The Universe and Planet Earth

ACKNOWLEDGEMENTS

The publishers wish to express their thanks to the institutes and scientists for kind permission to include the following photographs in this book.

Astronomy: Dr. D. E. Blackwell, Oxford University, Great Britain, 112; CERN, Geneva, Switzerland, 57; Dudley Observatory, Shenectady, USA, 111; European Space Agency, Paris, France, 389, 390; Enrico Fermi National Laboratory, Batavia, USA, 16, 17; Dr. Eleanor Helin, California Institute of Technology, Pasadena, USA, 106; Harvard College Observatory, Cambridge, USA, 175, 179; High Altitude Observatory, Boulder, USA, 174, 192; JPL/NASA, Pasadena, USA, 73, 75, 76, 78, 80–92, 95–104, 384, 385; Kiev Astronomical Observatory, USSR, 190; National Radioastronomy Observatory, Charlottesville, USA, 137, 138; Dr. Gordon Newkirk, Boulder, USA, 174, 192; Dr. Gennadi Nikolski, IZMIRAN, Moscow, USSR, 190; Ondřejov Observatory, Czechoslovakia, 169, 180; Oxford University Astronomy Department, Great Britain, 112; Palomar Observatory, Pasadena, USA, 113, 117, 125–127, 129–136, 139, 140, 148, 149, 154–156; Philips, Eindhoven, the Netherlands, 402; Dr. C. F. Powell, Great Britain, 43; Dr. A. Rots, Radiosterrenwacht, Dwingeloo, the Netherlands, 137, 138; Sacramento Peak Observatory, Sunspot, USA, 161, 172, 176, 178; Dr. E. H. Schröter, Kiepenheuer Institute, Izaña, Tenerife, 167; Solar Physics Group, American Science and Engineering Inc., Cambridge, USA, 173; Dr. J. Suda, Ondřejov, Czechoslovakia, 169; Westerborg Synthesis Radio Telescope, the Netherlands, 137, 138; Mt Wilson Observatory, Pasadena, USA, 166, 177, 191.

Geology: World Data Center for Rockets and Satellites, Houston, Goddard, USA, 204, 206, 238, 239, 242, 244, 247, 255, 264; JPL/NASA, Pasadena, USA, 243, 253, 254, 257–259; Scripps Institution of Oceanography, Deep Sea Drilling Project, 232–234; Agency APN, 169; Hawaii National Parks, 272.

Astronomy (chapters 1–4, 10–11)
Text by Josip Kleczek
Illustrations by Theodor Rotrekl, Pavel Rajský
and Vladimír Rocman

Geology (chapters 5–9)
Text by Petr Jakeš
Illustrations by Adolf Absolon, Theodor Rotrekl
and Hedvika Vilgusová

Translated by Stephen Finn
Graphic design by Karel Vilgus

This edition first published in Great Britain in 1987
by Octopus Books Limited
59 Grosvenor Street
London W1

ISBN 0 7064 2919 2
Printed in Czechoslovakia by TSNP Martin
1/19/02/51-01

THE UNIVERSE AND PLANET EARTH

Written by Josip Kleczek and Petr Jakeš

Translated by Stephen Finn

Octopus Books

Contents

MAN AND THE UNIVERSE

We are bound to the planet Earth by a force we call gravity. Only astronauts are, for a short time, free of this bond. They have seen and photographed our planet from space – half illuminated by solar rays and turning slowly in a black void, as it silently continues its year-long pilgrimage around the Sun. Through the icy endlessness of space it carries along 4000 million intelligent beings – creatures with their joys and sorrows, and with a persistent thirst for knowledge.

The Earth is one of nine planets which orbit the Sun. In terms of the size of the Universe, it is just a tiny speck. The Universe comprises everything on the Earth and beyond it. It includes the distant stars, the nearby planets and the infinitely remote galaxies. Everything around us belongs to the Universe, from the largest whales to the smallest bacteria, and from the stones on a beach to the dewdrop on a flower. Our own bodies are of course part of the Universe, and they take part in its development. We are made up of the same elementary particles as the stars, the stones and the flowers. Our bodies are subjected to the same forces as those which act upon everything else.

This book will tell you about all sorts of different things in the Universe, from elementary particles to supergalaxies, and about their origins, the changes they undergo, and the way they end their existence when the time comes. But above all, it will tell you something of the unity of the Universe, its structure, the processes that go on in it, and the role played in it by life, including intelligent life. Man is an intelligent animal, which means that he is able to understand and choose between things. (The Latin word *intellegere,* to discriminate, comes from *legere,* meaning to read, and *inter,* meaning between.)

When man learns about the Universe, he forms a 'picture' of it in his mind, rather like a drop of water reflecting the image of its surroundings. The joy of understanding is one of the most precious gifts man possesses.

NOTES TO READERS

- In this book, references to months and seasons for observations etc., relate to Europe, unless otherwise stated.
- Metric units are used throughout. To help readers unfamiliar with this system, Imperial equivalents are shown, but these should be taken as approximate, only.
- To deal with very large numbers, the mathematical 'shorthand' form is used. For example: 1000 which is $10 \times 10 \times 10$, is written as 10^3 and 1,000,000, which is 1 followed by six 0s, is written as 10^6. So, 5 million is 5×10^6.

  In a similar way, for very small numbers, one-millionth, which is 1 divided by a million, is written 10^{-6}.

  It should also be noted that 'per second' may be abbreviated per sec, or/sec, or sec^{-1}; and 'grams per cubic centimetre' may be abbreviated as g/cm^3 or $g\,cm^{-3}$.
- Where large numbers are named, the British system has been used. So,

  million = 1,000,000 = 10^6
  billion = 10^{12}
  trillion = 10^{18}
  quadrillion = 10^{24}
  quintillion = 10^{30}

1 The human organism is a system of organs **(1).** Organs are made of tissues, tissues of cells **(2),** cells of molecules **(3),** and molecules are systems of atoms **(4).** An atom is made up of a nucleus **(5)** and electrons **(6).** A nucleus is a system of protons and neutrons **(7).** Living organisms make up the biosphere **(8),** the biosphere is part of the planet Earth **(9),** which

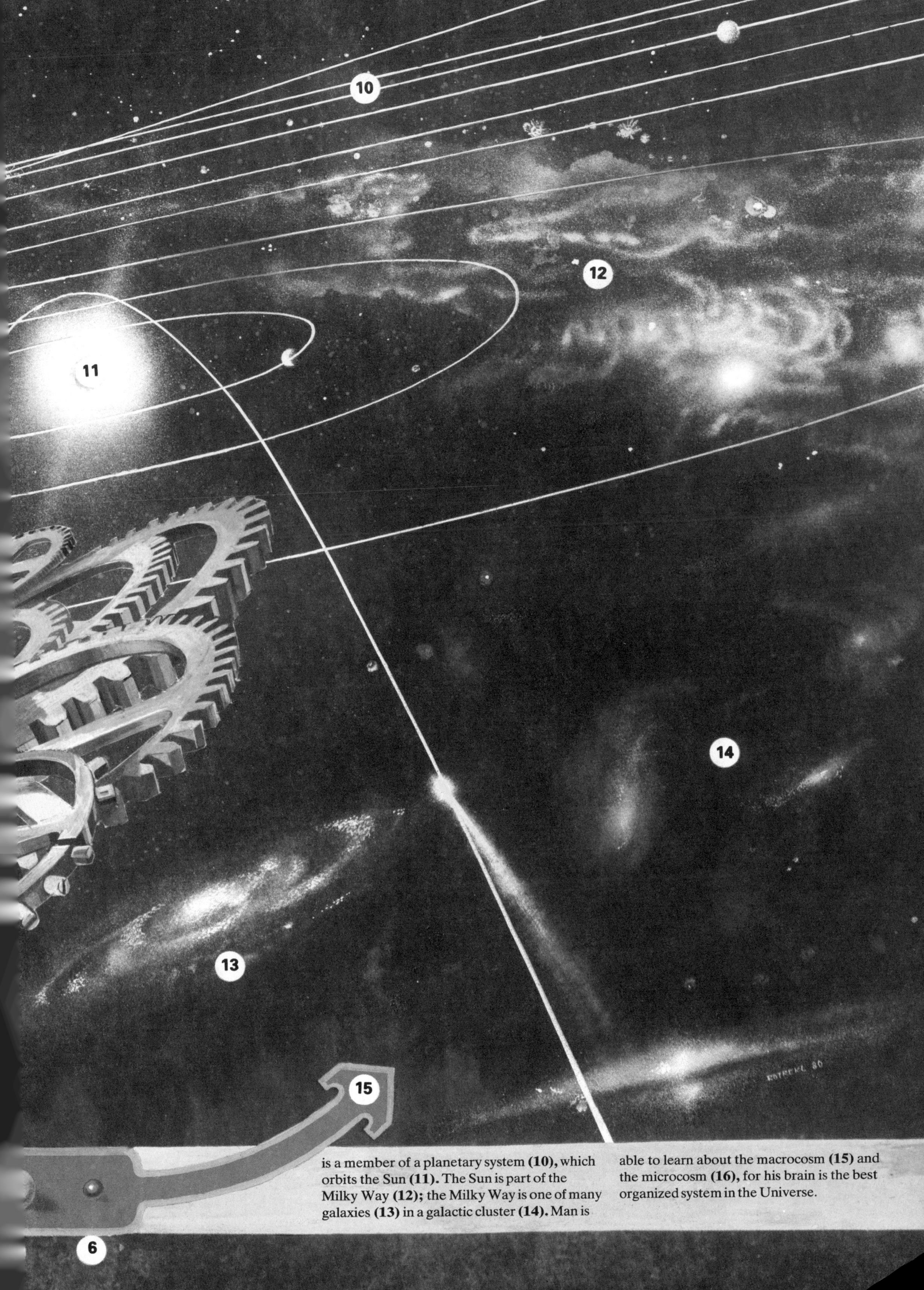

is a member of a planetary system **(10),** which orbits the Sun **(11).** The Sun is part of the Milky Way **(12);** the Milky Way is one of many galaxies **(13)** in a galactic cluster **(14).** Man is able to learn about the macrocosm **(15)** and the microcosm **(16),** for his brain is the best organized system in the Universe.

2 Our body is made up of organs, organs of tissues, tissues of cells, cells of organelles, organelles of macromolecules, macromolecules of atoms, atoms of electrons and nuclei, and nuclei of protons and neutrons. Man is a very complex system of elementary particles.

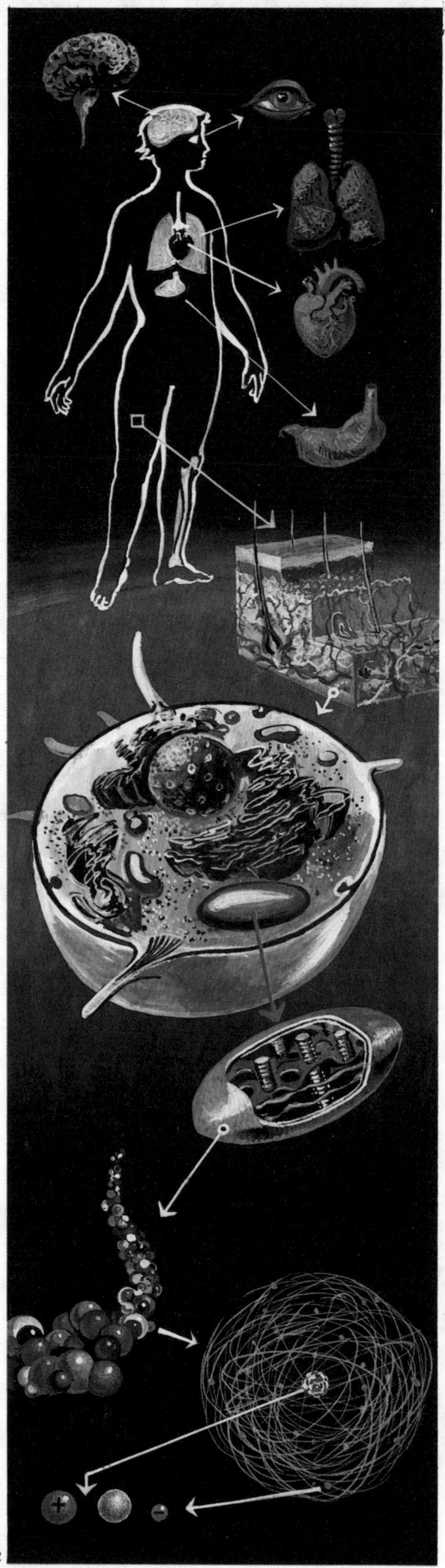

Man and the microcosm

The Universe is organized according to the laws of nature; it has order and hierarchy. The Greek word for world or universe, *kosmos*, also means order. The word *mikros* means small, so that a *microcosm* is a small world. The microcosm we shall look at is that consisting of tiny objects much smaller than man himself, such as elementary particles, atomic nuclei, atoms and molecules. The microcosm cannot be seen with the naked eye. The *macrocosm*, on the other hand, is everything we can see around us with the naked eye, and everything in the Universe which is bigger than man.

Our bodies are made up of various organs, such as the brain, the heart, the lungs and the stomach. These organs are complementary to each other, and incapable of working naturally in isolation. But in combination they make up an independent whole. Such a whole is called a system. Our body is a system of organs.

Each of the organs in our bodies is made up of tissues. Tissues are formed by cells which have similar functions. Cells are composed of tiny structures called organelles, and organelles are composed of molecules. A molecule is a system of atoms. The number of atoms in a particular molecule varies so that, for instance, a water molecule has three atoms, while a protein molecule may be composed of millions of atoms.

An atom is a system consisting of a nucleus, around which there are orbiting electrons. The atomic nucleus is the smallest known system. It consists of protons and neutrons. Protons, neutrons and electrons are the simplest building blocks of the human body and of everything else in the Universe, so we call them elementary particles. As yet we cannot say whether elementary particles are really the simplest things in the Universe, but they are the smallest discovered so far. As yet, they are indivisible. What we do know for certain is that atoms, molecules, cells, plants, animals and man himself – together with all he has ever made – and the Earth and the Sun, are systems of elementary particles, which means harmoniously arranged numbers of protons, neutrons and electrons (Figures 1 and 2).

Things differ from one another in the

number of elementary particles they contain and the way they are arranged. A quadrillion (10^{24} – 1 followed by 24 noughts) of protons, a similar number of electrons and a rather smaller number of neutrons go into the making of a grain of rice or an ant. However, the arrangement of the protons, neutrons and electrons is of course vastly different in the two entities.

The degree of organization of elementary particles is a fundamental factor which determines the nature of things. A piece of rock weighing about 1.5 kg ($3\frac{1}{4}$lb) contains 10^{27} elementary particles. The very same sort of particles could, however, be found in a highly sophisticated system – the human brain. Let us look at how these simple particles can produce an organ as complex as the brain. Estimates indicate that the human brain contains around a hundred thousand million nerve cells, interconnected by a complex network of thread-like axons and dendrites. The total length of these connections, which serve to carry information in rather the same way that the wires of a telephone network carry information, is roughly equivalent to the distance between the Earth and the Moon. The energy supply to the brain amounts to about 25 watts, whereas a computer capable of fulfilling its function, if one existed, would need the entire output of a power station. But after all, it did take around 3000 million years for man, with his perfectly organized brain, to develop from a single-cell organism (Figure 3).

Man is the most complex system of elementary particles known to have developed in the Universe. Though we are trying hard, we have not yet managed to find a more highly developed race of creatures anywhere in the Universe – indeed, so far we have discovered no other forms of life at all.

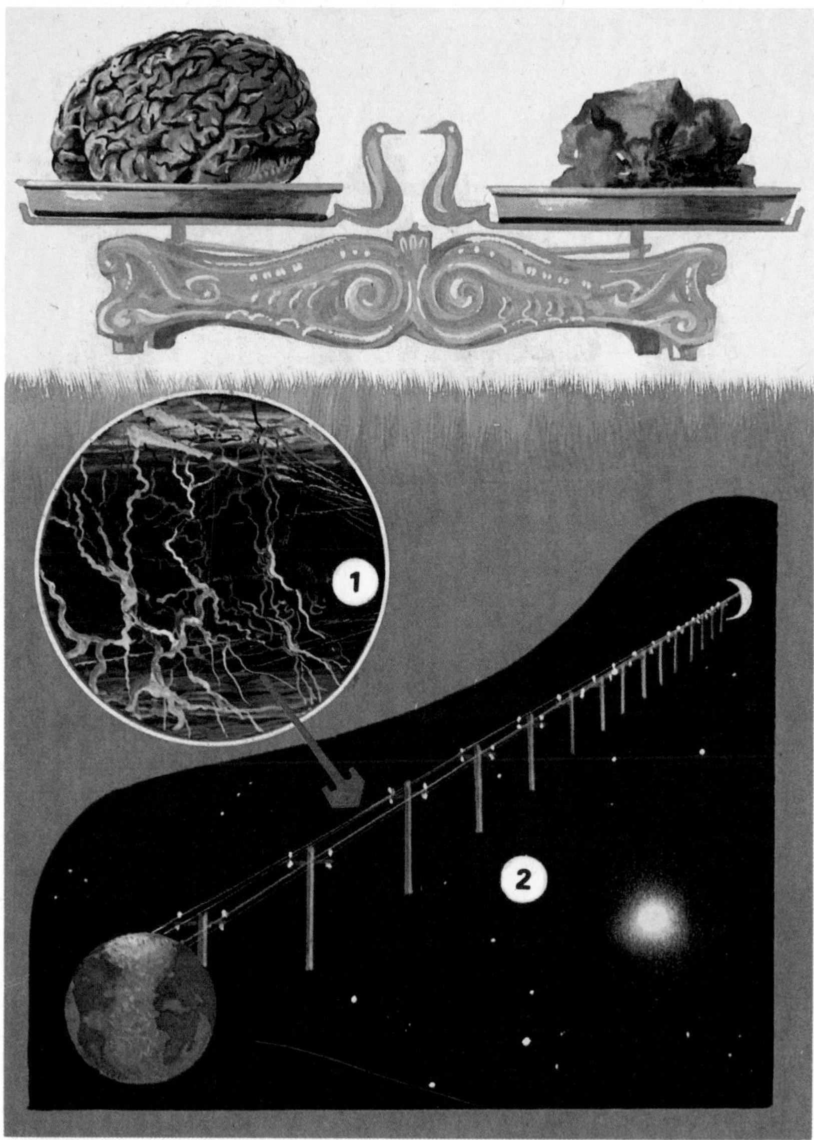

3

3 The human brain and a piece of rock of the same weight have the same number of protons, neutrons and electrons. But the brain has much better organization. The fibres linking the nerve cells of man **(1)** are together about equal to the distance from the Moon to the Earth **(2)**.

Man and the macrocosm

Man comes somewhere in size between the elementary particle and a star. To us, the particle is unimaginably small, and in comparison with a star a man is also infinitely small. Our own bodies consist of a few tens of thousands of quadrillions ($10^{28}-10^{29}$) of elementary particles. A similar number of human beings would correspond to the size of the Sun, since the number of elementary particles in it is about that much greater than the number in our bodies (Figure 4).

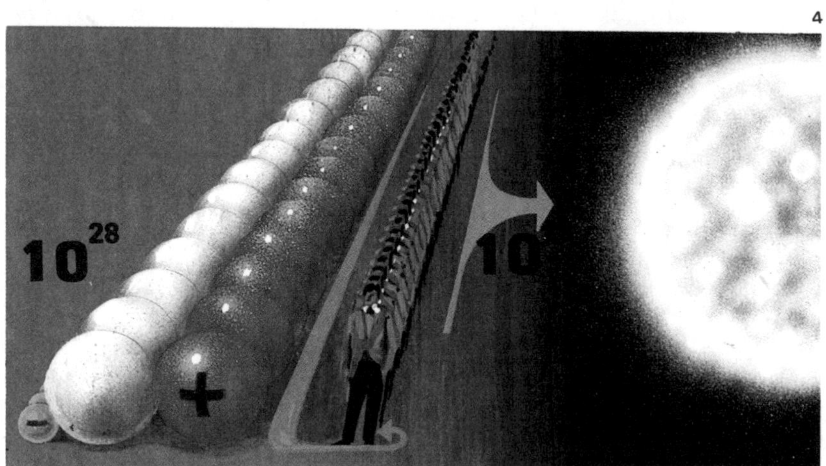

4

4 The human organism is composed of several tens of thousands of quadrillions of protons and electrons. The same number of human bodies would equal the mass of the Sun.

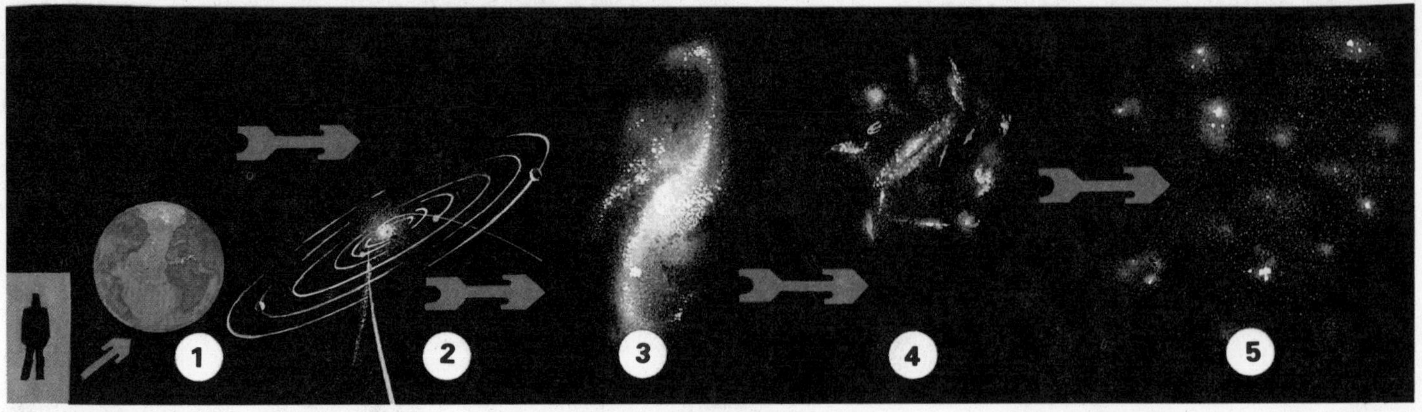

5

5 Man is a part of the biosphere, the biosphere is a layer of the Earth **(1),** the Earth is a part of the Solar System **(2),** the Solar System is part of the Milky Way **(3),** and the Milky Way is part of a galactic cluster **(4),** which is a part of a supergalaxy **(5).**

Though our bodies are very tiny compared with a star, they are much more sophisticated systems of elementary particles. That infinitely complex, yet perfectly organized system, the human brain, allows man, through his senses, and thanks to his capacity for thinking, to learn about himself, the microcosm and the macrocosm.

Man and the other animals and the plants are a part of the biosphere, which is that layer of the Earth's surface on which there is life. The biosphere, like the lithosphere (hard crust), the hydrosphere (waters) or the atmosphere, forms a part of the Earth. Inside the Earth there is an inner core and an outer core, and between the core and the crust there is the mantle. More details are given on this in Chapter 7. The individual parts of the Earth interact with each other. Our bodies are made of its atoms. We breathe the molecules of its atmosphere, exploit its mineral and energy resources, and are bound to it by its gravity. On the other hand we influence its land, water and air by our own activities.

The Earth itself is part of a higher system – the planetary system, which in turn is part of the Solar System. In addition to the nine major planets the Solar System contains comets, minor planets and innumerable smaller bodies ranging in size from specks of dust to huge lumps of rock (the 'meteoroid complex'). The centre of the Solar System is a star we call the Sun, whose gravity holds all these bodies close to itself. The structure of the Earth and the whole of the Solar System will be dealt with in Chapters 2, 4, 6 and 7. We are only interested here in forming a general picture of the Universe around us, to get a better picture of our position.

Our Sun is just another star in the Milky Way. The Milky Way appears as a bright strip that stretches right across our sky. On a clear, moonless night you can see it quite clearly with the naked eye. In fact it is an enormous system of about a billion (10^{12}) stars, one of which is our Sun. Such huge star systems are called galaxies. The Milky Way is our Galaxy, seen from within, as it were, and it is one of a local group of 21 galaxies. Along with other such groups it belongs to a cluster of galaxies containing thousands of galaxies in all. Several hundred such clusters of galaxies make up a supergalaxy, which is a system so huge that it takes light over 150 million years to cross it. The Universe contains many supergalaxies, all built of elementary particles: mainly protons, neutrons and electrons.

The Universe we know of to date is made up of something like 10^{82} elementary particles, most of them being protons, electrons and neutrons. From atomic nuclei to supergalaxies, the structure and evolution of the Universe are due to three forces that hold these particles together: nuclear, electrical and gravitational. One small link in the chain of cosmic evolution, and one tiny part of this immense system, is man.

6 The proton (+), the electron (−) and the neutron (no electric charge) are the elementary particles of which our bodies and all other things are built. Apart from ordinary matter there also exists antimatter made of antiprotons (−), positrons (+) and antineutrons. The whole of the space between particles is filled with radiation (photons) of various sorts: light of all colours **(1),** radio waves **(2)** and X-rays **(3).** The Universe is composed of matter and radiation, and the two components act on each other. Radiation is emitted from matter and is absorbed by it **(5),** is reflected from it **(4),** passes through it **(6),** is refracted **(7),** split into different colours (dispersion) **(8),** and its oscillations assume order within it (polarization) **(9).**

1. THE NATURE OF THINGS

The complexity of things

All things are made up of simpler parts. We shall remind ourselves of this important fact by looking at a simple example.

Every machine consists of a certain number of components. Thus a car has an engine, a body, wheels, steering linkage, brakes, suspension, transmission, and so on.

7 Order is the most important property of a system. The heap in the picture contains all the parts necessary to make up a car, but it is not a car. It lacks the required order.

Each of these parts consists of a number of smaller and simpler components – in the case of the engine there are cylinders, pistons, shafts, rods, valves, and so on. The parts of the car are joined together by items such as the universal joints, rivets and bolts to form a complete whole – a car. The way in which the parts are joined is functional, so the steering wheel is not attached to the exhaust, or the wheels screwed on to the roof. A whole such as a car, whose parts are organized to work together, is called a *system.*

Almost anything can be broken down into ever smaller and simpler parts. But where does it all end? Let us answer this question by considering a very simple system such as the drops of water that form the dew on a flower or go into making a cloud.

The drops, or droplets, of water which make up clouds are very small indeed. In fact each of them is about 0.01 mm in diameter. A thousand million of the droplets in a cloud weigh no more than a single gram (0.04 oz). But though these droplets are so small, they are each composed of an enormous number of molecules (Figure 8). A molecule is the smallest possible unit of water. Figure 8 shows a drop of water enlarged about 10 times. Its mass is around one-hundredth of a gram. Only a few molecules are shown, and in reality they would be very much smaller than they appear in the picture, since there are about 100 million billion (10^{20}) of them in a drop of water. In fact, even if more than 50 million people were to help draw these molecules, and each of them drew a molecule a second, night and day, it would take them about 50,000 years to draw them all!

Figure 8 shows a molecule of water magnified enormously. It consists of an atom of oxygen and two atoms of hydrogen. The hydrogen atoms are joined to the oxygen atom by an electrical force.

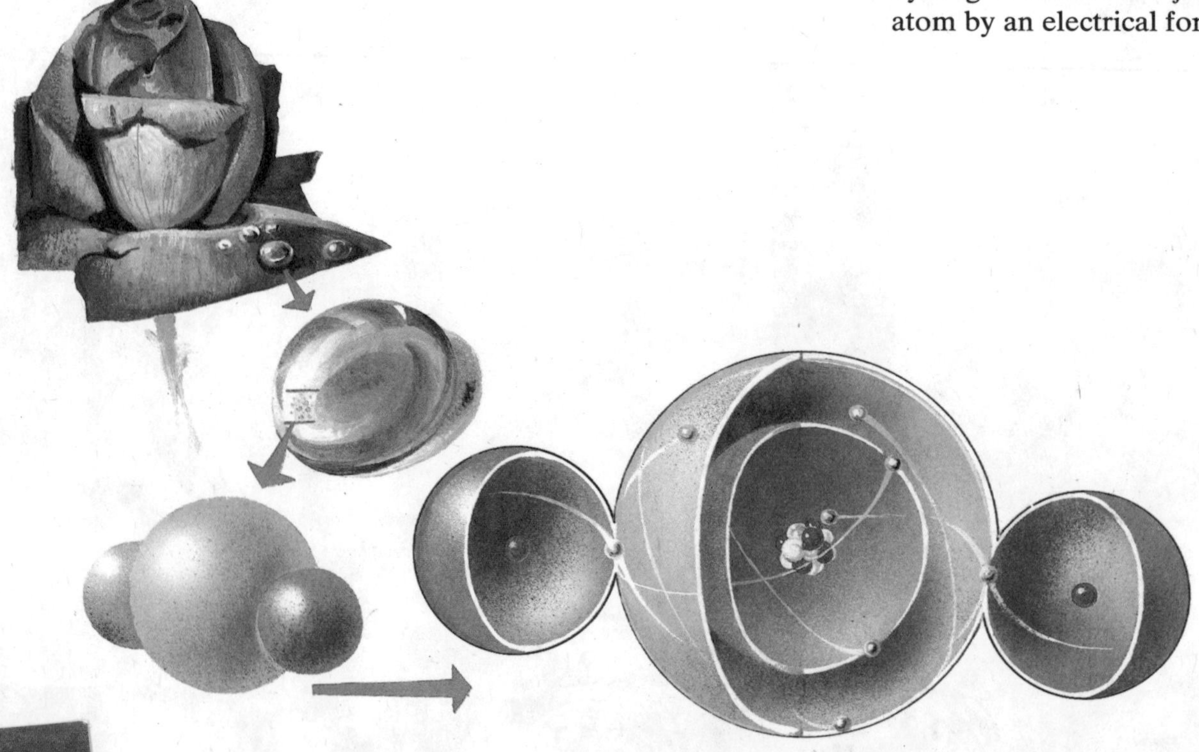

8 A drop of water is composed of an immense number (10^{20}) of molecules. A molecule of water (enlarged about one thousand million times) is shown at the bottom.

The hydrogen atom is the simplest and smallest atom of all. Ten million of them placed side by side would form a row 1 mm (0.04 in) long. At one time scientists thought that the atom was the smallest particle that exists, in other words it could not be divided any further. That is how it got its name, which in Greek means indivisible. But in this century research has shown that every atom is made up of two parts: the infinitely small nucleus and the electrons that orbit around it. The arrangement within atoms is similar to that of the Solar System, where lighter planets orbit the heavy Sun. But the force that binds the planets to the Sun is quite different from that which holds the electrons in their orbits around the atomic nucleus. The Solar System is held together by gravitation, while electrons are kept in place by an electrical force; they are negatively charged, while the nucleus is positively charged (Figures 9, 10 and 12).

The nucleus of the simplest of all atoms, the hydrogen atom, contains only a single proton. It is so simple that it cannot be divided any further. The nuclei of the other elements (helium, carbon, nitrogen... right up to uranium) were formed in the hearts of stars, and the material that went into their making was protons. So, for instance, the helium nucleus comes into being in the Sun and in many other stars by the joining together of four protons. Two of these lose their positive charge (Chapter 4) and such an uncharged proton is called a neutron. Thus the nucleus of helium is composed of two protons and two neutrons. All atomic nuclei heavier than the hydrogen nucleus are made up of protons and neutrons. They are held together in the nucleus by an enormous force called the nuclear force. Under normal conditions there is the same number of electrons orbiting the nucleus as there are protons in the nucleus itself.

The nucleus of an atom is many thousands of times heavier than its electrons. Almost the entire mass of the atom is concentrated in its tiny nucleus. About 100,000 nuclei would fit into the diameter of the atom. Though protons are so infinitely tiny, they are decisive in the life of the Sun and all the other stars. We might say figuratively that the proton is the 'hero' of the Universe (Figure 11).

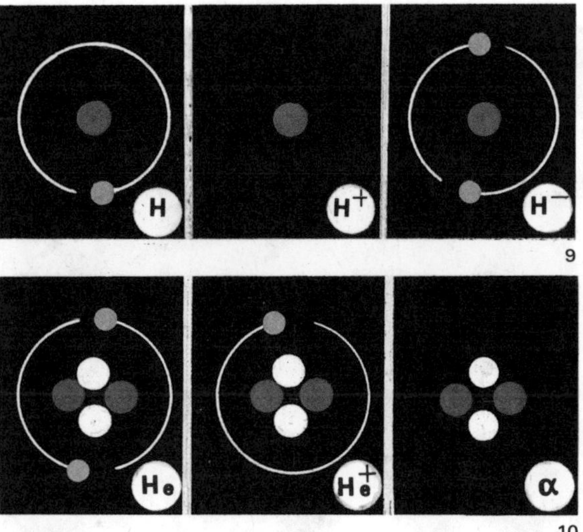

9 The neutral atom of hydrogen **(H)** consists of one proton and one electron. $\mathbf{H^+}$ is the positive ion of hydrogen, and has no electron. $\mathbf{H^-}$ is the negative ion of hydrogen, and has two electrons.

10 The alpha particle is the nucleus of the helium atom. $\mathbf{He^+}$ is the helium ion and **He** the neutral atom of helium.

Let us remember that all atoms are made up of only three types of elementary particles: protons, neutrons and electrons. Protons and neutrons make up the nuclei; nuclei and electrons form atoms, atoms form molecules, and molecules form the drop of water we have already mentioned, plus everything else on Earth. The Sun, too, consists of an enormous number of protons (altogether 10^{57}), neutrons (10^{56}) and a corresponding number of electrons (10^{57}).

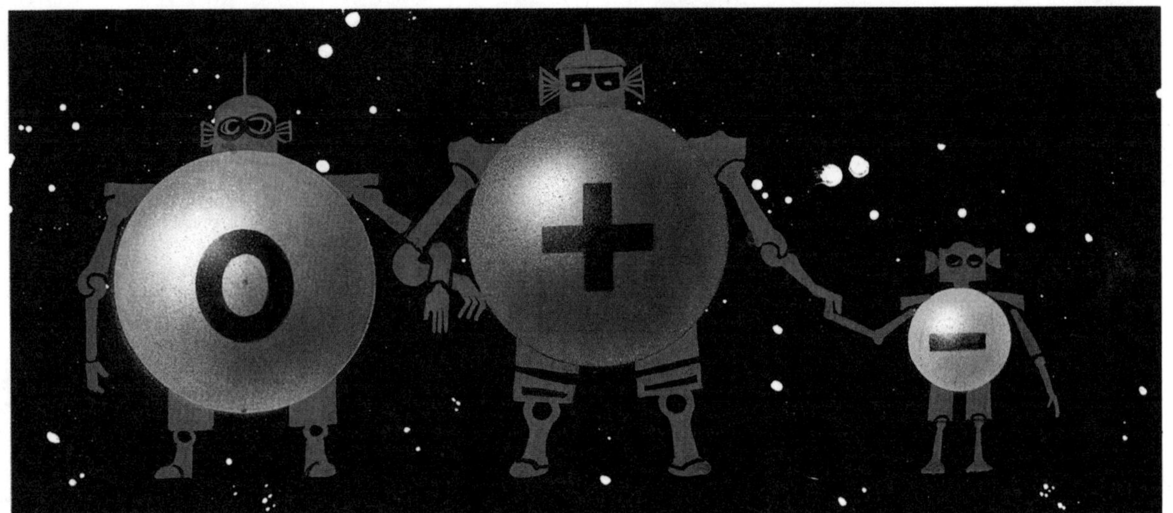

11 The proton is the true hero of the Universe. It is equipped with nuclear, electrical and gravitational forces. The neutron is obliged to live in the nucleus with a proton, since on its own it would disintegrate. The electron has only electrical force; its gravitation is tiny. It tries to stay close to the proton.

Elementary particles

As far as we know today, the proton, the neutron and the electron are so simple that they cannot easily be divided into anything simpler. But you might ask whether or not the future will bring a way of splitting them, as it did with the atom. After all, scientists in the last century were convinced that there was nothing smaller than the atom, whereas we now know that it is a system of elementary particles. Could not elementary particles be made up of some sort of subelementary particles, such as quarks or partons? This is quite a reasonable point of view, and we shall be considering it in Chapter 2. But for the time being we shall take a look at the basic properties of protons, neutrons and electrons, and of the other elementary particles we know.

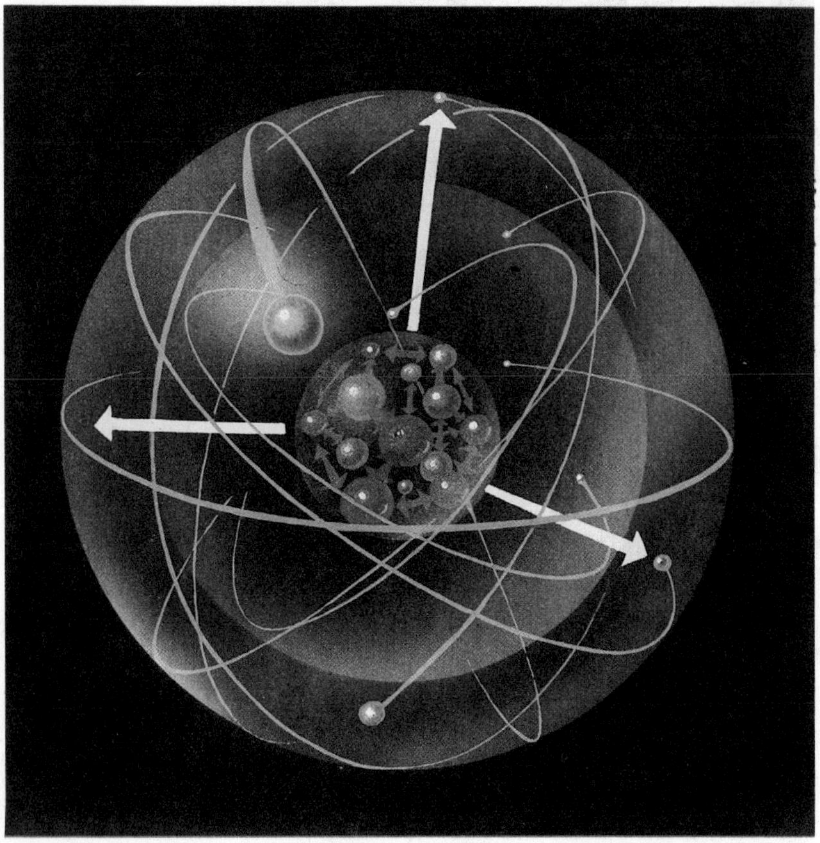

12

12 Protons and neutrons are held together in the nucleus by the nuclear forces (brown arrows). The electrons are bound to the nucleus by electrical force (yellow arrows).

The size of elementary particles

How big are elementary particles? It would be more appropriate to ask how small they are. A billion (10^{12}) elementary particles arranged one beside the other would make a row 1 mm (0.04 in) long. The diameter of an elementary particle is thus one-billionth of 1 millimetre (10^{-12} mm). That is the sort of dimension which is difficult to remember and difficult to pronounce. For this reason scientists named the unit 1 'fermi', in memory of the Italian physicist Enrico Fermi, who helped to discover what elementary particles are all about.

The mass of elementary particles

At the moment no one knows what elementary particles are made of. But we can determine how much of that 'something' there is in one particle. We call this amount the mass of the particle.

The greater the mass of a body, the greater is its weight. The greater its mass, the more difficult it is to set it in motion or to stop it if it is already moving. We call this unwillingness of a body to change its state of motion *inertia.* Elementary particles have a very small mass, and therefore also a very low inertia. This means it is easy to accelerate them almost to the speed of light, the fastest possible speed. Some particles even have zero mass – for instance the photon (Figure 14) – and these travel at the speed of light as soon as they come into being.

The mass of an elementary particle is one of the most important of its properties. Here we are speaking of its mass when it is not moving, which is known as its *rest mass.* Later on we shall see that the actual mass of a particle depends on the velocity at which it is moving. The faster it moves, the greater its mass, whereas its rest mass always remains the same. Thus, for instance, the rest mass of an electron is the same on Earth as in the most distant galaxy.

Other elementary particles

When protons, electrons and neutrons are investigated in pieces of equipment called accelerators, for short periods other particles appear. The accelerator (Figures 16 and 17) is a sort of huge 'microscope' where scientists study the properties of elementary particles. While they are doing this they keep discovering new sorts of particles, just as zoologists discover new species of animals and botanists find new plants. Figure 14 shows the best-known particles which are able to exist in the Universe at least for a short time. (Here 'short' means about one 100-millionth of a second.) Apart from these, a whole range of particles has been discovered with an infinitely short life, one billion times shorter than that of some of those in the table. These particles live for a mere 10^{-20} to 10^{-24} seconds, and they are called resonances. We are not yet sure whether they really are elementary particles at all. They do not seem to be of any importance in the Universe, so they have not been included in the table.

The particles in our table are arranged in order of rest energy. The greater the rest energy, the higher the position in the table. Let us consider the right-hand side of the table (the left-hand side will be discussed on page 26). The proton and the neutron are relatively heavy particles, so they are near the top of the table. The electron, on the other hand, is a fairly light particle, so it will be found at the bottom. The numbers at the side indicate rest mass multiplied by the square of the speed of light, i.e. c^2 (9×10^{20} cm^2/sec^2, or 9×10^{16} m^2/sec^2). The rest mass of an electron is denoted m_e, and it is around 10^{-27} g (more precisely 9.1×10^{-28} g, or 9.1×10^{-31} kg). Its rest energy m_ec^2 is 0.51 megaelectronvolts (0.51 MeV) – as indicated in the table. The electron is the lightest of the particles which have a mass. There are some indications that the neutrino also has a rest mass – about 50,000 times smaller than that of an electron m_e – but scientists are still not sure. It can be seen from the table that the heaviest elementary particle, known as the omega particle, is more than 3000 times heavier than the electron. But not even one billion (10^{12}) omega particles would have any effect on the most sensitive scales in the world, since one quadrillion (10^{24}) protons only have a mass of about 1 g (0.04 oz).

Rotation of elementary particles

Many of the bodies in the Universe turn about their own axes. This movement is called rotation. Planets, stars, moons and galaxies all rotate, but so do molecules, atoms and elementary particles. The rotation of elementary particles is known as *spin*. In some ways spin is similar to the rotation of large bodies, but in other respects it differs a great deal. We can imagine spinning particles rather as tiny tops. Unlike the rotation of large bodies, the spin of particles can neither be decelerated nor accelerated. Spin is an unalterable property of all elementary particles.

According to the value of spin they possess or do not possess, elementary particles fall into two groups. Particles which either do not spin at all or spin rapidly (i.e. have zero, or integer, spin) are called *bosons* after the Indian physicist Jagadis Chunder Bose. On the other hand, those particles which spin at medium speed (i.e. have half-integer spin) are called *fermions* after Enrico Fermi. There are major differences between the two types of particles (the division of particles can be seen clearly in Figure 14).

For the sake of clarity we may make a comparison with a train (Figure 18). In the first carriage, whose conductor is called Fermi, there is a rule that no more than two passengers can sit in any one compartment, one facing the engine and one with his back to it. But in the second carriage, where the conductor's name is Bose, the situation is

13

13 Peas circling the tip of the Eiffel Tower – that is what the electrons orbiting the nucleus are like.

14 Table of elementary particles. The bosons W^+ and W^- are called heavy or intermediate. Their rest mass and rest energy are nearly one hundred times larger than those of a proton and therefore they should be placed far above the Ω particles. ▷

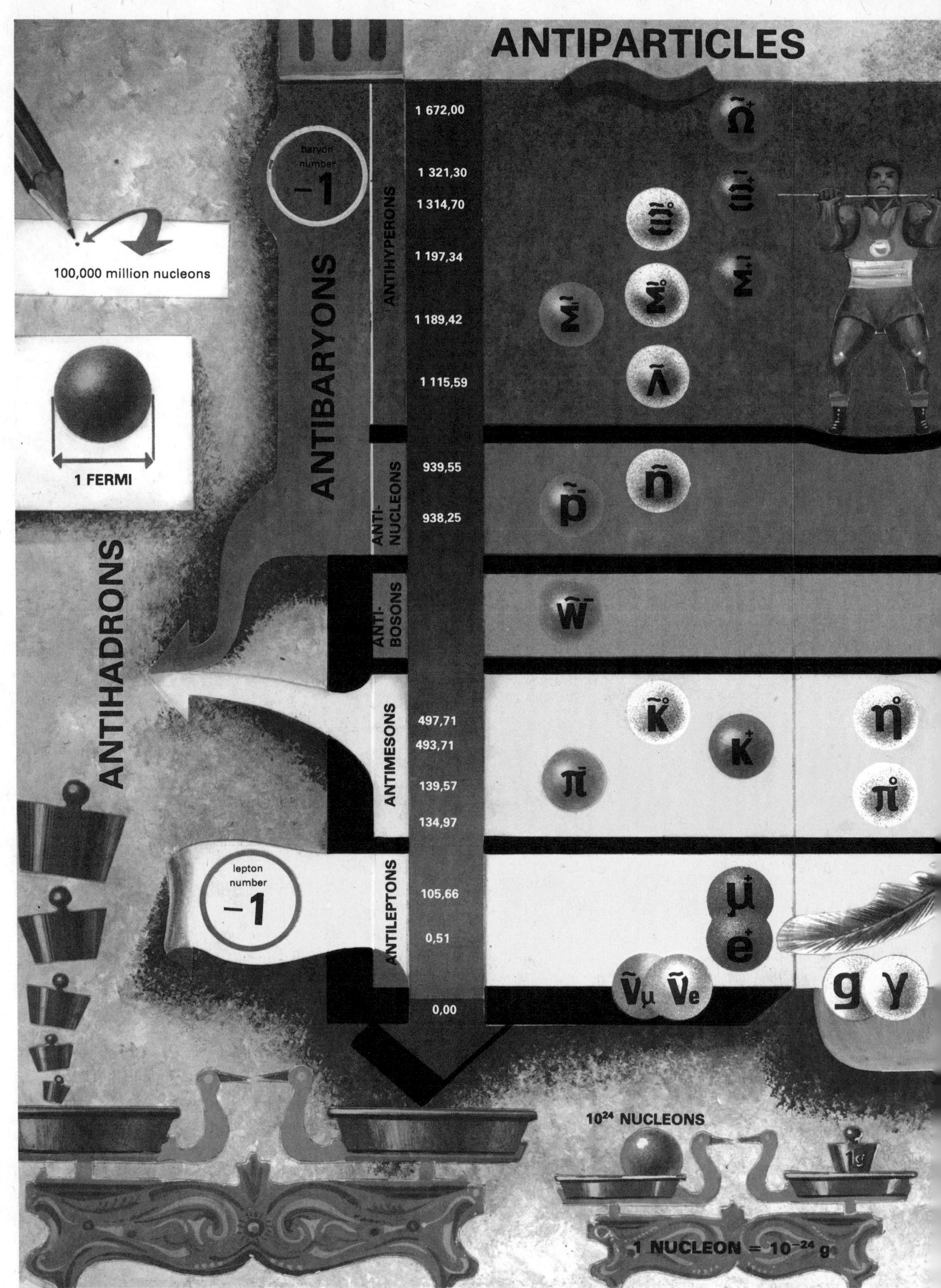
ANTIPARTICLES
100,000 million nucleons
1 FERMI
baryon number
−1
ANTIHADRONS
ANTIBARYONS
ANTIHYPERONS
ANTI-NUCLEONS
ANTI-BOSONS
ANTIMESONS
ANTILEPTONS
lepton number
−1
1 672,00
1 321,30
1 314,70
1 197,34
1 189,42
1 115,59
939,55
938,25
497,71
493,71
139,57
134,97
105,66
0,51
0,00
10²⁴ NUCLEONS
1g
1 NUCLEON = 10⁻²⁴ g

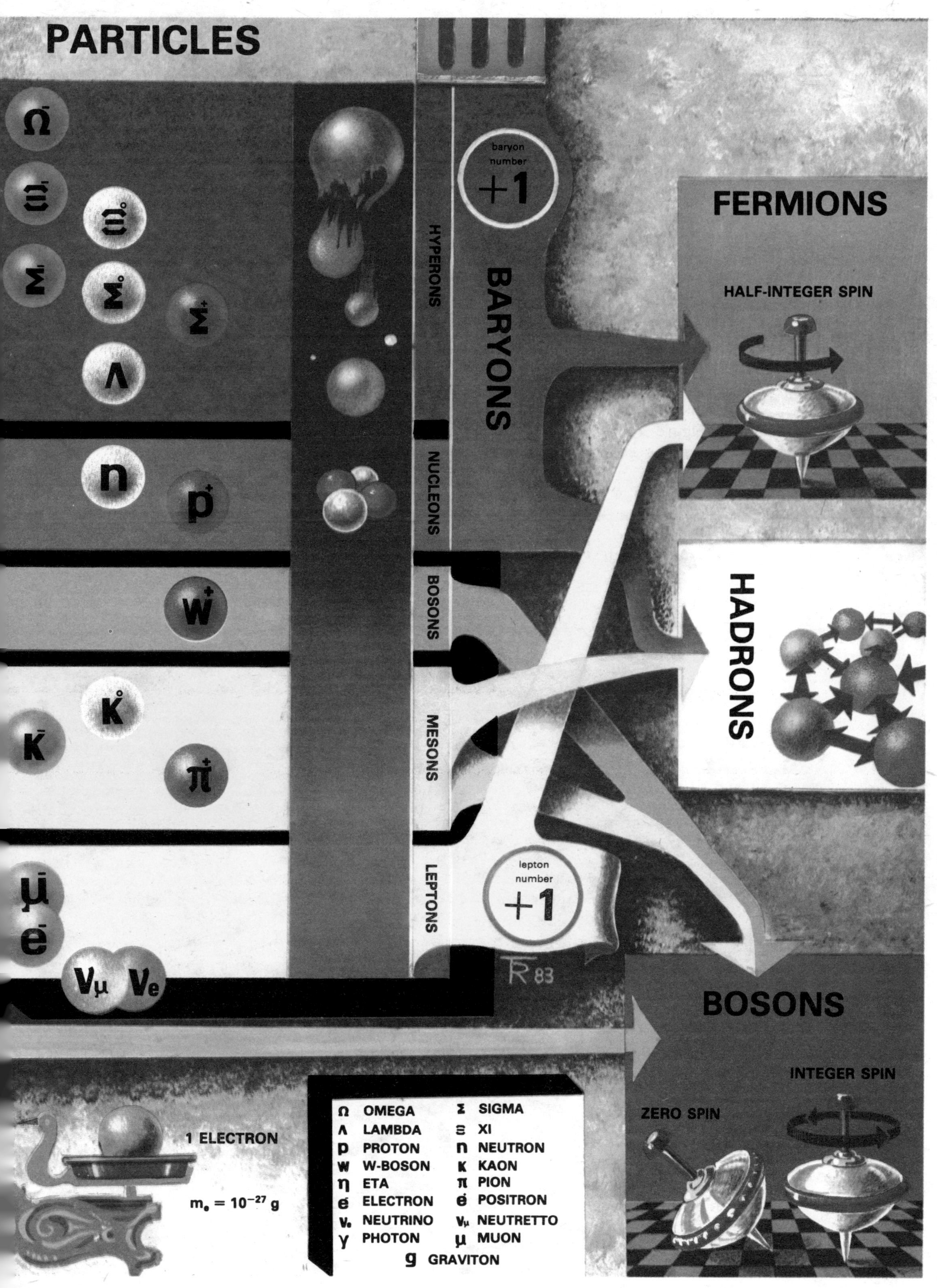
PARTICLES
baryon number
+1
HYPERONS
BARYONS
NUCLEONS
BOSONS
MESONS
LEPTONS
lepton number
+1
FERMIONS
HALF-INTEGER SPIN
HADRONS
BOSONS
INTEGER SPIN
ZERO SPIN
1 ELECTRON
$m_e = 10^{-27}$ g
Ω OMEGA
Λ LAMBDA
p PROTON
w W-BOSON
η ETA
e ELECTRON
ν_e NEUTRINO
γ PHOTON
Σ SIGMA
Ξ XI
n NEUTRON
K KAON
π PION
e POSITRON
ν_μ NEUTRETTO
μ MUON
g GRAVITON

15 The mass of a proton $m_p = 10^{-24}$ g.

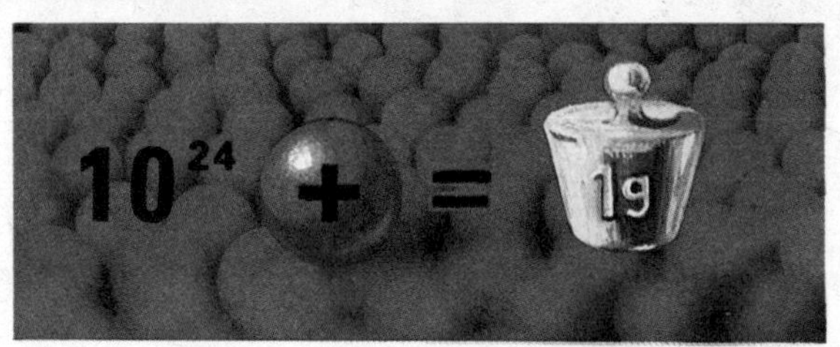

15

quite different. There is no restriction on the number of passengers that can sit in a compartment. The gregarious passengers here are like the particles which have zero or integer spin – a very large number of bosons, such as photons, will fit into a very small space.

The electrical charge of elementary particles

The elementary particles which are shown in red in Figure 14 have a positive electrical charge, and those shown in blue have a negative charge. Those particles which are white are neutral, with no charge at all.

The movement of electrically charged electrons is what we call an electric current. A charge of 10^{19} electrons is called a *coulomb*. A current of 1 coulomb per second is 1 ampere (Figure 19). Every second about 10 trillion (10^{19}) electrons pass through the filament of an electric light bulb.

The charge of 1 electron is 10^{-19} coulombs. This is the smallest electric charge which can exist. It cannot be divided up into anything smaller. All particles with a negative charge have the same amount of electricity as the electron (Figure 14). The charge of protons (and all positively charged particles) is of the same magnitude as that of electrons, though they have the opposite sign.

An electrically positively charged body has a shortage of electrons and a predominance of protons. A negatively charged body has an excess of electrons. A body has no electric charge when the number of protons in it is the same as that of electrons (Figure 20).

Electrically charged particles give rise to forces in the space around them. They affect other electrically charged particles which are close to them. The space around electrically charged particles in which their electric force operates is called an *electric field*. If a charged particle is in motion, it induces in the space around it a *magnetic force* (magnetic field). A charged particle which is accelerated or

16

16 A particle accelerator (at Batavia, Illinois, USA), photographed from the air. The particles (protons) are accelerated in a round underground tunnel (largest circle). The main building is at the bottom left. Three experimental channels run from it (dark green lines running towards the top left). The protons which are led from the tunnel into these channels have an energy of 400,000 MeV (4×10^{11} eV)

decelerated emits *electromagnetic radiation* (photons). Electric charges cannot be created or destroyed. This is the *Law of the Conservation of Electrical Charge.*

You can see from Figure 14 that the lightest particle with an electric charge is the electron. This means that an electron cannot disintegrate into lighter particles to which it might pass on its charge. Thus the electron is a stable particle.

Baryon charge or baryon number

Another important group of elementary particles is the *baryons.* Their basic property is the baryon charge (baryon number). The Greek word *barys* means heavy or massive, for baryons are heavy particles. They are at the top of Figure 14. The baryon group is a 'closed family'. If a baryon is transformed, it can only become another baryon – a neutron can change into a proton, for example, and vice versa. (Neutrons and protons are the basic particles in the atomic nucleus. For this reason they are called the nucleons.)

Baryons heavier than nucleons are denoted by means of Greek capital letters (lambda Λ, xi Ξ, sigma Σ and omega Ω). They are above the nucleons in the table and are called *hyperons.* All hyperons are unstable, disintegrating shortly after coming into existence. Even a lone neutron disintegrates (after about 10 minutes) into a proton. Every baryon tries to 'fall' as low as possible in the table, and the lowest baryon is a proton.

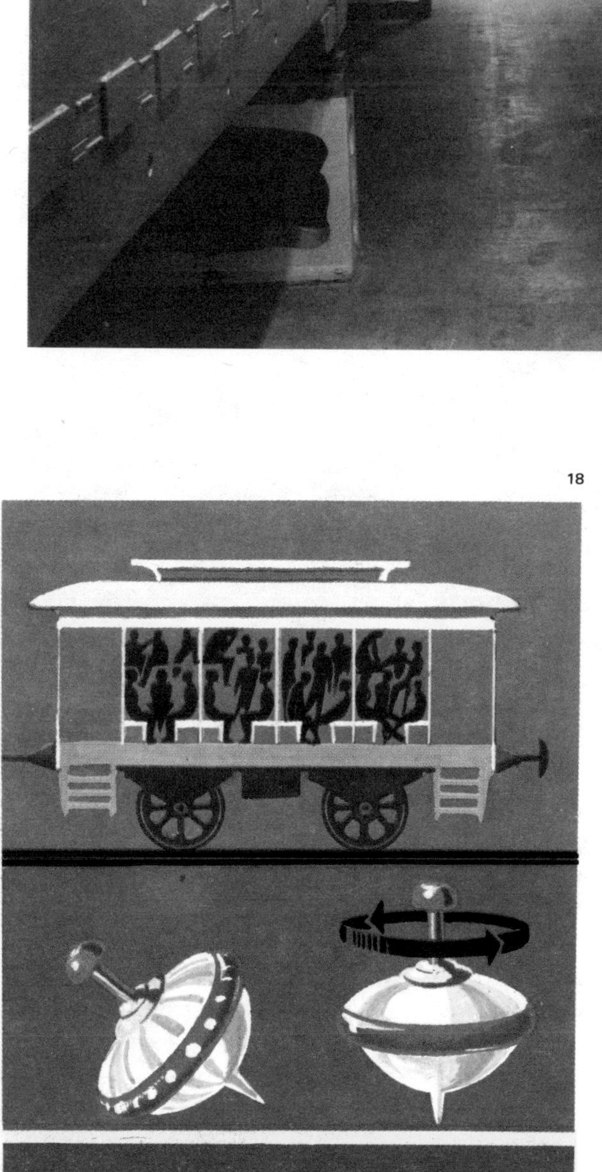

17

17 The inside of the accelerator tunnel from the previous picture. Inside the blue housing is a quadrangular tube 5 × 12 cm (2 × 4.7 in), from which the air is pumped. Along the tube is a total of 1014 magnets (on the yellow stands). These are water-cooled from pipes at the top. The tube makes a closed circle and protons run inside it; they make the circuit around 50,000 times per second. Each time, each proton acquires an energy of 2.8 MeV.

18

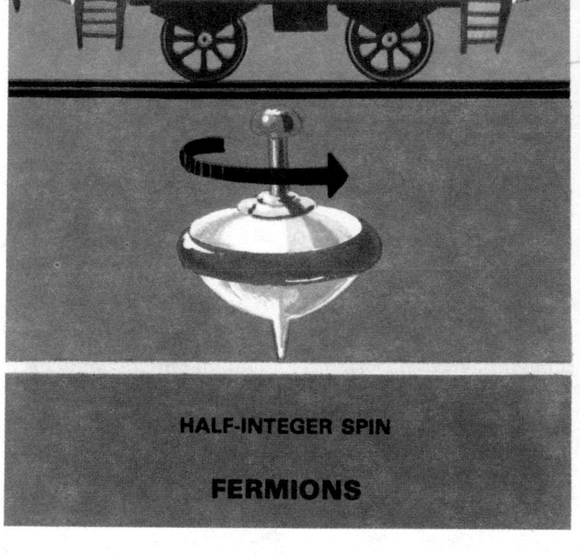

18 The behaviour of fermions and bosons can be imagined as passengers on a train.

Protons cannot disintegrate, since if they did the Universe would lose a baryon. And nature ensures that this never happens.

Physicists have assigned to every baryon a baryon charge of +1. All other particles (for instance electrons, photons and other bosons) have a zero baryon charge. The sum of all baryons which undergo any change is denoted N, and is called the total baryon

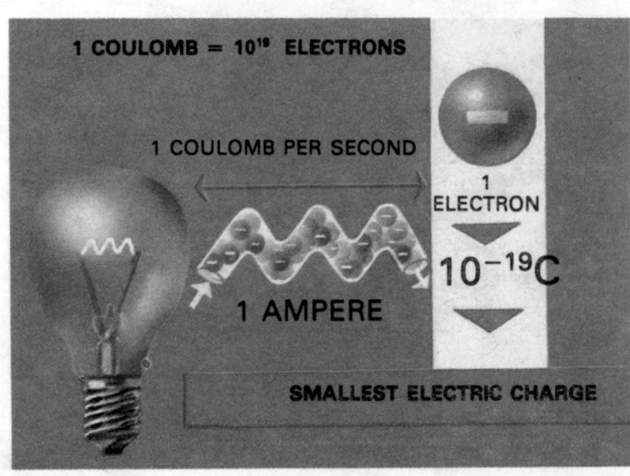

19 The electron (and the proton also) has the smallest electrical charge which exists in the Universe. Quarks are supposed to have smaller charge (e. g. one-third or two-thirds of that of an electron or a proton), but they can exist only in groups of two or three.

charge or baryon number. The total baryon charge remains unchanged regardless of the processes going on in the Universe. This is the *Law of the Conservation of Baryon Charge.*

A proton cannot disintegrate of its own accord, since that would break the Law of the Conservation of Baryon Charge. Nor can an electron disintegrate, which would break the Law of the Conservation of Electrical Charge. Both these *universal laws* (the conservation of electrical and baryon charges) apply without exception. The whole appearance of our Universe is based on the stability of the proton and the electron. If these two quantities (electrical and baryon charge) were not strictly maintained, the proton and the electron would disintegrate into lighter particles. Atoms and molecules would not be able to exist, and there would be quite a different sort of Universe.

Lepton charge or lepton number

The lightest particles of all in our table are the *leptons* – light particles. The Greek word *leptos* means light or tiny. The lepton family includes the electron (denoted e^-), the negative muon (μ^-), and two types of neutrino (the electron neutrino ν_e and the muon neutrino or neutretto ν_μ). As in the case of baryons, nature never loses any leptons. Every lepton has a lepton charge of +1. All other particles (baryons and bosons) have a zero lepton charge. The sum of lepton charges remains the same whatever changes take place, since they are subject to the *Law of the Conservation of Lepton Charge.*

For the sake of completeness, it should be mentioned that there are three more properties of elementary particles: isospin, strangeness and parity. However, since they are not important for our further considerations, we shall not examine them further.

This brings to an end our survey of the properties of the elementary particles on the right-hand side of Figure 14. We have seen that each of them has its own characteristics (mass, spin, electric charge, baryon charge and lepton charge).

The left-hand side of Figure 14 is a mirror-image of the right-hand side. Each of the particles on the right has its counterpart on the left; these are the *antiparticles,* which will be discussed on page 26.

Groups of elementary particles

Elementary particles are divided into several categories (see Figure 14).

Particles form two major groups, according to spin: fermions with half-integer spin; and bosons with zero or integer spin. This has been discussed on page 15 and in Figure 18.

Fermions

Fermions fall into two further groups, according to their mass. The group of heavy fermions, the baryons, can be seen in the top half of the table. The lightest of the baryons is the proton (p). The neutron (n) is only a little heavier. The nucleus of helium is made up of two protons and two neutrons, a total of four nucleons; the nucleus of carbon has 12 nucleons, and so on. Baryons which are heavier than nucleons are called *hyperons* (from the Greek word *hyper* meaning over). Thus the hyperons are the particles over the nucleons. They are unstable and quickly disintegrate into nucleons.

The second group of fermions is the light particles or leptons. Among these are the electron (e^-), the negative muon (μ^-) – also called a heavy electron, the electron neutrino (ν_e) and the muon neutrino (ν_μ) – also called the neutretto. Each of these four leptons has a lepton charge of +1. (Thus the antileptons on the left of the table have a lepton charge of −1, and they include the positron – or positive electron (e^+), the positive muon (μ^+), the electron antineutrino ($\tilde{\nu}_e$) and the muon antineutrino ($\tilde{\nu}_\mu$) – also known as the antineutretto.)

20 When you comb your hair, electrons pass into it from the comb. The comb becomes positively charged, and your hair negatively charged. The electrons jump back into the comb as sparks.

Bosons

The bosons can be seen on the extreme right of the table. They are sometimes referred to as the field particles or field quanta. They transmit the forces through which particles act on each other. The forces between particles (nuclear, electrical, weak and gravitational) will be dealt with on page 27. The bosons which correspond to these forces are the pions and kaons, the photons, the W-bosons and the gravitons. The gravitons are still only hypothetical, since they have yet to be discovered experimentally.

Pions or π-mesons and kaons or K-mesons are very important for nuclear forces. All particles which act on each other through nuclear forces are called *hadrons*. Thus the baryons and mesons belong to the group of hadrons. (The hadron group is indicated in Figure 14.)

A very important boson is the photon. Its rest energy is zero, but the photon can never actually be at rest. From the moment of its birth it rushes around the Universe at the speed of 300,000 km (186,400 miles) per second. Photons are very important for the transfer of energy and information. The next section is devoted to them. The symbols used for photons are either γ (the Greek letter gamma), or $h\nu$ – h being Planck's constant and ν (Greek nu) the frequency of photons.

Photons – particles of electromagnetic radiation

It is light which allows us to see the Universe, just as it lets us see our surroundings here on Earth. Light is electromagnetic waves (or electromagnetic radiation), similar to the waves on the surface of water. While water moves up and down as the waves cross it, in an electromagnetic wave the electric force oscillates in one direction and the magnetic force oscillates perpendicular to it (Figures 21 and 22). Light is the electromagnetic radiation (electromagnetic waves) which is visible to us. The wavelength of light is between 400 and 700 nm. These wavelengths are so short that a special unit called the nanometre (nm) had to be introduced to express them: 1 nm = 10^{-9} m = 10^{-6} mm. One million nanometres make 1 mm (0.04 in). Electromagnetic waves 400–450 nm in length appear as blue light, while waves 650–700 nm are seen as red. In between blue light and red light there is green, yellow and orange light (Figures 23 and 161). Two and a half thousand waves of blue light and 1400 waves of red light fit into the span of 1 mm (0.04 in).

Apart from electromagnetic light waves there are many other sorts of waves that affect our eyes, often unfavourably, but which we cannot see (Figure 23). Waves which are shorter than visible light, between 10 and 400 nm, are called *ultraviolet radiation*. Even shorter waves, from 0.001–10 nm, are called *X-rays*. And electromagnetic radiation with rays shorter than 0.001 nm is known as *gamma radiation.*

Radiation with a wavelength longer than that of red light, from 700 nm–0.3 mm, is called *infra-red radiation.* We cannot see it, but we can feel it in the form of heat. Electromagnetic waves from 1 mm (0.04 in) to many kilometres in length are called *radio waves.* They are transmitted by every radio or television station. Many bodies in space also emit radio waves, including the Sun and many of the other stars. Lightning also sends out radio waves, which are called *atmospherics* (Chapter 10, page 286). It is important to bear in mind that all these types of radiation – gamma rays, X-rays, ultraviolet rays, light rays, infra-red rays and radio waves – are electromagnetic waves that differ only in their respective wavelengths.

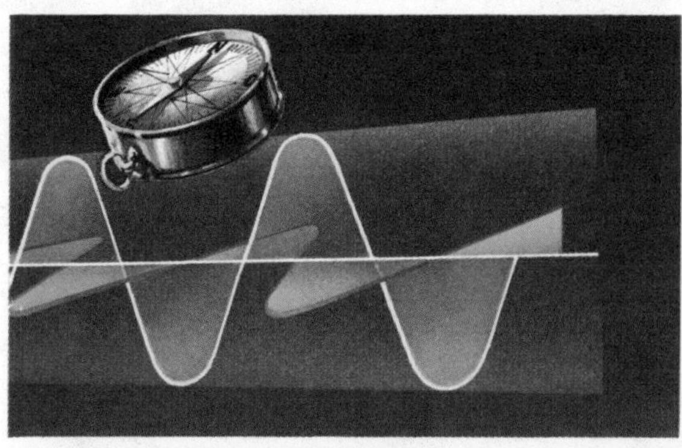

21 Electromagnetic waves.

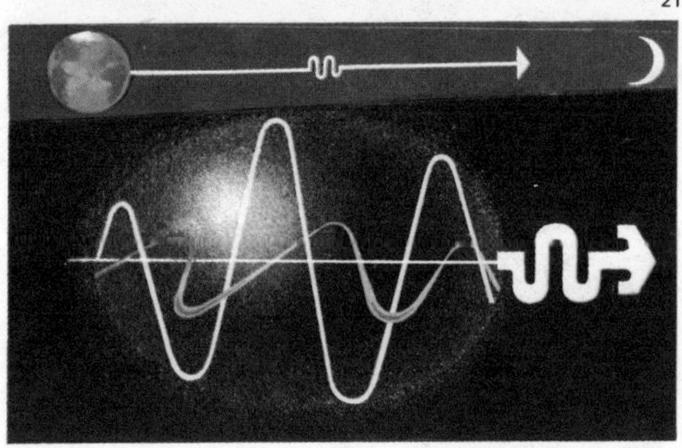

22 Electromagnetic waves consist of photons. They are clumps of energy in which an electrical force oscillates perpendicular to a magnetic one. The speed of photons in a vacuum is 300,000 km (186,400 miles) per second, which is almost the distance from the Earth to the Moon.

We have already seen that all bodies are composed of elementary particles (protons, neutrons and electrons). Similarly, electromagnetic radiation is made up of photons. These are small amounts of energy in which an electric force oscillates, with a magnetic force perpendicular to it (Figure 22). We know from earlier pages that a photon cannot stop, but continually moves in a straight line at the maximum possible speed, oscillating as it does so. In a vacuum all photons move at the same speed. Every second they cover 300,000 km (186,400 miles), which is about seven times round the Earth before you can say 'Jack Robinson'. In one second light can travel from the Earth's surface to the Moon. This immense speed of 300,000 km (186,400 miles) per second is called the speed of light in a vacuum, and is denoted c. In glass, water, air and other media light and other forms of electromagnetic radiation travel more slowly, at a speed less than c. This means that photons also move more slowly when not in a vacuum. This is the cause of the refraction of light (and of other forms of electromagnetic radiation) which occurs on passing from one medium to another (Figure 27).

Electromagnetic radiation can be imagined in two ways – either as waves of electrical and magnetic forces (Figure 21) or

23

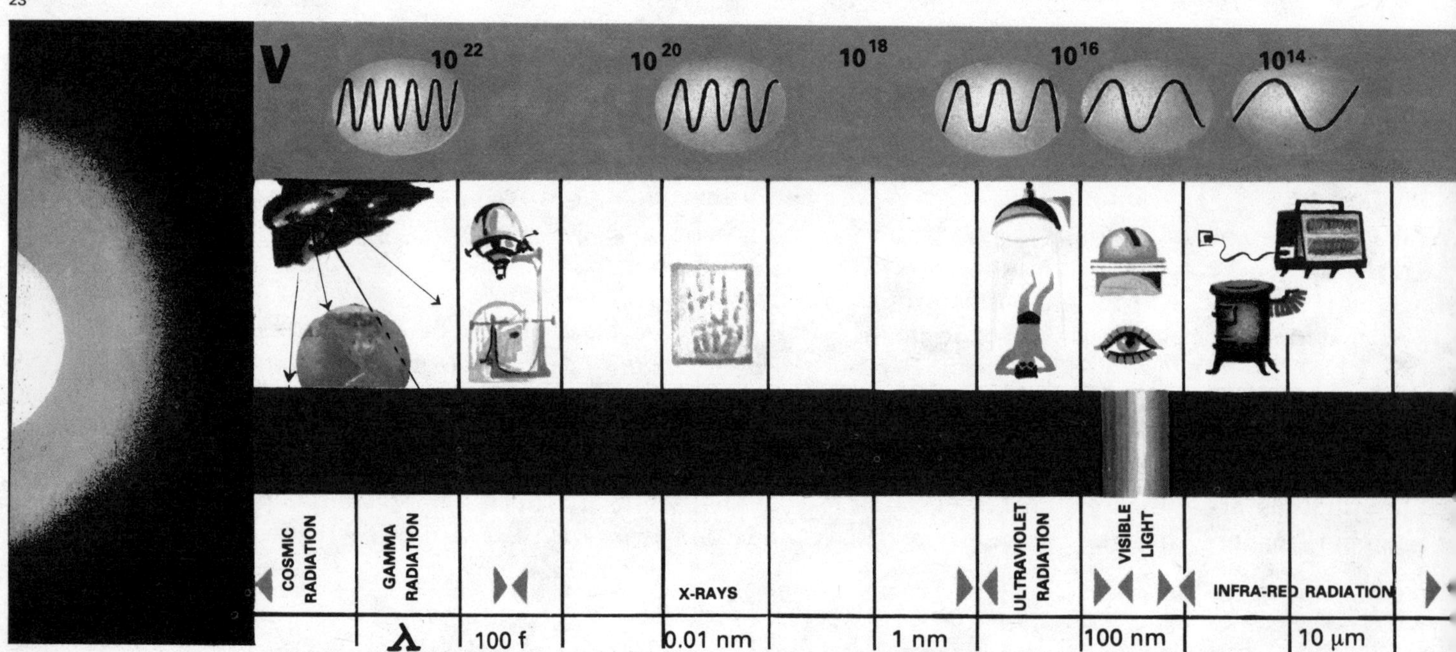

as a stream of photons (Figure 22). In the first case, that of the waves, we give the wavelength of the radiation. It is denoted by the Greek letter lambda (λ). In the case of the photons, we give the number of oscillations per second, which is denoted by the Greek letter nu (ν). An orange photon, for instance, has a wavelength of λ = 600 nm, and oscillates 5×10^{14} times a second (or at a frequency ν of $5 \times 10^{14}\ s^{-1}$). Every oscillation of the photon makes one wave – it covers the distance of one wavelength λ. Because it oscillates ν times a second, it covers a distance ν times longer than a wavelength, or $\nu \times \lambda$. But we know that in a vacuum every photon travels at 300,000 km (186,400 miles) per second (c = 300,000 km s^{-1}). This means that:

$$\nu \times \lambda = c$$

Let us test this in the case of an orange photon:

$$5 \times 10^{14}\ s^{-1} \times 600\ nm = 3 \times 10^{17}\ nm\ s^{-1} = 300{,}000\ km\ s^{-1}$$

It is easy to see that the higher the frequency of the electromagnetic radiation the shorter is the wavelength. A photon of violet light oscillates twice as fast as a photon of dark red light. Therefore the wavelength of violet light is only half that of red light. Radio photons, on the other hand, oscillate about one million times more slowly than light photons, so that radio waves are about one million times longer than light waves. The frequency of photons is rather like the different tones of the instruments in an orchestra. The high notes of the piccolo correspond to the gamma photons; the sound of the violin can be compared with light; the deep notes of the double-bass are like radio waves (Figure 24).

About 50 years ago astronomers were still observing the Universe by means of light alone. It was just as if they were listening to a melody played on the violin alone. But in the course of the last decade astronomers have learned to observe the Universe at all wavelengths of electromagnetic radiation. The whole range of radiation, from gamma to radio, gives us a complete picture of the Universe. It is just as though the whole of a philharmonic orchestra had joined in the piece.

Every body on Earth and everywhere in the Universe gives off electromagnetic radiation (photons) of various wavelengths. The higher the body's temperature, the higher the frequency with which it emits photons.

24 Observation of the Universe at all wavelengths of electromagnetic radiation can be compared to listening to the various instruments in an orchestra.

24

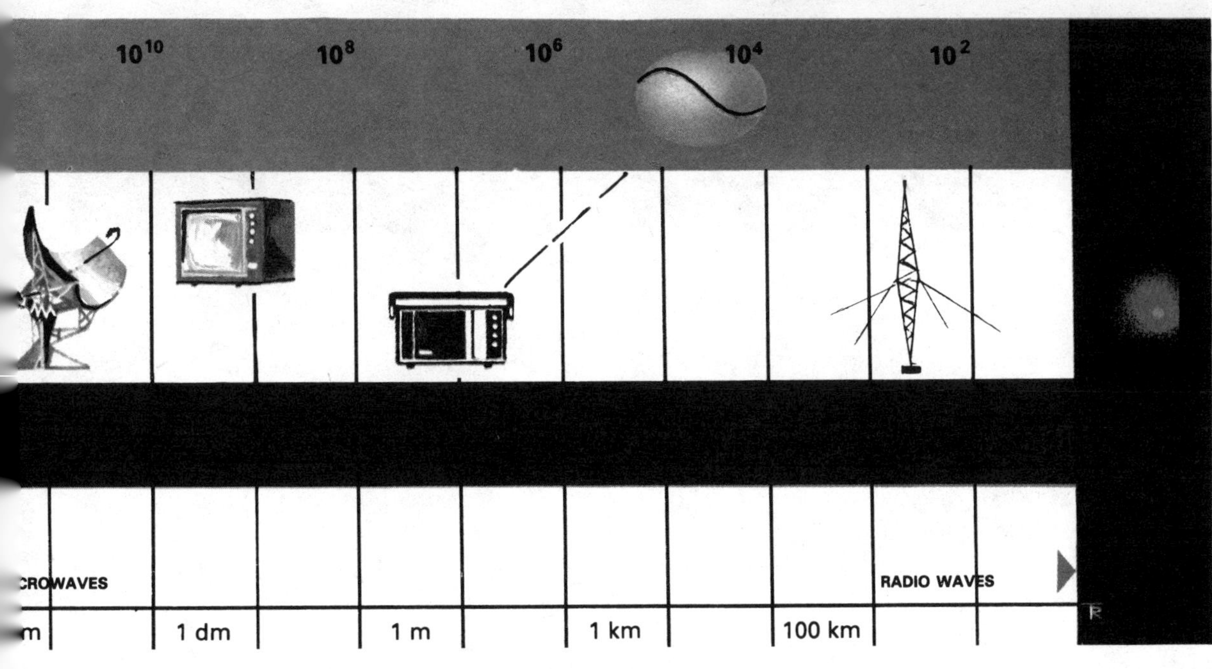

23 It is only in the last few decades that astronomers have been able to observe the Universe at all wavelengths of electromagnetic radiation. The upper scale gives the frequency of radiation in hertz. At the bottom are the corresponding wavelengths.

Figure 25 shows how much radiation is emitted by bodies at various temperatures, and what sort of radiation the body mostly emits – or at which frequencies the graph reaches its highest. This frequency is called the frequency of the maximum. The higher the temperature of the body, the higher is the frequency of the maximum. This dependence is called *Wien's Law of Displacement.* As-

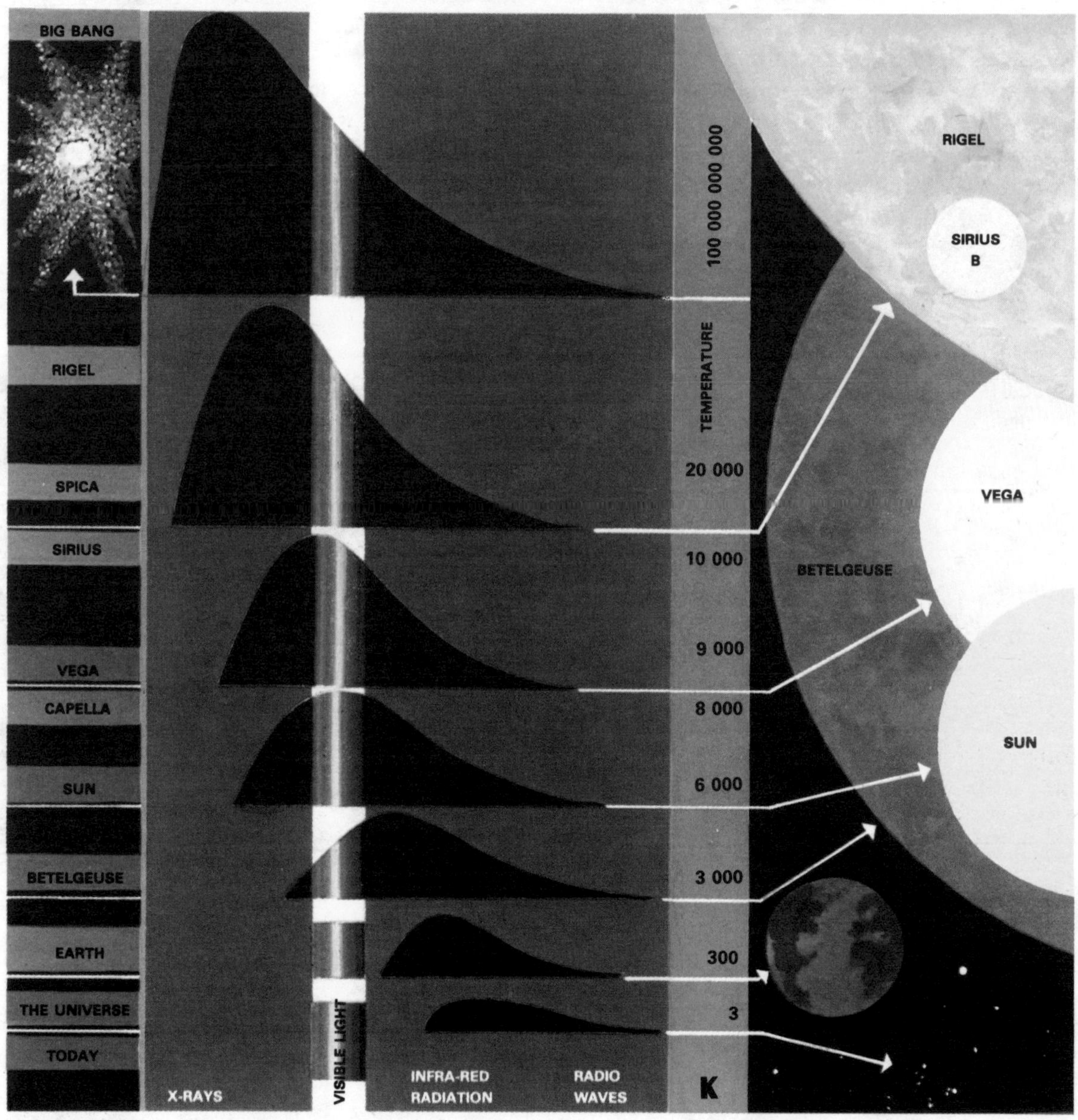

25 The radiation of cosmic bodies depends on their temperature. Temperature is given in kelvins. Here the size of various stars is shown; the graph shows how much radiation the star emits at various wavelengths.

tronomers measure the frequency of the maximum of a star's radiation by means of a spectrograph, and on the basis of Wien's law they can work out quite simply the temperature on that star's surface.

The radiation of stars (but also of other bodies) contains photons of all sorts of different frequencies. The spectrograph is an instrument which classifies the photons from a star according to their frequency. This 'signature' of the radiation is called a star's spectrum (Figures 25, 161 and 180). The spectrum of a star tells us not only its temperature, but also the velocity with which it is approaching us or receding from us, its chemical composition, speed of rotation, size, mass, and the amount of radiation energy it puts out. The spectra of stars and other bodies are of great importance for an understanding of the Universe.

If photons fall on a substance, one of several things can happen (Figure 6, on the right):

• Radiation falling on a smooth, shiny surface (such as a mirror) is *reflected*. When this happens the radiation changes direction and does not produce any change in the substance.

• Most radiation falling on a black, rough surface is *absorbed*. This means that the radiation is changed into heat, and the black object gets warmer.

• In the case of a transparent substance (such as glass), radiation *passes through* without being absorbed; if, however, the glass or similar material is coloured, some sorts of radiation are absorbed. Thus, for instance, only red light passes through red glass, while all other colours are absorbed and transformed to heat.

• *Diffusion* is the dispersion of light in all directions, and it occurs, for instance, on the molecules of air or on dust. Diffusion by the molecules of air disperses blue light in particular, whereas red light remains almost unaffected. This explains the blue sky and the redness of the Sun at sunset (Figure 26).

• If radiation falls on the surface of a transparent substance at an angle, the course it is following is *refracted*. The amount of this refraction increases the more slowly the radiation is transmitted through the transparent material in question.

• If white light falls on the wall of a glass prism, it is refracted. White light is composed of all the colours of the rainbow. Red light is transmitted in glass faster than blue light, so that blue light is refracted more (Figure 6). This is the reason white light is split up as it passes through a prism, a fact discovered by Isaac Newton around the year 1700. The band of colours into which a prism decomposes white light is called the *spectrum*. Something similar occurs in the crystals of ice which high clouds contain, giving rise to the magnificent sight of a solar halo (Figure 411). Another example is the rainbow, caused by the decomposition of sunlight in rain (Figure 412).

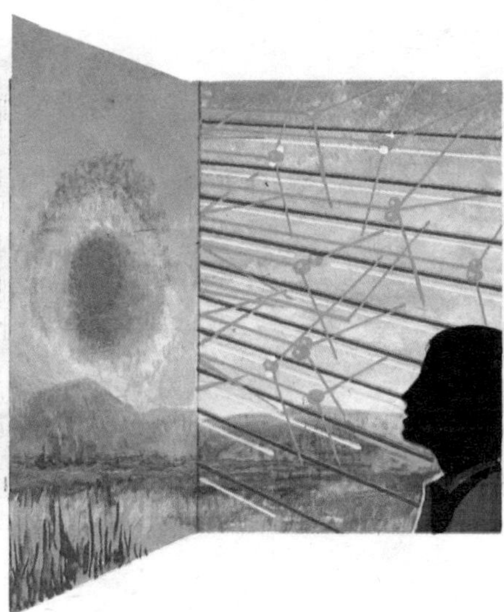

26

26 The diffusion of light on the molecules of air. Blue light is diffused most, while the red passes straight through. That is why the sky and the Earth are blue, while the Sun on the horizon is red.

27

27 The refraction of light in passing from one material to another. From experience the Indian knows he must aim a little lower than the place the fish appears to be.

• In passing through some substances (such as crystals), light is *polarized*. The light of the Sun or of an electric light bulb is unpolarized, or natural. This means that its photons are oscillating in different directions. But when light passes through a crystal only those photons which are oscillating in a certain direction can get through. Thus polarized radiation is organized and unidirectional, whereas unpolarized, natural light is disorganized.

Photons of various energies fly through the Universe. They occur not only in the endless expanses between the stars and galaxies, but also in stars themselves. In the heart of our Sun, for example, there is a vast quantity of X-ray photons. Inside heavy stars, whose mass is 10 or more times that of our Sun, there are countless gamma photons. It can be said that photons saturate the whole Universe and all the bodies in it, making up what is called the *photon component* of the Universe. At one time, right at the start of the existence of the Universe, photons were far more important than all the other particles. We shall be discussing this in Chapter 3. But photons have an important role even in today's Universe: they transmit energy, for example from the Sun to the Earth, and provide information about the bodies they have left and the space through which they have passed. It is thanks to solar photons that we are able to live here on the Earth, and our knowledge of the Universe is based on the photons which strike our planet from all directions.

All photons move through space at the same speed. For this reason they arrive from the nearest stars sooner than from distant galaxies. Thus the photons which are now reaching the Earth from all over the Universe are of various ages – the older they are, the farther away is the body which emitted them. Photons from meteors are about one-thousandth of a second old, those from the Moon $1\frac{1}{4}$ seconds old, those from the Sun take eight minutes to get here, and those from the nearest stars several years. The photons which are just arriving from the middle of the Milky Way are already 30,000 years old, those from the galaxy in Andromeda 2 million years, and the photons from distant galaxies have taken a few thousand million years to make the trip. The oldest photons of all are 10,000 million years old, and are called *fossil photons*. They were present at the very start of the Universe (Chapter 3).

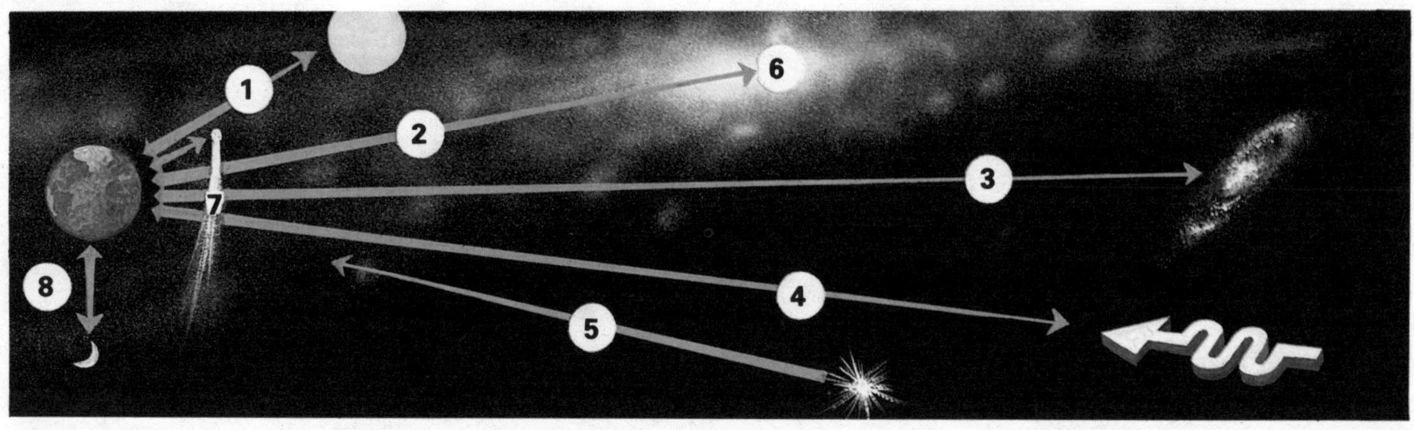

28 The photons which arrive on the Earth are the older, the farther away the body which emitted them. **(1)** the Sun (8 minutes), **(2)** the spiral arm of the Milky Way in the constellation of Sagittarius (10,000 years), **(3)** the galaxy in Andromeda (2 million years), **(4)** fossil photons (10,000 million years), **(5)** quasar (7000–10,000 million years), **(6)** the centre of the Milky Way (30,000 years), **(7)** a comet (5–30 minutes), **(8)** the Moon (1 second).

A look into space is a look into the past (Figure 29). And that past is the more distant, the farther into space we look. This is because of the finite speed of light. Whatever is happening, for example, in the neighbouring galaxy M 31 in the constellation of Andromeda, could be observed by our descendents after two million years, if they survive that long. For that is how long it takes a ray of light to get from galaxy M 31 to our own galaxy. In spite of this vast distance, galaxy M 31 is visible with the naked eye in the autumn sky (Figure 133). It is the most distant object in the whole of the Universe which we can see without the aid of a telescope.

Antiparticles

In order to get a better understanding of the word 'anti', you should take a look in the mirror. The person you see there is not exactly you, however like you he or she may be. If you part your hair on the left, then he or she parts it on the right. If you have a mole on one cheek, then he or she has it on the other. If you put your right hand to your heart, then that other person puts his or her left hand to the right-hand side of his or her chest. The creature hasn't even got its heart in the right place! Everything is the wrong way round; you might say that you were looking at your antiperson.

Not only your antiperson, but also everything else is the other way round when you look in a mirror: the hands of a clock move anticlockwise, right-handed people become left-handed, and so on. The world we see in a mirror is an *antiworld.* But it can exist quite well enough. There are even people who really do have their heart on the right-hand side, their appendix on the left, and so on.

Now we can understand what antiparticles are: they are elementary particles with the opposite properties to protons, neutrons, electrons and the others, which we discussed in detail on page 14. Antiparticles have opposite electrical, baryon and lepton charges. So, for example, an antiproton has an electrical charge of -1, a baryon charge of -1 and a lepton charge of 0.

Now we know what antiparticles are, we are left with the question of how they arise, what happens when they meet their opposite numbers (for example what happens when an electron meets a positron), and what their importance is to the Universe. Research in accelerators has shown that when antiparticles meet their counterparts they change into gamma radiation. This process is called *annihilation* (see page 40). Antiprotons, antineutrons and positrons make up anti-atoms and these in turn make up *antimatter* (Figure 30). Antimatter played a very important role in the creation of our Universe (Figure 157).

29 Let us imagine that it was possible, using a perfect telescope, to observe a planet 80 light years away, and to see a five-year-old boy on it. In fact the 'boy' would be 85. We should be seeing what he looked like 80 years ago, when the rays (which are reaching our eyes now) set off from such a planet.

Forces between particles — what holds the Universe together

We have said quite a lot about elementary particles in the preceding sections. The whole Universe and everything in it is made of them, from atoms to supergalaxies. We have looked at the properties of individual particles, and now we are going to consider the 'mortar' of the Universe, which holds the particles together.

There are four forces that hold elementary particles together: the nuclear, the electrical, the weak and the gravitational. It is this foursome that makes up the 'mortar' of the Universe, without which nothing at all could exist — not an atom, not a blade of grass, not a single star.

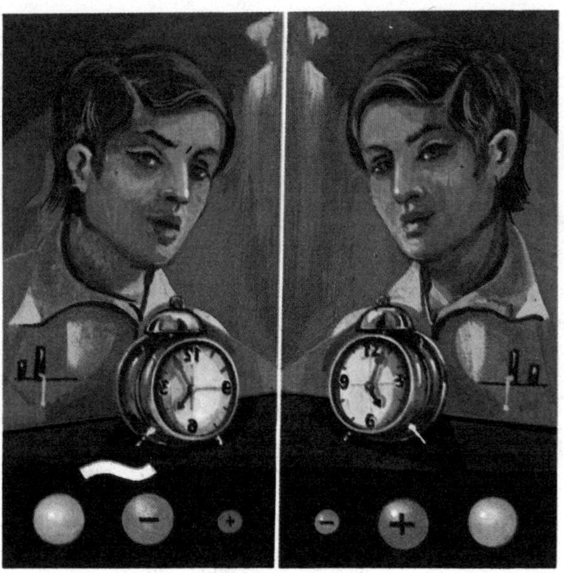

30 The antiworld is symmetrical to our own.

Every elementary particle has at least one force with which it acts on the others around it. The space around a particle in which its force acts is called its *force field.* The size of a force field depends on the type of force involved. The nuclear field is small, the nuclear force acting only in the closest vicinity of the particle (up to a distance of 1 fermi – Figure 33). The gravitational field of a particle, on the other hand, stretches an enormous distance. Though it is extremely weak, the gravitational force of a particle reaches out many million light years.

It is as though each force had its source in elementary particles, like a spring. Elementary particles are therefore said to be the source of a force or the source of a field. Particles with an electrical charge are surrounded by an electrical field, which means that the electrical charge of a particle is the source of its electrical force. Similarly hadrons (baryons and mesons) are a source of a gravitational force.

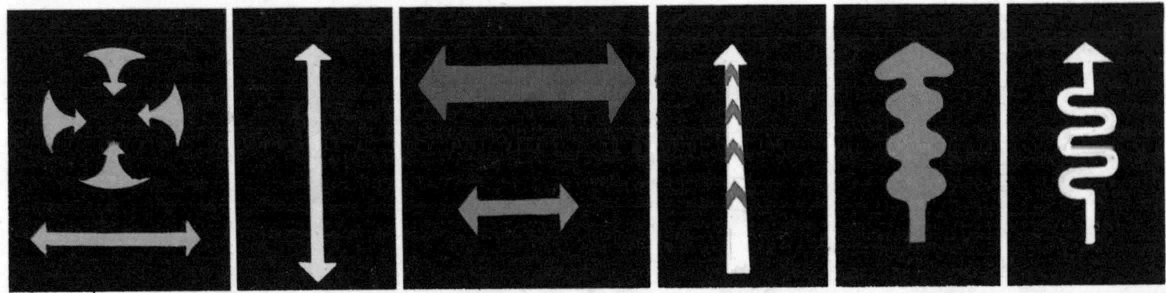

31 The figure shows the symbols which will be used throughout the book. The colour of arrows shows the type of force or energy which is involved: green = gravitation, yellow = electromagnetic force, brown = nuclear interaction, white arrow = centrifugal force, orange arrow = energy, white wavy arrow = photon and radiation.

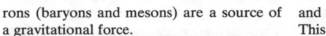

Every field around a particle contains energy. Let us suppose that we could suddenly remove the particle. For an instant the field would be on its own, and then would spread, independent of the particle which was its source. The field would gather into lumps which we call the quanta or particles of the field. A nuclear force gives rise to mesons, an electrical force gives rise to photons, a weak force gives rise to W-particles, and a gravitational force gives rise to gravitons. Note that all four field quanta are bosons (Figure 14). Thus field quanta have either zero or integer spin. They may have an electrical charge (mesons π^+, π^-, K^+, K^-) or no charge at all (such as the mesons π°, K°, the graviton and the photon). Their rest mass is either zero (photon, graviton) or many times that of an electron (mesons, W-particles). We should again point out that the graviton has yet to be observed, though we have every reason to suppose that it really does exist.

The basic properties of the four forces that make up the Universe are summarized in Figure 32. This illustration will give you an idea of the various forces (interactions) that occur between elementary particles:

• The *nuclear force* is the strongest of all, but it acts only over a very short distance – 10^{-15} m, which is 1 fermi, or one-billionth of a millimetre. Its source is the hadrons (baryons and mesons). As soon as two protons, for example, approach to within one fermi of each other (a distance equal to their diameter), they start to attract each other by means of the immense nuclear force (Figure 33). At such a short distance the attraction of the nuclear force between protons is much greater than their repulsion by the electrical force.

The nuclear force binds nucleons together and forms them into the nucleus of an atom. This ensures the stability of the nucleus against outside influences such as collisions. It was the nuclear force which constructed the nuclei of all the more complex elements from protons (i.e. the nuclei of hydrogen atoms). The formation of the heavier atomic nuclei as far down as iron (in the periodic table) is still taking place in the interiors of stars. The amount of hydrogen in the Universe is thus decreasing and the amount of heavier elements increasing. This change, brought about by the nuclear force, is called the *chemical ageing of the Universe* (Chapter 3).

When protons are converted into heavier nuclei energy is released, which is the source of the radiation of stars. Stars are, in effect, huge nuclear reactors. Almost all the energy on the surface of our own planet comes from the interior of the Sun. It was released in the solar interior by the nuclear force in the course of the conversion of hydrogen into helium (Chapter 4).

• The *electrical force.* The most important particles from which the Universe is made, the protons and electrons, have an electrical charge. But as you can see in Figure 14, other particles and antiparticles are also charged. Their electrical charge can have only one of two values: either 1.6×10^{-19} coulombs (for example the charge of protons or positrons), or -1.6×10^{-19} coulombs (for example the charge of electrons or antiprotons). Particles with the same sized charge may have widely differing rest masses. It is strange how precisely the same elementary charge has been given to all the charged particles in the Universe, without regard to their mass.

Electrons can be transferred from one body to another (Figure 20). So, for example, if we rub a glass rod with a silk scarf, electrons from the rod pass into the scarf. The rod then has a shortage of electrons – more protons than electrons. It is positively charged. The silk, on the other hand, has more electrons than protons, and is negatively charged. But if we then place the scarf against the glass rod, the electrons run back into the rod, so that each proton again has its own electron. Now neither the rod nor the scarf is charged – they are neutral.

The electrical charge of elementary particles is a source of electrical forces and electrical fields. Charged particles act on each other by means of an electrical force. If they have the same charge, they repel each other. If they have opposite charges, they attract each other (Figure 35). If a charged particle moves, it forms alongside itself a magnetic field in which a magnetic force acts. Every magnetic force is caused by the movement of charged particles, usually electrons or protons (Figure 36). If an electrically charged particle is rapidly accelerated or decelerated, it emits photons (quanta). A stream of electrically charged particles, such as electrons in a copper wire, is called an electric current.

Electrical forces play a vital role in the structure of things. A positive nucleus draws to itself negative electrons by means of an electrical force, thus giving rise to an atom (Figure 12). An electrical force binds atoms together to form molecules; molecules combine to form crystals, grains of dust or drops of water; and also living organisms, from single-celled organisms to man himself. Larger bodies, such as rocks, comets, small planets and lesser moons (up to 500 km/310 miles in diameter) are also held together mainly by electrical force. For this reason they are often irregular in shape (Figure 48). Though their molecules are also attracted by gravitational force, in the case of these bodies it is insufficient to mould them into a globe shape.

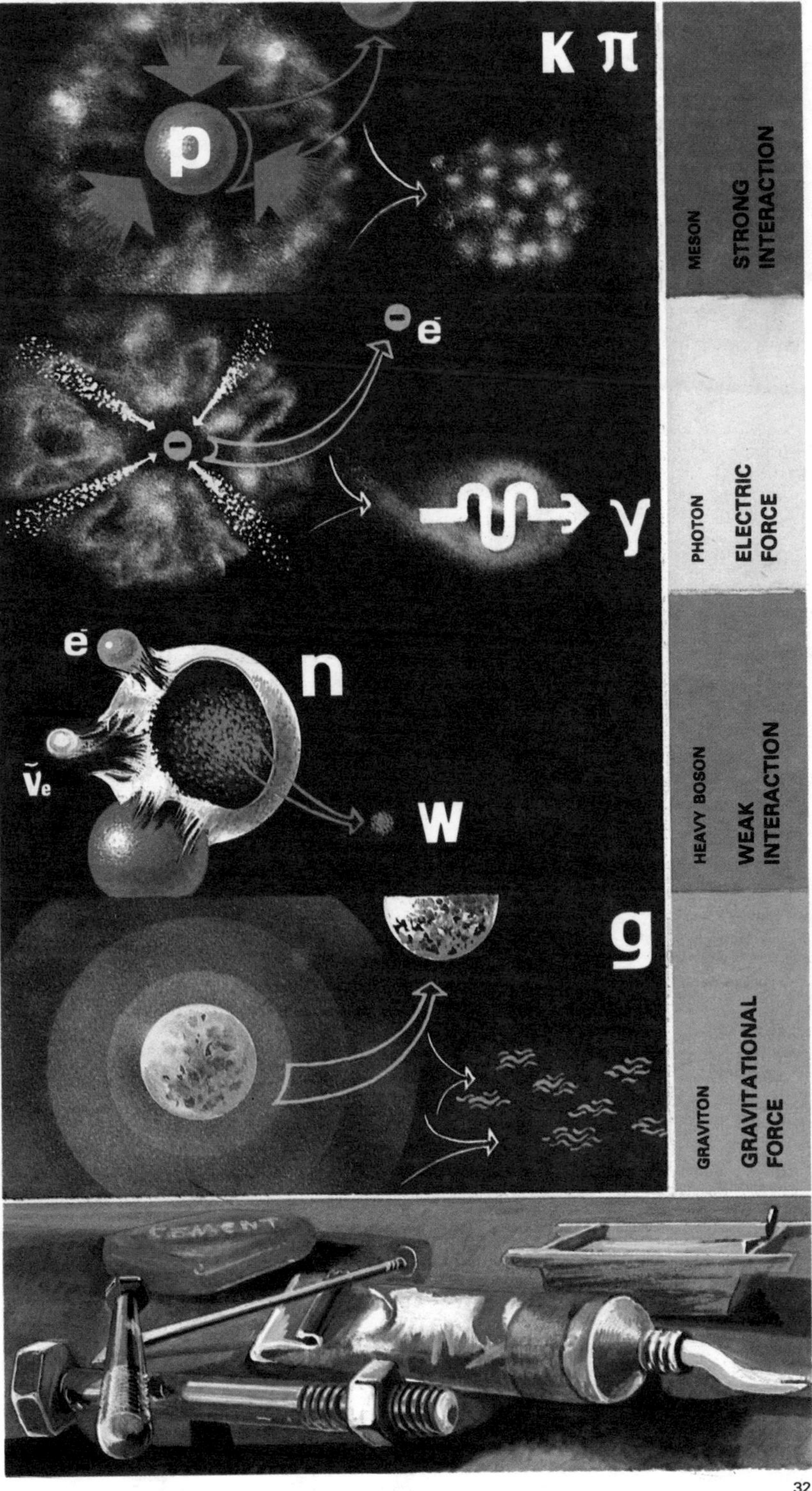

32 The basic forces (interactions) which hold the Universe together. The figure shows the field particles (K, π, γ, W, g) of individual interactions.

The electrical force and the associated magnetic force take effect in all sorts of ways. The structure of the electron envelope of the atom, the properties of molecules and the cohesion of molecules into crystals, and various chemical and physical properties of gases, liquids and solids, are all the result of the electrical force between the electron and the proton. Hardness, transparency and opa-

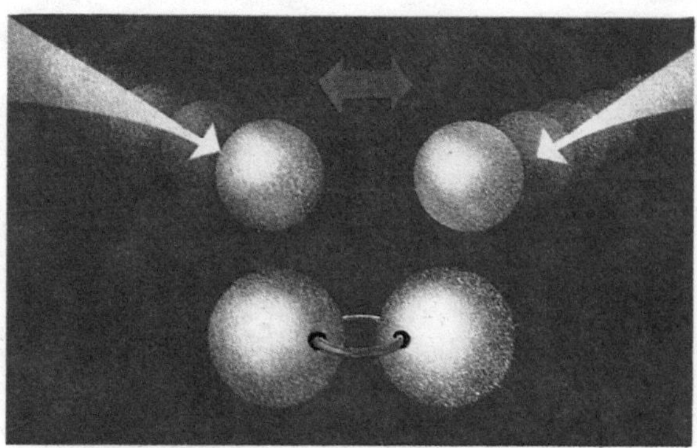

33 The nuclear force acts at a distance of 1 fermi (the length of the brown arrow).

queness of materials, perception by the senses, nervous excitement, the work of our brains and muscles and a host of other phenomena and processes on Earth and throughout the Universe can all be put down to this same electrical force.

• *Weak interactions.* While nuclear forces are confined to the close vicinity of hadrons, weak interactions take place inside particles. Their range is one hundred times less (10^{-17} m) than that of the nuclear forces (10^{-15} m). Because the size of elementary particles is about 10^{-15} m, it is clear that weak interactions take place in their very centres.

The best-known weak interactions are those in the course of which neutrinos and the antineutrinos come into being. Let us give here two examples known as beta-radioactivity.

A neutron on its own, i.e. a free neutron not bound in a nucleus, disintegrates as a result of weak interactions into three particles: a proton (p), an electron (e^-) and an antineutrino ($\tilde{\nu}_e$). This disintegration is expressed as follows (Figure 37):

$$n \rightarrow p + e^- + \tilde{\nu}_e$$

Electrons were previously known as beta-particles (or β^--particles), so this disintegration is known as beta disintegration (β^--disintegration). A free neutron disintegrates in this way in about 10 minutes. A neutron in an atomic nucleus can also disintegrate, and can do so in a fraction of a second or over a very long period (depending on the type of nucleus).

Some radioactive nuclei emit positrons (or β^+-particles). In these the weak interaction brings about the disintegration of a proton

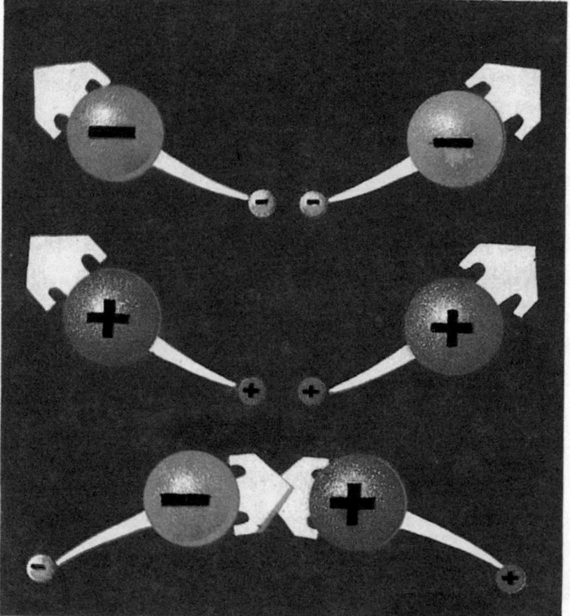

34 Every electron and proton in nature has the same amount of charge.

35 Electrical forces may be either of the repelling or attracting variety.

into a neutron (n), a positron (e^+) and a neutrino (ν_e). Thus (Figure 38):

$$p \rightarrow n + e^+ + \nu_e$$

A solitary proton (not bound to a nucleus) cannot disintegrate, since its mass is too low. But in a nucleus a proton can disintegrate, since it acquires the required mass and energy from the other nucleons. An example of the disintegration of a proton in a nucleus is beta disintegration (more precisely β^+-disintegration) of a carbon isotope:

$$^{11}_{6}C \rightarrow {}^{11}_{5}B + e^+ + \nu_e$$

$^{11}_{6}C$ is the nucleus of carbon, in which there are 11 nucleons (upper figure) and six of these are protons (lower figure). $^{11}_{5}B$ is the nucleus of boron with five protons and six neutrons, i.e. 11 nucleons in all. In the beta disintegration of the nucleus of carbon one proton is transformed by the weak interaction into a neutron, a positron and a neutrino.

Let us return to the two transformations caused by the weak interaction. Note that the neutrino is denoted ν_e and its antiparticle, the antineutrino, in the same way, but with a wavy line (tilde), i.e. ~. The tilde is also used to denote other antiparticles.

These two disintegrations are a good example of the conservation of the electrical, baryon and lepton charges. The neutron has no electrical charge, and the particles produced by its disintegration (shown on the right-hand side of the arrow) have an electrical charge of +1 (proton p), −1 (electron e^-) and zero (antineutrino $\tilde{\nu}_e$). Thus the sum of the electrical charges after the arrow is zero. When a proton disintegrates the same principle applies. In all transformations in the Universe the sum of electrical charges remains constant.

Similarly, let us verify the conservation of the baryon and lepton charges. Remember that the neutron and proton have a baryon charge of +1 and a lepton charge of zero. The electron and the neutrino have a baryon charge of zero and a lepton charge of +1. Both the antileptons $\tilde{\nu}_e$ (the antineutrino) and e^+ (the positron) have a baryon charge of zero and a lepton charge of −1. In both of the radioactive disintegrations the baryon charge is thus +1 before as well as after disintegration. The lepton charge before disintegration is zero, and the sum of the lepton charges of the resulting particles is also zero.

In the case of the disintegration of the carbon nucleus the three charges of the elementary particles are also conserved. It is easy to see that there are 11 baryon charges before the arrow, six positive electrical charges and no lepton charges. After disintegration there are still 11 baryon charges (in the nucleus of the boron), six positive electrical charges (five in the boron nucleus and one on the positron), and the total lepton charge (−1 on the positron and +1 on the neutrino) is zero.

Note one other point. When the neutron and the proton disintegrated, an electron and a positron were produced together with, very importantly, an electron neutrino and an electron antineutrino. Whenever an electron or a positron either appear or disappear, an electron neutrino or an electron antineutrino is involved. When a muon (μ^-, μ^+) appears or disappears, on the other hand, a neutretto or an antineutretto (ν_μ, $\tilde{\nu}_\mu$), i.e. the muon

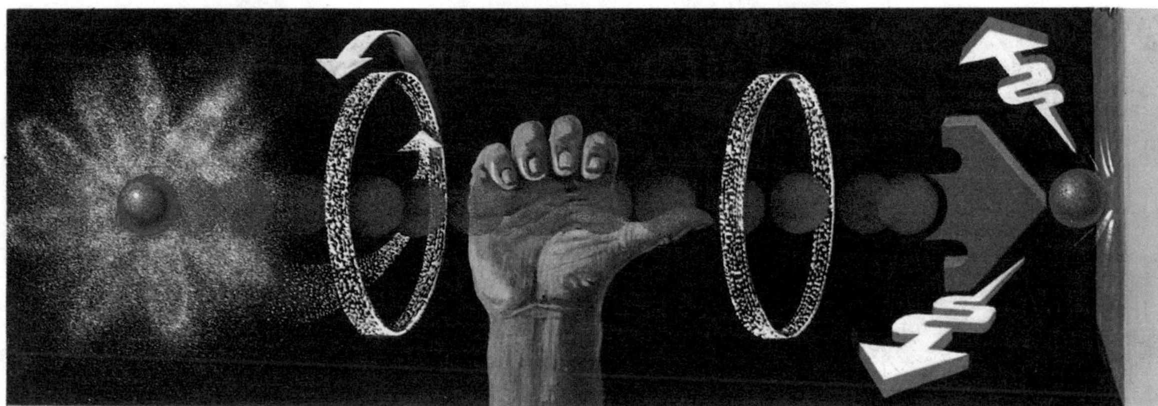

36

36 The creation of an electromagnetic field. A charged particle (such as a proton) is surrounded by an electrical field (shown only for the proton on the left as a white background). In the course of movement a magnetic field is formed (only two magnetic force lines are shown as white circles). If a charged particle is stopped, it emits photons (shown as the wavy arrows to the right).

neutrinos, is to be found. Let us take the example of the disintegration of π-mesons (Figures 39 and 41):

$$\pi^- \rightarrow \mu^- + \tilde{\nu}_\mu$$

and

$$\pi^+ \rightarrow \mu^+ + \nu_\mu$$

The lepton charge is conserved in these transformations, as we can easily show. The mesons π^- and π^+ have a lepton charge of zero; μ^- and ν_μ have a lepton charge of +1, while μ^+ and $\tilde{\nu}_\mu$ are antileptons with a lepton charge of −1.

An even more interesting case is that of the disintegration of muons. When this occurs either a neutretto (ν_μ) or an antineutretto ($\tilde{\nu}_\mu$) must appear. Since this disintegration gives either an electron or a positron, an electron neutrino (ν_e) or electron antineutrino ($\tilde{\nu}_e$) must occur. Thus, as illustrated in Figures 40 and 43,

$$\mu^- \rightarrow e^- + \tilde{\nu}_e + \nu_\mu$$

and

$$\mu^+ \rightarrow e^+ + \nu_e + \tilde{\nu}_\mu$$

Here, too, it is simple enough to show that the lepton charge is conserved when muons disintegrate. In fact it is conserved in each of the two groups, the electron group (e^-, e^+, ν_e, $\tilde{\nu}_e$) and the muon group (μ^-, μ^+, ν_μ, $\tilde{\nu}_\mu$), separately.

All four types of neutrino occur on Earth: in the atmosphere, through the disintegration of mesons and muons in cosmic radiation, in the thermonuclear reactions in the Sun and other stars, and in the extremely incandescent cores of heavy and very heavy stars (page 118). But they also arise in the vast expanse of interstellar space through collisions between cosmic radiation and interstellar gas.

We have already given several examples of weak interactions (weak forces) in the course of which a neutrino appears. We shall be looking at some more later (page 118). The neutrino behaves very indifferently towards the other particles. This is both a handicap and a boon. Once a neutrino is born, through a weak reaction, it flies through space on a practically endless journey, passing through stars, planets and galaxies. But for this very reason the neutrino is also our one hope of ever being able to observe the inside of the Sun and other stars.

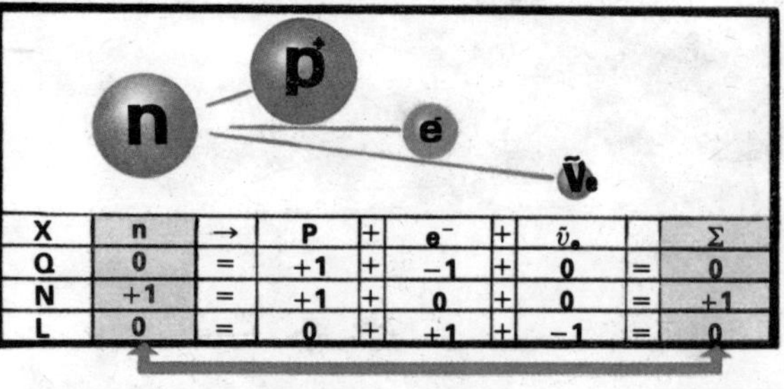

X	n	→	P	+	e^-	+	$\tilde{\nu}_e$		Σ
Q	0	=	+1	+	−1	+	0	=	0
N	+1	=	+1	+	0	+	0	=	+1
L	0	=	0	+	+1	+	−1	=	0

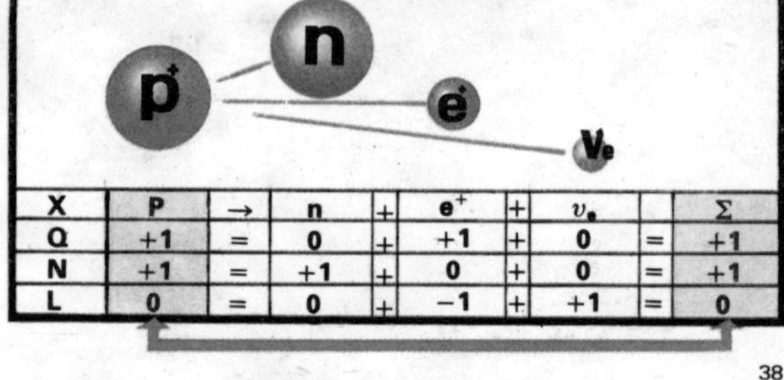

X	P	→	n	+	e^+	+	ν_e		Σ
Q	+1	=	0	+	+1	+	0	=	+1
N	+1	=	+1	+	0	+	0	=	+1
L	0	=	0	+	−1	+	+1	=	0

37 The disintegration of a neutron (beta disintegration). The electrical charge Q and the baryon number N are conserved, as is the lepton number L. The letter X shows the particle in question. The symbols apply to Figures 37–42.

38 Disintegration of a proton by the weak interaction. Such a disintegration can proceed only in a nucleus. Why?

Only when scientists at some future date have at their disposal sensitive detectors of neutrinos and antineutrinos will it be possible to say how much antimatter and how many antistars and antigalaxies there are in the Universe. Stars made up of ordinary particles (protons, electrons and neutrons) are a source of electron neutrinos, since protons disintegrate into neutrons (Figure 38) in them. But antistars, composed of antiprotons, antineutrons and positrons, emit electron antineutrinos, because in these negative antiprotons ($\tilde{p}$) disintegrate into antineutrons ($\tilde{n}$), electrons and antineutrinos (Figure 42):

$$\tilde{p} \rightarrow \tilde{n} + e^- + \tilde{\nu}_e$$

Here we again see the laws of the conservation of the electrical, baryon and lepton charges. A photon can never reveal whether it comes from a star or an antistar, since the photons emitted by an antistar are the same as those emitted by a star. Thus the photon and the antiphoton are completely identical particles. Only the neutrino and the antineutrino can tell us whether we are observing a star or an antistar.

• The *gravitational force* acts between all particles and antiparticles. It is always a force of attraction, but under the conditions of the microcosm it is exceptionally small (about 10^{40} times smaller than the nuclear force). For this reason it takes no part in the structure of the nuclei of atoms, molecules, rocks, small planets or small moons. It only plays a major role in the make up of planets, stars and all bodies with more than 10^{46} elementary particles. Such bodies have a diameter of about 500 km (310 miles) or more. Each particle in a body is attracted to all the others (this is called the *self-gravitation* of a body). By means of self-gravitation bodies seek to assume the shape of a globe. In the case of smaller bodies, perhaps only a few tens of kilometres across, self-gravitation is not sufficient to break up the crystalline structure of the body. But as soon as the body is larger than about 500 km (310 miles) in diameter the self-gravitation is able to overcome the electrical force in the crystals, and the solid matter inside becomes pliable, so that the body takes on the shape of a ball. For this reason the Moon (diameter 3476 km/2159 miles) is round, while small planets and small moons (such as Phobos and Deimos, the moons of Mars) are irregular lumps of rock (Figures 48, 75 and 77).

Gravitation is the most widespread force, since it acts between any two particles. On Earth we also speak of gravity (from the Latin *gravis,* meaning heavy) since it is gravitation that gives things weight. It is gravitation which keeps us on the surface of the Earth (Figure 46). Without it we would go flying off into space. But every particle of the Earth (there are about 10^{51} of them) attracts every particle of our bodies (which have about 10^{29} particles). This makes a total of 10^{80} very weak forces holding us down (Figure 44, on the right).

On Earth, gravitation is a very important force. Warm air is light and rises. Cold air flows in to take its place – which is how winds occur. Water vapour in the rising air condenses and forms clouds. The droplets of condensed vapour gather together into larger drops, and these fall to the ground in the form of rain. It is the force of gravitation which makes the raindrops fall. When they reach the ground the raindrops form rivulets, which run into streams. The streams run into rivers and the rivers into bigger rivers, which flow down to the sea. All because of the Earth's gravitation.

Without gravitation fire would not be able to burn, since hot air would not rise. The hot air which rises from the fire is replaced by cold, heavy air – air which is, even more importantly, rich in oxygen, enabling fire to continue to burn.

You may have noticed that bricklayers use a plumbline to ensure their walls are perfectly vertical. This length of twine with a weight on the end shows the exact direction of the force of gravity. But nature has always determined that plants grow in the direction

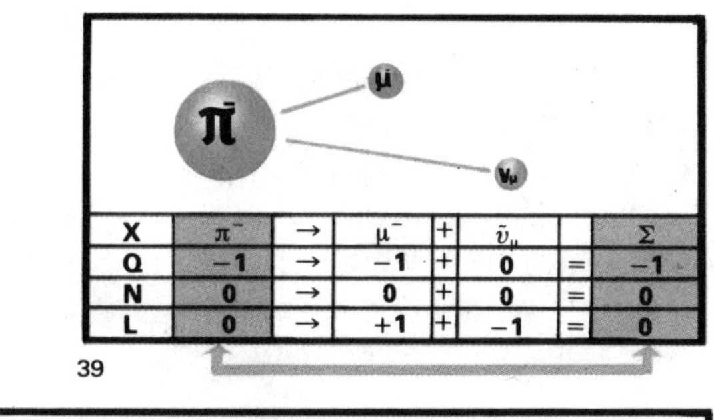

X	π^-	→	μ^-	+	$\bar{\nu}_\mu$		Σ
Q	−1	→	−1	+	0	=	−1
N	0	→	0	+	0	=	0
L	0	→	+1	+	−1	=	0

39

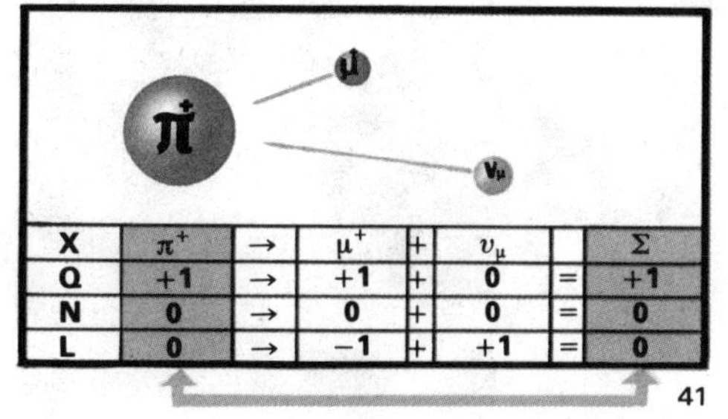

X	π^+	→	μ^+	+	ν_μ		Σ
Q	+1	→	+1	+	0	=	+1
N	0	→	0	+	0	=	0
L	0	→	−1	+	+1	=	0

41

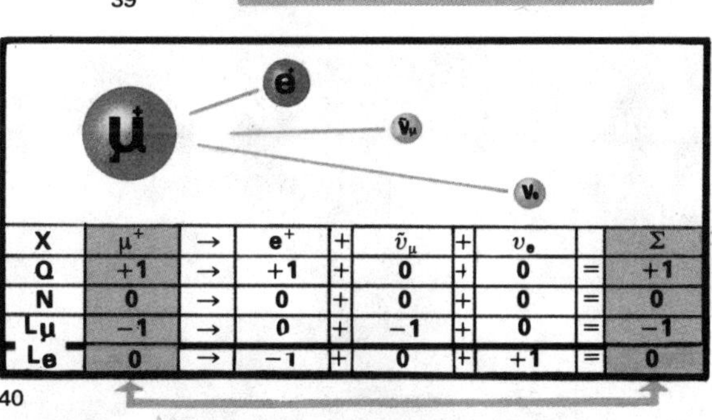

X	μ^+	→	e^+	+	$\bar{\nu}_\mu$	+	ν_e		Σ
Q	+1	→	+1	+	0	+	0	=	+1
N	0	→	0	+	0	+	0	=	0
L_μ	−1	→	0	+	−1	+	0	=	−1
L_e	0	→	−1	+	0	+	+1	=	0

40

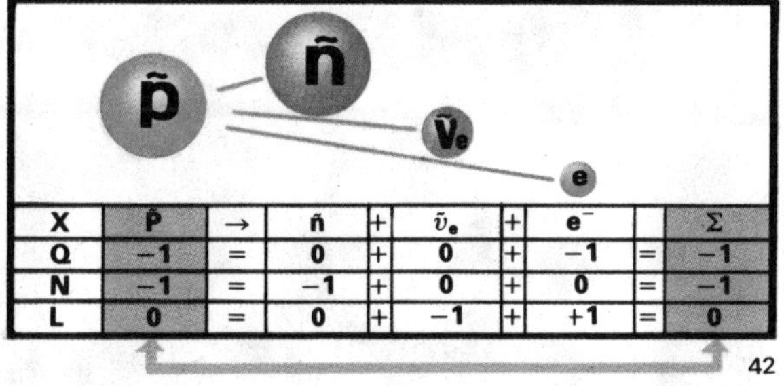

X	$\bar{p}$	→	$\bar{n}$	+	$\bar{\nu}_e$	+	e^-		Σ
Q	−1	=	0	+	0	+	−1	=	−1
N	−1	=	−1	+	0	+	0	=	−1
L	0	=	0	+	−1	+	+1	=	0

42

39 Disintegration of π^- mesons.

40 Disintegration of muons.

41 Disintegration of π^+ mesons.

42 Disintegration of antiprotons.

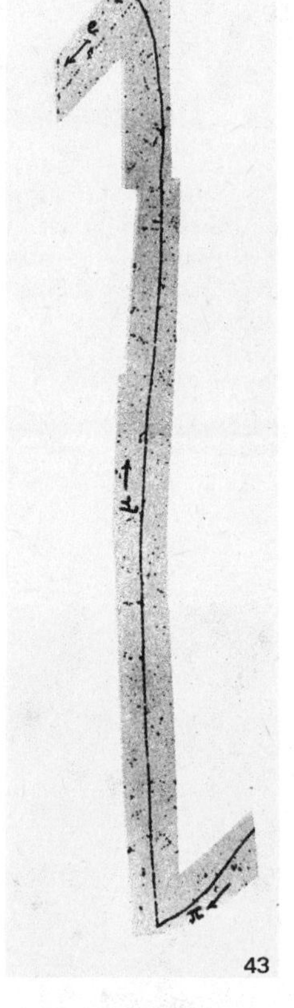

43

43 The π meson at the bottom disintegrated into a μ^- (upward path) and an antineutretto $\bar{\nu}_\mu$. At the top the μ^- disintegrated into an electron e^-, a neutretto ν_μ and an antineutrino $\bar{\nu}_e$. They have no electrical charge and left no trace in the photographic emulsion.

44 The action of gravitation. Every nucleon in our body is attracted by every nucleon in the Earth (right-hand picture). Fire, rain, water in a reservoir, the direction in which houses are built and the growth of trees are all the result of gravitation.

of gravitation (a phenomenon which is called geotropy). Nature has also equipped us with small canals in the inner ear (the part called the labyrinth) which, due to the force of gravity, help us to walk and stand upright.

For the sake of precision, it should be said that gravity is not quite the same thing as gravitation. If the Earth did not rotate, gravitation and gravity would be exactly the same. But, as a result of this rotation, there is a slight centrifugal force which marginally offsets gravitation. The difference is so small, however, that we are unable to perceive it (Figure 45).

Gravitation is the most important force in the Universe. It has formed galaxies, stars, planets and large moons, and given them their regular shapes (Figures 48 and 79). It was gravitation that warmed the newborn stars (see page 105), so that thermonuclear reactions could take place in them. The orbiting of the planets and comets around the Sun and of moons around their planets, and the flight of artificial satellites and space probes are all due to gravitation. And the motion of the Solar System and the stars, clouds, interstellar material, and star clusters around the centre of the Milky Way, are caused by the gravitational force of all the elementary particles (about 10^{69} in all) of which our Galaxy is made.

The first to calculate the force of gravitation accurately was Isaac Newton. Let us imagine two bodies of respective mass m_1 and m_2 kg at a distance from each other of r metres (Figure 46). According to Newton's Law the two bodies are attracted by a force F, expressed in units of N (newtons), for which the following is the equation:

$$F = 6.7 \times 10^{-11} \times \frac{m_1 \times m_2}{r^2}$$

According to Newton's Law the gravitational force is proportional to the product of the mass of the two bodies, and inversely proportional to the square of the distance between them. The force which attracts the first body to the second is equal to that which attracts the second to the first. Both forces act along a line joining the two bodies (Figure 46). Newton's Law allows us to work out the mass of the Sun (see page 124) and other heavenly bodies.

Einstein's general theory of relativity explains gravitation as the curvature of the space around bodies. The Earth has a relatively small mass, and the space around it is only slightly curved, so that Newton's law of gravitation can be applied. But in the case of very high-density stars the curvature of the surrounding space is so great that the general theory of relativity, or Einstein's equations of gravitation, must be used instead of Newton's Law. They are complicated equations which express the dependence of the curvature of the space (space-time) on the density of matter and its energy in these super-dense bodies (Figure 47).

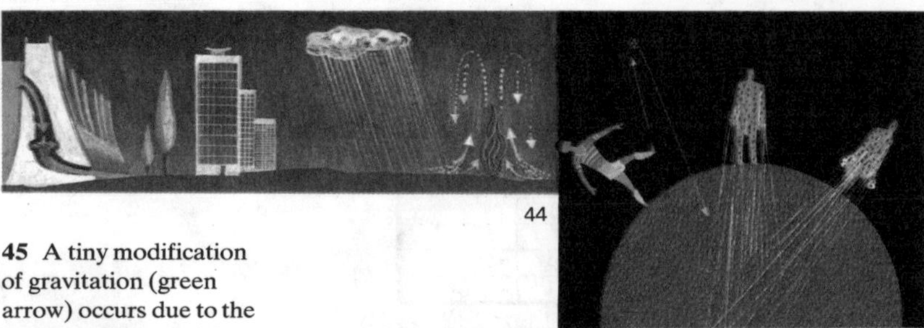

44

45 A tiny modification of gravitation (green arrow) occurs due to the centrifugal force of the Earth's rotation (red-and-white arrow). The resulting force (gravity) is marked by the thin white arrow.

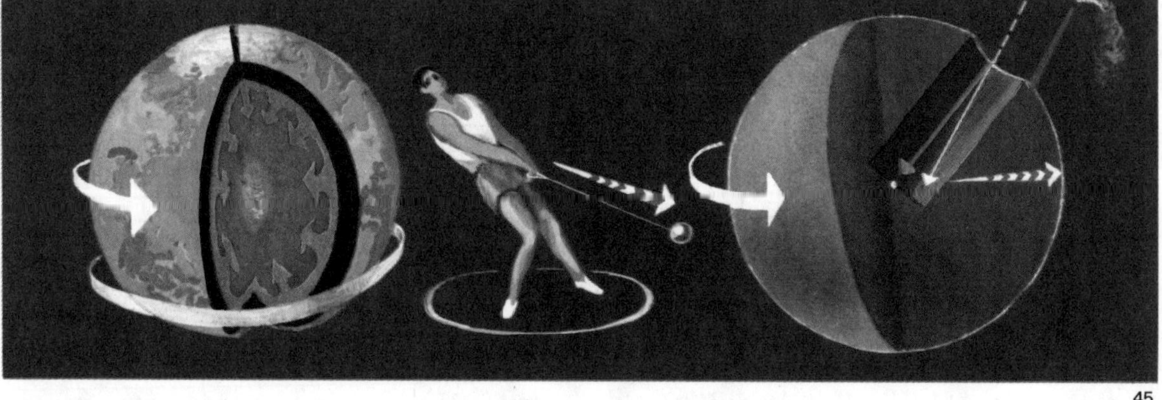

45

46

46 The Earth attracts an apple with the force F. The apple attracts the Earth with the same force. The magnitude of the force is determined by Newton's Law.

Gravitation is of immense importance not only for the structure of individual large bodies and of systems of large bodies, but also for their evolution. It was decisive in the formation of supergalaxies, galaxies and stars. In the chapter on the evolution of the Universe (Chapter 3), we shall see that gravitation also plays a part in the end of a star's life. The fate of the whole Universe is in its hands, since gravitation is slowing down its expansion to the point where it will eventually come to a complete stop. As a result of the gravitational attraction between all the galaxies a shrinking of the Universe will occur, leading to a huge collapse, in which all the galaxies, stars, planets and moons and all the molecules, atoms and atomic nuclei will cease to exist. Huge pressures and temperatures will lead to the disintegration of all things in the Universe to individual elementary particles (Chapter 3).

Finally, let us summarize the basic facts about forces between particles. In the very centre of particles the weak interaction is at work, causing the disintegration of particles, giving rise to neutrinos of all four types. The nuclear force (also called the strong interaction) holds hadrons together. It makes atomic nuclei out of neutrons and protons. It operates only in the close vicinity of hadrons (baryons and mesons), but is extremely strong.

The source of the electrical force is the electrical charge of particles. The electrical force forms atoms from nuclei and electrons and crystals, rocks and the smaller heavenly bodies, up to 500 km (310 miles) in diameter, from molecules. It also forms molecules into organelles, cells, colonies of cells and organisms. While the weak interaction acts at a tiny distance from the centre of particles (10^{-17} m), and the nuclear force acts at a distance of 1 fermi (10^{-15} m) and forms nuclei measuring several fermi across, the electrical force forms systems from atoms to small planets and small moons. These can measure from 10^{-10} m (one ten-millionth of a millimetre) to 500 km (310 miles) in diameter. In other words, from a system of two particles (the proton and the electron in the hydrogen atom) to systems of 10^{46} particles (in small planets and satellites).

For a larger number of particles the main force is gravitation. The structure of the large planets (such as the Earth with 10^{51} particles), stars and other large systems in the Universe (with more than 10^{46} particles) is conditioned and governed by gravitation alone (Figure 66).

Energy

Energy is a very important property of elementary particles. Every particle and antiparticle without exception contains energy, which they may lose by transferring it to other particles. On the other hand, a particle sometimes gains energy from other particles.

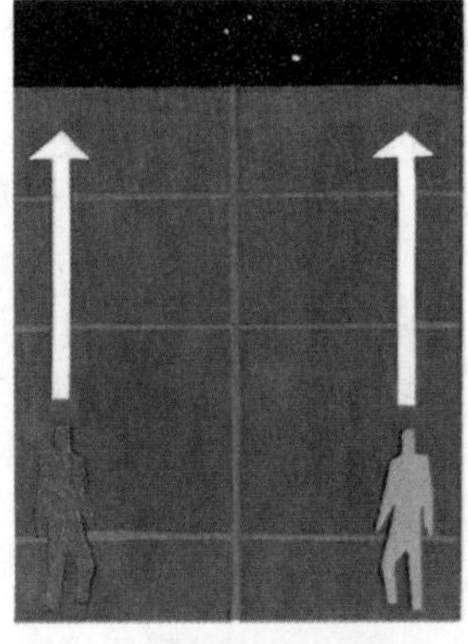

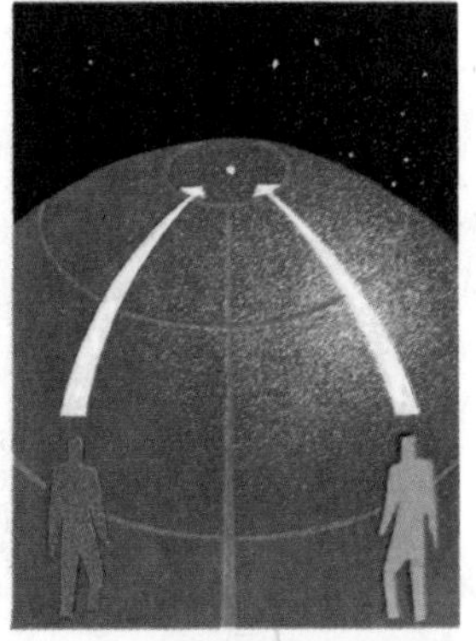

47

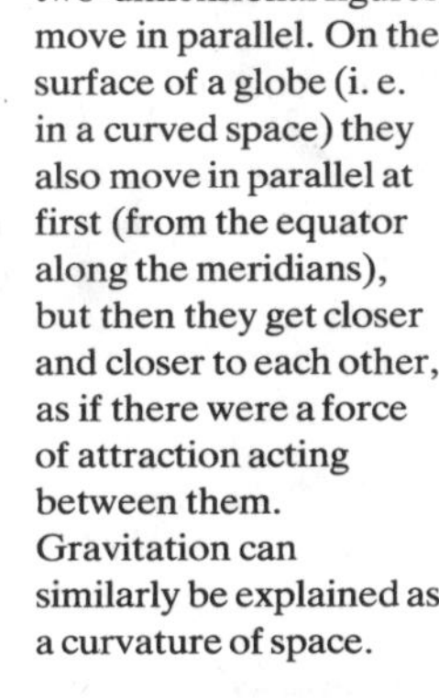

47 On a plane, both two-dimensional figures move in parallel. On the surface of a globe (i. e. in a curved space) they also move in parallel at first (from the equator along the meridians), but then they get closer and closer to each other, as if there were a force of attraction acting between them. Gravitation can similarly be explained as a curvature of space.

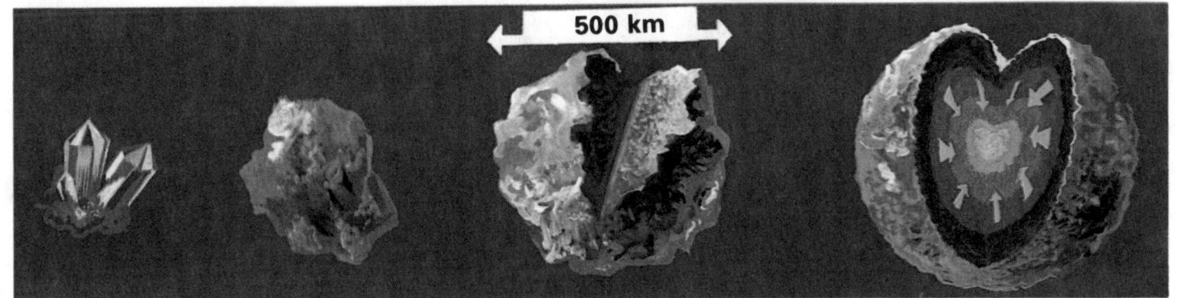

48

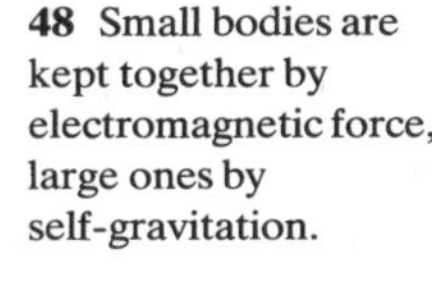

48 Small bodies are kept together by electromagnetic force, large ones by self-gravitation.

In this respect, energy differs from other properties, which remain unchanged. The energy of an elementary particle depends on its rest mass m_o (the rest energy of a particle), on the speed at which the particle is moving relative to its surroundings (kinetic energy) and on whether the particle is free or belongs to a system (binding energy) – Figure 49.

• Figure 14 gives the *rest energy* of each particle, m_oc^2. For instance, the rest mass of an electron, m_e, is 9×10^{-28} g (9×10^{-31} kg), that of a proton, m_p, is 1.67×10^{-24} g (1.67×10^{-27} kg), and so on. One of the great discoveries of this century is that a particle with a rest mass of m_o has an energy of m_oc^2. Because m_o is the rest mass of the particle, we call m_oc^2 its rest energy (Figure 50).

The figure c^2 is an enormous one. It means the square of the speed of light. The speed of light – 300,000 km (186,400 miles) per second – is the greatest speed which exists. No particle which has a rest mass can move at that speed. Only photons without rest mass can move at that speed in a vacuum. The square of the speed of light means that the speed of light is multiplied by itself, i.e. $c^2 = 90,000,000,000,000,000 \text{ m}^2/\text{s}^2$, or $9 \times 10^{16} \text{ m}^2/\text{s}^2$. The relation between the mass and the energy E_o of a solitary particle at rest is:

$$E_o = m_oc^2$$

where m_o is the rest mass of the particle and c^2 is $9 \times 10^{16} \text{ m}^2/\text{s}^2$. If we express the m_o of a particle in kilograms, we get its rest energy in joules. If m_o is in grams and $c^2 = 9 \times 10^{20} \text{ cm}^2/\text{s}^2$, the energy of the particle is in ergs. But the rest energy of particles is most often expressed in electronvolts (1 eV = 1.6×10^{-12} ergs) or megaelectronvolts (1 MeV = 1.6×10^{-6} ergs). (Joules are converted as follows: 1 eV = 1.6×10^{-19} joules, 1 MeV = 1.6×10^{-13} joules, 1 erg = 10^{-7} joules.)

As an example of the equation $E_o = m_oc^2$ we can calculate the rest energy of an electron m_ec^2 and the rest energy of a proton m_pc^2. Their rest masses m_e and m_p have already been mentioned. After multiplying by the square of the speed of light we get 0.5 MeV as the rest energy of the electron and 938.26 MeV as the rest energy of the proton. The rest energy of a photon is zero, since its rest mass is zero.

All we know of the neutrino's rest energy for the time being is that it is very small, maybe 50,000 times smaller than the electron mass m_e or less; we are not able to say what it is. It is possible that, like the photon, it has a zero rest energy, in which case the neutrino would, like the photon, move at the speed of light. But as yet we simply do not know.

The rest energy of a particle is always the same as that of the corresponding antiparticle (see Figure 14).

• Particles can acquire energy from neighbouring particles. If they are accelerated, they acquire *kinetic energy* (Figure 49). The total energy of a particle is then made up of its rest energy and its kinetic energy. We can express this relation as follows:

$$E = m_oc^2 + \frac{1}{2} m_ov^2$$

where v is the velocity with which a particle moves in relation to its surroundings. The faster a particle moves, the greater its kinetic energy, and thus also its total energy.

Particles which move very fast (whose velocity is greater than 100,000 km/62,000 miles per second) are called relativistic. The total energy of relativistic particles is greater than that given by the expression set out above. In the special theory of relativity there is a precise expression for the calculation of the total energy of a particle (Figure 49, above):

$$E = \frac{m_oc^2}{\sqrt{1 - \frac{v^2}{c^2}}} = \frac{m_o}{\sqrt{1 - \frac{v^2}{c^2}}} c^2 = mc^2$$

We can see from the last two expressions that the mass of a fast-moving particle, m, depends on its velocity v:

$$m = \frac{m_o}{\sqrt{1 - \frac{v^2}{c^2}}}$$

This relation says that with increasing velocity the particle increases its mass. The smallest mass is that of a particle which is at rest; its velocity v = 0 and its mass m is m_o (rest mass). The more rapidly the particle moves, the greater the mass (m) and the total energy (E) of a particle. The greater the velocity, the nearer the fraction $\frac{v^2}{c^2}$ approaches unity.

Thus the difference $1 - \frac{v^2}{c^2}$ and the entire denominator $\sqrt{1 - \frac{v^2}{c^2}}$ approach zero. But this means that the fraction's value rises greatly, since its numerator is the constant quantity m_o. The above relations for E and m show clearly how the mass and total energy of a particle change with increasing velocity. The nearer the velocity of a particle gets to the speed of light c, the closer the mass of the particle approaches infinity. In order to accelerate a particle to the speed of light, an infinite amount of energy is required. But such an amount of energy is unknown in the whole of the Universe. Therefore no particle which has a rest mass can move at the speed of light.

Kinetic energy can thus be thought of as the 'putting on of weight' by elementary particles. It is calculated by subtracting from the total energy mc^2 of a particle its rest energy m_oc^2 (Figure 49). If the velocity of the particle is low, its kinetic energy is $\frac{1}{2}\, m_ov^2$.

Particles acquire energy from other particles by various means: by collision, through an electrical force (for instance in an X-ray lamp), in a gravitational field (for instance free fall), in a variable magnetic field, in an accelerator (Figures 16 and 17), in an explosion of supernovas (Figure 152), in pulsars (Figure 153 number 6), in collisions of particles with an interstellar cloud ('Fermi mechanism') and in radioactive decay, etc.

• *Binding energy.* So far we have spoken of how particles can increase their energy. But the rest energy of a particle can also be decreased. In this case the rest mass of the particle also decreases. One might say that the particle 'loses weight', or decreases its mass, and therefore also loses part of its rest energy. The part of its rest energy which is lost is called the binding energy of the particle. The particle gives out the energy to its surroundings. This is therefore energy we can obtain, or release from a particle.

A particle releases part of its rest energy m_oc^2 if it forms a system with other particles (such as an atomic nucleus, an atom or a molecule, etc.). A free particle has to pay its 'entry fee' for becoming part of a system. But it can do so only from its rest energy, i.e. from its rest mass. This means that the forces acting between elementary particles and forming the system are able to release from them part of their rest energy. How much they release depends on whether the forces involved are nuclear, electric or gravitational.

1 kg (2.2 lb) of a substance such as coal, oil, water, uranium etc. has a rest energy:

$$E_o = 1\text{ kg} \times 9 \times 10^{16}\text{ m}^2/\text{s}^2$$
$$= 9 \times 10^{16}\text{ joules}$$
$$= 25{,}000{,}000{,}000\text{ kWh}$$

This huge rest energy contained by 1 kg

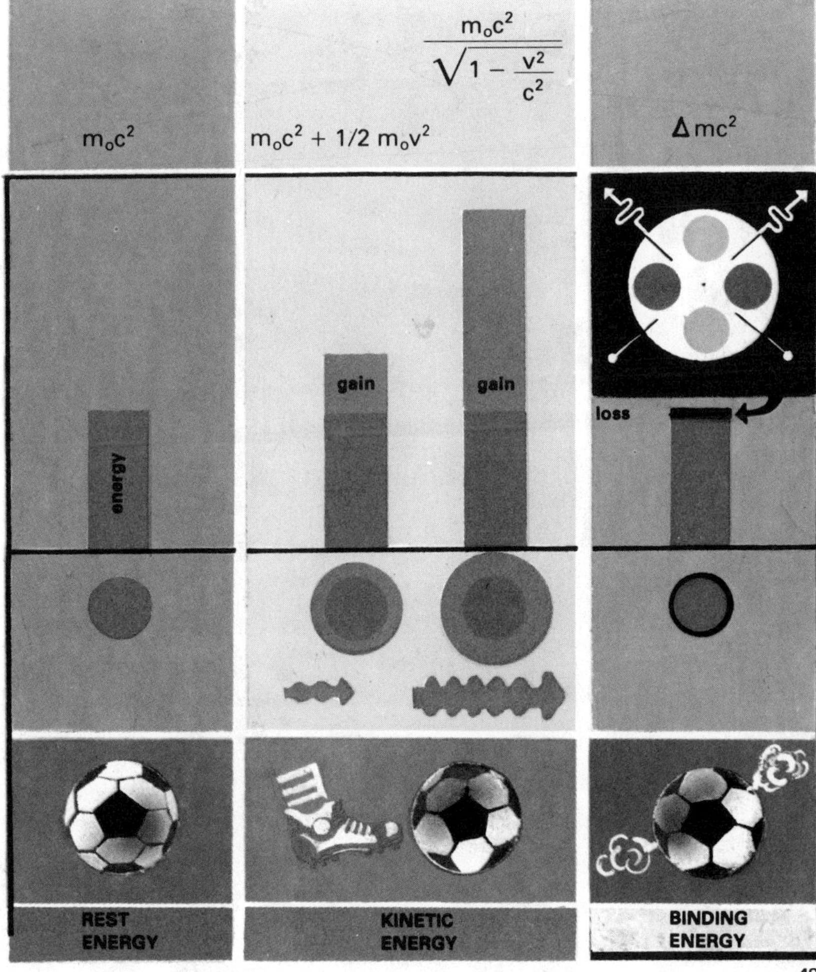

49 Particles at rest have a basic energy (rest energy). In motion they acquire in addition kinetic energy. In a system, on the other hand, they have to give up their binding energy.

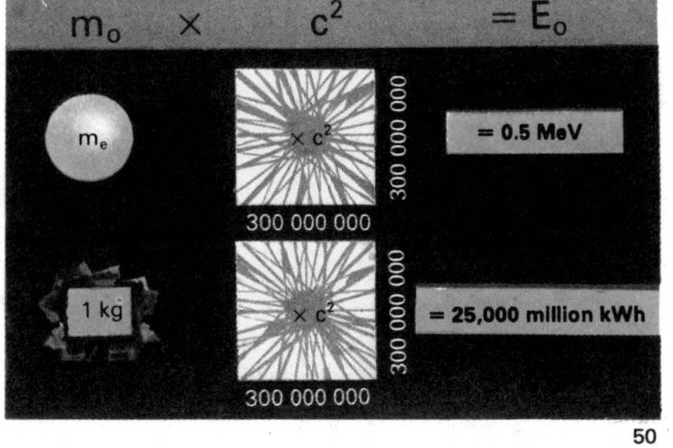

50 The mass m_o multiplied by the square of the speed of light c^2 gives rest energy E_o. This is true for a single particle as well as for massive bodies or any piece of matter.

(2.2 lb) of any material cannot as yet be released by man. He is able to release only a tiny part of it using electrical, nuclear and gravitational forces. We shall consider how the various forces release energy from matter.

Every chemical reaction results in a rearrangement of atoms in the molecules. This is brought about by the electrical forces between atoms. The chief chemical reaction for all living organisms is oxygenation. Animals obtain oxygen from the air by breathing. In their food they obtain carbon and hydrogen bound in organic molecules (sugar, proteins, etc.). When carbon and hydrogen are oxygenated, the energy necessary for all the life processes of the organism is released. Similarly, combustion (oxygenation) is as yet the main means of obtaining energy from materials such as coal, petroleum, wood and other fuels. But it is a highly inefficient way of doing it, since it releases less than one thousand-millionth (10^{-9}) of the rest energy of the material. So, for instance, about 5000 kcal of heat is released by burning 1 kg (2.2 lb) of coal, which is approximately 5 kWh of energy. But we have already seen that 1 kg (2.2 lb) of any substance (and therefore also of coal) contains a total energy of 25,000 million kWh. So, burning uses less than one thousand-millionth of the total of the rest energy, while all the rest remains in the ash and smoke. This gives some idea of just how wasteful burning, still mankind's main way of obtaining energy, really is. The other means of releasing energy using the electrical force (such as joining electrons with nuclei – recombination as in Figure 182, or molecules in crystals – crystallization) are also very inefficient.

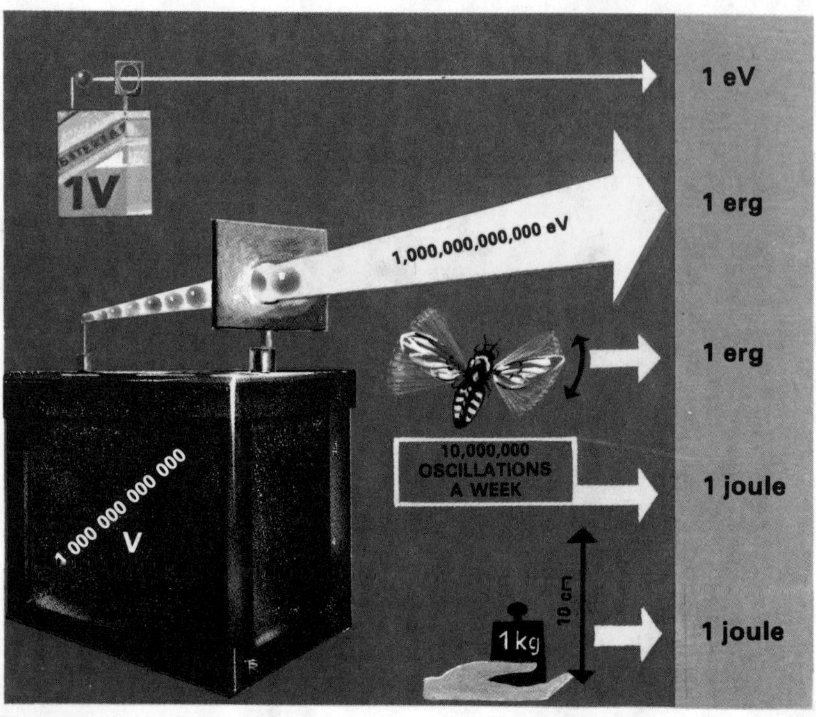

51

51 In a field of 1 volt an electron accelerates and acquires kinetic energy of 1 electronvolt (1 eV). This is the smallest unit of energy, and the erg and the joule can be derived from it.

Nuclear forces are much more powerful than electrical ones. They are able to release from a material several million times more of its rest energy. An atomic power station uses the nuclear forces to obtain around one-thousandth of the rest energy of uranium. Stars are even more efficient at doing this than we are (see page 130). The conversion of hydrogen to iron which goes on inside stars releases almost one-hundredth of the rest energy of hydrogen.

The Sun releases energy in a similar way to a hydrogen bomb. The difference is that the Sun is much more efficient in this work, and it maintains life, rather than destroy it.

The gravitational force can also release energy, but only in the case of bodies in the Universe which have immense mass (such as heavy stars, the compact nuclei of galaxies, and so on). In such heavy bodies gravitation is able to 'squeeze' out of a material up to half its rest energy. The Earth is a relatively small body, and so it is not possible to exploit gravitation here in order to obtain energy.

Figure 52 shows another, the most efficient, way all the rest energy can be released from a material. This is the annihilation of particles with antiparticles, or of a piece of matter and a piece of antimatter. Annihilation was important at the start of the existence of the Universe (Figure 157). In the next section we shall take a closer look at annihilation, and at the opposite process, which is the materialization of particles and antiparticles from energy.

So far we have spoken of the energy of particles which have rest energy and rest mass. A photon can never be at rest, and has neither a rest mass nor a rest energy. Its energy depends only on its frequency. The higher the frequency of a photon, the greater its energy. If we denote the frequency of oscillation of a photon in cycles per second with the Greek letter nu (ν) and the energy of the photon as E, the relationship is expressed as follows:

$$E = h\nu$$

where the letter h is Planck's constant, and has a value of 6.6×10^{-34} joule seconds ($h = 6.6 \times 10^{-27}$ erg seconds). While the frequency may vary, the value of h is always constant. It is named after the German physicist Max Planck, who discovered it, and plays a very important role in the Universe. It can be used to calculate the energy of radiation of various frequencies and wavelengths. So, for instance, radio waves with a wavelength of 2000 m (6562 ft) have a photon energy of 6×10^{-9} electronvolts. The greatest energy of all has been found in gamma photons from space: 10^{17} eV.

There are many more photons in the Universe than there are particles with rest energy (protons, electrons and neutrons) – about 2000 million times more. Instead of elementary particles with rest energy, we shall call them simply particles. The photon is a particle, too, of course, but without rest energy. For the sake of simplicity we shall speak in the further text of photons and particles. For every particle there are about 2000 million photons. For the most part they are photons with an energy of from 10^{-4} to 10^{-3} eV, or 'fossil' photons. With the expansion of the Universe, the frequency, and therefore also the energy, of photons is decreasing. Long ago, early on in the existence of the Universe (the so-called Big Bang), photons had much more energy, many thousands of millions of electronvolts. This prevalence of the photons lasted around 300,000 years.

As we have said, the Universe consists of particles and photons. We speak of the photon component and the particulate component of the Universe. There is a constant exchange of energy between them. For example, matter continuously radiates photons, but at the same time absorbs them. Other processes of the exchange of energy between the two components are annihilation and materialization, which we shall discuss next. Whatever the transformations of energy in the Universe, no energy can be lost, nor can it appear from nowhere. Energy changes its form, but never its quantity. This important fact is known as the *Law of the Conservation of Energy.*

Energy has an important role in every transformation which takes place in the Universe. It is constantly changing its form; so, for instance, when a petrol and air mixture is burnt in a car's engine, a very small part of its rest energy is converted into heat, which results in the movement of particles. By means of pistons this heat is converted into the energy of movement of the car. In the car's alternator or dynamo movement is converted into electrical energy. In the bulbs of the headlights electrical energy is converted into the energy of photons – into

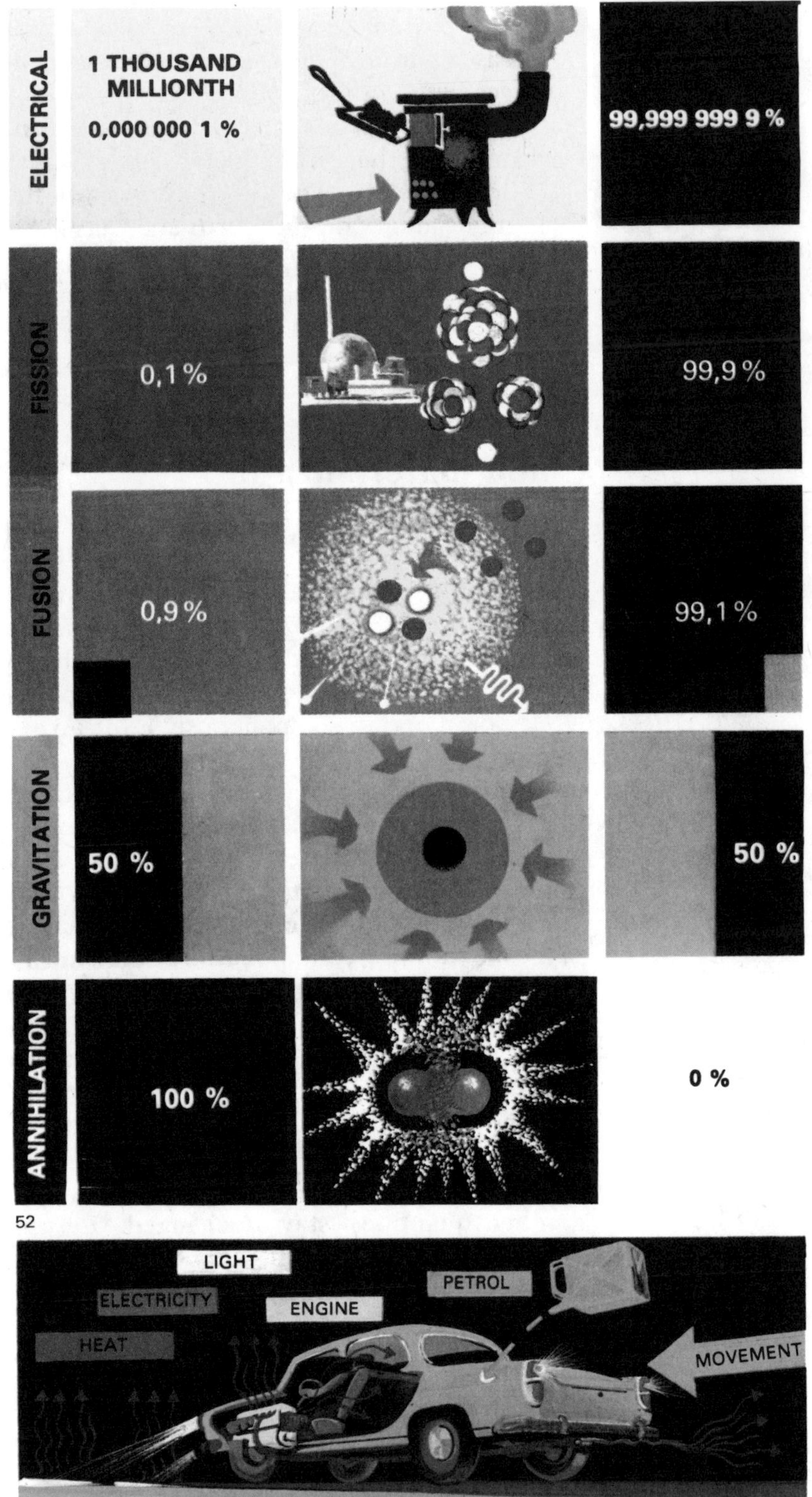

52 The release of rest energy from matter by the electrical, nuclear and gravitational forces and annihilation. On the left, the percentage of rest energy released, on the right the unexploited remainder.

53 The transformation of the chemical energy of petrol into movement, electricity, light and heat.

light. Light is absorbed by the road surface and turned into heat, i.e. the kinetic energy of particles. And so we might go on.

The Law of the Conservation of Energy states that every form of energy around us arose from another form of energy. So, for instance, the heat of our bodies and the work of our muscles are both transformed *chemical* energy which is taken into our bodies in the form of food (Chapter 10). The energy in food is of solar origin, trapped by green plants through photosynthesis. Solar radiation acquired its energy from the rest energy of the protons inside the Sun. There it is released through the action of the nuclear forces between protons (see page 130). The rest energy of these protons has been conserved since the start of the Universe (Figure 163 bottom left). Thus we can follow all the energy changes on the Earth and outside it. In any case it is a very fascinating history, and one closely connected with the creation and evolution of the Universe.

The annihilation and materialization of particles

The proton and the electron are stable particles. The antiproton and the positron (antielectron), too, do not decay by themselves, whereas most particles and antiparticles disintegrate spontaneously after a very short time.

The only way the proton and the electron can disappear is by meeting an antiproton or a positron. In that case annihilation takes place, and both particles are changed into gamma photons. Since the Latin word *nihil* means 'nothing', annihilation should mean that particles are destroyed, but actually only a transformation is involved. Particles with rest mass (protons, antiprotons, electrons and positrons) are changed into particles without it – photons. But the total amount of energy remains the same. Also the total electrical, baryon and lepton charges and other quantities stay unchanged. Let us examine some cases of annihilation:

$$e^- + e^+ \rightarrow 2\,\gamma$$

where the Greek letter γ (gamma) denotes a gamma photon with an energy of $\frac{1}{2}$MeV, which is the rest energy of an electron or a positron. Similarly, a proton and an antiproton are annihilated thus:

$$p + \tilde{p} \rightarrow 2\,\gamma$$

where $\tilde{p}$ denotes an antiproton and γ is a photon with a rest energy of 938 MeV, which is the rest energy of a proton or an antiproton. If a neutron is annihilated with an antineutron $\tilde{n}$ and both are at rest, then:

$$n + \tilde{n} \rightarrow 2\,\gamma$$

where the energy of the gamma photons is 939.5 MeV. But if the particles are moving rapidly their total energy is greater, and the energy of the resulting gamma photons will also be greater.

Now it is easy to understand why the antiprotons, the positrons and the antineutrons which appear on Earth or on the Sun have a very short life. The Earth and the Sun are made of ordinary matter, i.e. of protons, electrons and neutrons, and the moment antiparticles meet particles, they are annihilated. Ordinary matter (known by the Greek name of *koinomatter*) is a very hostile environment for antiparticles. Thus koinomatter and antimatter cannot exist in each other's close vicinity.

As yet we do not know where in the Universe antimatter is to be found. As has already been said, light rays cannot tell us, since the photons from matter are exactly the same as photons from antimatter.

Materialization is the opposite process to annihilation. Both these processes, annihilation and materialization, were very important in the beginning of the Universe (Figure 145). We can show what materialization is by means of a simple example. If a gamma photon with an energy of at least 1 MeV passes an atomic nucleus, it turns into an electron and a positron. This is the best-known case of materialization, and it is written thus:

$$\gamma \rightarrow e^+ + e^-$$

Note that the electrical charge is conserved; the charge of the photon is zero and the sum of the charges of the two new particles is also zero. The lepton charge is also conserved in materialization. If the energy of the photon is greater than the sum of the rest energies of the two particles, the excess energy is converted into their kinetic energy.

Similarly for a proton and an antiproton to arise:

$$\gamma \rightarrow p + \tilde{p}$$

the photon must have an energy of at least 1876 MeV, since the rest energy of a proton and an antiproton is 938 MeV.

Materialization means, generally speaking, a transformation of energy into particles with a rest mass. The energy may be in the form of a photon which passes close to the nucleus of an atom. But the kinetic energy of a proton in cosmic radiation can also materialize. A proton arriving from the depths of space can have a kinetic energy up to a billion times greater than its rest energy. And this huge amount of energy gives rise to a large number of particles in the Earth's atmosphere. Such high-energy protons (called primary cosmic ray protons) collide with a nitrogen or oxygen nucleus in the atmosphere. In this collision the nucleus is split into individual nucleons, and the enormous kinetic energy produces many millions of new particles and antiparticles of various sorts (baryons and antibaryons, leptons and antileptons, mesons and photons). All these particles together are called a shower of cosmic radiation. These showers are an example of materialization on a large scale.

But various materializations also take place in accelerators on the surface of the Earth (Figure 16). For instance, a very fast proton collides in a hydrogen chamber with a hydrogen nucleus, i.e. with a proton, and its kinetic energy is converted into a neutron, an antiproton and a meson (Figure 61):

$$p + p \rightarrow p + p + n + \tilde{p} + \pi^+$$

where the electrical charge is:

$$(+1) + (+1) = (+1) + (+1) + 0 + (-1) + (+1)$$

and the baryon charge is:

$$(+1) + (+1) = (+1) + (+1) + (+1) + (-1) + 0$$

We can see that the electrical and baryon charges were conserved in this materialization. The kinetic energy of the proton entering the hydrogen chamber must be the same as, or greater than, the sum of the rest energies of the particles which are produced (neutron, antiproton and positive pion). Thus it must be equal to or greater than:

$$939.5 \text{ MeV} + 938.2 \text{ MeV} + 139.6 \text{ MeV} = 2017.3 \text{ MeV}$$

This kinetic energy of the fast proton from the accelerator materialized (i.e. was transformed) into the rest energy of the neutron, the antiproton and the pion. Any excess of kinetic energy of the proton from the

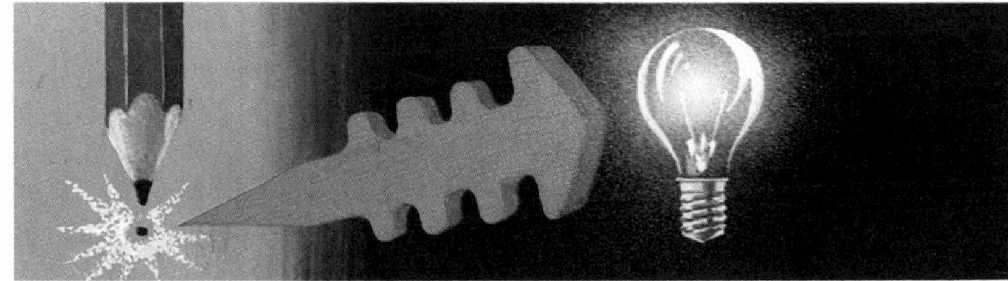

54 The rest energy in the dot would be enough to power a 100 W bulb for 100 hours.

X	e^-	+	e^+	→	γ	+	γ		Σ
Q	−1	+	+1	=	0	+	0	=	0
N	0	+	0	=	0	+	0	=	0
L	+1	+	−1	=	0	+	0	=	0

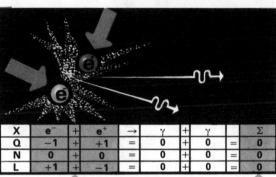

55 Annihilation of an electron and a positron at rest. The energy of the gamma photons which arise is equal to the rest energy of the electron (and therefore also of the positron).

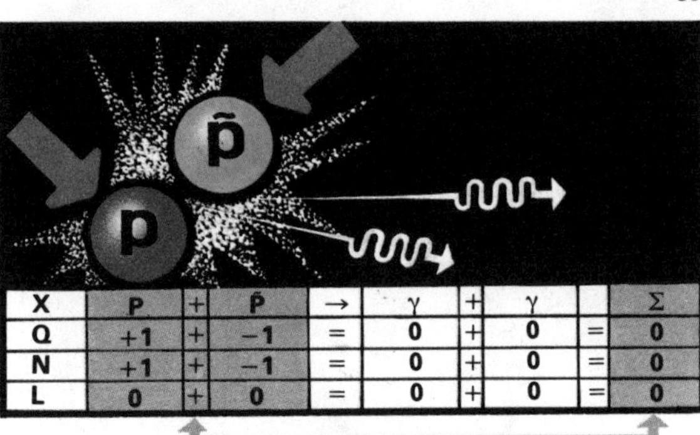

X	p	+	p̃	→	γ	+	γ		Σ
Q	+1	+	−1	=	0	+	0	=	0
N	+1	+	−1	=	0	+	0	=	0
L	0	+	0	=	0	+	0	=	0

56 Annihilation of a proton with an antiproton. The rest and kinetic energies of both particles are transformed into the energy of gamma photons. But they may be transformed into rest energy of other particles (see Figure 57).

accelerator is divided between the new particles as kinetic energy.

Materialization played a vital role in the early period of the Universe (the Big Bang), about 10,000 million years ago (Figure 145). Today, it occurs wherever high-energy particles are to be found whose kinetic energy can materialize: in collisions of cosmic radiation in interstellar space, in cosmic radiation showers in the atmosphere of Earth and the other planets, in particle accelerators, etc.

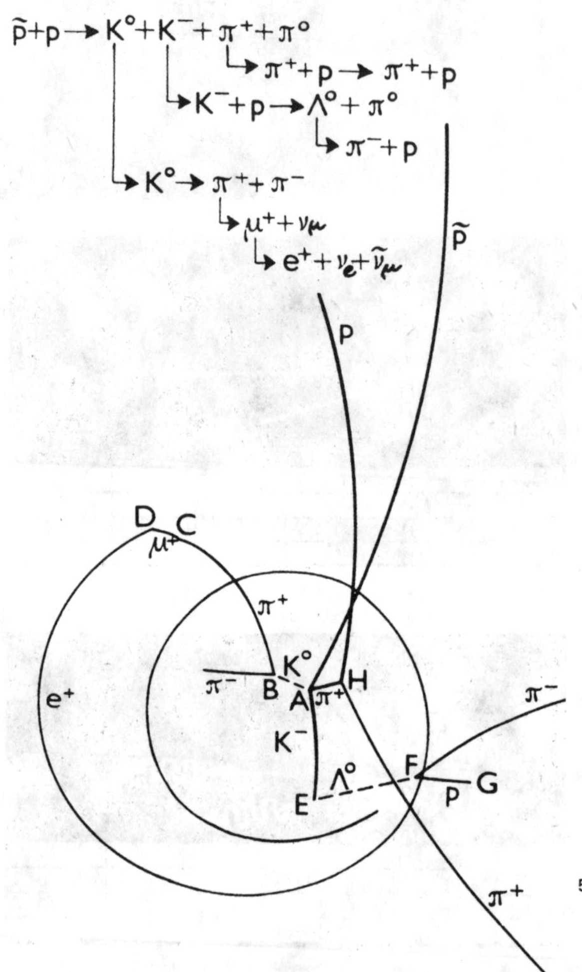

57 Annihilation of a proton p and an antiproton $\bar{p}$ (at point A). The antiproton $\bar{p}$ was injected downwards. The proton was at rest at point A. Four mesons resulted, K^O, K^-, π^+, and π^O. K^O and π^O left no trace, since they have no electrical charge. The kaon K^O disintegrated at point B into π^- and π^+; π^+ disintegrated at point C into μ^+ and ν_μ. μ^+ disintegrated into e^+, ν_e and $\bar{\nu}_\mu$ (at point D). The negative kaon collided with a proton (at point E): $K^- + p \rightarrow \Lambda^O + \pi^O$; Λ^O then disintegrated at point F into a proton p and a pion π^-. The pions π^O disintegrated after a short time into two gamma photons and left no trace. The pion π^+ collided with a proton p at point H and pushed it upwards, itself falling downwards. The photograph was taken in liquid hydrogen at CERN (Geneva).

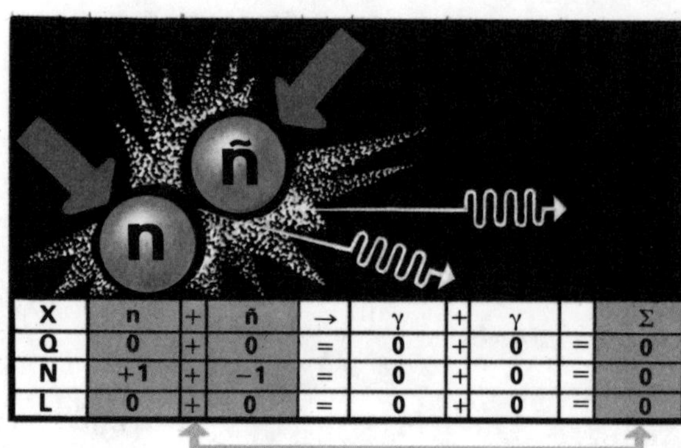

X	n	+	ñ	→	γ	+	γ		Σ
Q	0	+	0	=	0	+	0	=	0
N	+1	+	−1	=	0	+	0	=	0
L	0	+	0	=	0	+	0	=	0

58 Annihilation of a neutron n with an antineutron ñ. Their rest and kinetic energy is carried off by the photons which arise from annihilation.

The life of elementary particles

Not only animals and plants, but also elementary particles, have their fate. They come into being, act on other particles and form systems with them, move and acquire kinetic energy, or, on the contrary, lose their own rest energy. In the end, after a certain time (which may be short or infinitely long), they cease to exist, passing on their properties (energy, spin, electric charge, baryon and lepton charges) to the descendants into which they disintegrate.

The origin of elementary particles

All particles and antiparticles can arise through the materialization of photons or the materialization of the kinetic energy of high-energy particles. A third way in which particles can come into being is by the disintegration of heavier particles. We can recall, for instance, the disintegration of a neutron into a proton, an electron and an antineutrino, the disintegration of pions into muons, neutrettos and antineutrettos or, finally, the disintegration of muons into electrons, positrons, neutrinos, antineutrinos, neutrettos and antineutrettos.

In all disintegrations the rest energy of particles (neutron, pion, muon) is divided among the rest energies of the two or three resulting particles. The remainder of the rest energy of a disintegrating particle is passed on to the new particles as kinetic energy. In all the ways in which particles can disintegrate or appear the energy is conserved.

The fate of elementary particles

In the Universe everything's days are numbered – including elementary particles. Most particles in Figure 14 have only a short life. Immediately after coming into existence they disintegrate, which is why they are called unstable particles. All unstable particles soon disintegrate into stable particles: protons, electrons, neutrinos, neutrettos and photons.

The most important particles in our close vicinity are protons and electrons. Some

59

59 A high-energy proton of cosmic radiation smashes the nucleus of nitrogen or oxygen in the upper atmosphere. This gives rise to a large number of particles and antiparticles, known as secondary cosmic radiation.

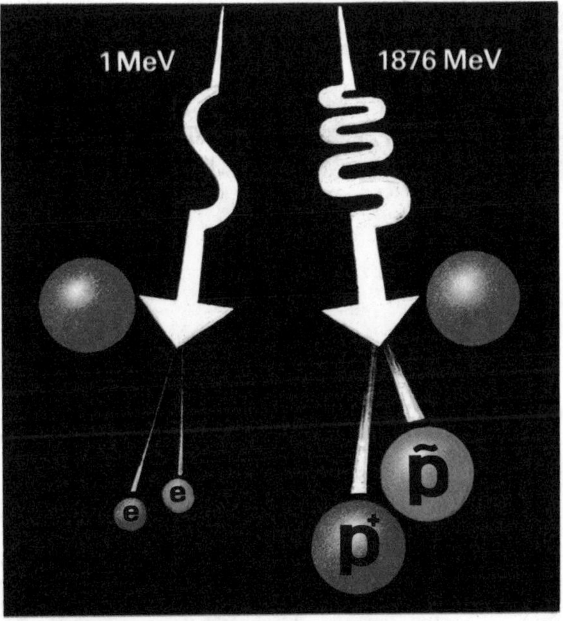

60

60 Materialization of gamma photons close to a proton (or other particle). The photon must have an energy of at least 1 MeV or 1876 MeV in order to change (materialize) into e^- and e^+ (with rest energy 0.5 MeV each) or into p and p̃ respectively (with rest energy 938 MeV).

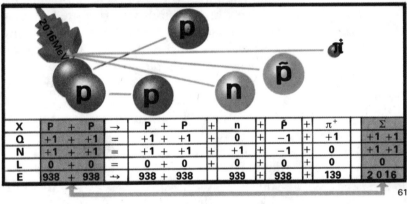

X	P + P	→	P + P	+	n	+	P̃	+	π^+		Σ
Q	+1 + +1	=	+1 + +1	+	0	+	−1	+	+1		+1 +1
N	+1 + +1	=	+1 + +1	+	+1	+	−1	+	0		+1 +1
L	0 + 0	=	0 + 0	+	0	+	0	+	0		0
E	938 + 938	→	938 + 938		939	+	938	+	139		2016

61

61 The materialization of the kinetic energy of two protons in a neutron n, an antiproton p̃, and a positive pion π^+. The total energy (rest and kinetic) of the two protons prior to the collision is equal to the total energy of all the particles after materialization.

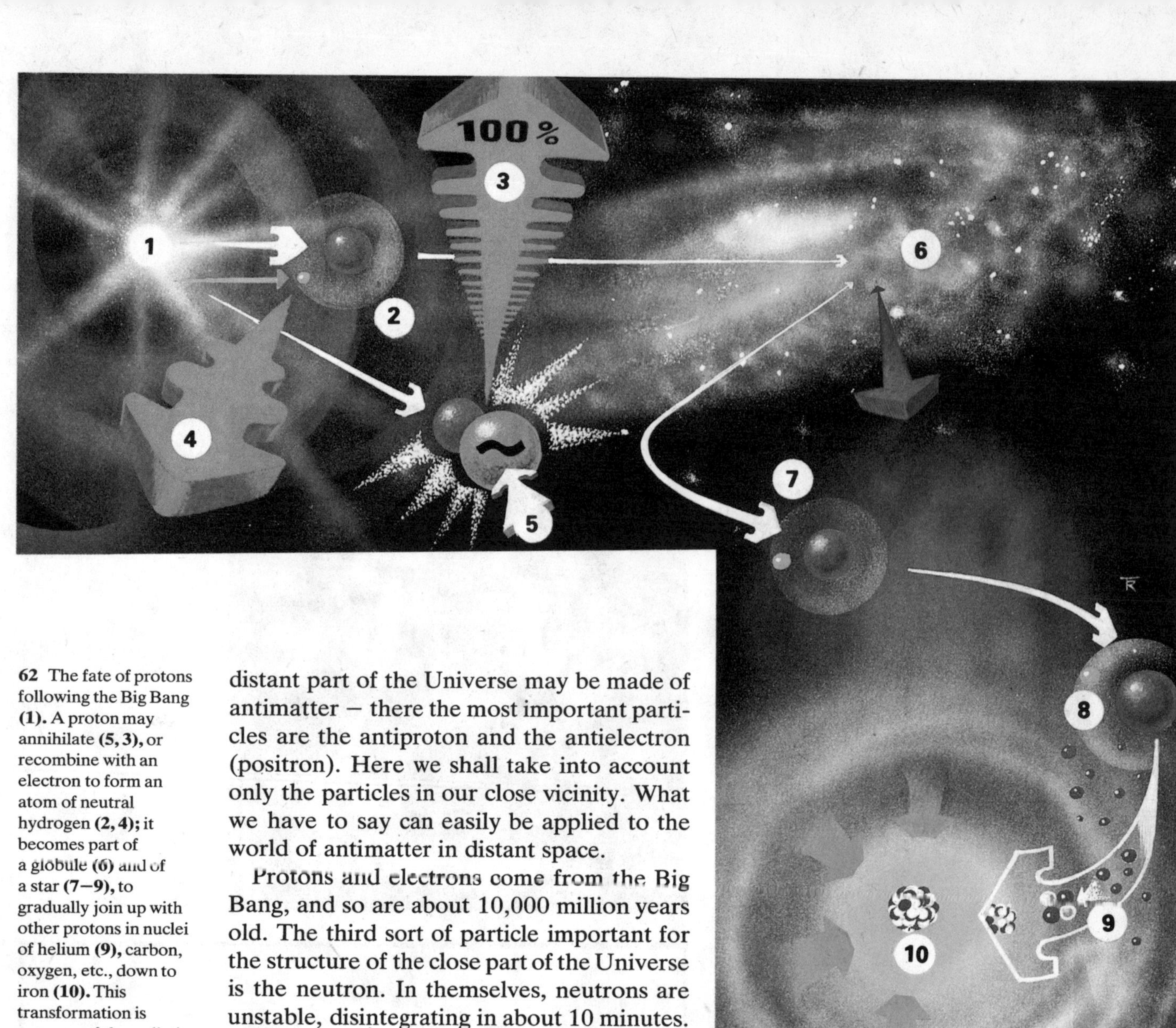

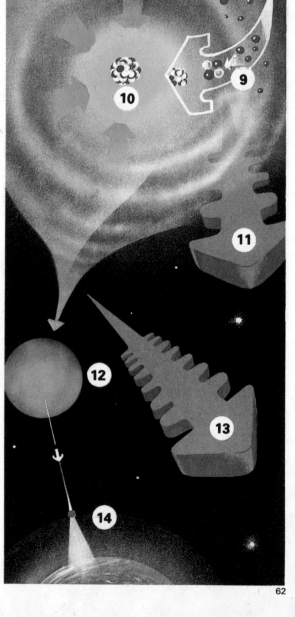

62 The fate of protons following the Big Bang **(1).** A proton may annihilate **(5, 3),** or recombine with an electron to form an atom of neutral hydrogen **(2, 4);** it becomes part of a globule **(6)** and of a star **(7–9),** to gradually join up with other protons in nuclei of helium **(9),** carbon, oxygen, etc., down to iron **(10).** This transformation is a source of the radiation of the star **(11).** Then the star collapses into a neutron star – a pulsar **(12),** and a large amount of energy is released – a supernova **(13).** In the process the protons change into neutrons **(12).** The neutron star rotates very quickly and accelerates protons into cosmic radiation **(14).**

distant part of the Universe may be made of antimatter – there the most important particles are the antiproton and the antielectron (positron). Here we shall take into account only the particles in our close vicinity. What we have to say can easily be applied to the world of antimatter in distant space.

Protons and electrons come from the Big Bang, and so are about 10,000 million years old. The third sort of particle important for the structure of the close part of the Universe is the neutron. In themselves, neutrons are unstable, disintegrating in about 10 minutes. They are only stable in the nucleus of an atom. Large quantities of them are constantly being produced in the hearts of stars, where atomic nuclei are formed from protons (Figure 151).

Neutrinos and neutrettos are being made in the Universe all the time: through the transformation of protons into neutrons inside stars, the materialization of photons in the hearts of very heavy stars (see page 118) and by the disintegration of a large number of unstable particles. They come into being in the cosmic radiation showers in our atmosphere, and in collisions of cosmic radiation in interstellar space. Huge numbers of neutrinos and neutrettos were produced in the Big Bang (Chapter 3). Because neutrinos are not absorbed by practically any matter, their number in the Universe is constantly increasing. Like the photons, they fill the whole of the Universe. This phenomenon is called the *sea of neutrinos.*

62

Photons come into being in the disintegration of some unstable particles. (Figure 14 shows them at the bottom, denoted γ). A huge number of photons remains from the time of the Big Bang. These are called fossil or relic photons. They also fill the whole of the Universe, and their frequency (and therefore their energy) is constantly falling, because the Universe is expanding. At present all bodies are contributing to the photon component of the Universe, but expecially the stars and nebulae. The photons arise at their surface from the energy of electrons (Figure 182).

At the start of the Universe all particles were free. At that time there were no atomic nuclei, no planets and no stars. These systems formed later, after the Universe was about 300,000 years old, and the hot matter had cooled sufficiently by expansion (Chapter 3).

Only the neutrinos, the neutrettos and the photons did not form any system, since their mutual attraction is too small. They remain free to this day.

Proton – the hero of the Universe (Figure 11), underwent many dramatic events due to its electric, nuclear, weak and gravitational forces. Still in the early days of the Universe, 300,000 years after it came into existence, the free protons and electrons joined together to form the atom of hydrogen (a system of one proton and one electron linked by electrical force). Their rest energy was reduced in the process by only 13 eV. This means that the electrical attraction of the particles released only about one 100-millionth of their rest energy. The protons in the heavy stars gradually transformed themselves into iron (Chapter 3). In doing so each proton released nearly 1 per cent of its rest mass (9 MeV out of 938 MeV). A proton in a very heavy star, which at the end of its life shrinks its mass into a very small volume, may lose up to one-fifth of its rest energy (and thus also one-fifth of its rest mass). Finally, in collisions between protons and antiprotons, no system is created, but all the rest energy is released in the form of photons (annihilation).

In the preceding paragraphs we have seen that elementary particles are subject to all kinds of changes. By means of movement they can gain energy, whereas when they enter a system their energy is reduced. Unstable particles (such as the neutron or the hyperon) spontaneously disintegrate into lighter ones, but at high densities they, too, become stable (such as neutrons in a nucleus or in a neutron star, hyperons in a hyperon star, page 48). On the other hand stable particles (such as the proton and the electron) are annihilated in collisions with their antiparticles. At high densities unstable neutrons change into stable hyperons. Nothing in this Universe has its existence assured, since not even the stable elementary particles, the most basic structural units, can be sure of their own survival.

Matter – agglomerations of elementary particles

Matter is a cluster of a large number of protons, neutrons and electrons. There need not be any order in this cluster, so that these elementary particles may be free and move 'as they please', independent of each other. The very hot and dense matter which was present at the start of the Universe, for instance, consisted only of elementary particles (Chapter 3). On the other hand, at low temperatures and densities particles have a tendency to form simple systems: nuclei, atoms, molecules and crystals.

The arrangement of elementary particles into the simplest systems is a natural result of the nuclear and electrical forces which particles exert on other particles in their vicinity. These forces are inborn to particles. Protons and neutrons are equipped with the nuclear force, protons and electrons with the electrical force. The action of the forces between particles is conditioned by the distance they are apart, i.e. by the density of the matter; it also depends on the speed at which the particles are moving in the matter – which means the temperature of the matter. Thus the observed properties of a material will depend on its density and temperature. This dependence is shown in Figures 63 and 64.

Heating up of matter

The heating up of matter means an increase of the kinetic energy of its particles. In our illustration heating up means movement towards the top. If we heat up crystals of water – snowflakes – their crystalline structure is broken down, and the matter (water) passes from the solid state to the liquid. The distance between molecules of water is very small, as it was in the crystal. The molecules are attracted to each other by a sufficiently strong force so that they remain near to each other and the liquid does not expand. But the thermal movement of molecules allows them to move in relation to each other, because they are not as rigidly bound to each other as they were in the crystal.

Through further heating the speed of the water molecules increases. When the molecules reach sufficient speed, they overcome the attraction of the neighbouring molecules in the liquid and escape from it. By evaporation the liquid changes into gas. The thermal movement of the molecules totally outweighs their attraction. The molecules of gas move quite independent of each other and their movement is quite chaotic, so that the gas has a tendency to expand. The molecules collide with each other and in doing so change their direction and speed.

If we continue to heat a gas, the collisions between molecules become more and more violent. At a sufficiently high temperature they are so violent that the molecules break each other down into individual atoms. The kinetic energy of the molecules is greater than the binding energy between their atoms. This break-up of molecules is called dissociation. So, for instance, if water is heated to over 2000 kelvin, it dissociates into hydrogen and oxygen. It loses its molecular structure and ceases to be water, changing into a mixture of the two gases.

If a gas is heated still further (to a temperature of many thousands of kelvins), the atoms move faster and faster. Their electron envelopes crash into each other violently. At temperatures over 10,000 kelvin the collisions between electron envelopes are so violent that the electrons are knocked out from them. We say that the atoms are ionized. Atoms which before the collision were neutral (since the number of electrons in the envelope was the same as that of protons in the nucleus) become atoms with an incomplete electron envelope; such atoms with one or more electrons missing are called *ions,* and are positively charged, since the number of protons in the nucleus is greater than the number of electrons in the envelope. In a very hot gas there are many positive ions and free electrons. Such gases, either partially or totally ionized, are called *plasma.* Stars are huge balls of plasma. In plasma the kinetic energy of the electrons is greater than their binding energy in atoms. That is why the atoms are split into ions and electrons, or ionized.

If plasma is heated further, the atoms are broken down completely into bare nuclei and free electrons. At temperatures of many thousand million kelvins such violent collisions occur in the plasma that the nuclei are broken down into individual nucleons. At such high temperatures the kinetic energy of particles is so great that it exceeds the binding energy of the nucleons in the atomic nuclei. Such matter is called *nucleon gas* or *nugas.* It is made up of protons, neutrons and electrons which are moving very fast. It occurs in the death-throes of heavy stars, where the star collapses under the force of its own gravitation (a phenomenon known as gravitation collapse) and heats up its plasma to temperatures of many thousands of millions of kelvins (Figure 152).

In conclusion we can say that with increasing temperature matter becomes more and more simple. At extremely high temperatures it consists only of free, very rapidly moving elementary particles. There is no trace of any structure such as nuclei, atoms or molecules.

Compression of matter

If the density of a material increases, each cubic centimetre contains more particles, and they are closer together. One cm^3 (0.06 cu in) can contain anything from a single particle (as in the case of interstellar space), to ten trillion particles (10^{19}), as in the case of the air we breathe. The stars called white dwarfs (Chapter 3) have more than 10^{27} particles per cm^3 (0.06 cu in), and the extremely dense neutron stars (see page 115) have more than 10^{35} particles in a single cm^3 (0.06 cu in). Let us return to Figure 64, and take a look at how the properties of matter change as it is compressed into an ever-decreasing volume. In the figure, compression (increase in density) is represented by movement from left to right.

If we were to compress a bucket of water into a thimble, the water molecules would be so crowded that the electrons would no longer know to which atoms they belonged, and would fly around in confusion between the nuclei of oxygen and hydrogen. At such great densities there are more than a thousand times more particles (electrons, protons and neutrons) in 1 cm^3 (0.06 cu in) than in ordinary water. In such crowded conditions the atoms cannot exist with an electron envelope, and the electrons are released from the envelope, just as they are in the case of collisions at high temperatures. Here the release of electrons (or ionization) is the result of high density, not of high temperature, and this process can occur at very low temperatures. The free electrons no longer belong to a particular nucleus, but are the common property of all the nuclei in the matter. Such matter as this is thus similar to perfectly ionized plasma, where there are also only bare nuclei and free electrons. The difference between hot plasma and dense matter lies in the motion of electrons.

Electrons are fermions, and we have already seen that there may only be two in a 'compartment' (Figure 18). This is called the *Pauli Exclusion Principle,* after the Swiss scientist Wolfgang Pauli. In a very small space there can only be two electrons of the same velocity, but turning in opposite directions (i.e. having opposite spin). Any other electron thrust into the same space must differ from the two which are already there. They differ in having greater speed, and therefore also greater energy. Even at a very low temperature, the electrons in very dense matter must move at very high speed. When they strike other parts of the material they exert a force we call pressure. This pressure of a very dense material depends only on its density. Such a very dense material is called *degenerate.* (The pressure of a normal plasma, for example in the Sun, depends not only on density, but also on temperature).

When our Sun reaches the end of its days, it will become a body known as a white dwarf the size of Earth, and with a density of about 1 million g/cm^3. The Sun's plasma will turn into degenerate matter. White dwarfs are the 'pensioners' of the Universe, living off the heat accumulated in their lifetime by means of thermonuclear reactions (see page 115).

If we were to compress degenerate matter even further, a piece of a white dwarf, for instance, its density would increase even further. At densities around 10^8 g/cm^3 (100 tonnes per cm^3), the kinetic energy of electrons reaches 0.8 MeV and their total energy 1.3 MeV. At this stage the energy of the protons in the nuclei (938.2 MeV) and the total energy of the electrons (1.3 MeV) is equal to the energy of the neutrons (939.5 MeV). The electrons join up with protons to form neutrons. With a further increase in density the total energy of the electrons further increases, so that when a proton

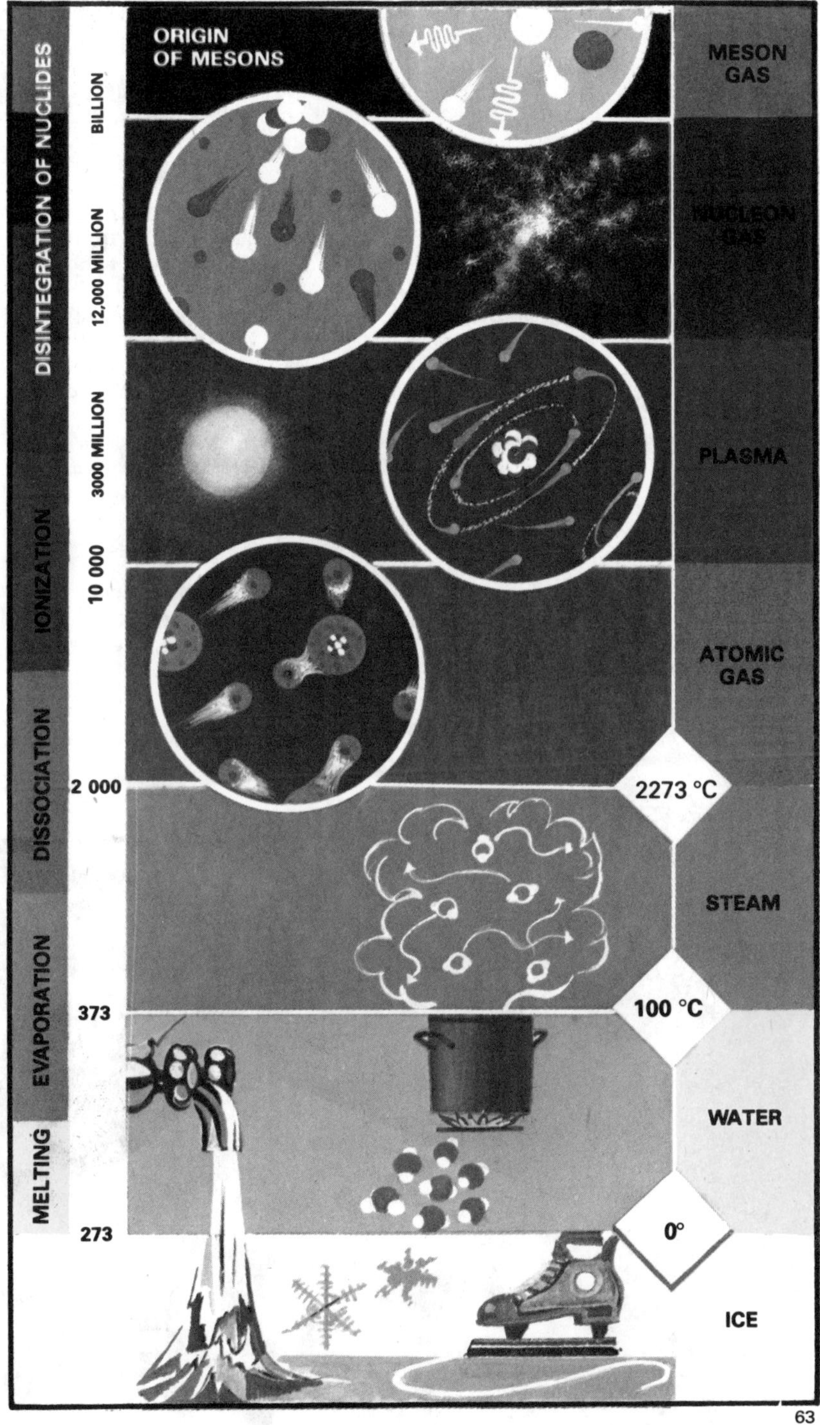

63 If a material is heated, its temperature rises (on the left in kelvins, on the right in degrees Celsius). Heating changes the arrangement of the particles in the material (centre pictures). On the extreme right are the names of states, and on the left the transitions between them.

63A The valley of stability represents the structure and binding energy of nuclids. It is made of many levels, or terraces. The depth of the terrace below the top of the valley gives the binding energy of one nucleon in the nuclid. For instance, Be8 is composed of four protons and four neutrons and the binding energy of one nucleon is −7.3 MeV. The nucleus of beryllium Be8 therefore requires an energy of 58.4 MeV in order to be broken down into protons and neutrons. The nuclid Be6 has a depth of −4.8 MeV, and after a very short time 'falls down' to a lower terrace. It is the endeavour of the protons (on the upper platform, i.e. with binding energy zero) and all nuclids to get to the bottom of the valley of stability. Thus in the interior of stars the nuclei of all the heavier elements down to iron are formed from those of hydrogen (Figure 151). In the explosion of a supernova (Figure 152) nuclei heavier than those of iron are created. In order to form a more accurate picture the valley of stability is cut in half between iron (Fe) and cobalt (Co). That is the deepest point of the valley of stability, around −10 MeV.

combines with an electron to form a neutron energy is released. Under such conditions neutrons become stable particles, since it would be necessary to add energy in order for them to disintegrate into a proton and a high-energy electron. There is no place for low-energy electrons due to Pauli's Principle. Thus the protons in the nuclei of the very dense material and the free electrons become neutrons. This process is called *neutronization.* At densities above 10^8 g/cm^3 matter turns into *neutron gas.* In the course of this the nuclei naturally cease to exist, since the

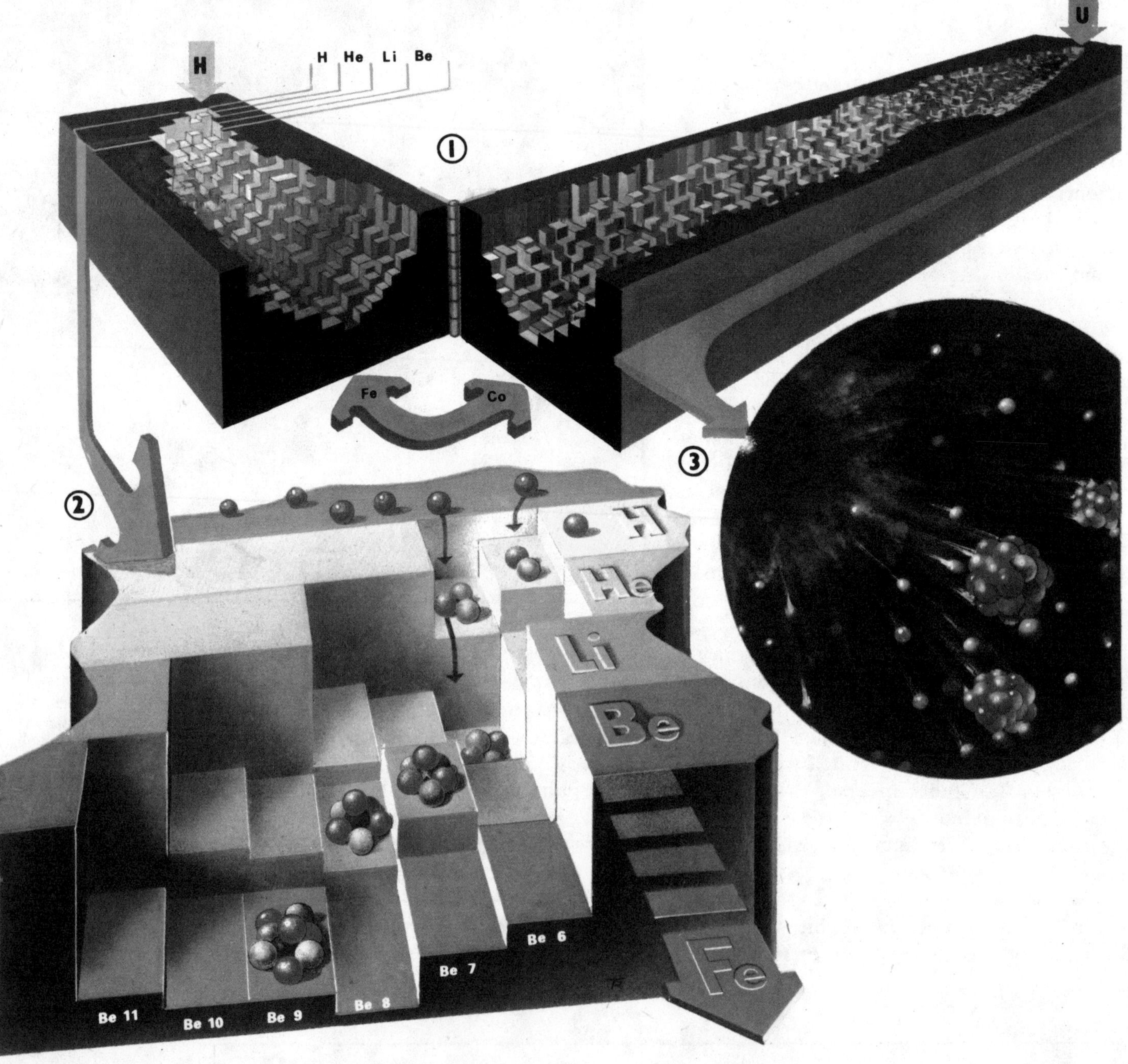

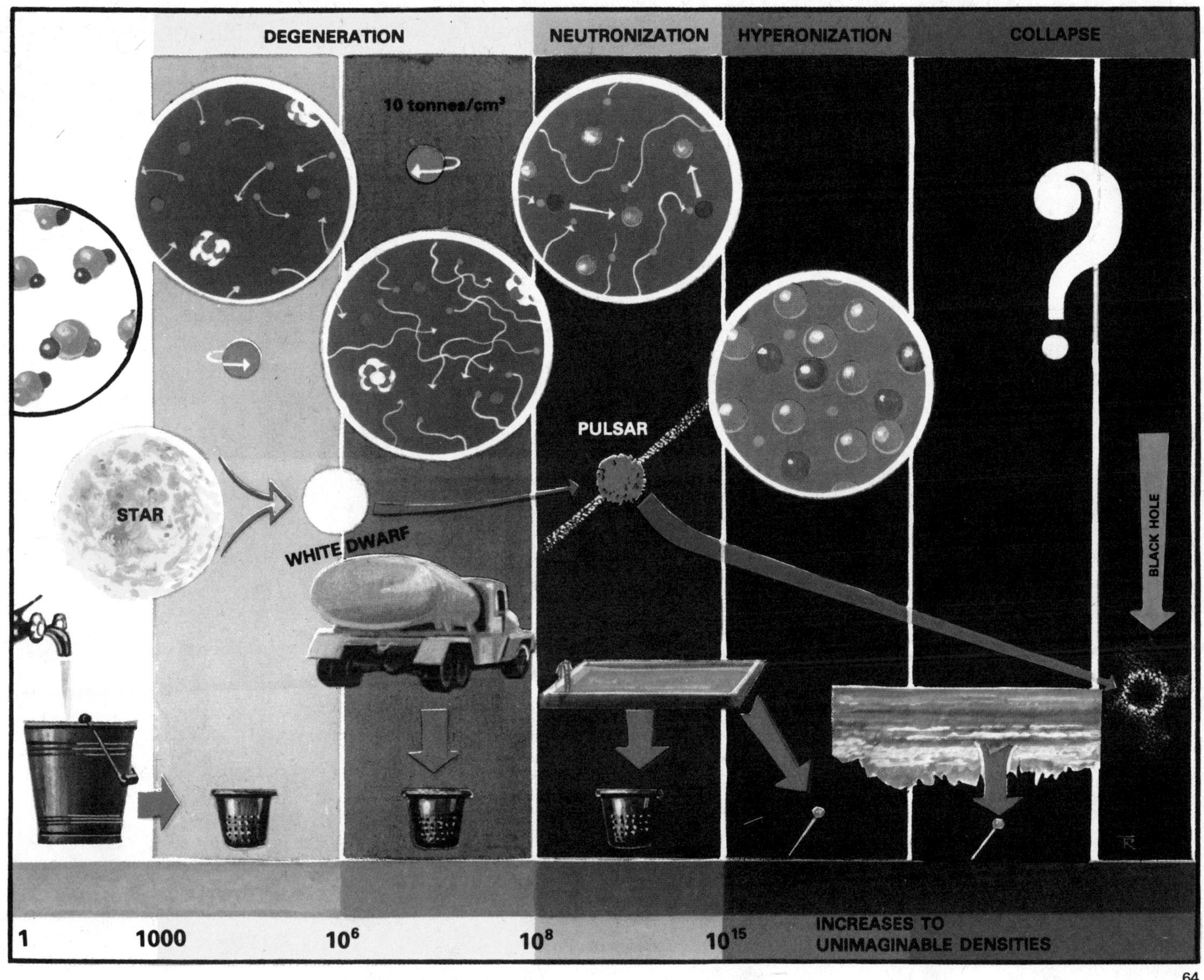

protons have turned into neutrons. *Pulsars* (Chapter 3) are made of neutron gas. If we were to compress a road-tanker full of water to the size of a pinhead, or to compress a whole swimming-pool into a thimble, we should end up with neutron gas – without molecules, without atoms, without nuclei. But we should need a very large amount of energy to achieve this.

If we were able to compress neutron gas still further, at densities of about 10^{15} g/cm^3 (1000 million tonnes per cm^3) – which would be like cramming our swimming-pool into our pinhead – we should obtain what is called hyperon gas, with hyperons in it as well as neutrons.

If we compressed hyperon gas to even higher densities, we should form a material whose properties we do not know. But such matter does indeed exist in the Universe, in 'black holes' (see page 118). We do not know what the matter in black holes is like, since not a single particle, not even a photon, is able to escape to give us a clue.

We have thus reached the conclusion that the denser a material is or the higher its temperature, the simpler it is. At high densities or temperatures matter can exist only in the form of free elementary particles. On Earth temperatures and densities are low, so elementary particles here can be arranged into systems. Life can only exist at low temperatures and densities such as those existing on Earth.

64 The properties of materials change with density. The water from a bucket, a tanker or a swimming-pool, when compressed into a thimble, would change into degenerate or neutron gas. The water from a swimming-pool compressed into the volume of a pinhead would change into hyperon gas. The densities at the bottom are in g/cm^3.

65 The Universe is made of three types of elementary particles: protons, neutrons and electrons. Protons and neutrons form nuclei, nuclei and electrons form atoms **(1),** atoms make up molecules **(2),** molecules crystals **(3),** crystals rocks, and rocks some planets and moons **(4).** Stars, too, are systems of atoms **(5),** in whose interiors the atoms of all heavier elements were formed from those of hydrogen. These later formed complex molecules **(6)** and living cells **(7),** and all the organisms of the biosphere **(8).**

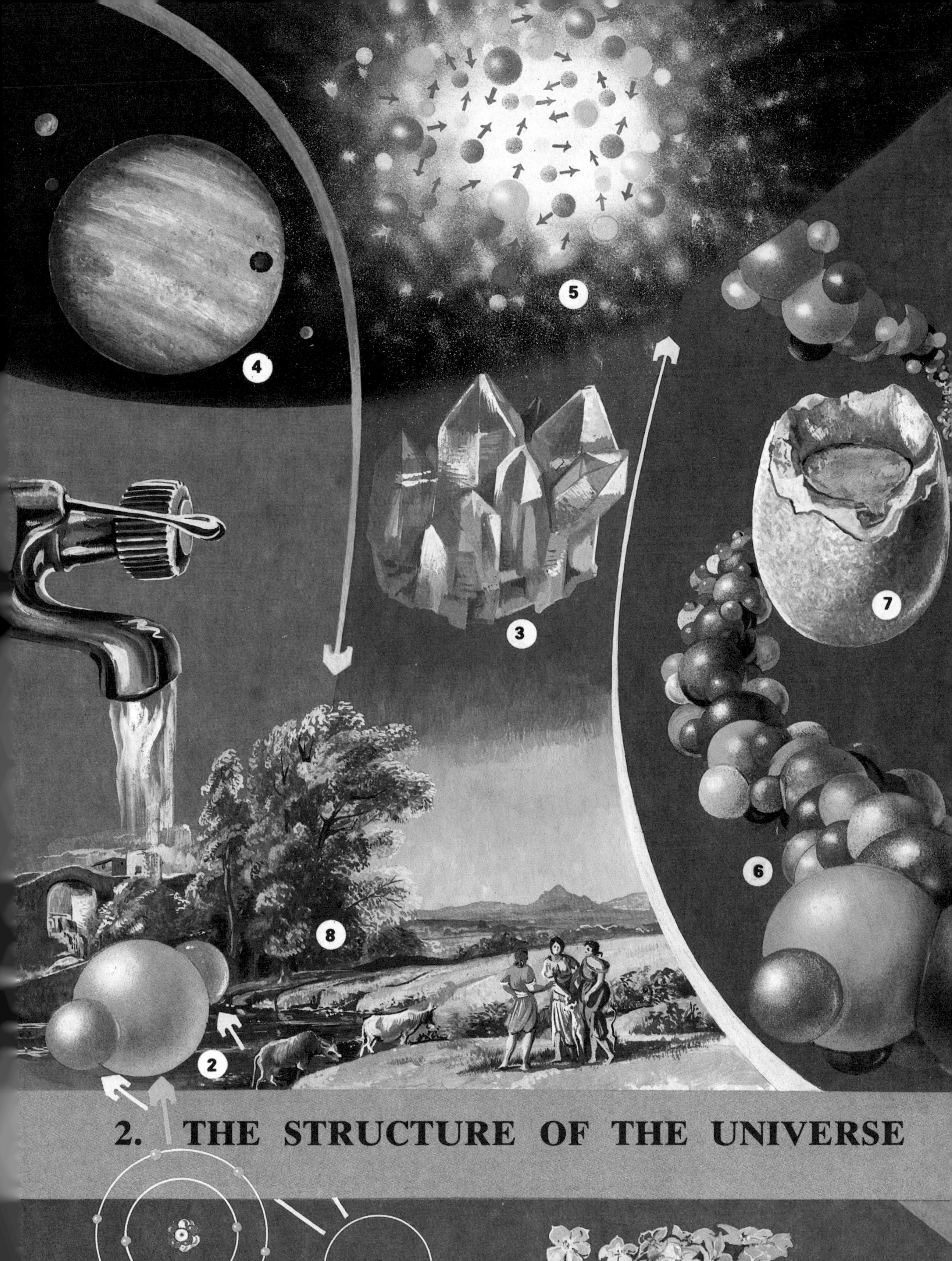

2. THE STRUCTURE OF THE UNIVERSE

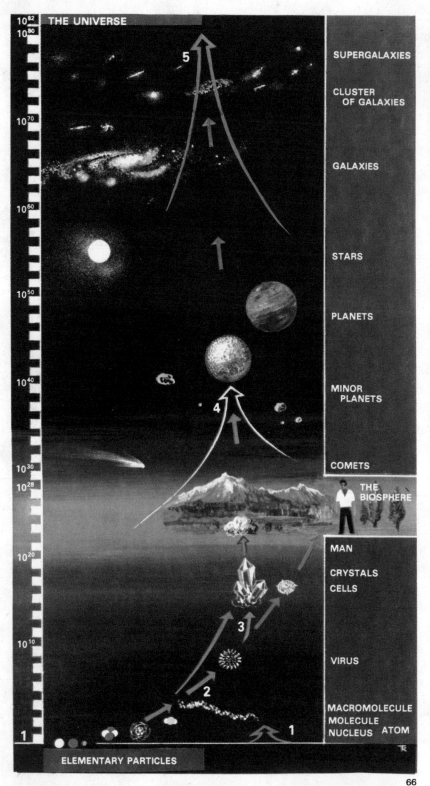

66 The nucleons in the atomic nucleus (neutron and proton extreme bottom) are bound together by strong interaction **(1)**. From the atoms to smaller bodies of the planetary system (yellow arrow) electromagnetic forces act **(2–4)**. Planets, stars and systems of them are bound by gravitation **(5)**. The vertical axis shows the number of elementary particles of which the system is composed. From left to right the organization (the orderliness) of the system increases.

From quarks to supergalaxies

In this chapter we shall examine various objects (i.e. systems) in the Universe, from quarks to supergalaxies. Figure 66 will help us get an idea of the size of these systems. The number of elementary particles constituting a system is indicated on the left-hand side. One step up means 10 times more particles, two steps mean 100 more particles, etc. The binding forces holding the systems together are marked by coloured arrows (1, 4, 5).

66

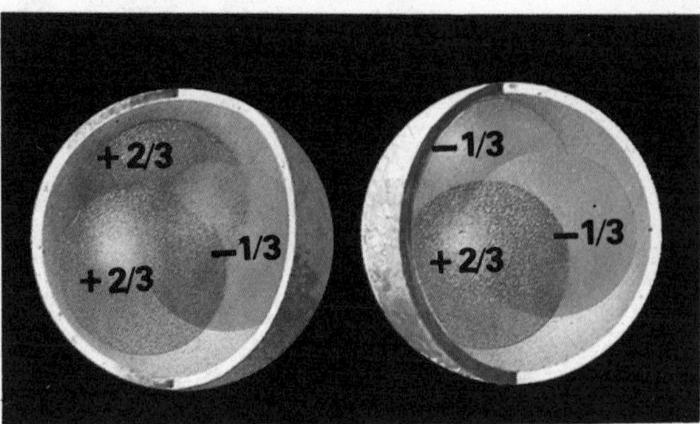

67

67 The composition of the proton and the neutron from the hypothetical quarks **d** $(-\frac{1}{3})$ and **u** $(+\frac{2}{3})$ might look something like this. The proton would be composed of the trio **uud**, the neutron of the trio **ddu.**

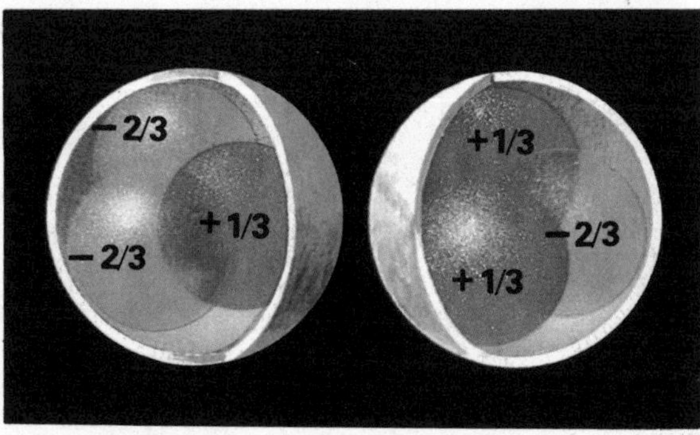

68

68 The composition of the antiproton and antineutron from hypothetical antiquarks $\bar{u}$ $(-\frac{2}{3})$ and $\bar{d}$ $(+\frac{1}{3})$ The antiproton would be made of the trio $\bar{u}\bar{u}\bar{d}$, the antineutron of the trio $\bar{d}\bar{d}\bar{u}$. All baryons and antibaryons consist of three quarks and antiquarks; mesons only of two of them.

The nucleon — a system of quarks?

In the first chapter we have seen that everything in the Universe around us is composed of nucleons and electrons. Up till now these have been considered the simplest elementary units, which are not capable of being broken down further. So far, no one has succeeded in dividing a proton, a neutron or an electron into simpler parts.

This idea of the indivisibility of particles has been somewhat shaken in recent times.

Many experts now believe that even elementary particles may be made up of something simpler, i.e. subelementary particles. If they are right, not only the proton and the neutron, but all hadrons, are made up of units called quarks. Baryons are supposed to consist of three quarks, mesons of only two (Figures 67 and 68).

Quarks are thought to have charges in thirds ($-\frac{1}{3}$ and $\frac{2}{3}$). They can only exist inside hadrons. But we must not forget that quarks are only hypothetical at the moment, and cannot be proved experimentally. (Neither the electron nor any other lepton can be explained as a system of quarks.)

The atomic nucleus – a system of nucleons

The smallest system of elementary particles is the nucleus of an atom. It is made up of nucleons, which are joined together by the huge nuclear force. The nucleus of an atom is a formation a few fermi across, but with immense density, 10^{14} g/cm^3. According to the number of nucleons we distinguish various sorts of nuclei, or nuclids. Every nuclid is described by two numbers: the number of protons, Z; and the number of neutrons, N. Thus the number of nucleons in a nuclid is Z + N, which is denoted A, and called the nucleon number or mass number. It indicates the mass of the nuclid.

Over 1400 types of nuclid are known. The number of protons in nuclids varies from one to 105, that of neutrons from zero to 157. Most nuclids are radioactive, and decay in time.

Scientists have constructed a map of nuclids. It is a network of squares where the number of neutrons N and the number of protons Z are marked. Each nuclid has its own pair of numbers (N and Z) and thus its own square. But there is not an actual nuclid for every square. We might choose, for instance, square 10,2 for 10 neutrons and 2 protons. But in nature there is no nuclid of helium (2 protons) with 10 neutrons. Nor is there a carbon nucleus with 20 neutrons, so square 20,6 is also empty. Such nuclei cannot exist, since the nuclear forces which hold the nucleus together do not permit it.

The nucleus is the 'heart' of the atom. For this reason the chemical designation of the atom (element) is used to denote it. Two small numbers are added: the lower one indicates the number of protons in the nucleus, Z, and the upper one the mass number, A. For example, $^{1}_{1}$H is the nucleus of hydrogen (proton), or $^{4}_{2}$He is the nucleus of helium, with two protons and two neutrons, i.e. four nucleons in all.

A nucleon is bound to its nuclid by a high binding energy. This energy was given up by

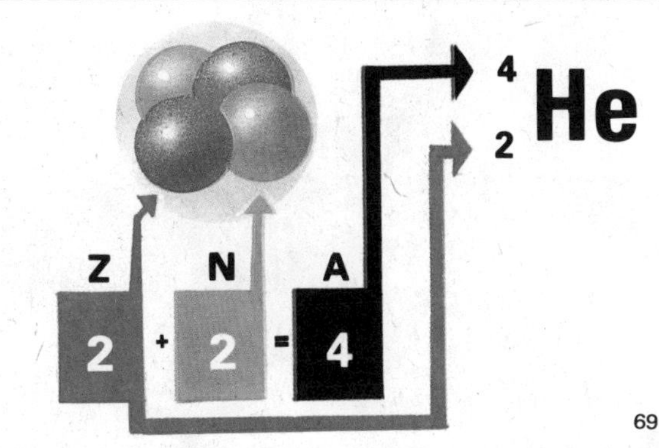

69 The nucleus of helium (alpha particle) is composed of two protons (Z) and two neutrons (N), i.e. a total of four nucleons (A).

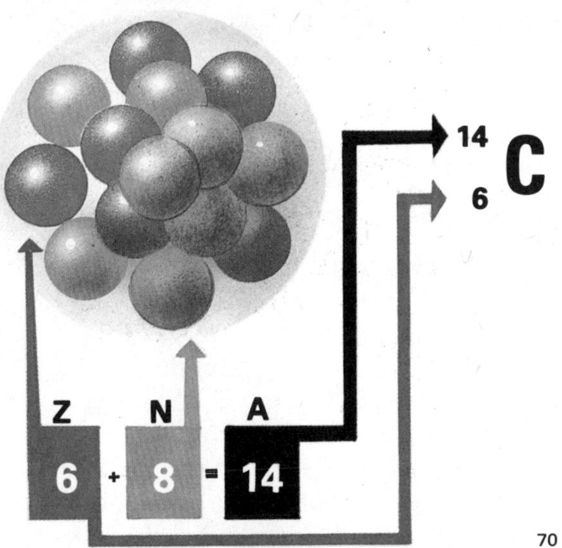

70 The nucleus of radioactive carbon with six protons and eight neutrons is unstable (6 + 8 = 14). After about 5000 years one of its neutrons disintegrates, giving rise to a nucleus of nitrogen $^{14}_{7}$N.

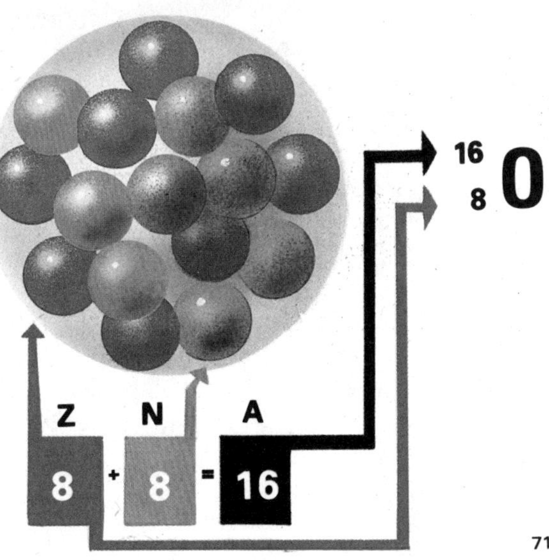

71 The nucleus of oxygen with eight protons (Z) and eight neutrons (N).

the nucleon when it entered the nuclid. The higher the binding energy of the nucleon in the nuclid, the more stable is the nuclid. We can use the model in Figure 63A to compare the binding energy of various nuclids. For every nuclid the binding energy of each of its nucleons is represented by the depth of its respective square. A nuclid with a binding energy of 3 MeV per nucleon is at a depth of 3 cm (1.2 in), one with a binding energy of 7 MeV at a depth of 7 cm (2.8 in), etc. In this way the binding energy of all nuclids is represented by a sort of terraced valley. There are three numbers corresponding to every square step: the number of neutrons, the number of protons, and the binding energy. We shall call this model the valley of stability. There should be over 1400 square steps in it, since that is the number of nuclids we know. The valley of stability serves as a very good model for understanding all nuclear reactions which take place in the Universe. It explains the origin of the chemical elements from hydrogen as it takes place in the stars (Chapter 3). A nuclid which moves to a lower step releases energy. The combination of four protons in the nucleus of helium, for instance, or the fission of the nucleus of uranium with 235 nucleons ($^{235}_{92}U$) into two lighter nuclids means that the nucleons in the nuclids must release binding energy, since they fall lower down.

At the start of the Universe there were only protons and electrons. Free protons have no binding energy and are the highest in our model, as if they were on the top of our model ready to roll down into the valley and come to a stop on one of the square steps or 'terraces'. An opportunity to do so occurred after the formation of the stars. In the hot cores of stars protons moved so fast that in collisions they overcame their mutual repulsive electrical force and came within 1 fermi of each other. Then the nuclear forces drew them together and a nucleus came into being; in other words, protons fell into the valley of stability. This happened at the moment when the nuclear force joined them in a nucleus.

The older the Universe gets, the more protons get drawn into nuclei. As time goes by the number of protons at the top decreases and the number of nuclei on the steps below increases. Stars live on the binding energy which is thus released. In these transformations electrons are unimportant, for this is a nuclear reaction. Instead of saying that nuclei of hydrogen combine in the nuclei of heavier elements, we say simply that hydrogen changes into heavier elements.

How many atoms (more exactly, how

72

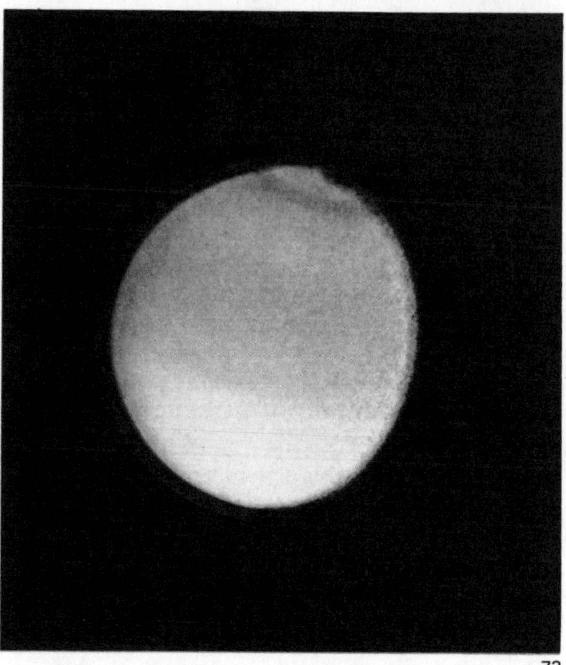

73

72 Saturn's rings are made up of a large number of lumps of ice, small stones and fine dust.

73 Saturn's largest satellite, Titan. It has a diameter of 5800 km (3600 miles), and is larger than Mercury (4850 km/3012 miles). The surface of Titan is not visible, being covered by an opaque atmosphere of nitrogen, hydrogen, methane, ammonia and various organic compounds.

many of their nuclei) of various elements have already been made in the Universe from hydrogen?

How abundant are the individual chemical elements in the Universe? Geologists have determined the chemical abundances for the Earth (Figure 224). Astronomers did the same for the whole Universe. What are the results of their research?

There are still more atoms of hydrogen than of anything else; they have not yet had the opportunity to change into heavier atoms (i.e. to fall into the valley of stability). There is relatively very little lithium, beryllium or boron – these atoms are changed into helium right at the start of the evolution of stars (Chapter 3). On the other hand iron atoms are very abundant because their nuclei are the most stable of all (they are at the bottom of the valley of stability).

In spite of the great variety of atomic nuclei or nuclids, all these systems are based on two sorts of elementary particles only – the protons and the neutrons bound by nuclear force. The electrical force of a proton acts over a much greater distance than its nuclear force. It attracts electrons and repels other nuclids. Thus the more complex system of the atom is built by the electrical force out of a nuclid and electrons.

The atom – a system of a nuclid and electrons

A positively charged nuclid (nucleus) binds to itself electrically the same number of electrons as it has protons. If the atom was enlarged one billion times (10^{12}), we should see in the centre the heavy nucleus, measuring a few millimetres across. Light electrons about 1 mm (0.04 in) in diameter would be orbiting the nucleus at a distance of a few hundred metres. The nucleus is positively charged (in our model it is red), while all the light electrons are negatively charged (shown in blue). The nucleus is several thousand times heavier than all the electrons put together.

As we have already said, the electrons are rather a long way from the nucleus. In the space between the nucleus and the electrons electrical forces are at work. The electrons are arranged around the nucleus according to certain rules. Together they are called the electron envelope. The dimensions of this electron envelope are about 100,000 times greater than those of the nucleus. The electrons have a binding energy of a few electronvolts. The electron on the inside of the envelope (nearest the nucleus) can have a binding energy of up to 1000 electronvolts. Even that is very small compared with the binding energy of a nucleon in the nucleus

74 A comet is a rock about 1 km (0.6 miles) across, covered in a thick layer of ice, snow and dust. When it comes within 200–300 million km (124–186 million miles) of the Sun, the surface of the comet sublimates, producing a large gaseous envelope (the coma) and a long tail.

(around 8 million electronvolts). This is the reason why the atom is not nearly as stable and resistant to external shocks as its nucleus.

Much of the matter in the Universe is in the incandescent cores of the stars. There, violent collisions disrupt the atoms (the phenomenon called ionization). In some stars (such as white dwarfs) densities are so great that their matter is degenerate (Figure 64). There are only a few places in the Universe where there is a sufficiently low density and temperature for nuclei to preserve their inflated envelopes. Luckily, the surface of our planet is one of those places – thanks to the magnetosphere and the atmosphere, which protect all the atoms on the Earth's surface from the destructive photons and fast-moving particles from space.

The molecule – a system of atoms

Atoms combine to form a higher system called the molecule. A molecule is the smallest sample of matter (for instance, the smallest sample of water is the water molecule, composed of two atoms of hydrogen and one of oxygen). Protein molecules are made up of many thousands of atoms of carbon, hydrogen, nitrogen, oxygen, phosphorus and sulphur.

From a little less than 100 different atoms it is possible to build an almost unlimited number of various sorts of molecule. Chemists know more than one million different sorts of molecules. Atoms in a molecule have a precise spatial arrangement. In a molecule of ethyl alcohol (C_2H_5OH), for instance, a carbon atom is joined by electrical force to three atoms of hydrogen and another atom of carbon. The second carbon atom is joined to two atoms of hydrogen and one of oxygen. Then the atom of oxygen is joined to another atom of hydrogen.

75 Phobos is the nearer and larger of Mars' satellites. It is potato-shaped, measuring 19 × 21 × 27 km (12 × 13 × 17 miles). Its long axis always points towards its parent planet. The craters measure from 10 m to 1.2 km (33 ft to 0.75 miles).

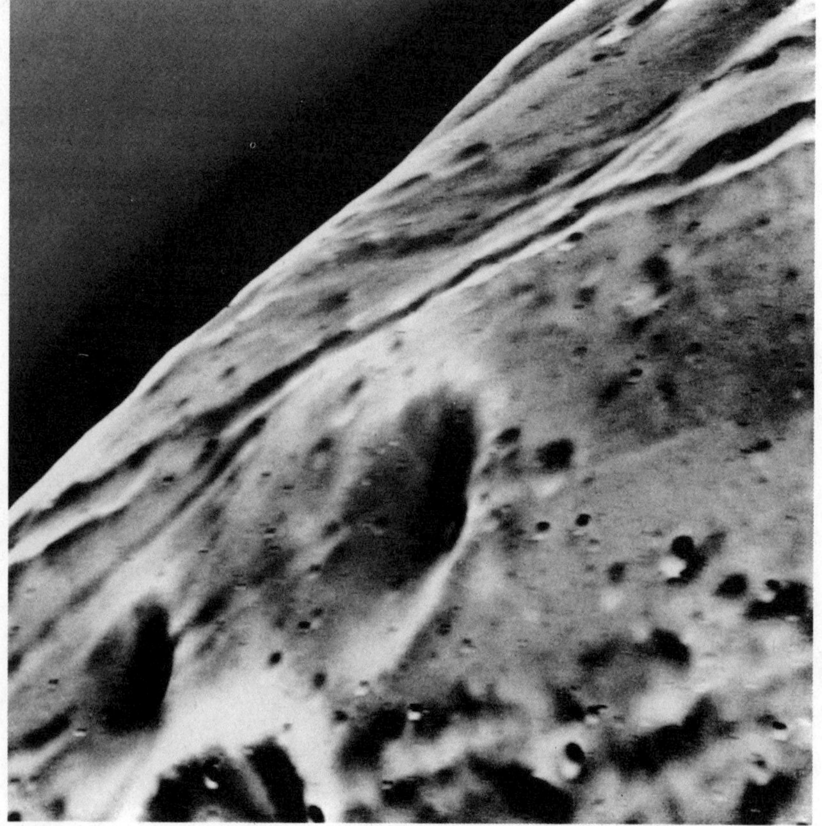

76 The surface of Phobos is covered with numerous craters which were caused by the fall of meteoroids, similarly to those on the Moon. Many of the craters were formed 3500–4000 million years ago. But the long, thin grooves have yet to be explained; they are 100–200 m (330–660 ft) across.

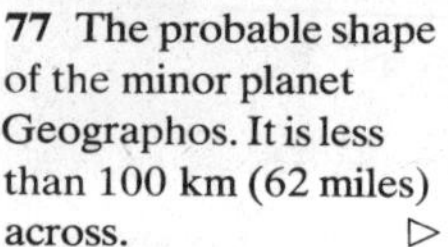

77 The probable shape of the minor planet Geographos. It is less than 100 km (62 miles) across. ▷

When a molecule is formed from atoms, a tiny part (about one thousand-millionth) of the rest energy of all its atoms is released. This energy is called the binding energy of the molecule. A few common binding energies are shown in the following table:

C – C	2.56 eV	(saturated carbon)
N ≡ N	6.8 eV	(in the N_2 molecule)
O – O	1.52 eV	(in the H_2O_2 molecule)
O = O	4.17 eV	(in the O_2 molecule)
O – H	4.78 eV	(in the H_2O molecule)
C – O	3.04 eV	(in the molecule of alcohols)

In chemical reactions the structure of molecules is altered. Either energy is gained from the new bonds (a process called exothermal reaction) or it must be supplied (endothermal reaction) for the reaction to take place.

There are molecules everywhere in the Universe where the temperature or density is not too great. As, for example, in the Earth's crust, in its atmosphere, on the other planets and satellites, in comets, and in the dust clouds of interstellar space. Simple molecules can also be detected (by analysing their spectra) on the cooler stars, though only on their cold surfaces, since there are not even complete atoms in their hot centres. Radio waves have revealed about 40 different molecules in the interstellar dust clouds. There they are protected not only from photons of ultraviolet and X-radiation, but also from the destructive particles of cosmic radiation.

The crystal and the cell – systems of molecules

So far we have looked at the electrical force which binds electrons to the atomic nucleus and atoms in molecules. There is also an attractive electrical force at work between molecules. It forms complex systems of them. In Figure 66 we saw how everything in the Universe is gradually built up of elementary particles by means of three basic forces. Here we shall examine one level of the hierarchy of the Universe – the transition from molecules to systems of them.

For more than 3000 million years the molecules on the surface of the Earth have been organizing themselves into more complex and more sophisticated systems: from molecules to macromolecules to giant molecules to organelles to cells to colonies of cells to multicellular organisms (Figure 66). The evolution of life meant that more and more sophisticated systems of molecules appeared on the Earth. Since one can expect suitable conditions for the evolution of life to exist in other planetary systems also, we may suppose that life also exists in many other parts of the Milky Way and other galaxies. Such questions are the subject of a new branch of science called exobiology.

The molecules of any material are in constant and disorganized movement. The higher the temperature of the material, the faster they move. When the material cools down, their movement becomes slower, and at a certain temperature the attractive force between molecules prevails over their movement and binds them into a regular structure. A crystal of solid matter is created. If their temperature falls sufficiently low, almost all materials solidify into crystals.

A crystal is a system of many molecules. Unlike liquids and gases, crystals have regularly arranged molecules. They form a regular network called the crystal lattice. The crystals of various materials differ in shape and colour. This variety in the world of crystals is due to the electrical force which acts between the molecules and their atoms (Figures 66 and 217–223). Crystals are much simpler than the simplest living system of molecules, the cell. They are also much more common; they are to be found everywhere in space where matter exists in solid form (the Earth's crust, interstellar dust, comets, satellites, etc.).

Crystals differ a great deal from living cells. But they do have one thing at least in

78 A comparison of Jupiter's satellites: Io (3600 km/2236 miles), Europa (3100 km/1925 miles), Ganymede (5270 km/3273 miles), Callisto (5000 km/3106 miles). They are called Galileo's satellites after their discoverer.

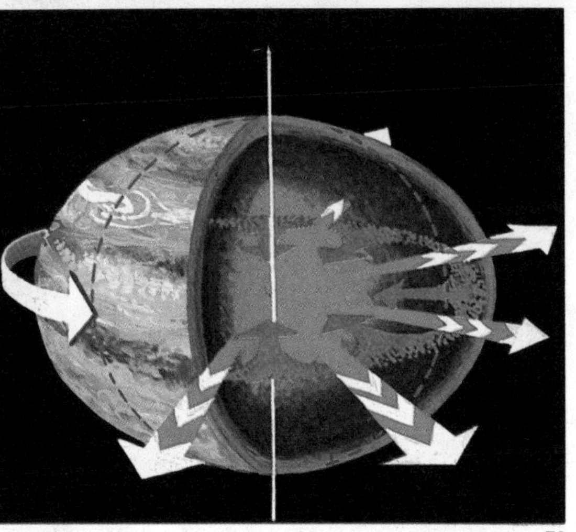

79 Large bodies (planets and stars) have, as a result of self-gravitation, a spherical shape. If they rotate rapidly, centrifugal force flattens them. An example is Jupiter.

common – both of them grow. Crystals grow from a solution, which means that they increase their volume in an ordered manner. In salt water, for instance, small salt crystals form, growing in size as the water evaporates.

A crystal attaches to its sides more and more molecules, extending its lattice. The moment a molecule becomes attached to the side of a crystal, it releases a tiny part of its rest energy. The electrical force arranges the molecules spatially in such a way that the energy released is as great as possible. This explains why various materials (differing from each other in their molecules) form crystals of different shapes. The energy released by the crystal during its growth is also called binding energy. The greater it is,

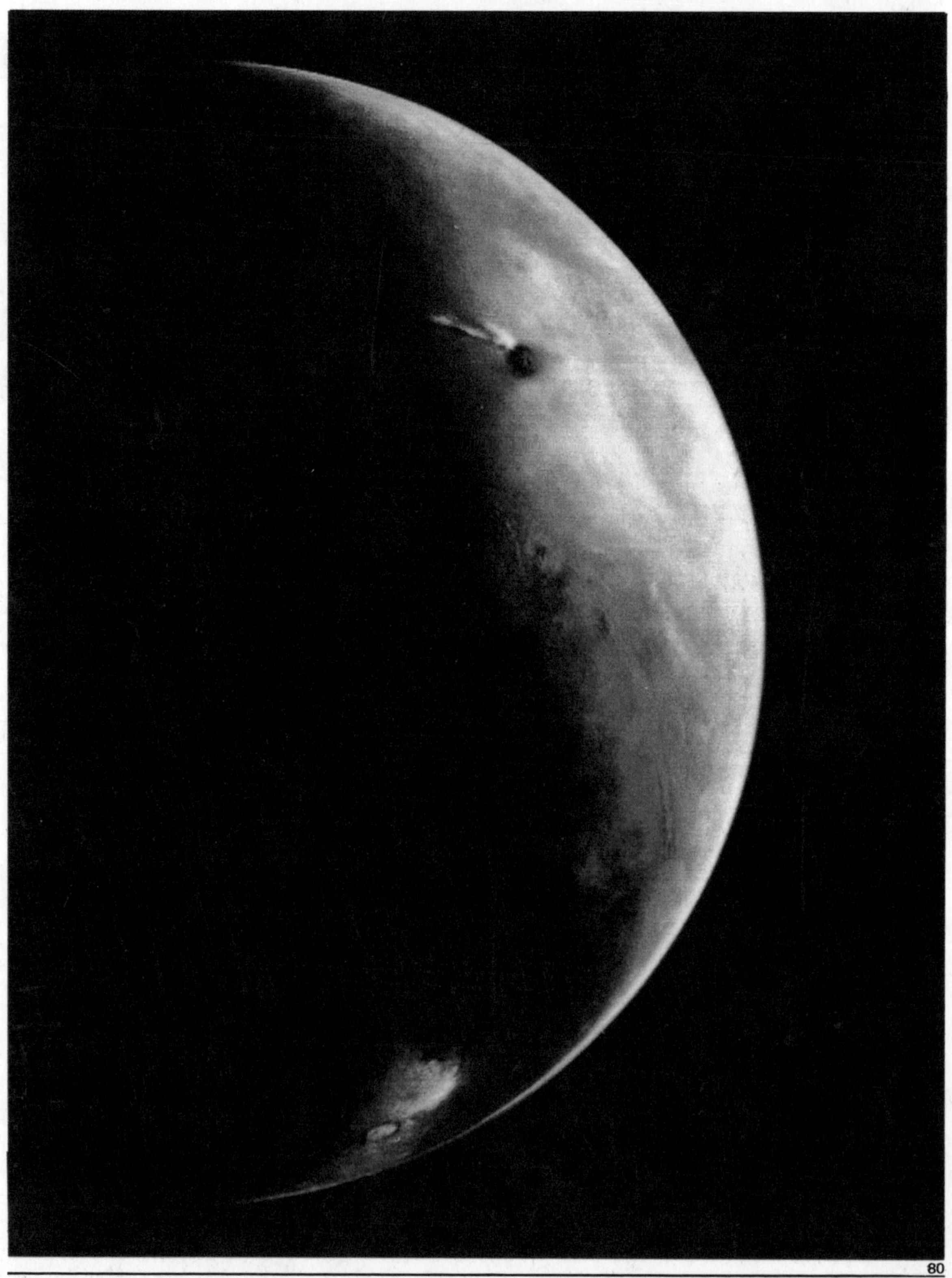

80 A crescent of Mars photographed by Viking 2 from a distance of 419,000 km (260,200 miles). Viking approached Mars from the dark side. From the centre to the south-east the long canyon Valles Marineris can be seen. The north pole is at the top, and the south pole is in darkness at the bottom (it was winter in the southern Martian hemisphere at the time). Close to the south pole is a large crater covered in ice and fog. The long white strips beyond the volcano are clouds of ice crystals.

the more stable the system of molecules – the crystal – is, and the more resistant to external influences. It must be heated to a higher temperature before it melts. Or a greater pressure must be exerted on it to crush the crystal lattice. In the interior of solid bodies consisting of more than 10^{46} elementary particles (Figure 48) self-gravitation is strong enough to achieve this.

body. Such a ranking is usually called a hierarchy of structures. Each item is made up of simpler parts, which are called subsystems. A crystal, for instance, is a system of molecules, so a molecule is a subsystem of the crystal. But the molecule itself is a system of atoms, so the atom is a subsystem of the molecule.

Rock is a grouping – a system – of various

81

81 A colour photograph of the surface of Mars, taken from the unmanned probe Viking 1. The orange-red material of the surface (coloured by limonite – an iron oxide) is found in most parts of the planet. Below it is a darker substrate (visible in the bottom right-hand corner).

Rock – a system of crystals

From the start of this chapter we have been considering how elementary particles are gradually built up into ever more complex systems. Figures 1 and 66 show this increasing complexity from elementary particles, through the nucleus, the atom, the molecule, the crystal and rock, right up to a planetary crystals, bound together by electrical forces. It may be large, making up an entire range of mountains. But a piece of a meteorite is also rock, made up of all sorts of different crystals (such as troilite, taenite, pigeonite, etc.).

Geologists distinguish rocks more by their origin than by the crystals (minerals) of which they are composed. Rocks which originated by the cooling and crystallization of hot magma are called volcanic. The

Earth's crust in particular is made of these, though they do not emerge on to the surface everywhere. Sedimentary and metamorphic rocks make up a smaller proportion. Rocks, their structure and their evolution are dealt with in detail in Chapter 7. Let us just add that the rocks of the Earth's surface on which we walk are mainly composed of crystals of various silicates. In the upper part there are chiefly aluminium silicates, in the lower part ferrous and magnesium silicates.

82 The sand dunes of Mars indicate high wind speeds.

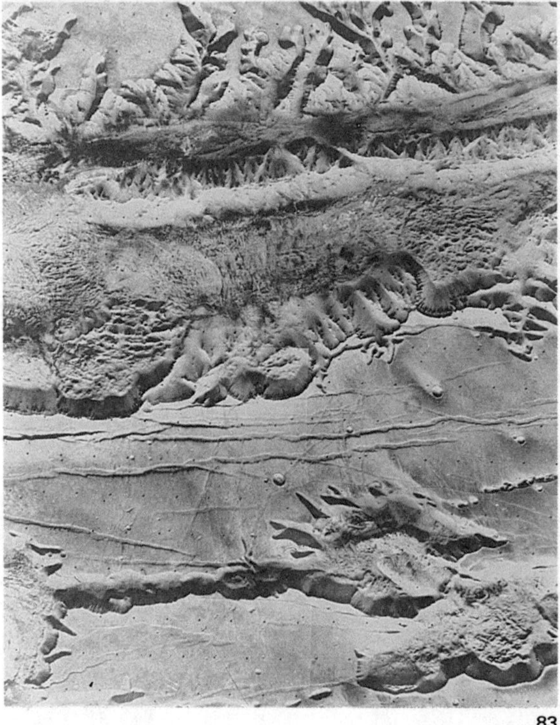

83 This huge canyon (called Valles Marineris) is 6 km (3.7 miles) deep. Its width is around 100 km (62 miles), and its length about 5000 km (3106 miles). It was formed long ago by the action of water, which occurred here in large amounts.

Planetary bodies – systems of rocks

The most complex systems existing in a solid state are minor planets, satellites and planets. These are systems of rocks. Some planets (and even one of Saturn's satellites, Titan) are surrounded by a gaseous atmosphere, but their central part is always made of rock. This is also true of the largest of the planets, Jupiter, whose central region (core) is probably made of ferrous silicates. The core of Jupiter and the other large planets (Saturn, Uranus and Neptune) is surrounded by an extensive layer of gas – mainly hydrogen (H_2), helium (He), methane (CH_4) and ammonia (NH_3). Between the gaseous envelope and the core is a thick layer of metallic hydrogen. The pressure and density in it are so great that electrons move freely among protons. At pressures greater than 2.5 million atmospheres hydrogen has the properties of a molten metal (rather like mercury). The inside of Jupiter and the other giant planets consists of metallic hydrogen. In the centre, beneath this metallic hydrogen, there is a small core made of iron and rocks (mainly silicates). The core is probably surrounded by a layer of solid metallic hydrogen, above which there is liquid metallic hydrogen. At enormous pressures even hydrogen can form rock, and this is what envelops the cores of the giant planets.

Not only hydrogen, but ice may also exist as a rock, and form part of the structure of planetary bodies. Titan, for example, is covered in ice, and its low density (1.6 times that of water) indicates that this, the largest of Saturn's moons, has a small core of silicate rocks, surrounded by this layer of ice. The polar regions of our own planet are, of course, covered by a thick layer of permanent ice.

A study of Saturn's rings from close up (offered by Voyager's instruments) shows that they are in fact a huge number of ice boulders some several metres across, which orbit their giant planet like little moons. They form rings because their orbits are in the same plane.

Much smaller bodies than Titan, the comets, have a similar rock structure. Titan has a mass of 9×10^{21} kg, while the mass of Halley's comet is 2.5×10^{16} kg. Comets, of course, come from the distant frozen frontiers of the Solar System, where there are many thousand millions of them. There the effect of solar radiation is negligible, so that while they are there comets have neither a head nor a tail. They are just huge balls of dirty snow and ice, roughly 10 km (6 miles) in diameter. When these big snowballs get closer to the Sun than the planet Mars, the Sun's heat sublimates their ice into gas. Such a snowball can release tens of tonnes of various gases (water vapour, CO, CO_2, OH, NH, O_2, etc.) and dust particles a second. It means that the snowball consists not just of water, snow and ice, but also of methane and ammonia ice. The gases and dust released form an extensive envelope around the snowball, known as a *coma*. The coma (from the Greek word for hair) is very rarified, much less dense than the air we breathe. It is much larger than our Earth, sometimes even bigger than the Sun itself. The solar wind blows the coma away from the Sun, thus forming the comet's tail, which is often many millions of kilometres long.

Small solid bodies made of rocks, such as meteoroids, the nuclei of comets, small satellites and the vast majority of minor planets, are held together by electrical force. Their mass is small, so their self-gravitation is also low. For this reason the shape of small bodies is irregular. An example is Mars' moon Deimos, measuring 11 × 12 × 15 km (6.8 × 7.5 × 9.3 miles). The larger Phobos, measuring 19 × 21 × 27 km (12 × 13 × 17 miles), is also 'potato' shaped. The minor planet Geographos has an irregular shape, and is elongated and cigar-like (Figure 77). Its dimensions are a few tens of kilometres.

The larger and heavier a solid body is, the greater the effect of its gravitation, which acts not only on other bodies, but also on its own. This is known as self-gravitation, and attracts the individual particles of a body to each other. Inside solid bodies more than 500 km (310 miles) in diameter the self-gravitation is so great that it breaks up the crystalline structure of the rocks. The solid rock thus becomes a pliable, dough-like material where the pressures in different directions are evened out. The irregularity of the body thus disappears: tall projections are heavy and sink downwards towards the middle, while light, thin parts of the body (depressions) are pushed through the doughy material to the top. This equalization of pressures is called *isostatic equilibrium*. The body tries to assume a round shape through its own gravitation (Figure 48). The round shape of rapidly rotating bodies (such as Jupiter) is flattened by centrifugal force (Figure 79).

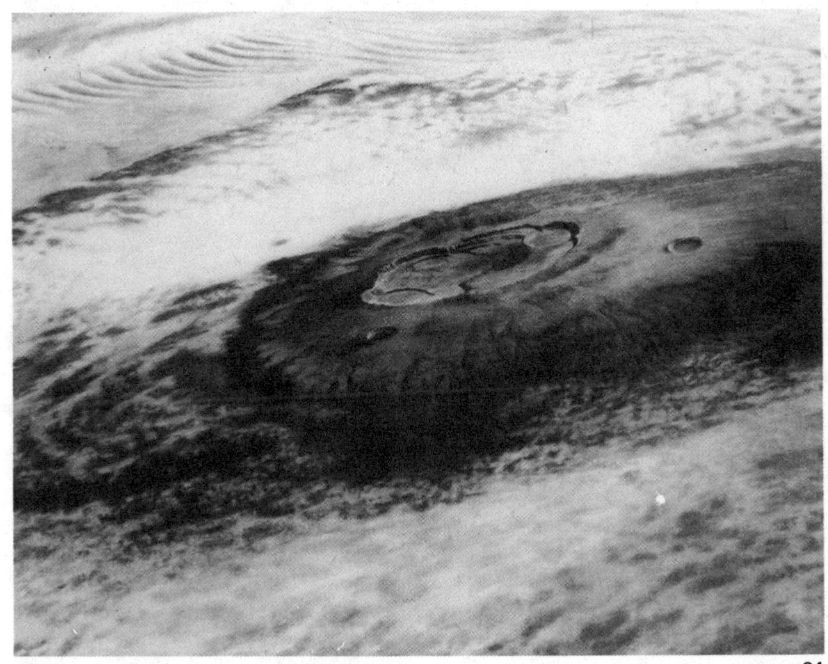

84

84 The largest and youngest of the Martian volcanoes is Olympus Mons. Its diameter at the base is 600 km (373 miles), and its height 25 km (16 miles). The round craters are the remnants of the volcano's activity. The largest of them is 80 km (50 miles) across. The clouds around the volcano cover its base.

The Solar System

Our Sun is surrounded by a large number of bodies of various sizes, from microscopic micrometeoroids (grains one-thousandth of a millimetre across) to the giant planet Jupiter. All these bodies together are known as the Solar System, because the Sun holds them close to itself by its gravitation, forcing them to travel around it in elliptical orbits.

The chief members of the Solar System are the Sun and the nine planets (Figure 108). Some of these planets (Earth, Mars, Jupiter, Saturn, Uranus, Neptune and Pluto) have orbiting satellites. We now know of the existence of about 60 of them in the Solar System, for thanks to interplanetary probes we are discovering new, smaller satellites. In recent times these natural satellites, which came into being along with the planets themselves, 4600 million years ago, have been joined by artificial ones, launched by man. They contain scientific instruments to help us learn more about the Sun, the planets and their natural satellites.

We can also consider as satellites the huge number of small bodies a few metres across which orbit Jupiter, Saturn and Uranus, forming their 'rings' (Figures 72, 88, 90, 95, 96 and 97). Satellites and rings move in the vicinity of their planets, which means in the region where the gravitational attraction of the planets is greater than that of the Sun. If a satellite escapes from the gravitational pull of a planet, it ceases to be that planet's satellite. It becomes a planet in its own right, and its movement is decided by the Sun (Figure 94).

Since the time of Nicolaus Copernicus it has been known that the planets orbit the Sun along orbits which lie in approximately the same plane – the plane of the ecliptic (Figures 1 and 107). The astronomer Johannes Kepler of Prague deduced from the observations of Tycho Brahe that the orbits of the planets are elliptical, with the Sun at one of their foci. The nearer a planet (or other body) is to the Sun, the more quickly it orbits. We see the planets from Earth against the background of the *constellations of the Zodiac,* which are distributed along the ecliptic (Figure 105). If we observe the sky for a long time (several days or even weeks), we find that the planets move in relation to the Zodiac, seeming to pass from one constellation to the other. That is why they are called planets – the Greek word *planetein* meaning to wander.

Much smaller bodies than planets are planetoids (called also asteroids or minor planets). You cannot see them with the naked eye, and some of them can be observed only with the world's largest telescope. Through a telescope the planetoids look like stars, and only after some hours' observation does it become clear that they are moving in relation to the stars. The total number of planetoids is estimated at some 400,000, though only a few thousand have actually been observed. Their total mass is about one-thousandth that of the Earth. The diameter of the largest, Ceres, is 1020 km (634 miles), while Adonis at 300 m (984 ft) is one of the smallest. Most of the planetoids move in the belt between the orbits of Mars and Jupiter (Figure 107). Only a few (such as Adonis) come nearer the Earth, or go beyond the orbit of Jupiter (such as Hidalgo). The smallest planet, Pluto, is sometimes considered to be a planetoid. With its satellite Charon it forms a double planetoid. Some planetoids have names given to them

85

85 A photograph of Jupiter from 47 million km (29 million miles) away. The clear white strips are clouds of ammonia crystals from gases which rise and then condense. The dark strips are warmer regions, where gases fall towards the planet's centre, are compressed and heat up.

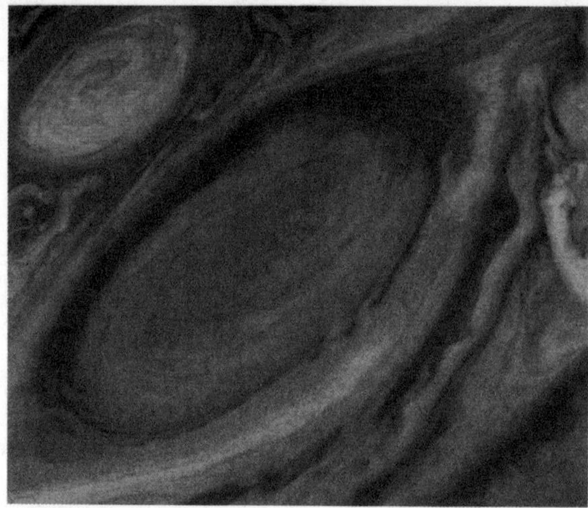

86

86 The red spot is the most conspicuous feature of Jupiter's atmosphere. The gases in its centre are almost motionless. On the edge they move in an anti-clockwise direction and make a circuit every six days.

by their discoverers. The planetoids which pass within the orbit of Mars are probably extinguished comets, whose envelope of ice and snow has been exhausted. They are in fact the cores of short-period comets (made of rocks), which can never form a coma and a tail again.

The smallest members of the Solar System are the meteoroids – various sized lumps of rock (Figure 111). They orbit the Sun in elliptical orbits, like the rest of the large bodies. The smallest meteoroids are composed of a few hundred molecules only, while the largest are measured in fractions of a kilometre and approach the size of the planetoids. The meteoroids are known collectively as the *meteoroid complex.* Its meteoroids are scattered in an ellipsoidal

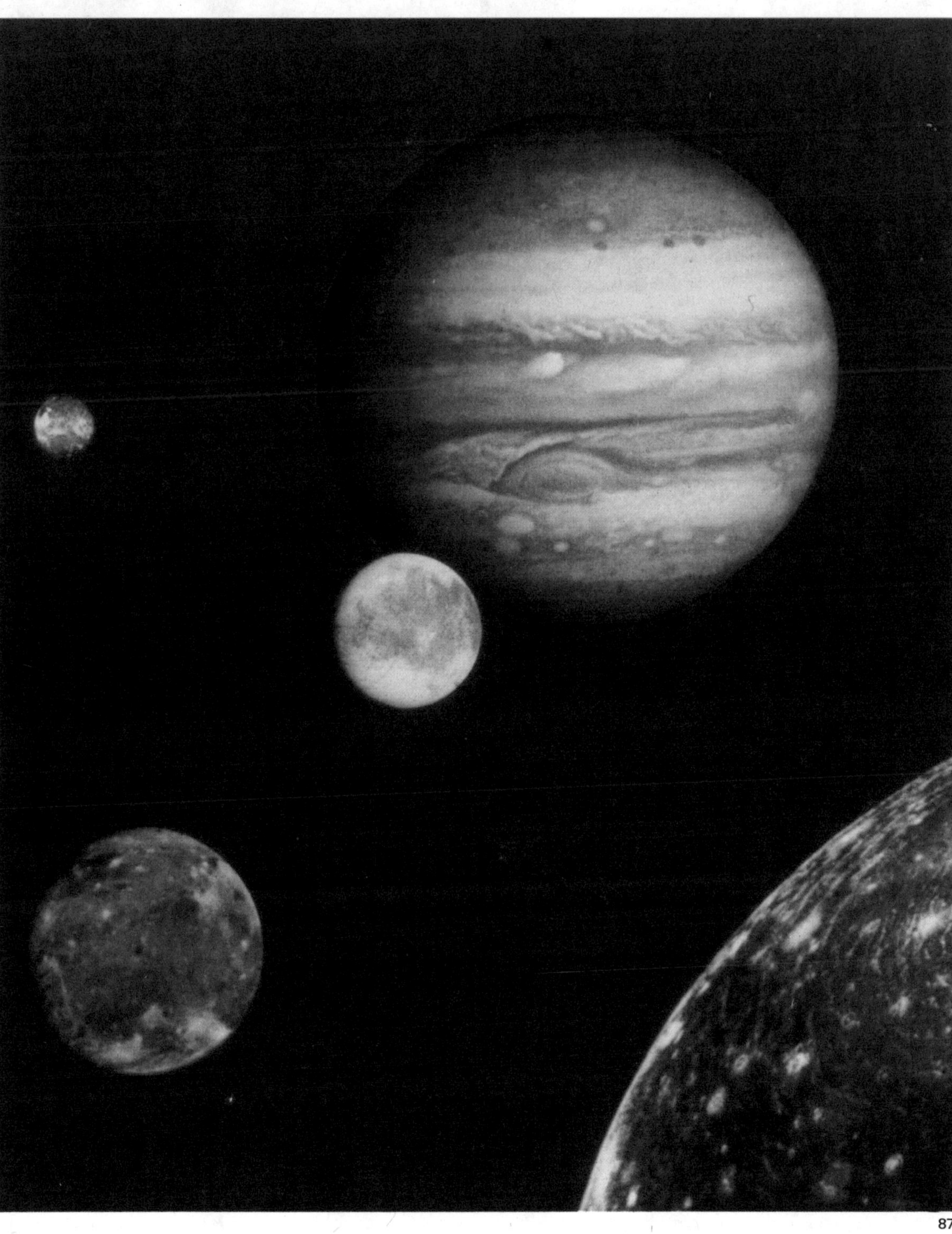

87

87 Jupiter and its four largest satellites: Ganymede, Callisto, Io (upper left) and Europa (in front of Jupiter).

88

88 Saturn and its satellites: in front is Dione, to the right are Tethys and Mimas, and to the left of the rings are Enceladus and Rhea. All of them (except for Enceladus) have crater-pitted surfaces. In the distance (top right) is the planet's largest satellite, Titan.

space around the Sun and it stretches along the plane of the ecliptic far beyond the Earth's orbit. You can observe the Sun's rays diffused and reflected by the meteoroid complex before sunrise in autumn or after sunset in spring. It looks like a shining cone in the constellations of the Zodiac. For this reason we speak of it as the *Zodiacal light* (Figure 112).

The furthest planet of the Solar System is Pluto. It is about 40 times further away from the Sun than the Earth. To use scientific terms, Pluto is about 40 astronomical units away from the Sun. An astronomical unit is the average distance between the Earth and the Sun: 150 million km (93,204,000 miles) – or more precisely 149.6 million km. Though Pluto is often considered to be the most distant planet this is not always the case. Its orbit is an elongated ellipse and Pluto can come nearer to the Sun than Neptune. From 1979 till 1999 the most distant planet is Neptune. But neither Pluto nor Neptune are on the edge of the Solar System. The Sun's attraction reaches out much farther. The solar wind ceases at a distance of some 100

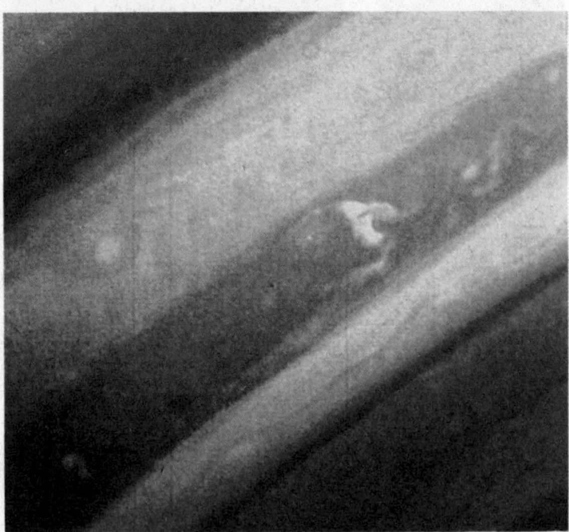

89

89 The northern part of Saturn photographed from a distance of about 9 million km (5.5 million miles). In the central dark band there are white clouds (similar to our storm clouds). The smallest details are 175 km (109 miles) across. This photograph was taken from Voyager 1, on November 5, 1980.

90

90 Voyager 2 encountered Uranus in January, 1986. This drawing depicts Uranus with its 10 rings. The axis of the planet is almost in the ecliptic, and at that time was pointing approximately towards the Earth, like the aerial of the probe. The long, ribbed arm (bottom left) carries a magnetometer; the cylinder at the bottom is an electricity generator, and at the top are various instruments. Ten new satellites were discovered.

astronomical units (Figure 114 and 185). A comet can occur several hundred times further out than that, at a third of the distance to the nearest star, Proxima Centauri, which is 4.25 light years away from the Sun, or around 269,000 astronomical units. The comets, between 30,000 and 100,000 astronomical units away, form the outer envelope of the Solar System.

Our own star, the Sun, is not the only one accompanied by a planetary system. It is likely that there are thousands of millions more in the Milky Way with such systems. But as yet not even the most powerful telescopes are able to observe their planets directly, since the light these reflect is too weak.

The star — a huge, yet simple system

A star is a heavenly body which shines with its own light. This distinguishes it from a planet, a comet, a moon or a nebula, which are illuminated by the Sun or a nearby star. The material of which stars are made is very hot gas — plasma. The highest temperatures on the surface of stars reach 150,000 kelvin (on the surface of newly born white dwarfs). These temperatures are measured by means of an analysis of the radiation from the surface of the stars. Not a single photon escapes from the interior of a star to give us any direct information on it, but we can make a reliable estimate of the temperature in each part of the inside of a star. The centre of the Sun, for instance, has a temperature of 13 million kelvin. Stars composed of a greater number of nucleons than the Sun (i.e. stars with a greater mass) have higher temperatures inside them — tens to hundreds of millions kelvin. The temperature in the centre of the stars with the greatest masses reaches over 3000 million kelvin prior to their explosion as supernovas (Figure 153).

A star is a huge, but nonetheless very simple system of elementary particles. The number of nucleons the average star is made up of is unimaginably high. It is expressed by a figure followed by 57 noughts. The number of nucleons in the Sun is 300,000 times

91

91 An artist's impression of the future meeting of Voyager 2 with the planet Neptune and its satellite Triton, which will take place on August 24, 1989.

92

92 The Voyager planetary probe, which contributed a great deal to the knowledge of Jupiter, Saturn, Uranus and their satellites. At the top is an aerial with a mirror, pointing earthwards. To the left the electricity generator, where heat is obtained from the decay of isotopes, and then turned into electricity by means of thermo-electric cells. The long pole is an aerial which receives the radio waves of planets. To the right are various measuring devices (taking measurements of plasma, particles, cosmic radiation, ultraviolet and infra-red spectra) and television cameras.

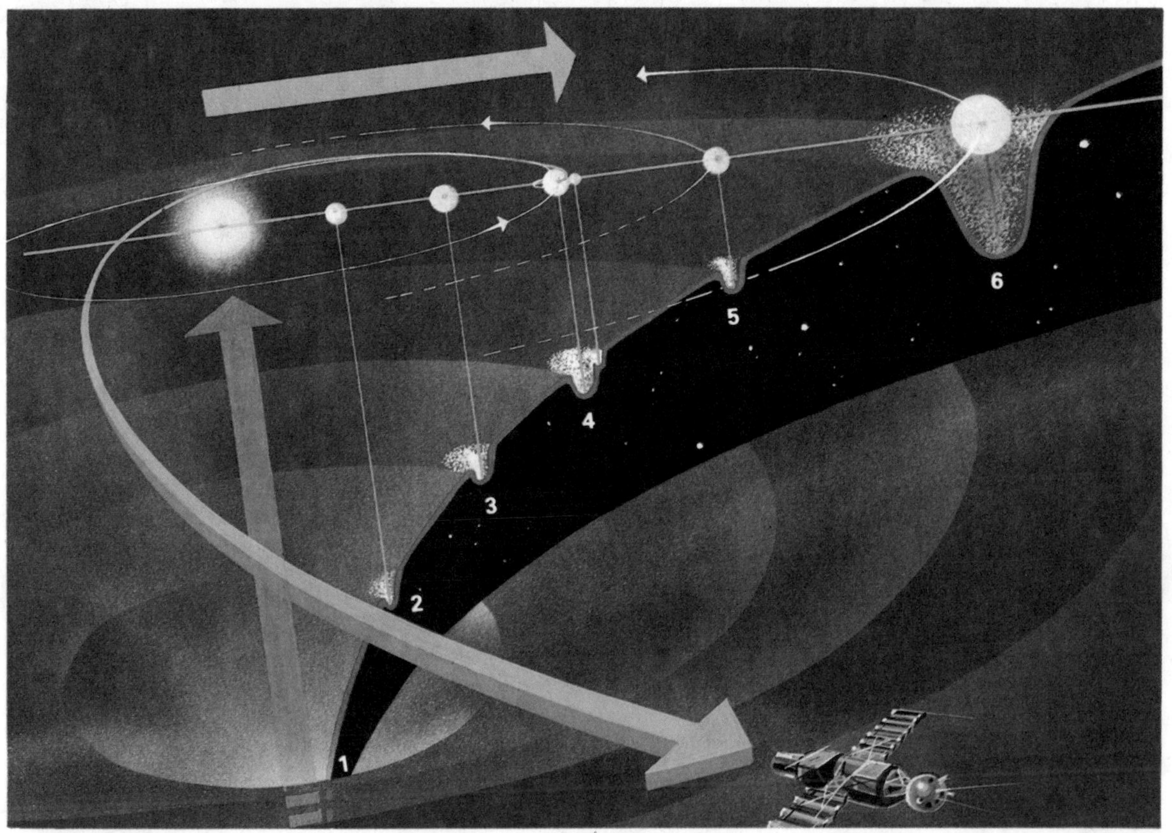

93 The planets orbit the Sun in a similar fashion to a cyclist riding on a banked track. If they were to cease to do so, they would fall into the Sun. Around the Sun **(1)** and the planets (**2** – Mercury to **6** – Jupiter) are 'gravitation pits'. The greater the mass of the body and the smaller its dimensions, the deeper the pit is. The deepest of all are the gravitational wells of black holes.

greater than the number of nucleons in the Earth. The amount of matter in a star, or its mass, is determined by the number of nucleons of which it consists.

Though the Sun is much larger than the Earth, it is a far simpler system. You can see this in Figure 66. Like the other planetary bodies, the Earth is composed of rocks; the rocks are composed of crystals, the crystals of molecules, the molecules of atoms, the atoms of nuclei and electrons. But you can also see in the illustration that a star is made only of atomic nuclei (chiefly of protons and alpha particles) and electrons. There are very few simple molecules and atoms, and no more complex systems at all. They are not able to exist in the hot interior of a star, for the atoms and molecules would disintegrate at once in violent collisions. So stars are made mainly of nuclei and electrons. It is easy for us to calculate the temperature, density, pressure and chemical composition at any depth below a star's surface. We are not able to do the same in the case of the Earth. Perhaps we should explain how it is that astronomers know the interiors of distant stars so much better than other scientists know that of the planet upon which we walk.

How does an astronomer know about the inside of a star? The answer is that, mainly from observations, he determines the mass, diameter and surface temperature. We cannot see the interior, but we know that it is made of plasma. The behaviour of plasma has been studied. We know, for instance, that the pressure in plasma is the greater the higher the temperature and density of the plasma. The pressure which exerts a force in a certain place in the interior is the same as the weight of all the layers above the place. If the pressure in the plasma was greater, the star would expand, and if it was less, the star would contract. But if the diameter of the star is constant, the weight and pressure are in equilibrium. This equality and several other relations can be expressed mathematically (using equations). Computers then help us to calculate a model of the star, its temperature, density, pressure and chemical compostion at any depth below the surface.

The basic force which keeps together all the nuclei and electrons in such a way that they form a star is gravitation. More precisely it is self-gravitation, since every nucleus in the star is attracted to every other nucleus by gravitational force. There is even a weaker gravitational attraction between the nucleus and the electron. Even the electrons in a star are attracted by gravitation, though it is very weak, since their mass is much less than that of the nuclei.

The smallest stars have a mass about 30,000 times greater than that of the Earth. The biggest have a mass of up to 10 million times that of the Earth.

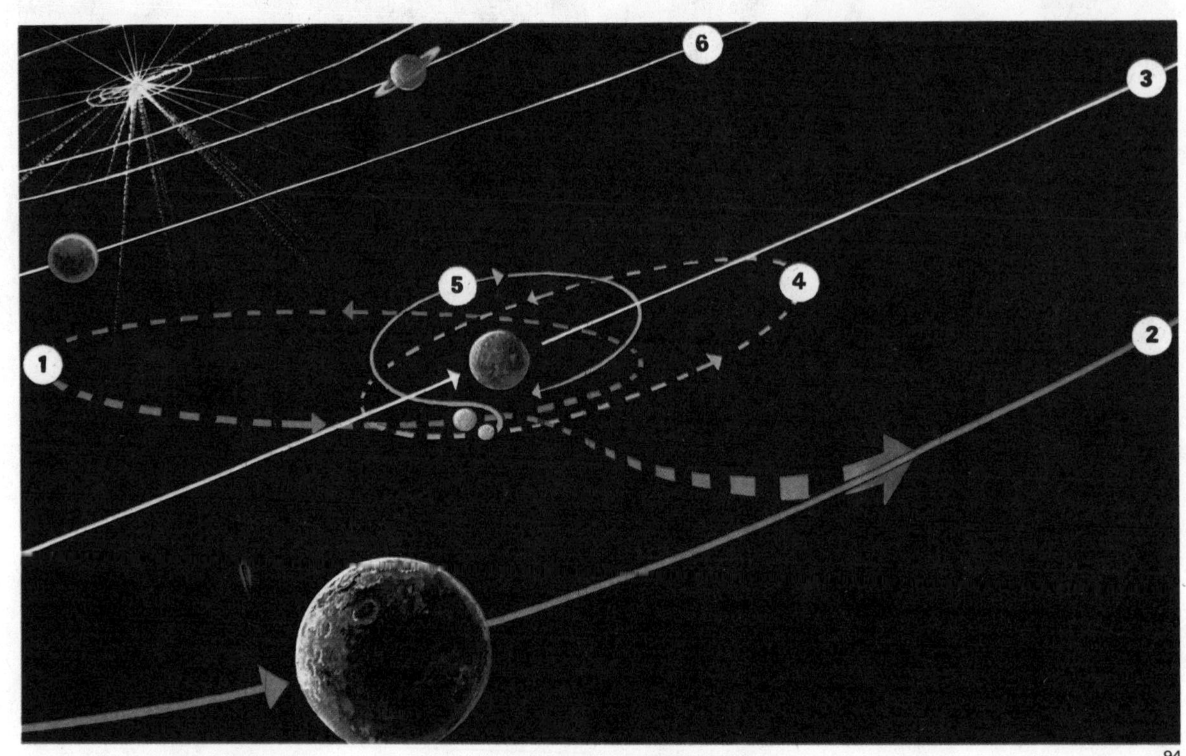

94 The most distant of the planets, Pluto, passes between Uranus **(6)** and Neptune **(3)** during part of its orbit. It is thought to have once been a satellite of Neptune **(1)** which received so much energy from Triton **(4)** that it left the gravitational field of Neptune to orbit the Sun on its own **(2).** Triton began to orbit in the opposite direction **(5).** The dotted lines show the paths of Pluto **(1)** and Triton **(4)** before the collision, the unbroken line after.

94

The dimensions of the stars are very varied. White dwarfs are the same size as the Earth, and their density is about a million times that of water. The matter in them is degenerate. Supergiant stars, on the other hand, have volumes many million times larger than the Earth.

The smallest stars that have been observed are the neutron stars. Their volume is 100 million times smaller than that of the Earth. In order for their enormous mass (the same as that of a normal star) to fit into such a small volume, neutron stars have an immense density. The matter in them exists only as neutrons. Neutron stars are observed as 'pulsars' by means of radio waves. Pulsars are the remains of stars more massive than the Sun (Chapter 3).

The distance of stars

Stars are an unimaginable distance away from us. The only exception is our own star, the Sun, which is many millions of times nearer than all but a few of the others. Nevertheless, it is difficult to imagine the distance from us of our own Sun (Figure 159), let alone the other stars. But astronomers are able to measure and calculate such distances, though they are so immense that a new unit of measurement, the *light year* had to be introduced. No one can imagine a light year, since it represents a distance beyond human experience. The light year is the distance a beam of light travels in a year. In a single second light travels 300,000 km (186,400 miles). As we have remarked, in the time it takes to say 'Jack Robinson', light can get from here to the Moon (it takes astronauts three days to do that). In a single minute light travels 18 million km (11,178,000 miles). Practically speaking, this means that an interplanetary probe at a distance of 18 million km (11,178,000 miles) from the Earth receives a signal transmitted

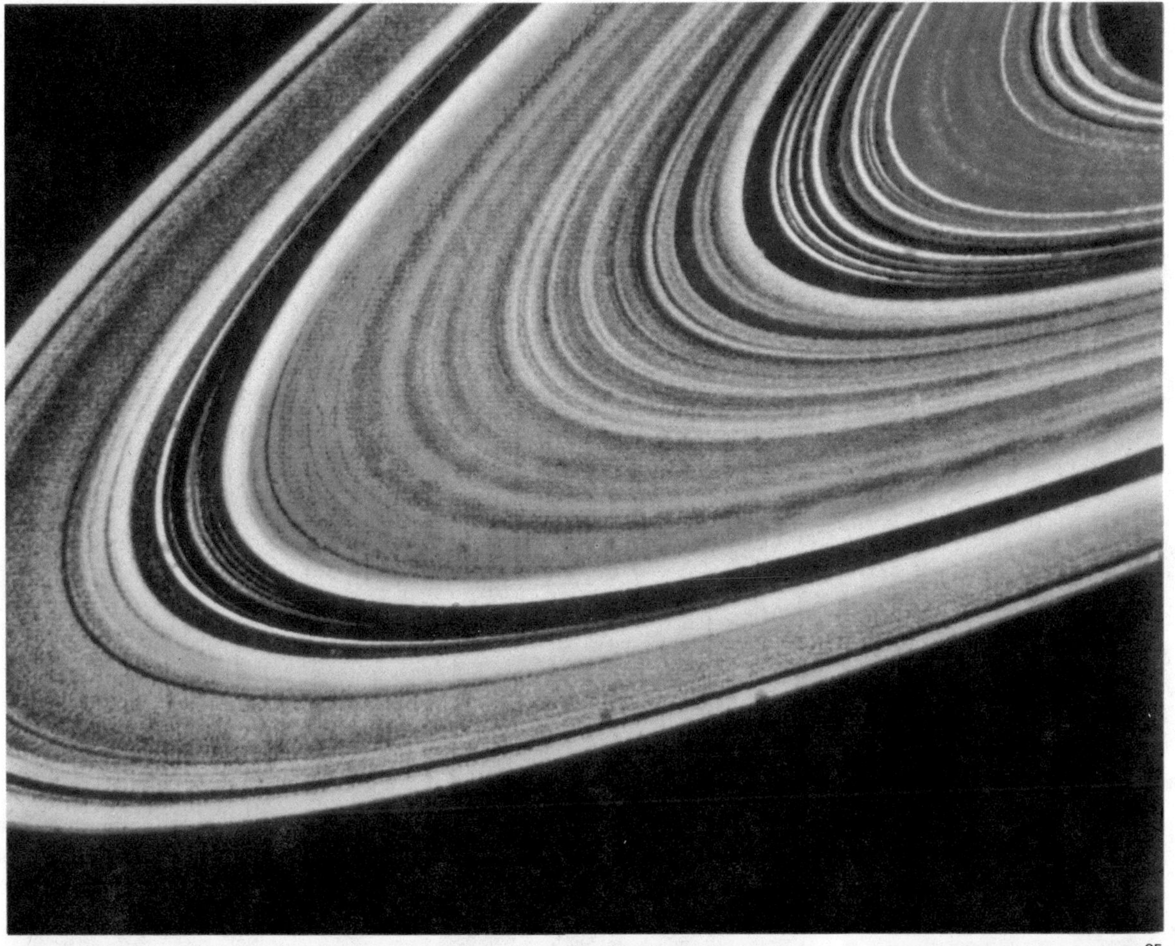

95 Part of Saturn's rings. This photograph was taken from a distance of 8.9 million km (about 5.5 million miles) in such a way as to emphasize the differences in ultraviolet radiation. Ring C is blue, ring B is brown inside and green outside, ring A is grey. The difference in colour is due to chemical composition.

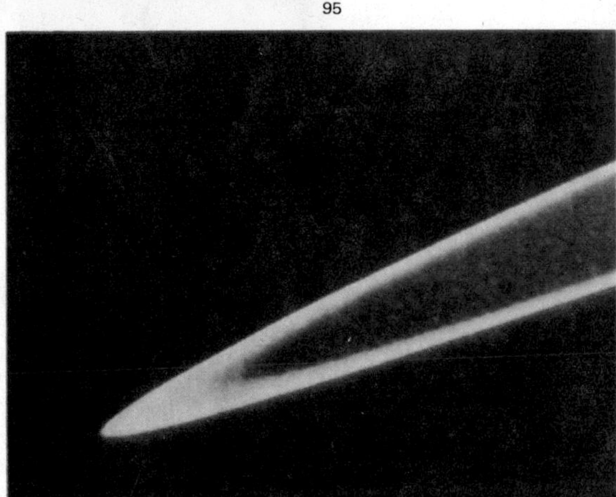

96 Jupiter's rings from a distance of 1.5 million km (about 0.9 million miles). Similarly to those of Saturn, Jupiter's rings also consist of large amounts of stones, dust and large lumps of ice. Their outer edge is 128,000 km (79,500 miles) from the centre of the planet, but the inner, dark part extends into its atmosphere.

97 Saturn's rings photographed on August 29, 1981, when Voyager was approximately 3.4 million km (2,111,000 miles) beyond the planet. The planet's surface is visible through the rings (seen from below), which proves that they are not monolithic, but are made up of a large quantity of boulders, rocks, pebbles, dust and ice.

from the Earth about one minute later. In one year light covers a distance of 10 billion (10^{13}) km (6 billion miles): that is what is meant by a light year.

A distance of 3.26 light years is called a *parsec*. It corresponds to 206,265 astronomical units, or 3.086×10^{13} km. Very often even the parsec is too small to measure the great distances in space, so the *megaparsec* is used, which is a million times larger, or 3,260,000 light years.

The distance of the stars is measured in a similar way to that used by scientists called geodesists when measuring inaccessible places on Earth. First of all they measure the distance between two arbitrary points A and B. The inaccessible point – let us call it C – forms with A and B a triangle. Its side AB has been measured and then the angles α and β. It is then a matter of simple calculation to arrive at the distance of point C from A and B. In measuring the distance of stars it is necessary to use points A and B as far away from each other as possible. The greatest distance which can be achieved on Earth is the diameter of the Earth's orbit (Figure 116). Thus the stars and their surroundings are photographed twice at an interval of half a year, with the Earth at two diametrically opposite points on its orbit. The stars on the second photograph are slightly shifted with respect to their position on the first; this shift is the greater, the smaller the distance between us and the star. It is this difference (the star's angle of parallax), a very small one, but measurable by means of accurate instruments, which allows us to calculate the distance of stars. There exist other ways of determining the distance of stars, but the method described is the one most often used.

98 The surface of Ganymede is covered with craters and grooves. The satellite is larger than Mercury, and is made of rocks and ice.

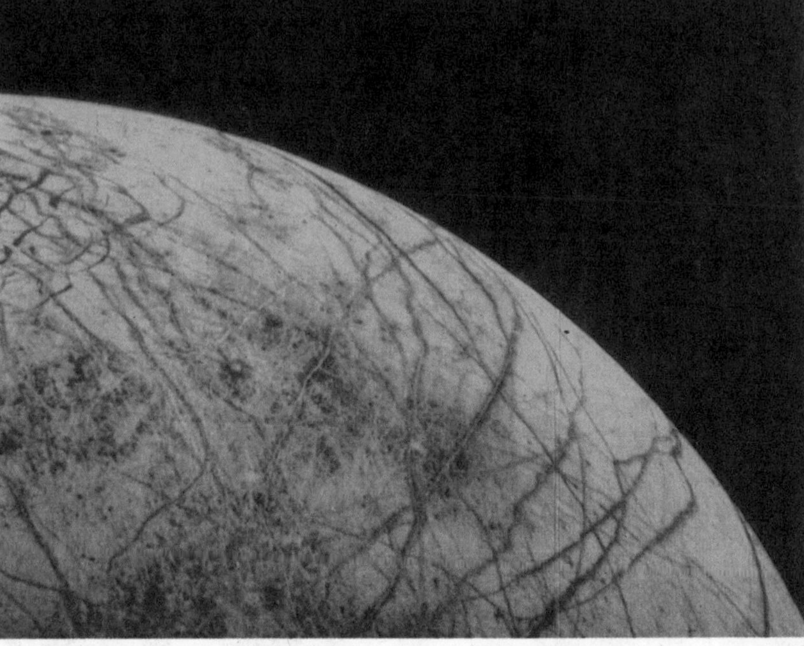

99 The surface of the satellite Europa is smooth. The dark lines on the surface are reminiscent of cracks in the arctic ice. Europa is enclosed in a shell of ice several tens of kilometres thick. Below the ice is an ocean several kilometres deep. Water from this seeps through the cracks in the ice and freezes.

The motion of stars

Nothing is static in the Universe. Stars, planets, comets, meteors, satellites, artificial satellites and spaceships all move, at speeds varying from a few kilometres per second to a hundred kilometres per second. Detecting the motion of stars can be compared with watching the movement of an express train which is passing through a station and whistling as it does so. The train moves in relation to its surroundings, so that we have to move our heads to keep our eyes on it. In the case of stars the same movement occurs in respect of the other stars, galaxies and stellar clusters. This movement is called the proper motion of stars.

The actual motion of stars is much faster than that of the train. But the stars are so far away that we are not able to detect any change of their position in a whole lifetime. For this reason our ancestors called them 'fixed stars'. They thought that the stars in the sky were motionless. It took huge telescopes, spectrographs and precise photographic equipment to show astronomers that the stars in the sky do actually change position slightly in the course of a lifetime. Astronomers photograph a region of the sky twice at an interval of several decades. By means of precise measurement of the two photographs they determine how much the stars have shifted. This shift is very small

100

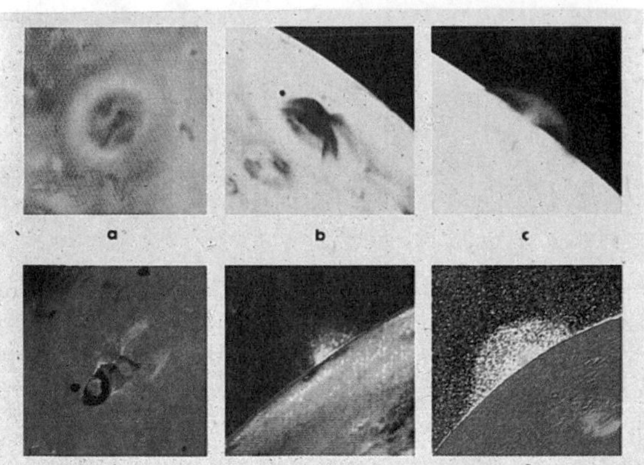

101

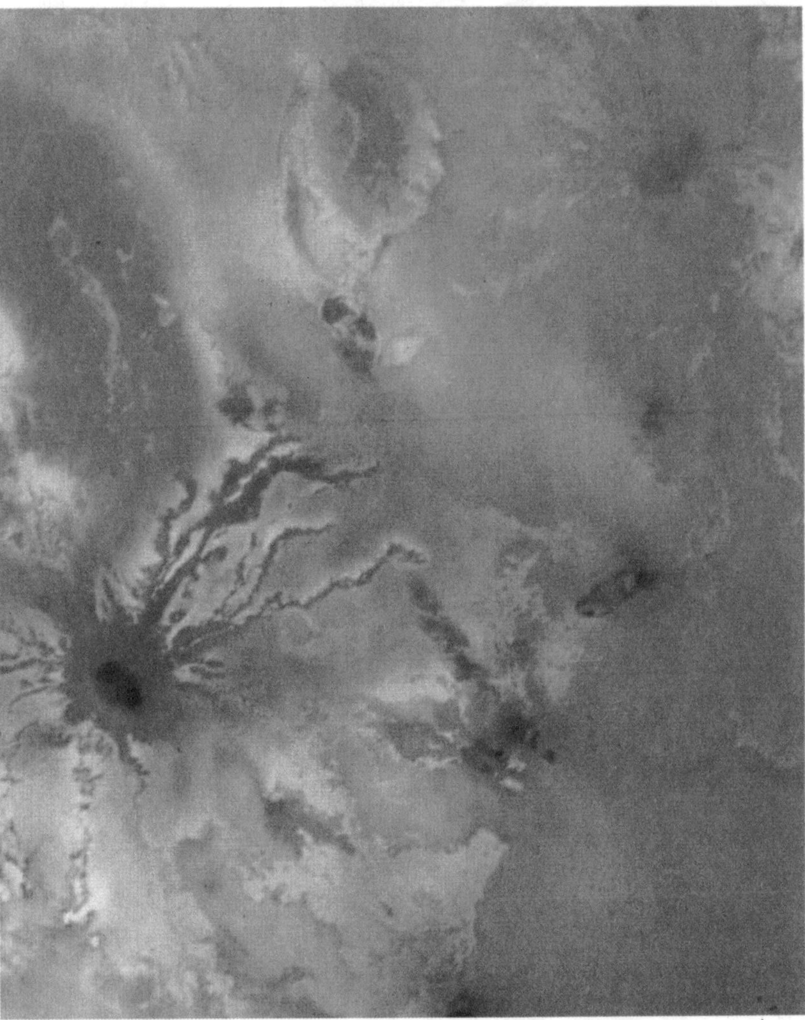

102

100 The greatest surprise during research into Jupiter's satellites was the discovery of volcanic activity on Io. It is due partly to the gravitational effect of the parent planet, and partly to the radioactive decay of Io's interior.

101 Several active volcanoes on Io. The speed at which gases are emitted is up to 1 km (0.6 miles) per second. By comparison the fastest rate at which Etna emits gases is 50 m per second.

102 A volcano on Io. The black crater contains molten sulphur. There are many such volcanoes there. The sulphur, oxygen and sodium thrown out by the volcanoes reach far into the space around Jupiter.

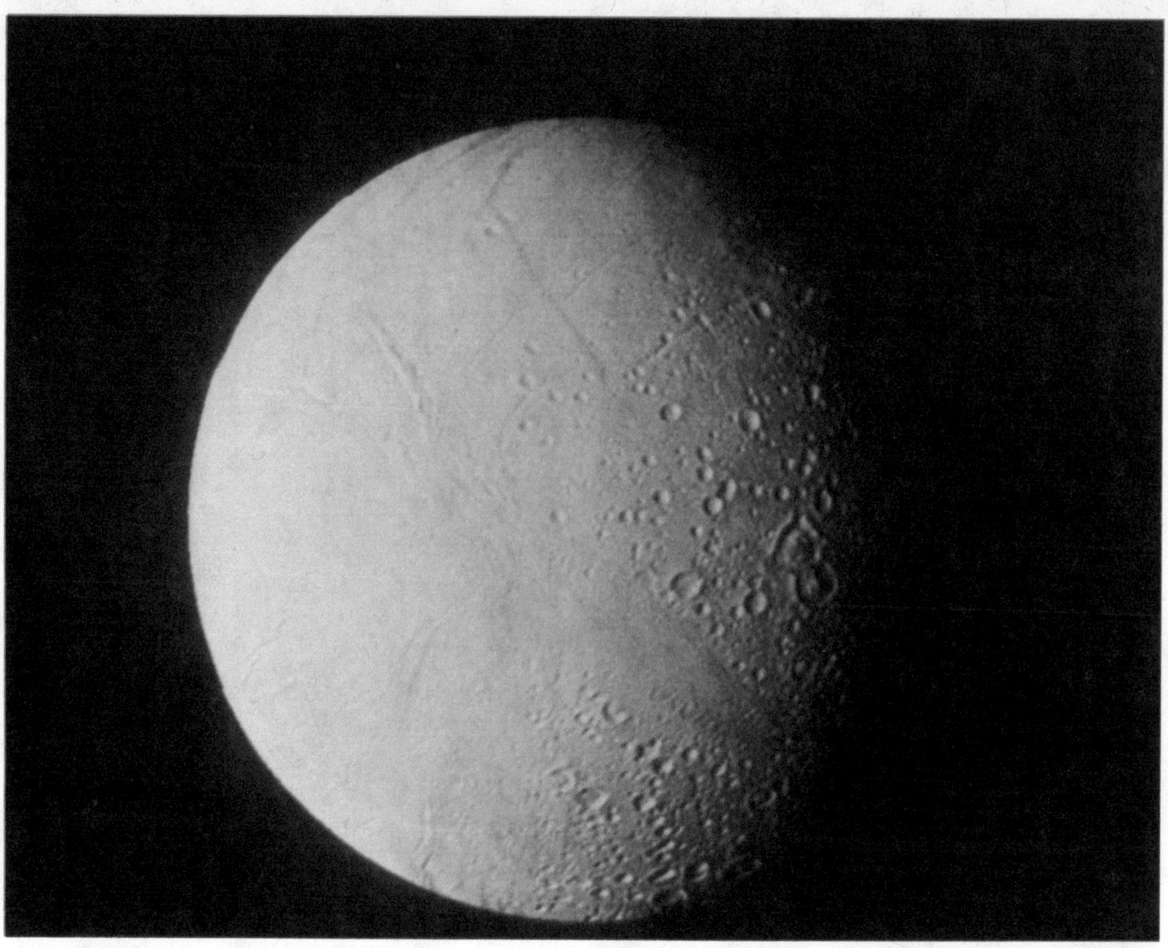

103 Enceladus is 500 km (311 miles) across. The smooth areas to the left are young, having been covered with molten material from the interior in the course of the last million years. The breaks in the crust are up to several hundred kilometres long.

103

indeed. The fastest-moving stars take several centuries to move across our sky by one-half degree, the distance equal to the apparent diameter of the Moon. In the course of millenia, the movement is more apparent. Because the direction of proper motion across the sky is different for different stars, in the course of tens of millenia the whole appearance of the constellations is changed (Figure 120).

You may have noticed while standing on a railway station that the pitch of a train's

Planet	Distance (ast. units)	Orbital period (years)	Rotation period	Mass (Earth = 1)	Equatorial radius (Earth = 1)	Mean density (kg/m^3)	Number of moons (1986)
Mercury	0.39	0.24	58.6 days	0.06	0.38	5600	0
Venus	0.72	0.62	243 days	0.82	0.95	5200	0
Earth	1.00	1.00	23 h 56 min	1.00	1.00	5518	1
Mars	1.52	1.88	24 h 37 min	0.11	0.53	3950	2
Jupiter	5.20	11.86	9 h 50 min	317.89	11.23	1314	16
Saturn	9.54	29.46	10 h 14 min	95.15	9.41	704	20
Uranus	19.18	84.01	10 h 49 min	14.54	4.06	1210	15
Neptune	30.06	164.79	15 h 48 min	17.23	3.88	1670	3
Pluto	39.44	247.70	6.4 days	0.002	0.20	800	1

Equatorial radius of Earth 6378 km
Astronomical unit 149.6 million kilometres
Mass of Earth 5.98×10^{24} kg
Sidereal year (360°) 365.26 days

whistle gets higher as it approaches, and falls as the train recedes from us. By means of accurate measurement of the tone of the whistle we could in fact ascertain whether the train was getting nearer or further away. The same goes for stars, except that instead of whistling they emit light of a certain frequency. This frequency can be measured by means of an instrument called a spectrograph, which is mounted on the end of a telescope instead of the eyepiece. The speed at which a star approaches or recedes measured in this way is called its *radial velocity*.

The radiation of stars

Stars are huge sources of light and other forms of radiation: infra-red, radio waves, ultraviolet and X-rays. Our eyes are sensitive to the light only and not to the other types of radiation. The radiation emitted by the stars has its origin in the thermonuclear reactions deep in their interiors. From there it reaches the cooler surface of the star and travels on to the surrounding interstellar space. The total amount of radiation a star emits per second (the star's output) is called its *luminosity*. The luminosity of a star does not depend on its distance from the Earth.

The Sun has a luminosity of nearly 400,000 trillion kilowatts (3.8×10^{23} kW). Only one 2000-millionth of this impressive amount of energy falls on the Earth each second, or about 200 billion kilowatts (180×10^{12} kW, or 180,000 TW). The luminosity of the Sun is used to express that of other stars. There are giant stars, and even supergiants (such as the stars Rigel and Deneb), whose luminosity is many thousand times that of the Sun. Dwarf stars, on the other hand, have luminosities one thousand times less than that of the Sun. The differences in the luminosities of dwarfs and supergiants can be compared with the difference in brightness between a firefly and a large searchlight.

The brightness of a star as seen from the Earth depends not only on its luminosity, but also on its remoteness. Of course, distance does not affect the luminosity of the star; it is simply that the inhabitants of the Earth see close stars brighter than more distant ones of the same luminosity.

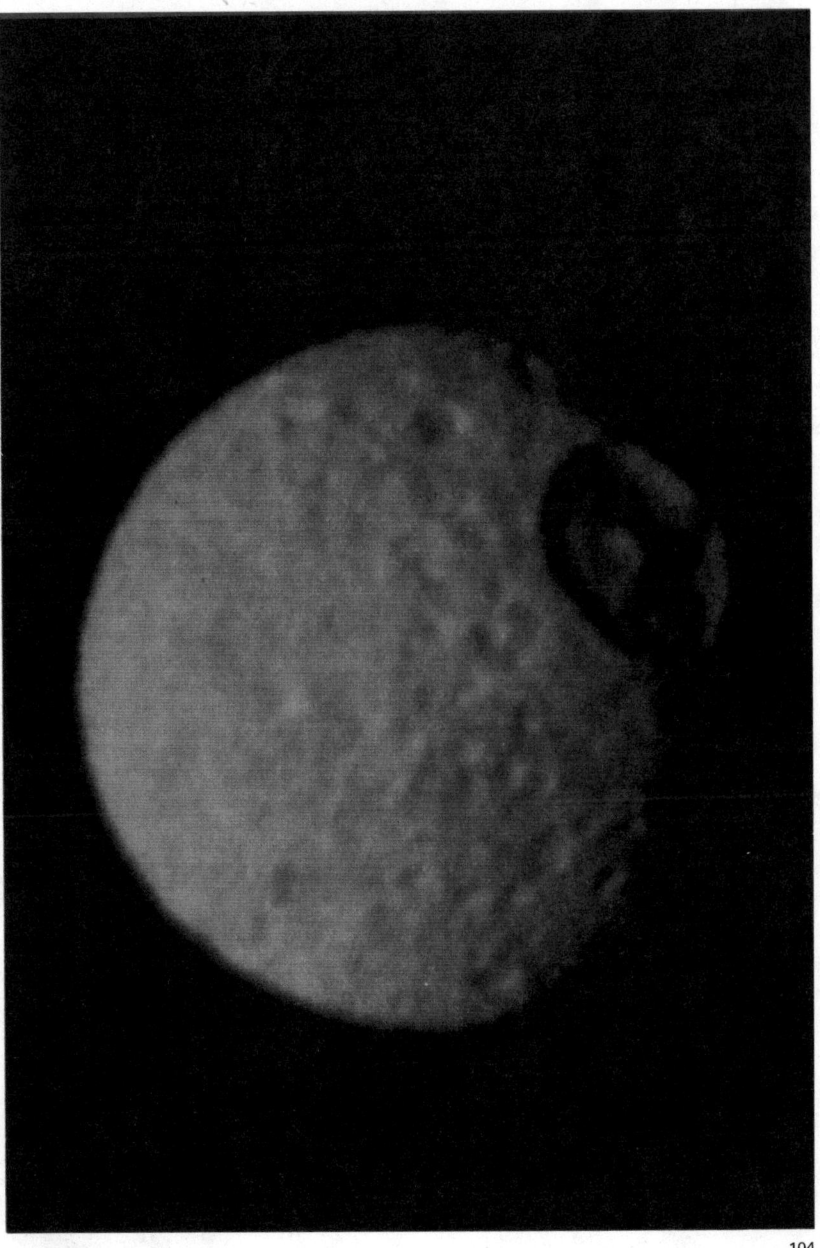

104

104 Mimas (a satellite of Saturn) is covered with many craters on its ancient surface. The largest craters are 130 km (81 miles) across. The majority of the craters were formed about 4000 million years ago.

Variable stars

Each star in the Universe is evolving in itself, changing its internal and external properties over very long periods. We cannot, of course, ever observe these gradual evolutionary changes, which take millions to thousands of millions of years to complete. But there are many stars whose brightness changes very quickly. We call them variable stars. Comparing the brightness of a variable star with the brightness of other stars, we can easily note its changes.

Stars change their brightness for various reasons. Some of them pulsate (rather like

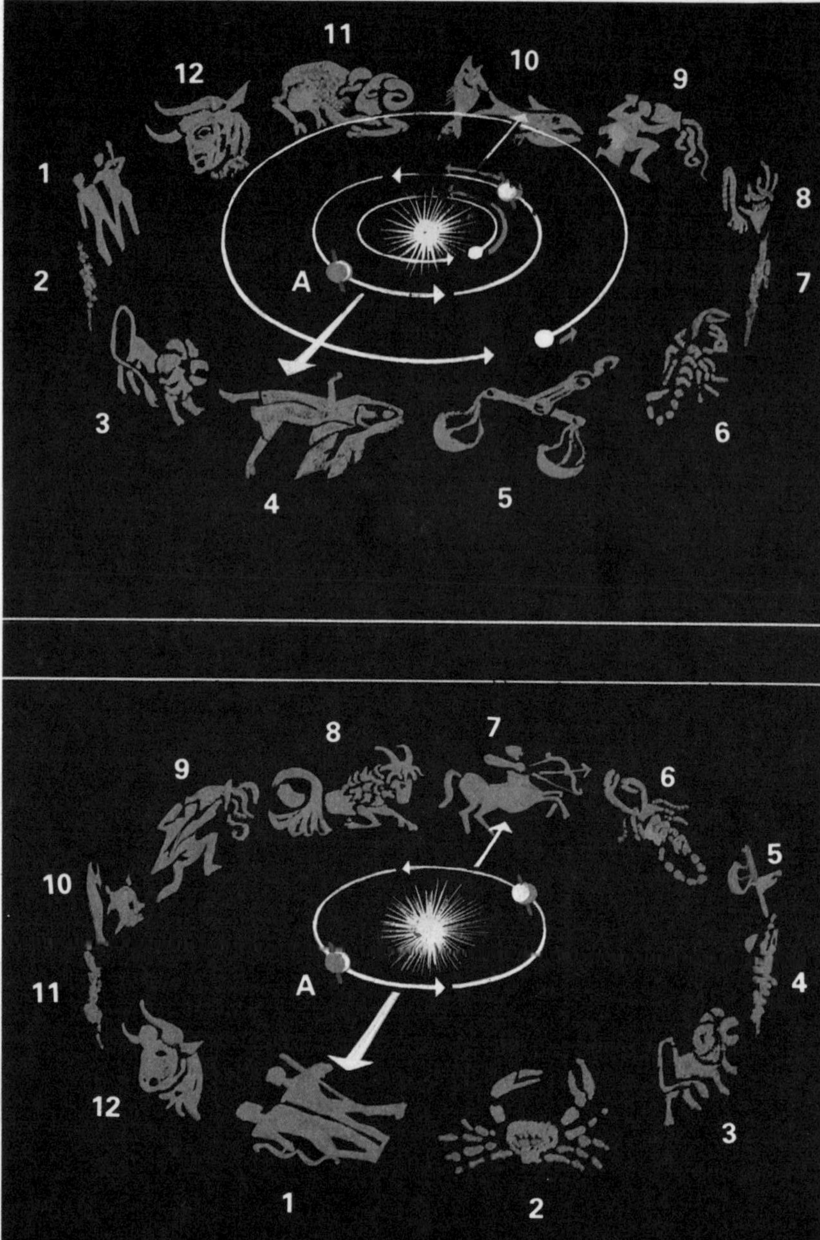

105 The constellations of the Zodiac form a band along the plane of the Earth's orbit (the ecliptic): Gemini **(1),** Cancer **(2),** Leo **(3),** Virgo **(4),** Libra **(5),** Scorpio **(6),** Sagittarius **(7),** Capricorn **(8),** Aquarius **(9),** Pisces **(10),** Aries **(11)** and Taurus **(12).** The solar rays which pass through the Earth (for example neutrino rays) always head towards the opposite sign of the Zodiac, which we can see best at midnight. The constellation which includes the Sun cannot be seen because of the clear blue sky. Spring begins when the Sun is in Pisces (top picture). Winter starts when it is in Sagittarius (bottom picture).

106 The minor planet called 1981 VA was discovered at Mount Palomar in this photograph. While stars appear as bright spots, the minor planet formed a line of light, since it moves across the sky. This is one of the minor planets which passes inside the Earth's orbit. In future it may collide with the Earth or the Moon.

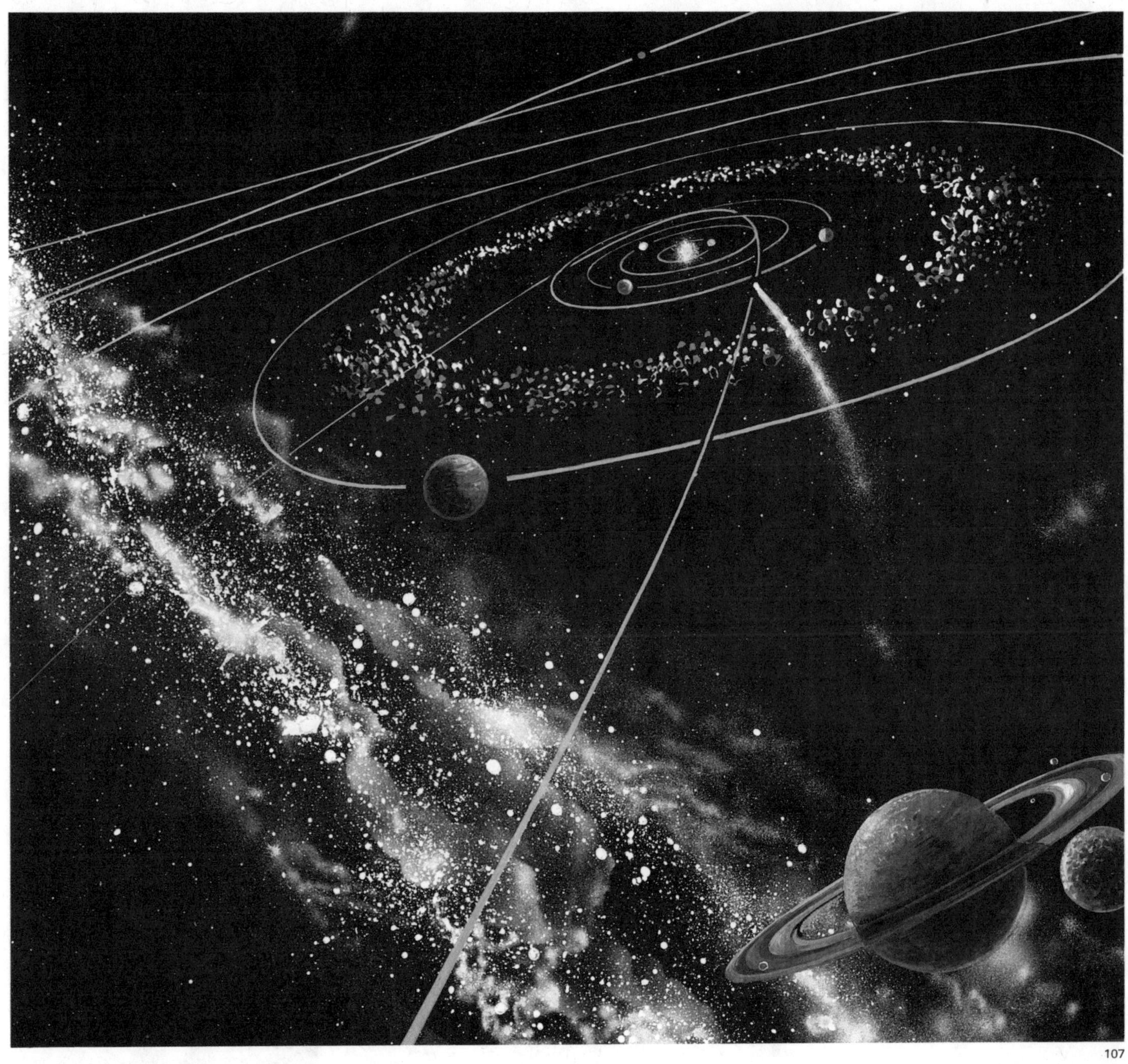

107

a heartbeat). Some (for example RR Lyrae) pulsate rapidly, one pulse taking only a few hours. The pulsations of other stars take several days (for example δ Cephei) or several months (for example Mira Ceti).

Close double stars are not spheres, but elongated ellipsoids, exhibiting radiating surfaces of various sizes as they orbit around each other. This sort of change in brightness is displayed by, for instance, ε Lyrae. In the course of orbiting the two components of a double star may cover one another, which means a change in their overall brightness. An example of this type of variable – called an eclipsing variable – is Algol, in Perseus. Exploding stars change suddenly in brightness. These include novas, supernovas, etc. Some rotating stars are variable because they have regions of different brightness in different places on their surface. On some stars, for example, there are very cool regions (similar to sunspots), or very hot places (similar to flares on the Sun). This group includes, among others, the pulsars, which are very rapidly rotating neutron stars. Their pulsations are exceptionally regular, and last approximately a second.

The most slowly, but also the most strong-

107 The orbits of the planets of the Solar System are approximately in the same plane – the ecliptic. Those of comets are at an angle to the ecliptic. The picture shows the orbit of Halley's Comet. Between the orbits of Mars and Jupiter is a band of minor planets.

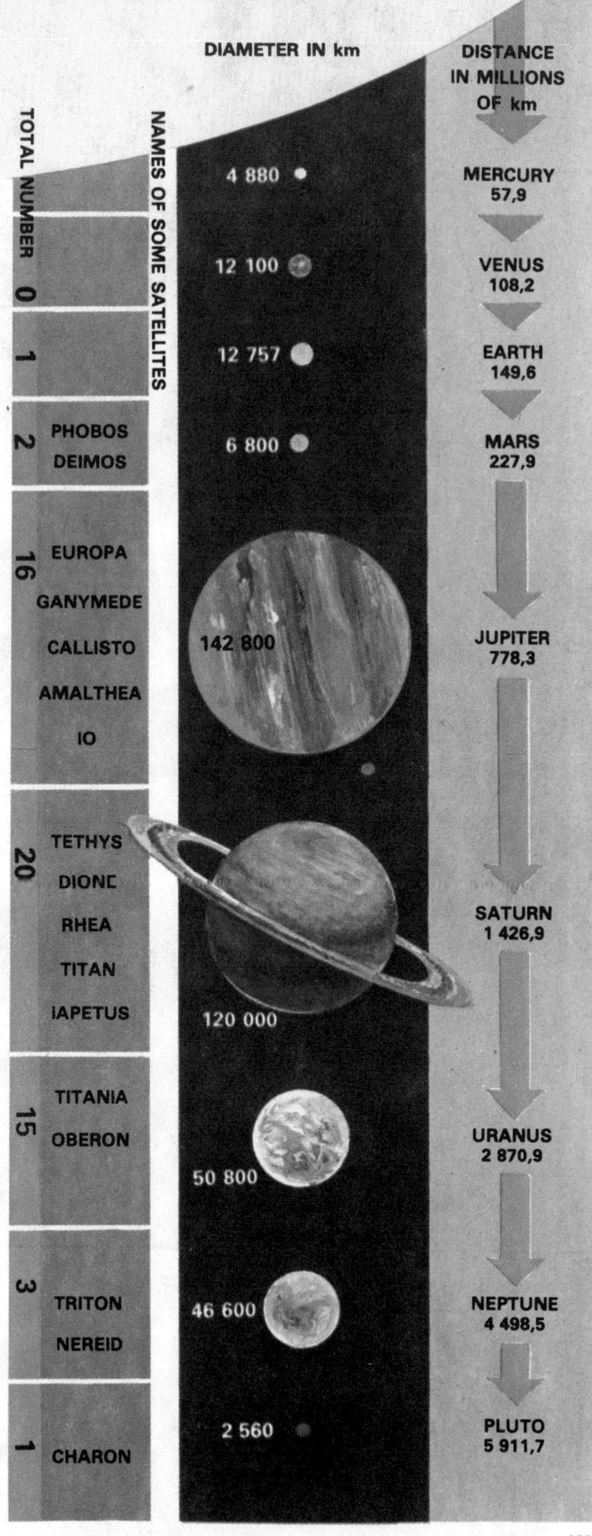

108 The relative sizes of the planets of the Solar System. The number of satellites is not final: the more observation methods improve, the greater the number of satellites discovered. Thus the Voyager's recent flyby of Uranus (January 1986) led to the discovery of 10 smaller satellites of Uranus.

ly, pulsating stars are those such as Mira Ceti. Planets associated with variable stars could not support life on their surface. Let us imagine that the Sun began to grow brighter until, after five months, it became a thousand times brighter than it is today; and then, in another six months, waned to its normal brightness. Life on Earth would, of course, be quite impossible under such circumstances.

Nebulae

In the space between the stars individual atoms of the elements and grains of dust are flying about. Their density is very small indeed. In 1 cm^3 (0.06 cu in) you would find only a few atoms, while in the same volume of the air we breath there are a trillion of them. In some parts of the interstellar space there are, however, densities of as much as one hundred to one thousand times as much, and it is here that the interstellar clouds form. If a cloud shines, we call it a bright nebula. But a nebula has no light of its own. The light comes from a nearby hot star, which happens to be in the vicinity. This is the case, for instance, of the large nebula in the constellation of Orion (Figure 148). Some bright nebulae have an irregular shape, and are called diffuse. Others have a regular ring shape (ring nebulae) or the shape similar to a planet (planetary nebulae). Dark nebulae can be seen against the background of bright nebulae or of the Milky Way as dark patches (Figures 130 and 149). Interstellar matter and nebulae are found particularly in the spiral arms of galaxies.

The major part of interstellar matter is the ancient hydrogen from which our Galaxy and other galaxies were born more than 9000 million years ago (Figure 138). A large part of it increased in density to the extent that it formed stars, where it changes into helium and all the other elements (Figure 151).

110 In the centre of a star there is rapid movement of free electrons (blue), protons (red), alpha particles, and a large number of photons (yellow). In this wild mixture there are very frequent collisions, in which particles and photons exchange energy.

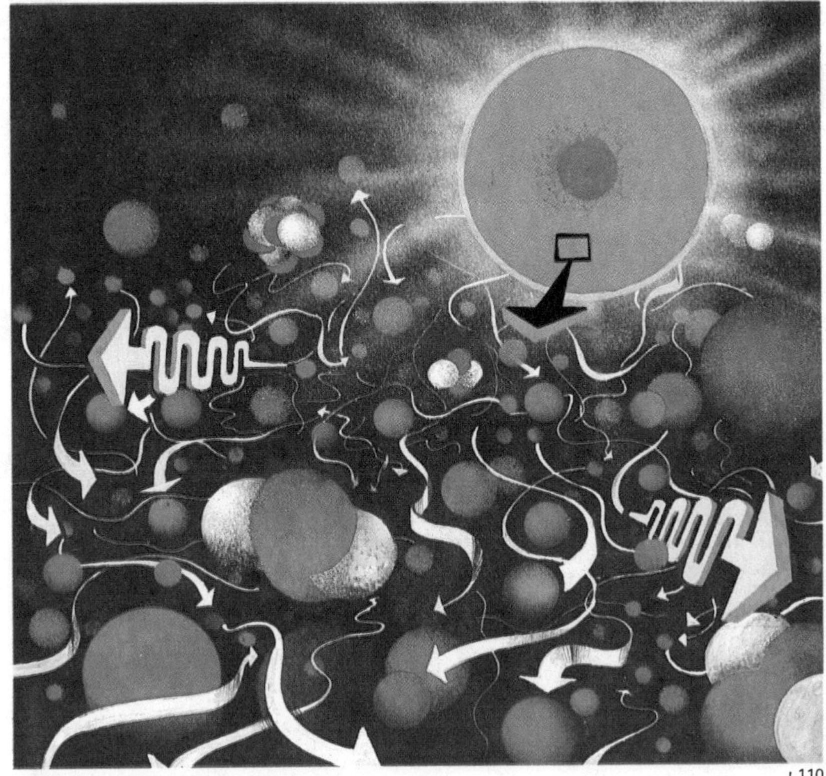

110

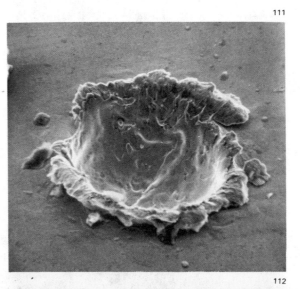

111

111 A microscopic 'crater' in a smooth steel plate from the outer casing of an artificial satellite. Such craters are made by the impact of small meteoroids (about $\frac{1}{10}$ mm across). On impact the metal heats up, melts, splashes, and at once 'sets' again. The crater in the photograph is $\frac{1}{6}$ mm across, $\frac{1}{15}$ mm deep.

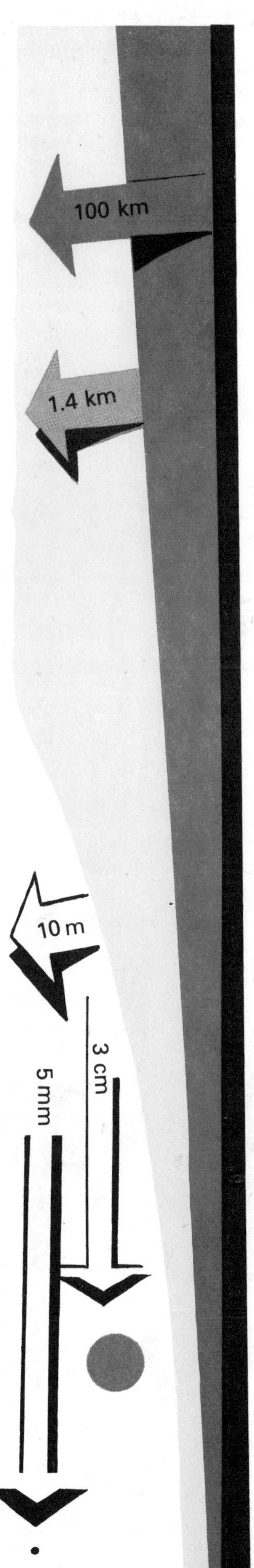

109

109 The diameters of stars reduced one million times (the size shows how large a 1 : 1,000,000 model of a star would have to be). For instance, a white dwarf measuring 10,000 km (6214 miles) across is represented by a ball 10 m (33 ft) in diameter, the Sun by one 1.4 km (0.9 miles) in diameter, a red giant by one 100 km (62 miles) in diameter, a neutron star by one 3 cm (1.2 in) in diameter and a black hole by one 5 mm (0.2 in) in diameter.

112

112 The Zodiacal light photographed on Mount Chacaltaya in the Bolivian Andes (5200 m/17,060 ft above sea-level). The short streaks of light are the paths covered by stars during the long photographic exposure.

113 The Mount Palomar Observatory in California. The huge telescopes detect the light from very faint stars. The diameter of the mirror of this telescope is 5 m (16 ft), and it captures several million times as much light as the human eye. Therefore it may 'see' several thousand times further into the Universe than can the naked eye.

Through the explosions of novas and supernovas (Figure 152) and through the solar wind, these heavier elements get into interstellar space, where they mix with the remnants of the original hydrogen. Through the contraction of nebulae and invisible interstellar clouds, further generations of stars are still being born. Thus a constant cycle is repeated: interstellar matter → star → interstellar matter → star, and so on, with a diminishing amount of the primordial hydrogen and an increasing amount of the heavier elements in the Universe.

Systems of stars

Solitary stars occur only rarely in the Universe. Most stars are grouped with others and form double, triple and multiple stars. Many of them are surrounded by planets. Larger systems of stars, containing from a few dozen to hundreds of thousands of them, are called *star clusters*. Systems of several million to a few billion stars are called galaxies. The force which binds stars into systems of various sizes, from double stars to giant galaxies, is gravitation. The greater the gravitational binding energy, the more stable the stellar system. But there exist stellar systems (called *stellar associations*) whose binding energy is small. These contain from a few dozen to several hundred stars. The kinetic energy of individual stars is then greater than their binding energy; it is then easy for stars to escape the association. These unstable associations of stars disintegrate soon after coming into existence.

A pair of stars which orbit each other is called a *binary*. More exactly, they both orbit around a common centre of gravity. If these stars can be seen through a telescope, we speak of a *visual binary*. But there are also many binaries which are so close to each other that even in the largest of telescopes we see them as a single point of light. But the spectrum of this point of light betrays the fact that it is a binary. The spectrum is an artificially formed 'rainbow' of the light from a star. But it is much more detailed than the rainbow of sunlight. The dark lines which interrupt the spectrum (see Figure 161) are the basic source of our knowledge of the stars. The binary nature of a star, for example, is given away by the fact that the dark

114

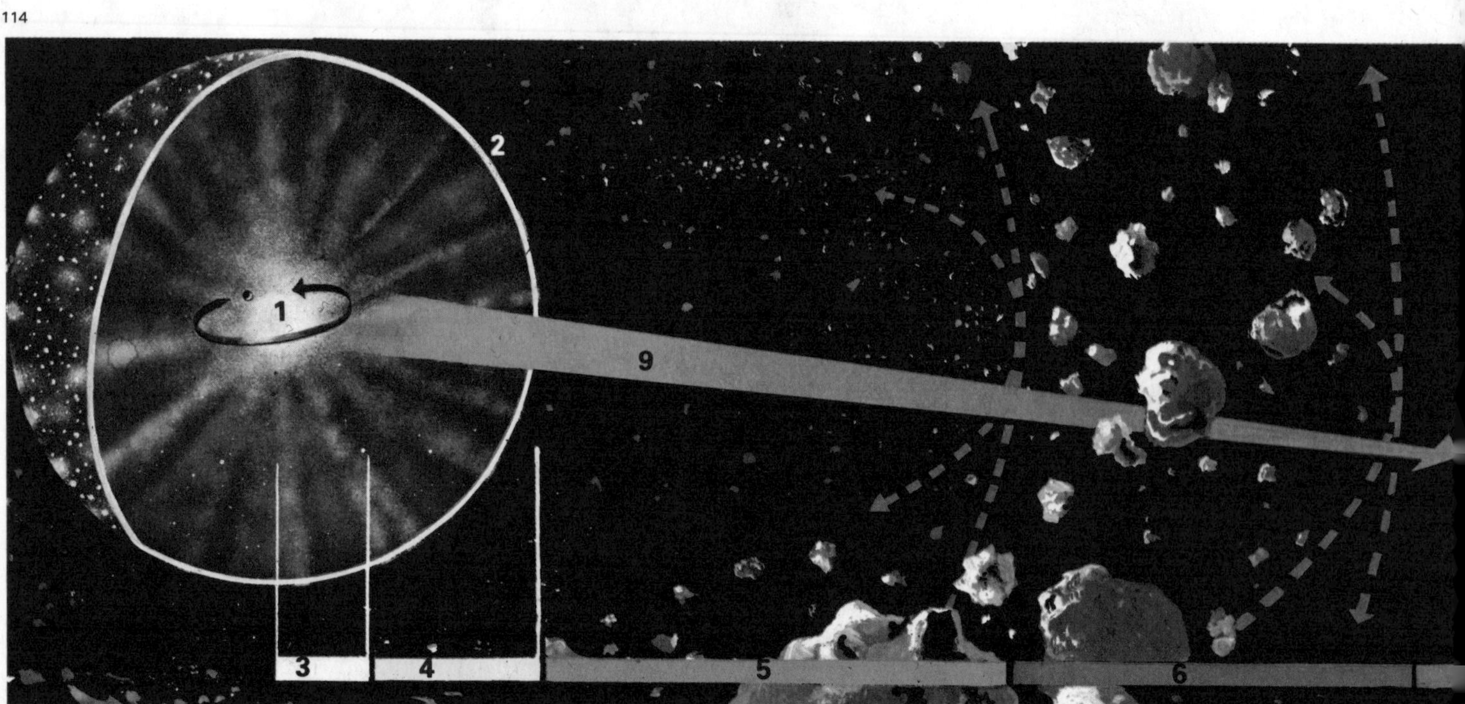

lines in the spectrum alternately separate and merge again. Such binaries as this are called *spectral binaries.* A star which is moving towards us has its lines shifted towards the blue end, while one that is moving away has them shifted towards the red end of the spectrum. This phenomenon is called a *Doppler shift.* The magnitude of the shift shows us the speed at which the stars are moving.

Some close binaries move in such a way that their orbital plane nearly 'passes through our eyes'. These stars alternately eclipse each other. Each time this happens their total brightness decreases. Such pairs of stars are called *eclipsing binaries* (Figure 123). An example is Algol in Perseus, or the beta star in the constellation of the Lyre.

A special sort of binary is the *X-ray binary* (called also the X-star, Figure 124). These are very close binaries. The main component is an ordinary star a few million kilometres across. The other component is a degenerate star – a white dwarf, a neutron star, or even a black hole. Hot gases from the main star fall on to its small, degenerate companion. These fast-falling gases (plasma) collide, giving off X-rays as they do so. (In a similar way as when fast electrons fall on to a metal plate in a medical X-ray lamp.) In X-radiation many thousands of times more energy is given off than the total radiation (luminosity) of our Sun. The source of energy for these binaries is the gravitation energy of the hot gases of the large star, which are attracted by the enormous gravitation of the small, degenerate component. Examples of X-ray binaries are the sources known as Cyg X-1 (meaning the first X-ray source in the constellation Cygnus – the Swan), Sco X-1 (in Scorpio), Cen X-3 (the third X-ray source in Centaurus) and Her X-1 (in Hercules).

The importance of binaries for astronomers lies in the fact that they allow us to determine the mass of the two stars. The two components of a binary are mutually attracted by gravitational force, which tries to bring them closer together and to join them

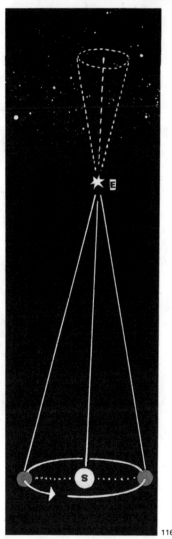

115

115 The cupola of the large telescope (Figure 113) at Mount Palomar.

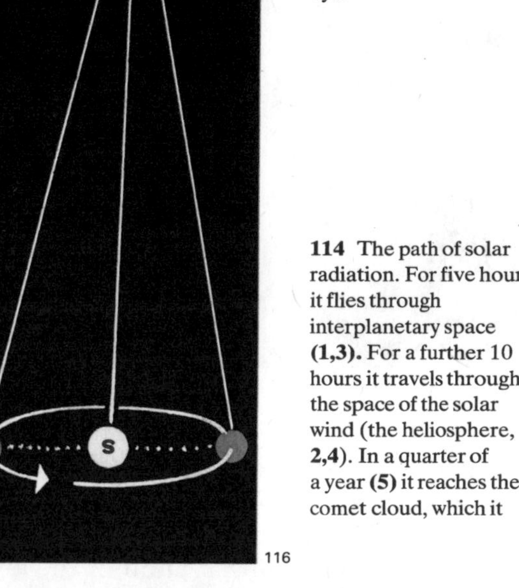

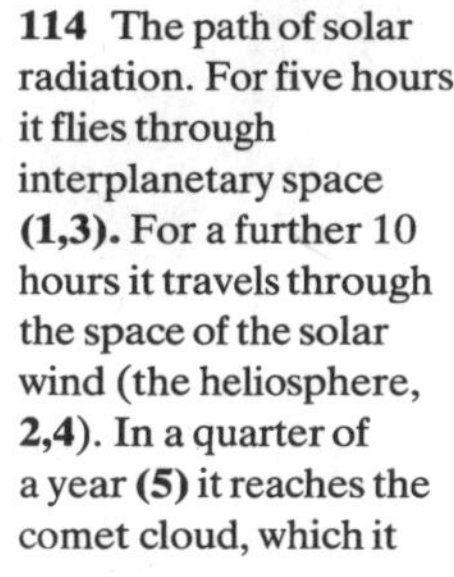

116

116 The distance of a star is determined by the angle at which we should see the radius of the Earth's orbit (150 million km/93,200,000 miles) from it. The smaller the angle (called parallax), the further away the star is. Parallax is determined by photographing the same part of the sky at half-year intervals. The star seems to have moved, like the finger of an outstretched hand if you look at it with first your left, then your right eye.

114 The path of solar radiation. For five hours it flies through interplanetary space **(1,3).** For a further 10 hours it travels through the space of the solar wind (the heliosphere, **2,4**). In a quarter of a year **(5)** it reaches the comet cloud, which it takes more than one year to fly through **(6).** There the Sun's attraction ends **(9)** and the attraction **(10)** of the nearest stars **(8)** begins. After that the solar light passes through the endless frozen waste of interstellar space **(7).**

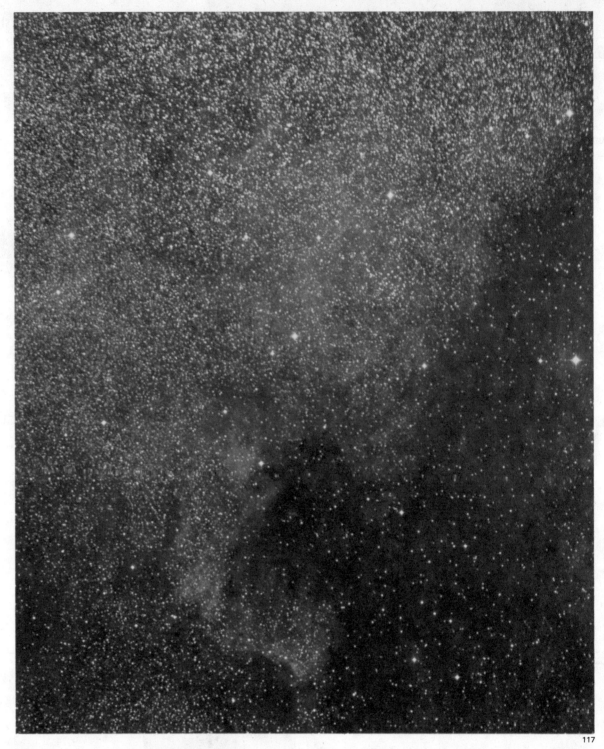

117 The nebula called, because of its shape, North America. The reddish colour is due to hydrogen (the H-alpha line).

in a single star. But centrifugal force acts against this gravitational force, since the two components are orbiting around a single centre of gravity. Observation of the binary can reveal the distance of each component from this centre of gravity. These distances show us the ratio of the masses of the two stars. From the period and the velocity we determine the centrifugal force, which is equal to the gravitational force. But the gravitational force depends on the masses of the two components. Thus we can calculate from the data ascertained for the binary the mass of each of its components. Figuratively speaking, the stars can be 'weighed' (Figures 119 and 121).

Clusters of stars are divided into the *associations* of small numbers of stars, of which we have already spoken, *open clusters,* and *globular clusters.*

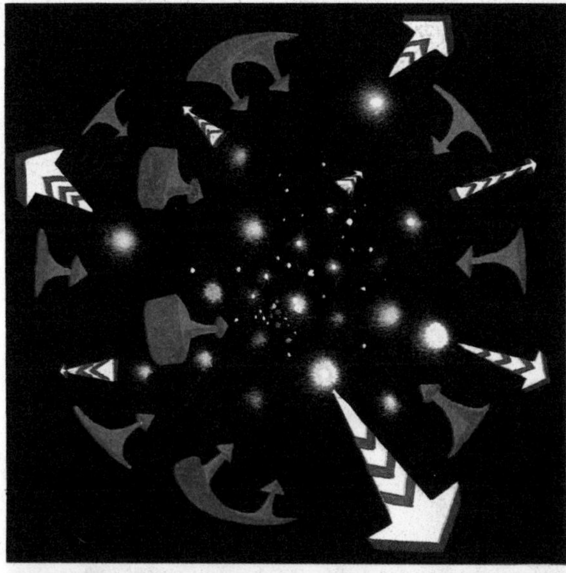

118

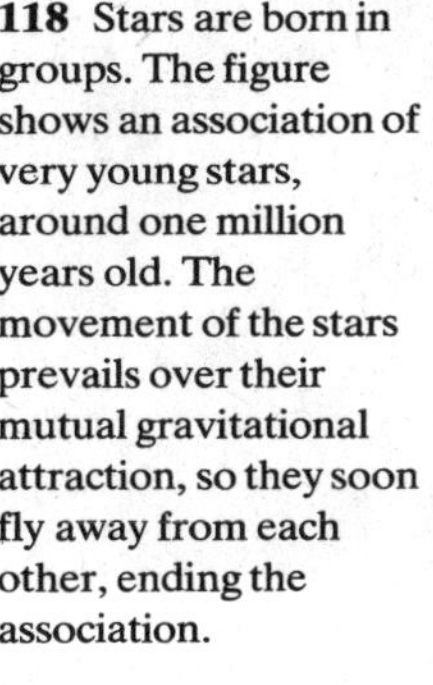

118 Stars are born in groups. The figure shows an association of very young stars, around one million years old. The movement of the stars prevails over their mutual gravitational attraction, so they soon fly away from each other, ending the association.

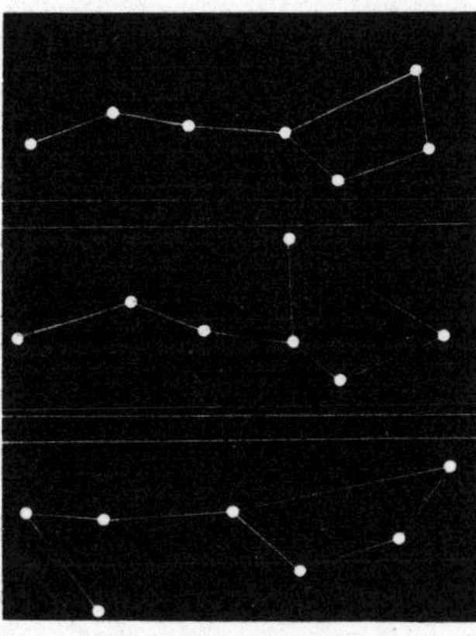

120

120 Stars move at speeds of many kilometres a second. Their position in the sky is therefore constantly altering (though extremely slowly), as can be seen from the example of the Great Bear. At the top you see its shape today, in the middle how it looked 200,000 years ago, and at the bottom what it will look like in another 200,000 years.

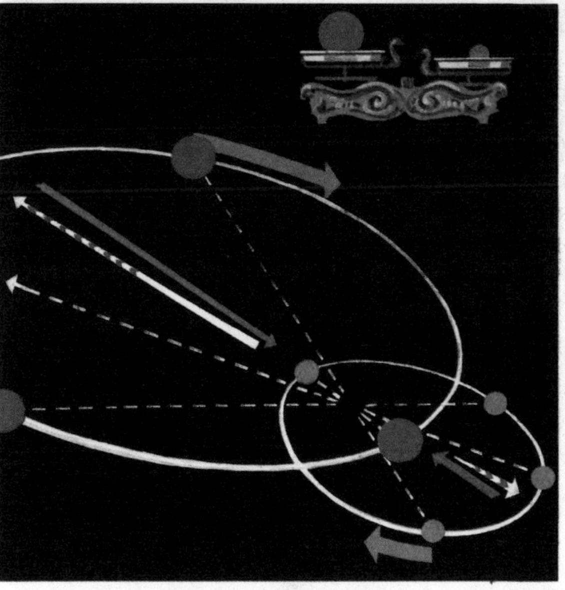

119

119 A binary consisting of a light (red) and a heavy (blue) star. Both orbit a common centre of gravity. The distances from this centre can be used to calculate their masses.

Open clusters do not have a precise shape. They vary in diameter from 20 to 100 light years. They can contain from a few dozen to several thousand stars. They occur close to the Milky Way, and are therefore sometimes called galactic clusters. We know about a thousand of them, and there are several thousand in the whole of the Galaxy. (References to 'the Galaxy' indicate our own galaxy.) Open clusters are also observed in other galaxies. They are young formations, tens to hundreds of millions of years old. A well-known open cluster is the Pleiades in the constellation Taurus, which is 50 million years old (Figure 125).

Globular clusters contain a much larger number of stars – from a hundred thousand to several million. In that part of the Galaxy which can be observed directly, we can see about 120 of them, and in the Galaxy as a whole there are about 300. The stars in globular clusters have a high binding energy, which is why these clusters do not spread apart. They are very stable formations, and the oldest in the Galaxy. They orbit the centre of the Galaxy in long ellipses. Globular clusters are much older than open clusters. The condensation of the Galaxy and all globular clusters probably took place simultaneously more than 9000 million years ago. Globular clusters contain the oldest stars, which have existed throughout the whole history of our Galaxy.

The Milky Way – a system of stars, stellar clusters, and nebulae

'...For the Milky Way is nothing other than an agglomeration of innumerable stars. Whichever direction we look, we see huge clouds of stars; many of them are large and exceptionally bright, but the number of small stars cannot be ascertained...'

Thus wrote Galileo, over 300 years ago, after pointing his simple telescope at the Milky Way. The truth of his words has been verified by modern research. A telescope reveals this silvery strip to be composed of countless stars.

The Milky Way runs from the constellation of Scorpio, northwards through Sagittarius, Scutum, Aquila, Sagitta, Vulpecula, Cygnus and Lacerta, with an arm reaching out to Cepheus. Then it passes through Perseus and Auriga, between Gemini and Orion, Monoceros and Canis Major. In the southern

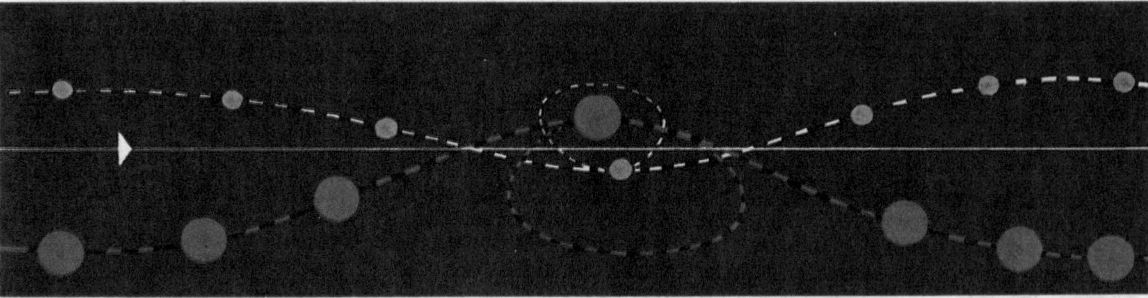

121 The centre of gravity of a binary star shifts very slowly across the sky (white line with arrow). In the process the two stars describe wavy lines which allow their masses to be determined.

122 Through a telescope we see the star Castor as two blue stars which orbit each other. A little further away is a reddish, faint star which orbits the other two once in several thousand years. Spectrographs show that each of these three components is a binary ('spectroscopic' binary). The figure shows the star Castor from an imaginary planet.

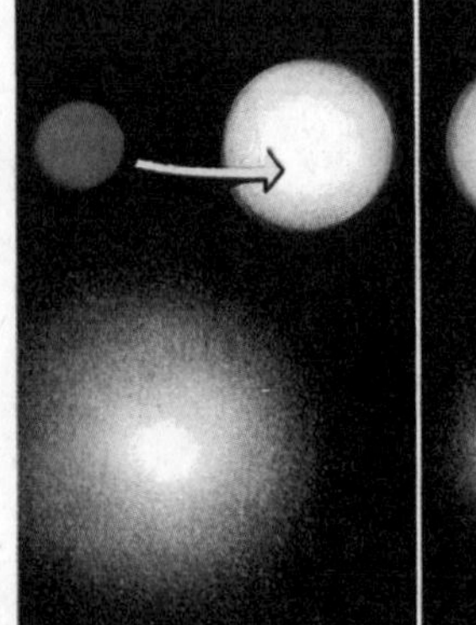

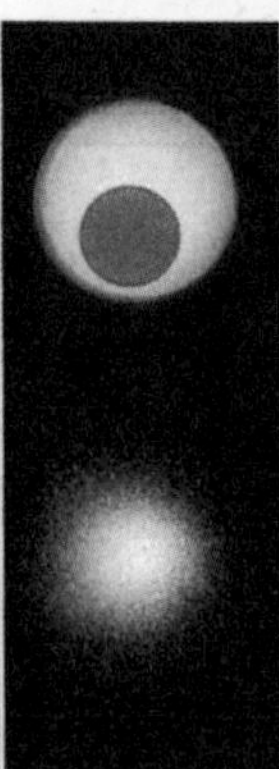

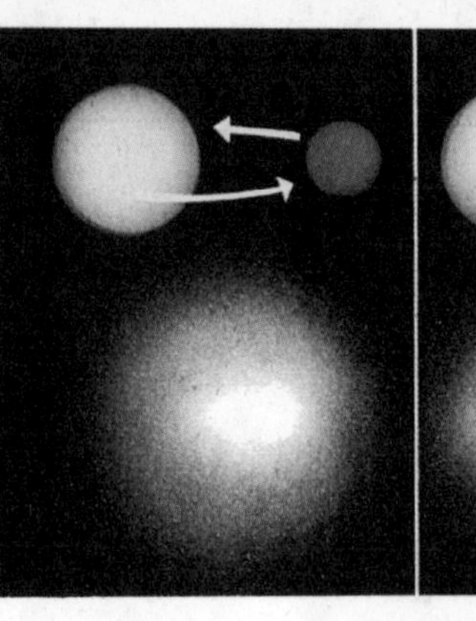

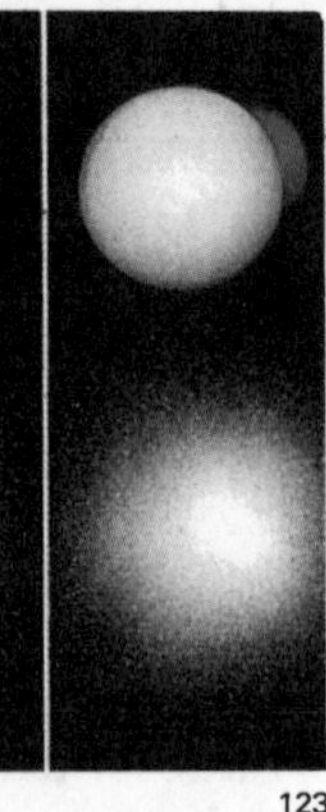

123 Eclipsing binary. The small red star and the large white one orbit around a common centre of gravity, sometimes covering each other. Because of the great distance the two stars are seen as one, whose brightness varies.

124 An X-ray binary. Gases leave the large star and are pulled in the direction of the small neutron star or black hole. They are accelerated to enormous speeds and collide, are heated to very high temperatures, and emit X-rays. ▷

hemisphere it passes through the constellations of Puppis, Pyxis, Vela, Carina, the Southern Cross, Musca, Centaurus, Lupus, Norma and Ara, and back to Scorpio.

The Milky Way is a huge system of about 300 globular stellar clusters, several thousand open clusters, about a billion stars, many nebulae and a great deal of interstellar dust and gas spread out in interstellar space. It is our Galaxy, and one of its billion stars is our Sun (Figure 128).

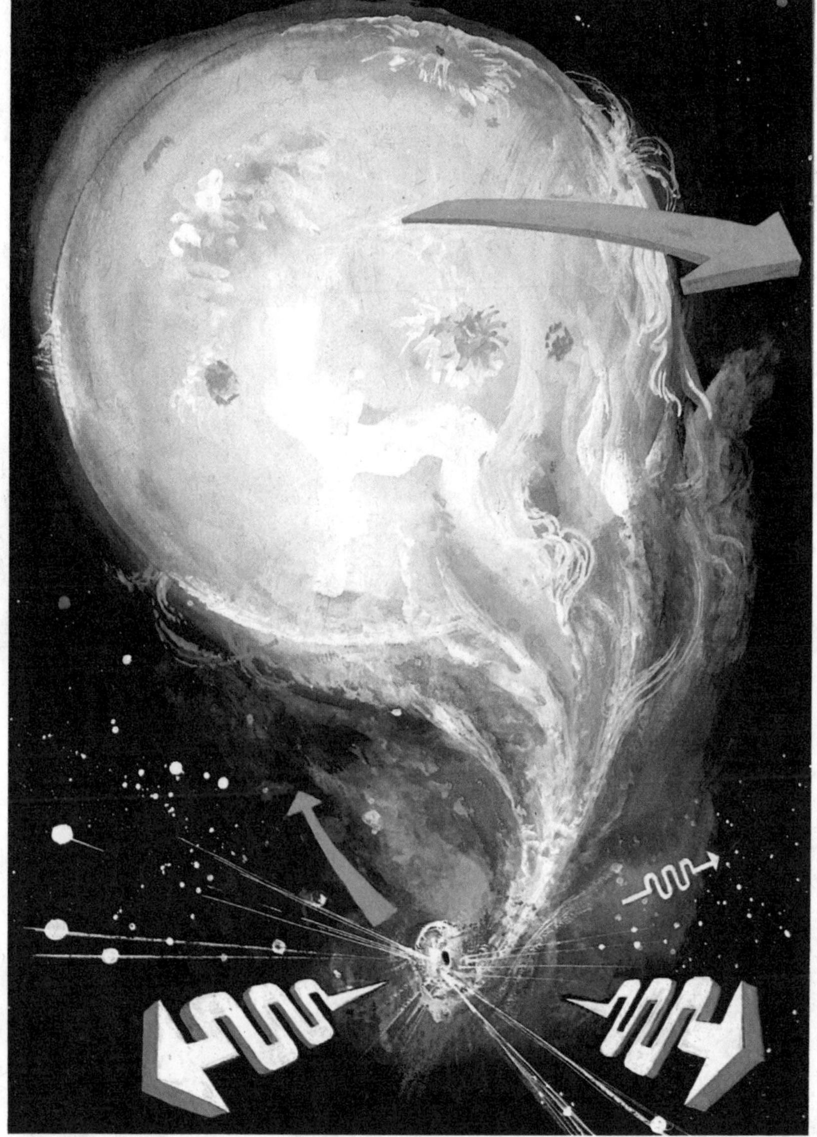

124

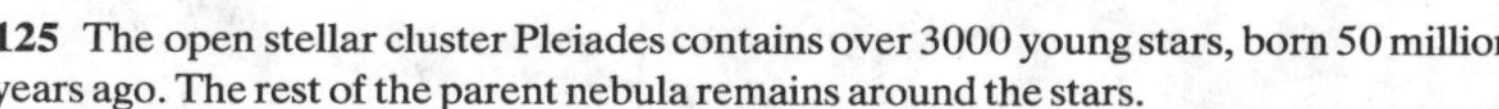

125 The open stellar cluster Pleiades contains over 3000 young stars, born 50 million years ago. The rest of the parent nebula remains around the stars.

125

126

126 The open cluster in Cancer. It contains about 80 stars and is 2500 light years away from us.

127

127 The globular cluster in Canes Venatici. It contains 250,000 stars and is 40,000 light years away. It is nearly 10,000 million years old. Globular clusters are the oldest systems of stars in our Galaxy.

128 If we were to travel 300,000 light years in the direction of the constellation Cancer and look back, we should see the Galaxy as it is depicted in picture **(A).** The Solar System is at point **X.** The stars of the Galaxy are for the most part distributed in a disc, and particularly in the spiral arms inside it (for example **Y** and **Z**). The disc is surrounded by a spherical halo (only the right-hand half, **13,** is shown). The galactic corona which surrounds the halo is much more extensive (reaching as far as the Magellanic Clouds) and is not shown.

The arms and the disc stretch along the galactic plane (the rectangle **A**). The Galaxy turns about its axis – **Q** – in a clockwise direction – **14.** The distances in the plane and perpendicular to it (left), and between galaxies, are expressed in light years. Below the galactic plane there are three galaxies: the large galaxy in Andromeda **(10),** and the Large and Small Magellanic Clouds (**12** and **11**). The flat cylinder **X** around the Solar System shows the distance in the galactic disc over which we can see (about 5000 light years). Inside this cylinder the red block has sides of 1000 light years. It is shown enlarged at the top left and some bright stars are marked: the Pole Star **(1),** the Pleiades **(2),** Antares **(3),** Spica **(4),** Mizar **(5)** and Betelgeuse **(6).** ▷

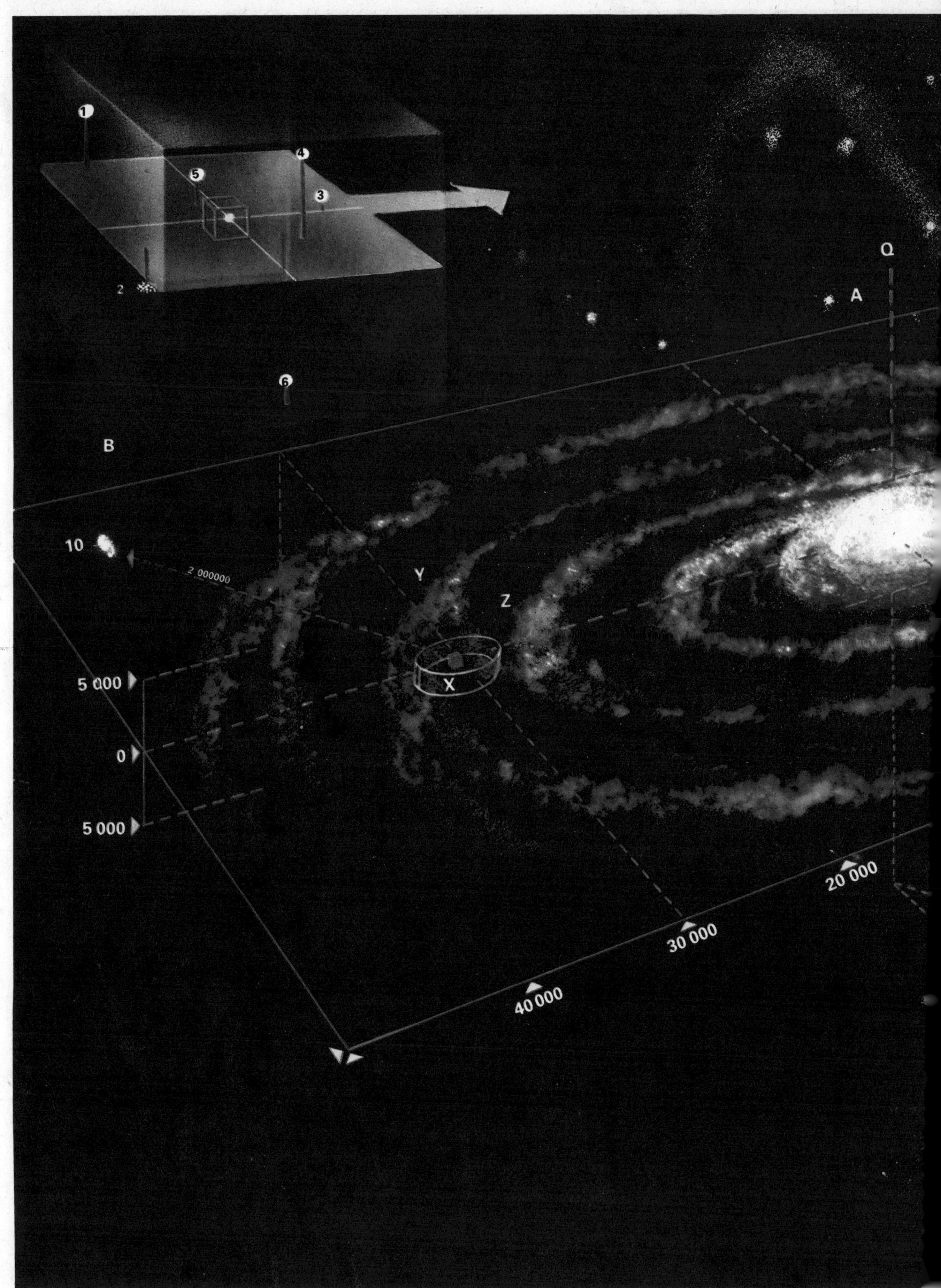
1
2
3
4
5
6
Q
A
B
10
2 000000
Y
Z
X
5 000
0
5 000
20 000
30 000
40 000

13
1
2
3
4
5
6
100 000
C
14
50 000
23
21
26
1
2
24
0
000
13
4
25
3
20
21
22
D
200 000
250 000
12
11

The diameter of the Galaxy is 100,000 light years. The distance between the Solar System and the centre of the Galaxy is 30,000 light years. Between us and the centre are two spiral arms. In these arms most of the stars and interstellar material are concentrated. New stars are being born there. At a distance of ten thousand light years from the centre of the Galaxy (in the direction towards us) is an arm which is moving away from the centre at a speed of 35 km (22 miles) per second. This is a result of an explosion which took place in the nucleus at the very centre of the Galaxy. At a distance of 20,000 light years from the centre (i.e. 10,000 light years from us) is another arm, called the Sagittarius arm. The Sun is at the inner part of the Orion arm (Figure 128). A large number of stars and interstellar matter hides the nucleus of the Galaxy from our view (Figure 130); otherwise it would light up our planet at night like the Moon.

Though the light from the galactic nucleus is entirely absorbed on its way to us, infra-red radiation and in particular radio waves from it do reach us, and we are able to observe the composition of the nucleus and the processes which are going on inside it. But there is still much work to be done before we properly understand its structure and the dramatic explosions which take place in it from time to time.

Our Galaxy is a rather flat system. Most of the stars, all the open clusters and all the interstellar matter are concentrated in a relatively thin disc. In that disc the stars and interstellar matter are grouped in spiral arms. These arms are the cradle of new stars. We

129

129 The spiral galaxy called the Whirlpool Galaxy, seen from above. It is accompanied by a small galaxy, just as the Magellanic Clouds accompany our own Galaxy. It is 12 million light years away.

130

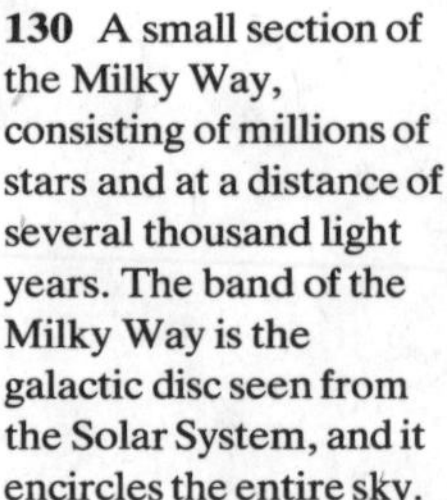

130 A small section of the Milky Way, consisting of millions of stars and at a distance of several thousand light years. The band of the Milky Way is the galactic disc seen from the Solar System, and it encircles the entire sky.

◁ The red cube from the centre of Figure B is enlarged, and its edges are 100 light years long **(C)**. In it the positions of the close stars are marked: Capella **(1)**, Castor **(2)**, Pollux **(3)**, Vega **(4)**, Arcturus **(5)**, Sirius **(6)** and Proxima Centauri **(7)**.

The position of the Solar System (i.e. of the ecliptic) in the Galaxy is shown in Figure **D** at the bottom right. The orbit of the Earth around the Sun (seen from below, i.e. from the south) marks the start of spring **(1)**, summer **(2)**, autumn **(3)** and winter **(4)**. The galactic plane **(20)** leans at an angle **(25)** towards the ecliptic and the galactic centre lies beneath the ecliptic at a small angle away from it **(26)**. The axis of the Galaxy is **(21)**, **(22)** is the axis of the ecliptic, **(23)** the axis of the Earth in the direction of the Pole Star, and **(24)** the direction of the galactic centre.

131

131 The spiral galaxy in the constellation Coma Berenices. This is how we should see our own Galaxy edge from a distance of 10 million light years.

can form an impression of what they probably look like by looking at other galaxies (Figure 129). Globular clusters and some old stars (called Population II) are situated in an extensive spherical space around the disc of the Galaxy (Figure 128) called the halo. The diameter of the halo is about 100,000 light years. The disc and the halo are surrounded by the enormous galactic corona, whose diameter is about 400,000 light years.

The galactic plane is the plane which bisects the disc. In Figure 128 it is indicated by the red rectangle A. The Sun and its system lie close to the galactic plane. That is why the Milky Way, which is the galactic disc seen from the Solar System, divides the sky into two equal halves.

The Galaxy – a huge system of stars, stellar clusters, nebulae and interstellar matter – is held together by the gravitational force between all the members of the system. The resulting gravitational force points towards the centre (the nucleus) of the Galaxy in the case of each member of the system. For this reason all members orbit the centre of the Galaxy along elliptical orbits, in the same way as the planets orbit the Sun. The stars in the disc orbit the nucleus along approximately circular orbits. The orbital speed of the Sun is 230 km (143 miles) per second. The members of the Galaxy in the halo, such as the globular clusters, orbit the galactic nucleus along very elongated ellipses, one of whose foci is in the centre. The galactic corona and the halo are the oldest parts of the Galaxy, while the flat disc came into being later.

Galaxies – huge systems of stars

According to their shape, galaxies are divided into three groups: elliptical galaxies (Figure 139), spiral galaxies (Figures 132 and 135), and irregular galaxies. The most common galaxies are elliptical ones (around 60 per cent of all galaxies). Spiral galaxies, such as our own, make up about 30 per cent of the total. About 10 per cent of galaxies are irregular.

We have spoken of why the nucleus of our own Galaxy is not visible. But we can see and photograph the centres of many other galaxies. They are compact, dense and, compared with the size of the whole galaxy, very small formations. They are not more than a few light years across. But the density of the nucleus is a million times greater than the average density of the galaxy. If we were to depict the Galaxy as a spiral 10 m (33 ft) in diameter, its nucleus would be no larger than a pinhead. Though the nucleus is so small, it is of basic importance. In the case of spiral galaxies spiral arms (usually two of them) stretch out from the nucleus. From some galactic nuclei huge clouds of hot plasma are thrown at a speed of several thousand kilometres per second. Very active galactic nuclei throw out clouds of relativistic electrons with magnetic fields, which then form the extensive lobes of radio galaxies.

The nuclei of highly active galaxies look like very bright stars. From them, galaxies radiate most of their energy. It is not yet known exactly how such a huge flux of energy is released in such a small volume as that of the galactic nucleus. These are not the thermonuclear reactions which take place in normal stars and are quite insufficient to account for such an intense process. It is either gravitation (such as plasma falling into a huge black hole in the middle of the nucleus), or the annihilation of matter with antimatter – a process which is 100 per cent efficient in releasing rest energy (page 40). The activity of the nuclei of some galaxies (for instance Seyfert's galaxy, the N-galaxy, some radio galaxies and quasars) is the most dramatic phenomenon in the whole of the Universe today. Some experts think that these are the local remains of the Big Bang, or perhaps something similar to delayed Big Bangs on a smaller scale taking place in the nuclei of galaxies. This is as yet a mere hypothesis, however, since the source of the immense radiation and very violent explosions of active galactic nuclei is something we are not yet able to explain.

The number of stars in a galaxy may vary from a hundred million (dwarf galaxies) to a few billion (giant galaxies). The distance between individual galaxies is usually several times their diameter (roughly hundreds of thousands of light years). Those galaxies nearest to the Milky Way are the Large Magellanic Cloud and the Small Magellanic Cloud. The Large Magellanic Cloud contains 10,000 million stars and is about 40,000 light years in diameter. The Small Magellanic Cloud contains 2000 million stars, has a diameter of over 30,000 light years, and is over 200,000 light years away from us. Both these galaxies are clearly visible with the naked eye in the southern hemisphere. The first to inform Europeans of their existence was Ferdinand Magellan, in 1519. They are small, irregular galaxies, considered to be companions of our own.

Systems of galaxies

In the hierarchy of the Universe, galaxies are merely units of higher systems. There are few solitary galaxies. It is more usual to find double, triple or multiple galaxies. Larger numbers of galaxies (up to several dozen) are contained in *groups of galaxies*. Their diame-

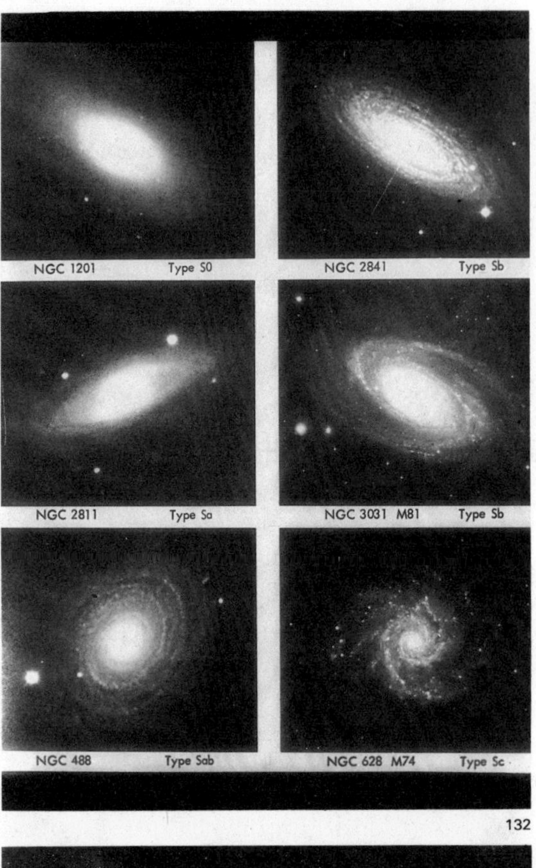

132 Spiral galaxies differ in shape, the size of the central bulge and the twist of the arms. They are divided into various types.

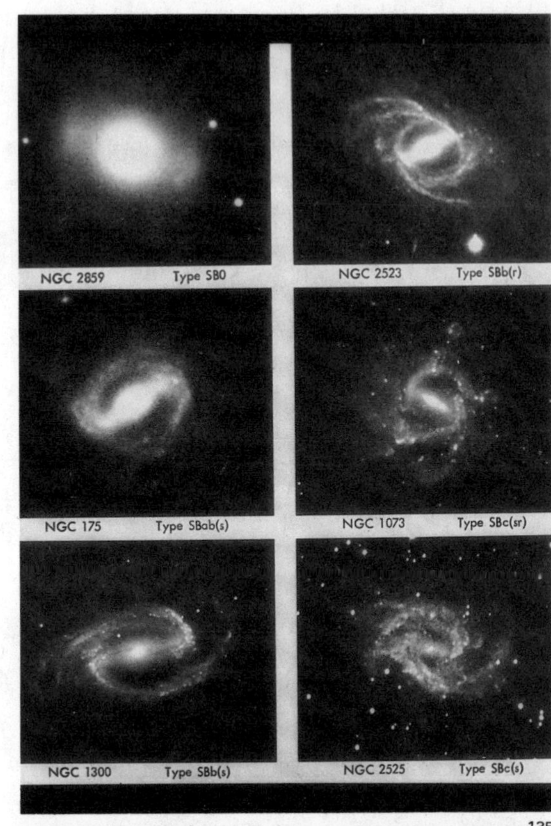

133 The large galaxy M 31 in Andromeda. Though it is 2 million light years away, it can easily be seen in autumn (in the northen hemisphere) with the naked eye. It is accompanied by two dwarf galaxies, and contains over one thousand million stars.

133

134 The arms of the large galaxy in Andromeda contain relatively young stars. Individual bright stars can be seen there.

134

135 Some spiral galaxies have a bar. Their arms run first radially from the galactic centre and then at a greater distance twist into a spiral.

136 The central bulge of the large galaxy in Andromeda. It is composed of very old stars (almost 10,000 million years old).

136

ter is 3 million to 10 million light years. In the closest vicinity, up to 50 million light years away, there are 55 groups of galaxies, but only a few isolated galaxies. Our *local group* is small, containing about 25 galaxies, and having a diameter of 3 million light years. The latest research shows that some small irregular galaxies in the local group are fragments of a large galaxy. They either broke away from it or were thrown out of it during a period of intense activity. The local group includes our own Galaxy, galaxy M 31 in Andromeda, the spiral galaxy M 33 in the constellation of the Triangle, the large galaxy called Maffei 1 in Cassiopeia, both the Magellanic Clouds, and others (Figure 141).

Systems of galaxies larger than groups are called *clusters of galaxies* (Figure 139). Their size varies from 5 million to 15 million light years. They contain hundreds to thousands of galaxies. A cluster containing a large number of galaxies has a regular spherical structure. In its centre there is usually a giant elliptical galaxy, which itself contains a few billion stars. Clusters of galaxies are systems held together by gravitational force. The galaxies in a cluster orbit around a common centre of gravity; the greater the total mass of a cluster, the greater the speed with which the galaxies orbit around its centre. Thus we can determine the mass of a cluster from the speeds of

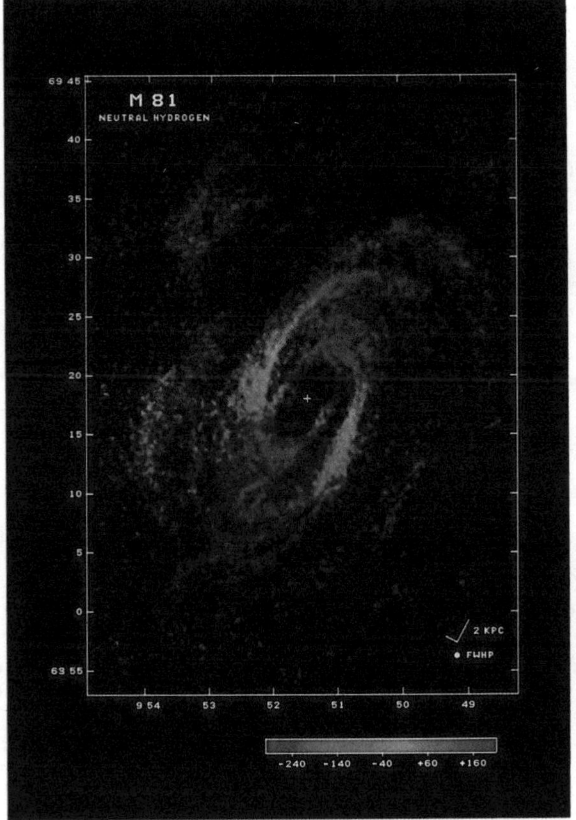

137

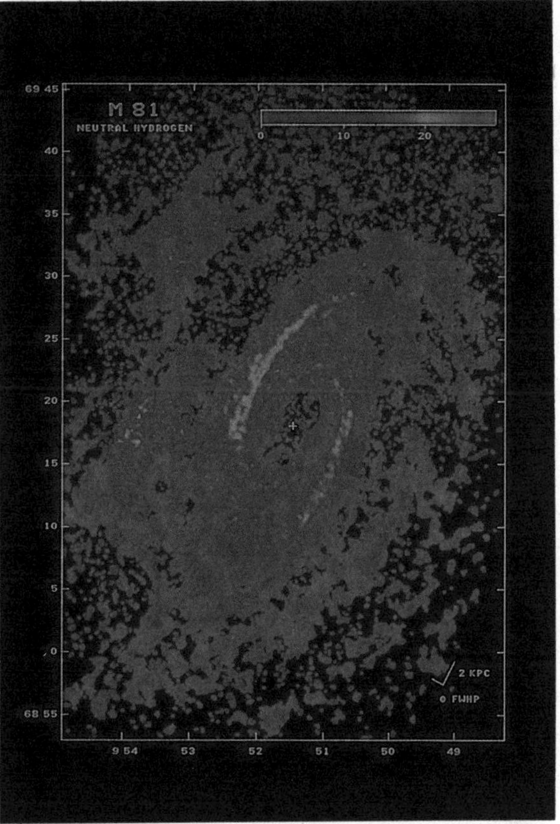

138

137 The rotation of galaxy M 81, determined according to the Doppler shift. The lower part is approaching (blue), the upper part receding (red). The speed of rotation in kilometres per second can be determined according to the scale at the bottom. Negative velocities are towards us, positive ones are away from us.

138 Galaxies are enveloped in a large amount of neutral hydrogen. This is the remains of the original hydrogen from which the galaxies formed around 10,000 million years ago. In this galaxy (M 81) the amount of neutral hydrogen is measured chromometrically (blue indicates a small amount of hydrogen, red a large amount).

the individual galaxies. Measurements show that the total mass of a cluster is much greater than the mass of all its luminous galaxies. This means that they contain matter in an invisible form. The actual amount of matter, i.e. of visible and invisible matter together, can be determined from the orbiting of galaxies around the centre of the cluster.

The Universe – a system of supergalaxies

Observation of galaxies indicates that there are systems even larger than clusters of galaxies. They are called *supergalaxies* or *clusters of clusters*. A supergalaxy is a system containing many clusters of galaxies, groups

140

of galaxies and multiple and individual galaxies.

The centre of our Supergalaxy, the one to which the local group and thus also our Galaxy both belong, is close to the cluster in the constellation of Virgo, about 60 million light years away. Our local group is close to the edge of our Supergalaxy, near its southern perimeter. The diameter of our Supergalaxy is about 150 million light years (Figure 142). Some supergalaxies are considerably larger, for example that in Perseus is about 750 million light years across.

The Universe is thus a system of supergalaxies. Supergalaxies rotate, so that the centrifugal force acting on the clusters and groups of galaxies acts away from the centre of the supergalaxy – i.e. in the opposite direction to the gravitational force of the supergalaxy. Galaxies also move relative to each other, so that each supergalaxy is moving away from the others (like raisins in rising dough, where each raisin also gets further away from all the others).

It is not only supergalaxies, but galaxies also that are moving away from each other. The further they are from each other, the faster the distance between them is growing (Figure 143). This fact has been expressed by

139

139 The cluster of galaxies in the Coma Berenices contains around 800 galaxies. It is 350 million light years away and is receding from us (as a result of the expansion of the Universe) at the rate of 6700 km (4160 miles) per second. The brightest of them is the giant elliptical galaxy containing several billion stars (just below the centre of the photograph).

140 A group of five galaxies in the constellation of Serpens. Like stars, galaxies seldom occur singly.

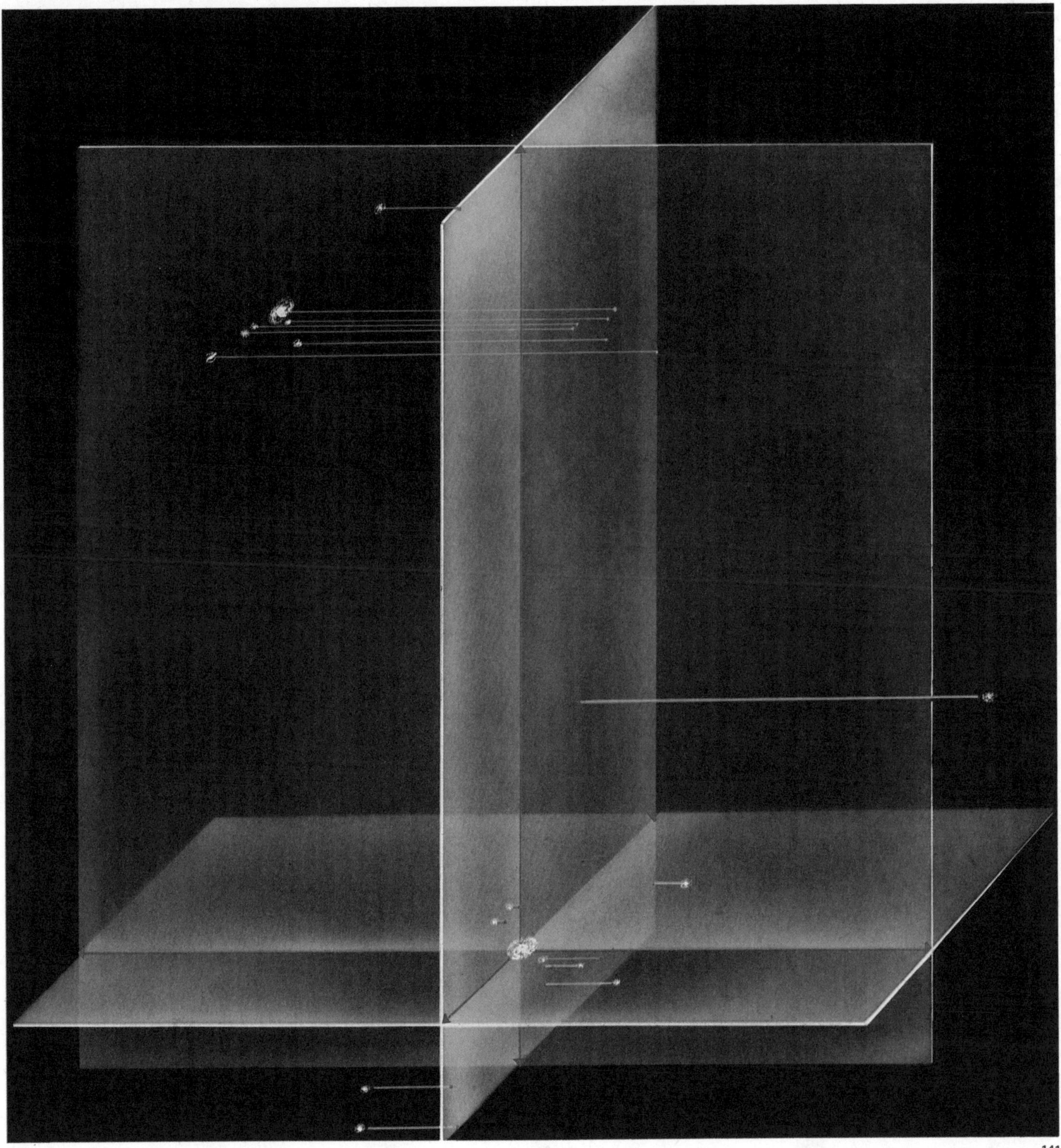
141

141 The local group of galaxies. The galactic plane is vertical, from front to back. The galactic axis is from left to right. The Magellanic Clouds are below the galactic plane (on the left, close to the Galaxy). The Galaxy is at the intersection of the planes. At the top left (i.e. below the galactic plane) is the galactic group around the large galaxy in Andromeda. It is 2 million light years away.

Edwin Hubble by means of the simple formula or relation:

$$v = Hr$$

In Hubble's relation v is the velocity with which two galaxies are moving away from each other (in kilometres per second); the letter r represents their distance apart in megaparsecs (Mpc); H is *Hubble's constant,* and has a value of 55 km (34 miles) per second per Mpc. This means that two galaxies 1 megaparsec away from each other move apart at the rate of 55 km (34 miles) per second, or the distance between them increases by 55 km (34 miles) every second. To give another example of Hubble's relation: two clusters of galaxies at a distance of 10

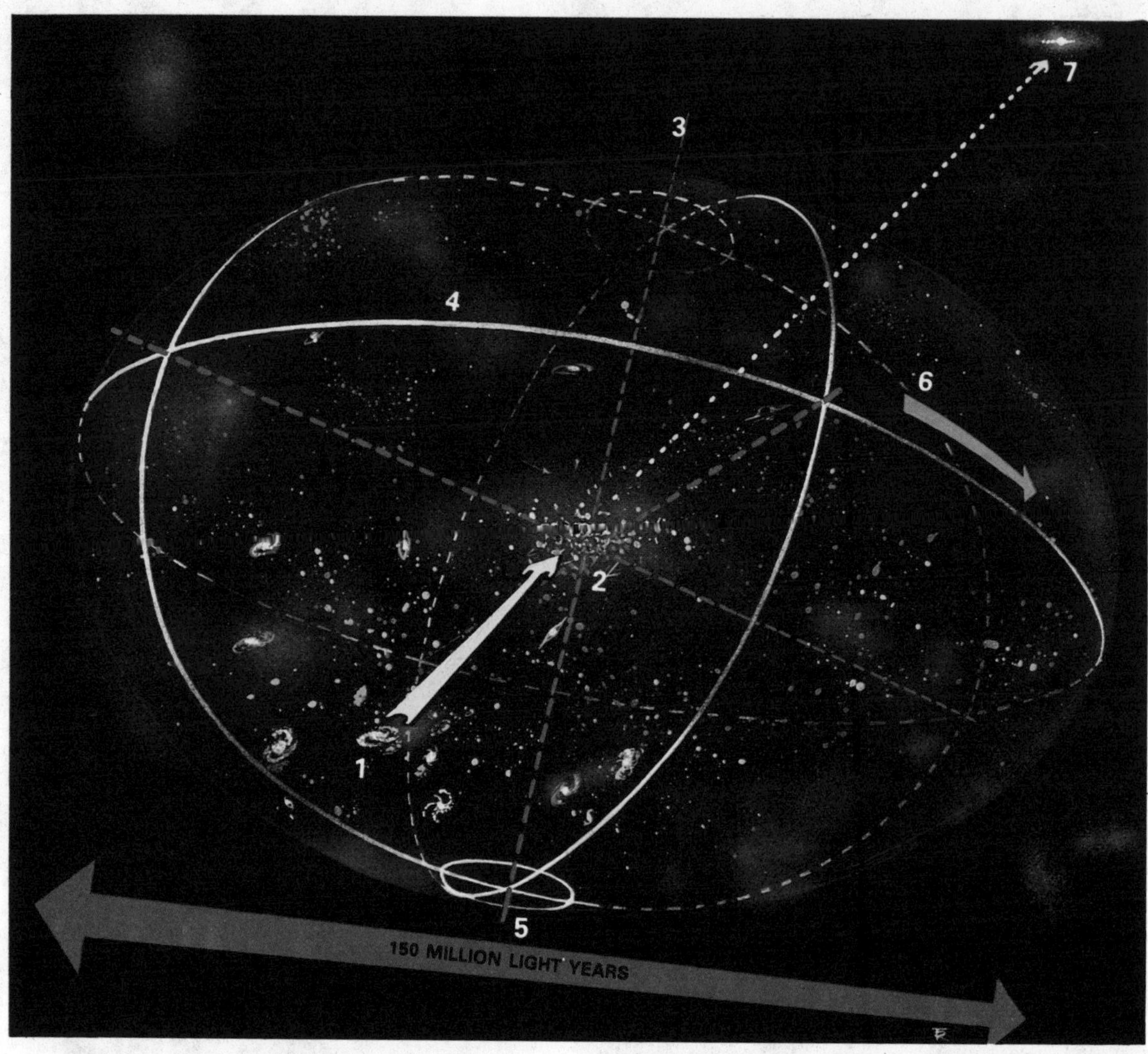

142

142 A diagramatic depiction of the Supergalaxy. It contains millions of galaxies, from dwarf galaxies to giant elliptical galaxies containing several million million stars. The approximate position of our own Galaxy is shown at **(1),** the centre of the Supergalaxy at **(2),** in the direction of the constellation of Virgo. The axis of rotation of the Supergalaxy is shown at **(3),** its equator at **(4),** its south pole at **(5),** and its direction of rotation at **(6).** Far beyond the centre of the Supergalaxy lies the best-known of the quasars **(7).** The size of our Supergalaxy is 150 million light years (red arrows). It is relatively small.

megaparsecs from each other move apart at the rate of 550 km (340 miles) per second. Supergalaxies 400 Mpc (1200 million light years) apart are parting company at a speed of 22,000 km (13,600 miles) per second. This motion of clusters of galaxies and supergalaxies away from each other is known as the *Expansion of the Universe.*

If it were not for the gravitation which attracts all supergalaxies, clusters of galaxies, groups of galaxies and galaxies to each other, the expansion of the Universe would continue to take place at the same rate according to Hubble's relation. But due to mutual attraction the expansion is slowing down. This means that Hubble's constant is getting smaller all the time. Observations show that it will decrease to zero in about 30,000 million years, when the expansion of the Universe will cease altogether. (This process can be compared to a stone thrown straight up in the air, whose motion slows down until at the highest point it stops altogether.)

In the past the expansion of the Universe was much faster than it is today. The rate of slowing down (i.e. the 'deceleration parameter') can be ascertained, and from it we can calculate when the supergalaxies and clusters of galaxies will be at their farthest from each other, which is when the expansion of the Universe will stop altogether. Similarly, we can count backwards and work out when the expansion of the Universe began. At that time the rate of expansion (Hubble's constant) was high. We therefore speak of a great explosion – the Big Bang. We find that this occurred some 10,000 million years ago (Chapter 3).

Thus the Universe we know today is about 10,000 million years old, and when it reaches the age of 40,000 million years its expansion will stop. Then contraction of the Universe will begin, when the distance r between supergalaxies and clusters of galaxies will decrease. Hubble's constant will be negative, and will fall to ever lower values. The supergalaxies and clusters of galaxies will fall towards each other at an ever-increasing rate (our stone falling back to the ground). In the end there will be a tremendous collision, a Big Collapse of all supergalaxies and clusters of galaxies, galaxies and stars, and a very hot and dense ball will be formed, like the one which existed at the start of the Universe in the Big Bang.

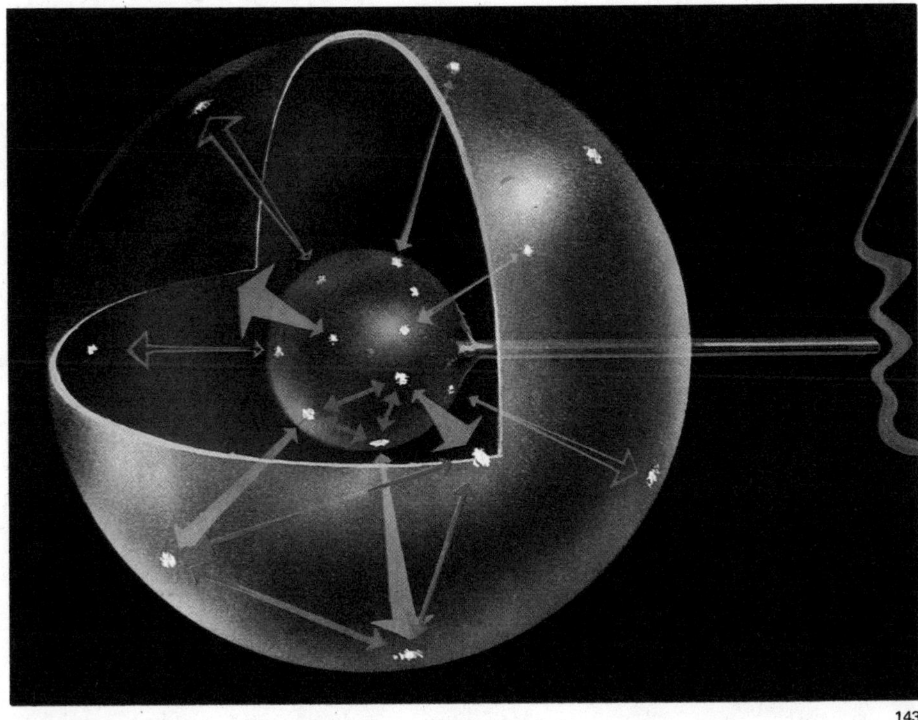

143

143 In the expanding Universe, the further galaxies are from each other, the faster they move away from each other. The figure shows the distances of three galaxies denoted by red arrows. They were previously close (small sphere), but now they are far away (large sphere).

The Big Collapse will be like the Big Bang in reverse. It will occur about 70,000 million years from now, when the Universe will be 80,000 million years old. In the inferno of the Big Collapse all systems will perish, since their binding energy is much smaller than the huge energy of the masses hurtling into the inferno will be. That will be the end of all systems – from the atomic nucleus to the supergalaxy. Everything will be demolished to elementary particles.

What will happen then? First, we must consider how all these systems came into being after the Big Bang.

144 In the course of 10,000 million years the shapeless, chaotic matter evolved into supergalaxies, in which galaxies formed, and protostars turned into stars and planets. Evolution on the Earth continued to the stage of intelligent creatures with an advanced culture. It took about 10,000 million years for the elementary particles of the Big Bang (top left) to develop into their most advanced system – the human brain.

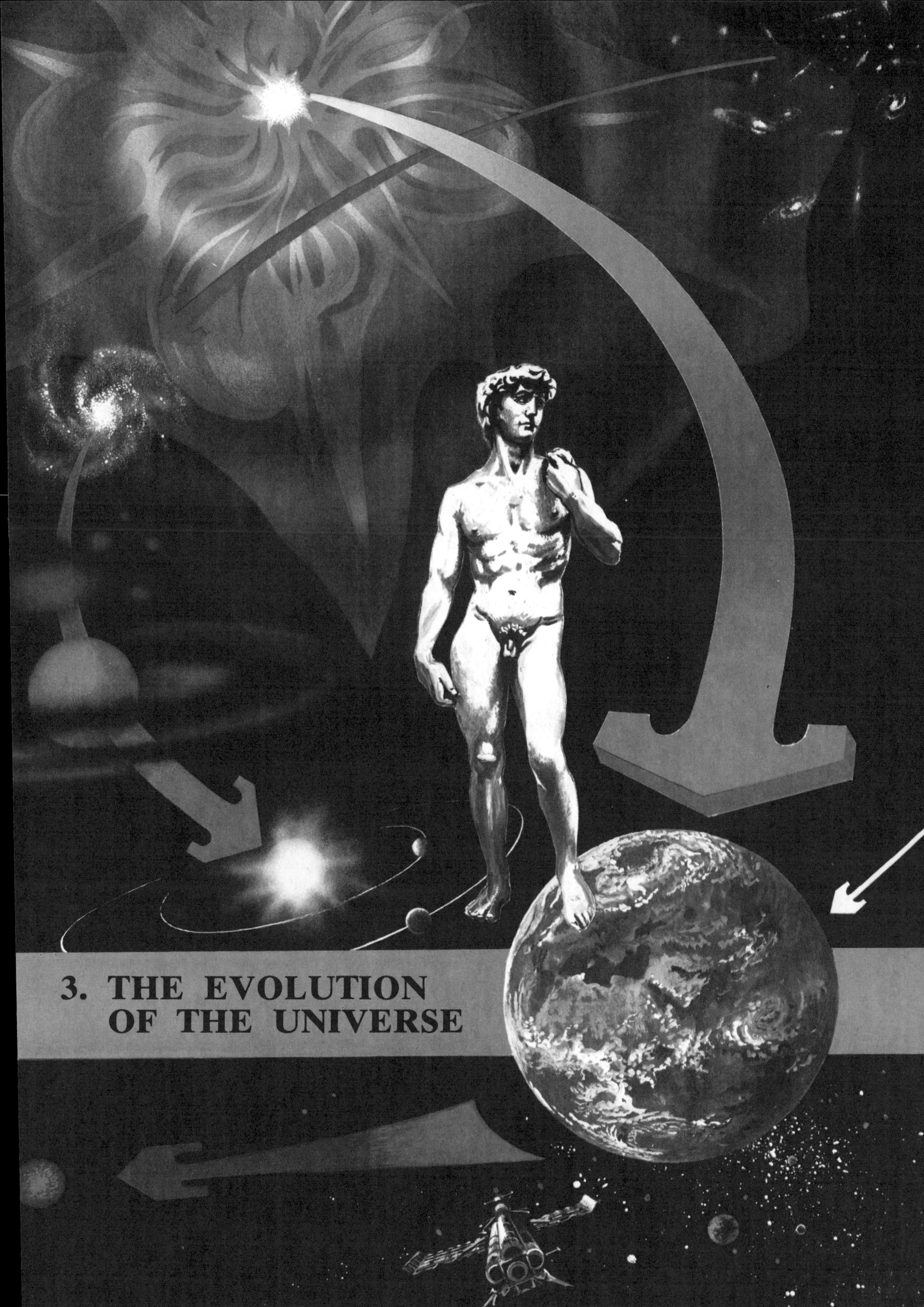

3. THE EVOLUTION OF THE UNIVERSE

From chaos to order

The structure of the Universe, as we have seen in the preceding chapter, is a result of events long past. But the whole of the Universe, and everything in it, changes with time. This process of change, or evolution, goes on very slowly; the Universe is, after all, many millions of times older than astronomy, and indeed the whole of human culture. The evolution of life on Earth forms only a tiny part of the evolution of the Universe itself.

Let us take a look at the main events in the history of the Universe. This chapter might be called a concise history of the Universe – and it is very concise indeed. That history begins with the Great Explosion or Big Bang, and ends with the present day. Its characters are all the elementary particles; we can truly speak of a 'theatrum mundi', in which the whole of history in the usual sense of the word (the history of the individual nations, of mankind itself, of the biosphere, and so on) makes up only a tiny episode.

Modern astronomical observations indicate that the Universe came into existence roughly 10,000 million years ago, from a huge ball of fire, tremendously hot and very dense indeed. Its composition was very simple. It was so hot that it was made up only of free elementary particles, which moved very violently, crashing into each other.

In the course of the 10,000 million years since the Big Bang this primitive, shapeless matter has gradually evolved into atoms, molecules, crystals, rocks and planetary bodies. Stars have come into being – systems of huge numbers of elementary particles with a very simple organization. On some planetary bodies similar to the Earth living organisms may have come into being.

For the time being the evolution of the Universe ends with man. We do not as yet know if and where the Universe has evolved further, or where evolution on Earth will lead to. In any case we are just a tiny fraction of the Universe, a tiny link in its evolution.

The beginning of the Universe

In Figure 143 we saw how our Universe is expanding. We regard the moment when this expansion began as the beginning of the Universe. It is the zero point in Figures 145 and 157. That was the start of the earliest and most dramatic period of the Universe's existence, known as the Big Bang.

The expansion of the Universe means that the same number of elementary particles and photons fill an ever-enlarging space. Thus with that expansion the average density of the Universe is falling. We may suppose that at a time long, long ago (around 10,000 million years ago) the Universe was very dense indeed. In addition it must have been very hot. So hot that the density of radiation completely predominated over the density of matter. In other words, the energy of all the photons in 1 cm^3 (0.06 cu in) was greater than the total energy (mc^2) of all the particles in 1 cm^3.

In the earliest period of all, in the very first moments of the Big Bang, all matter was a very hot and immensely dense mixture of all particles, antiparticles and high-energy gamma photons. The energy of the particles was much greater than that of particles accelerated in the world's biggest accelerators (Batavia, CERN, Serpuchovo). Particles and antiparticles collided and annihilated, but the resulting gamma photons immediately materialized in particles and corresponding antiparticles (see page 40). We can express this extremely rapid formation and extinction of elementary particles as follows:

$$\text{particle} + \text{antiparticle} \rightleftarrows \text{gamma photons}$$

where the top arrow represents annihilation and the bottom one materialization.

A detailed analysis shows that the temperature of matter T fell with time t according to the simple relation:

$$T = \frac{10^{10}}{\sqrt{t}}\ K$$

The formula allows us, for instance, to say that at the time when the Universe was one ten-thousandth of a second old (i.e. $t = 10^{-4}$ s), its temperature was 10^{12}K, which is 1 billion kelvin. We can easily show this, since if we substitute 10^{-4} for t, then $\sqrt{10^{-4}}$ is 10^{-2}. But dividing by 10^{-2} is the same as multiplying by a hundred, so that we get: $T = 10^{10} \times 10^{2}\ K = 10^{12}\ K$, as mentioned above.

Another important relation, allowing us to see into the earliest events in the Universe, is:

$$h\nu = kT$$

The expression on the left is the energy of a photon which is oscillating ν times a second. The letter h is Planck's constant, which has already been mentioned ($h = 6.6 \times 10^{-27}$ erg seconds or 6.6×10^{-34} joule seconds). The expression on the right is the product of Boltzmann's constant k and the temperature T ($k = 1.38 \times 10^{-23}$ $JK^{-1} = 1.38 \times 10^{-16}$ erg $K^{-1} = 8.17 \times 10^{-5}$ eVK^{-1}). The product kT expresses the mean kinetic energy of the particles in matter at a temperature of T. The relation thus tells us that the energy of photons in hot matter was approximately equal to the kinetic energy of particles and antiparticles.

The temperature T of hot, dense matter at the beginning of the Universe fell in time t, as is expressed in the relation given. This means that the average kinetic energy of the particles, kT, decreased. Thus, according to the relation, the photon energy $h\nu$ also had to fall. The only way this is possible is through a fall in the frequency, ν, of the photons. The fall in photon energy with time had serious consequences for the appearance of particles and antiparticles by materialization. We have already seen in Chapter 1 that a photon can change into a particle and an antiparticle only if it has sufficient energy. For a photon to change (materialize) into a particle and an antiparticle with a mass of m_o and a rest energy of m_oc^2, it must have at least an energy of $2m_oc^2$. This condition is expressed by the relation

$$h\nu \geq 2m_oc^2$$

With the expansion of hot matter the energy of photons $h\nu$ fell, and as soon as it decreased below the sum of the rest energies of the particle and the antiparticle ($2m_oc^2$), photons were no longer able to give rise to particles and antiparticles of mass m_o. For instance, a photon with an energy less than 2×938 MeV $=$ 1876 MeV, cannot materialize into a proton and an antiproton, since the rest energy of the proton and the antiproton is each 938 MeV.

We can substitute the kinetic energy of particles kT for the energy of photons $h\nu$ in the inequality above, arriving at:

$$kT \geq 2m_oc^2$$

or:

$$T \geq \frac{2m_oc^2}{k}$$

Expressed in words, this inequality means that particles and antiparticles materialized in hot matter only up to the point where the temperature of the matter did not fall below the value of

$$\frac{2m_oc^2}{k}$$

A glance at the table of elementary particles shows how they gradually ceased to come into existence from the top downwards (i.e. with falling m_o or m_oc^2). But antiparticles and particles were able to cease to exist without restriction; whenever they collided, they annihilated into photons.

In the earliest phase of the Universe all types of particles and antiparticles were produced from high-energy photons. This process gradually slowed down, which led to an extinction of most of the particles and antiparticles. Since annihilation takes place at any temperature, the following process occurred:

particle + antiparticle ⟶ 2 gamma photons

whenever matter and antimatter came into contact.

The process of materialization:

gamma photon ⟶ particle + antiparticle

on the other hand, was able to take place only at sufficiently high temperatures. Only then did gamma photons have a high enough energy to materialize (since $h\nu = kT$, as shown above). According to the cessation of materialization (as a result of falling temperature), the evolution of the Universe is divided into four periods: the hadron era, the lepton era, the photon era and the stellar era.

• *The hadron era.* At the very high temperatures and densities which prevailed at the start of the Universe, matter was made of all the particles and antiparticles which are shown in Figure 14. It can be shown that all the particles were present in almost equal numbers. There were, for example, as many

RADIATION
10^37
erg/cm³
PARTICLES
10^23
erg/cm³
ENERGY
TIME
00
BIG BANG
photons
particles
density
temperature
HADRON ERA
10^37
10^23
10^16
10^13
LEPTON ERA
10^12
10^12
10^24
10^33
SECONDS
10^-4
1 SECO

145 This figure shows diagramatically the prestellar period of the Universe, which lasted about 300,000 years. The age of the early Universe is indicated at the left and at the bottom. The parallel lines give the temperature of the Universe (in kelvins), density (g/cm^3), the total energy of particles per cubic centimetre (erg/cm^3) and the energy of all photons in one cm^3. According to the Big Bang theory, at first the Universe was extremely dense and hot. The first ten-thousandth of a second (10^{-4} s) is called the hadron era. At that time all particles existed in approximately the same numbers. Since the hadrons are the most numerous group in Figure 14, there were more of them than anything else in the Big Bang; they came into existence and passed out of existence in large numbers. As soon as the temperature fell to 10^{12} kelvin, hadrons annihilated and disintegrated and went out of existence; there was no longer sufficient energy for them to be formed by materialization of gamma photons. In the lepton era (10^{-4} s to 1 second) hadrons passed out of existence. The temperature was still favourable for the formation (materialization) of leptons. They were formed (by materialization) and disappeared (by annihilation) in large numbers. When the temperature of the expanding Universe fell to 10^{10} kelvin, the electron-positron pairs ceased to form. The energy of the gamma photons became less than 1MeV – i.e. insufficient for electron-positron formation. This meant the end of the lepton era. The long period that followed was one of radiation, lasting 300,000 years. With the expansion of the Universe not only temperature, but also density fell. The density of radiation fell more rapidly than that of matter. The radiation era thus ended when the two densities were equal. All three initial periods of the Universe together are called the prestellar era. At its end free electrons joined up with protons and formed atoms of hydrogen. The gamma photons of the hadron era degraded into light photons at the end of the radiation (i.e. prestellar) era. (Continued in Figure 146.)

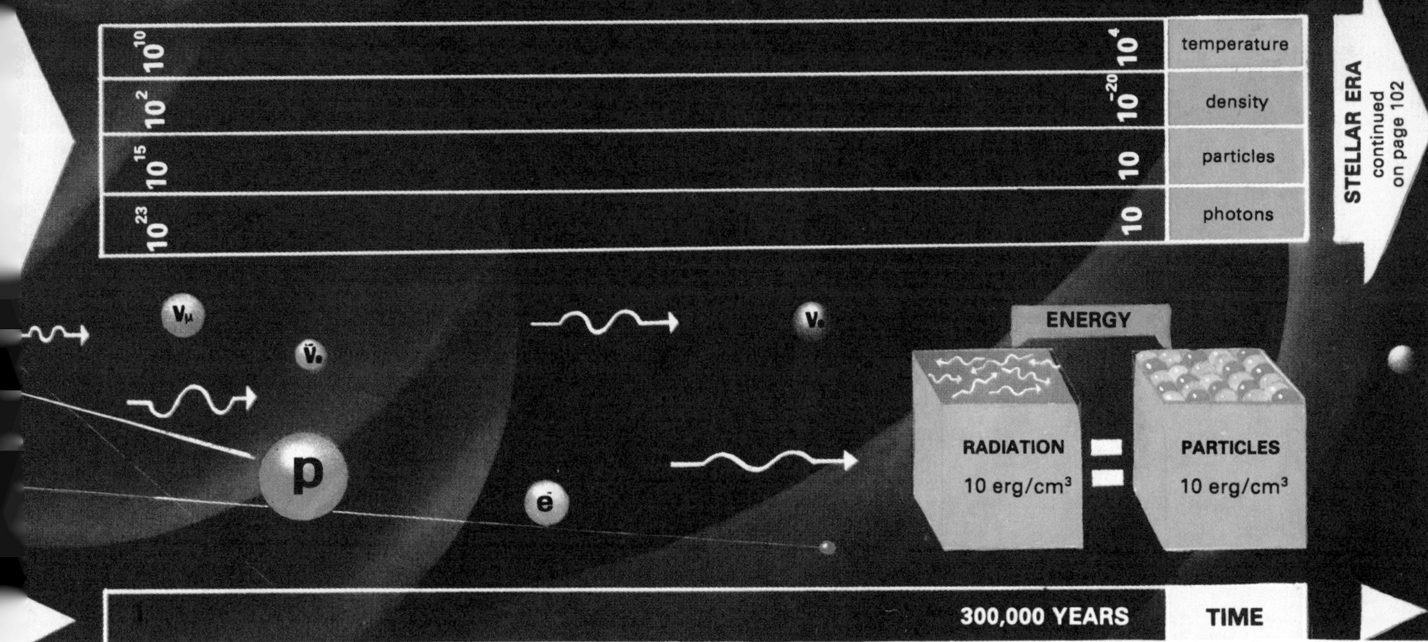

neutrons or omega particles as protons. It is clear from a glance at the table that there were much greater numbers of hadrons than leptons. Thus in the earliest days of the Universe matter chiefly comprised hadrons. The earliest era is therefore called the hadron era, though leptons existed as well.

After the first millionth of a second (when the Universe was 10^{-6} s old), the temperature T fell to 10 billion kelvin (10^{13} K), as follows from the expression on page 96. The average kinetic energy kT of particles and hν of photons was around 1000 million eV (10^3 MeV), which corresponds to the rest energy of baryons. In the first millionth of a second of the evolution of the Universe, materialization of all baryons took place at great speed, just as often as annihilation. But after the first millionth of a second had elapsed materialization of baryons came to a halt, since at temperatures lower than 10^{13} kelvin photons no longer had the necessary energy for it. But the process of annihilation of baryons and antibaryons continued until the moment when the pressure of radiation drove matter and antimatter apart. The hyperons (the heaviest of the baryons) are unstable, and through spontaneous disintegration they all changed into the lightest baryons (the protons and neutrons). This meant that the most numerous group of baryons, the hyperons, disappeared from the Universe. Though neutrons could have continued to disintegrate further (into protons), protons themselves could not decay any further, since this would be against the Law of the Conservation of Baryon Charge (see page 20). The disintegration of hyperons took place in the interval from 10^{-6} to 10^{-4} seconds.

At the time when the Universe was one ten-thousandth of a second (10^{-4} s) old, its temperature had fallen to 10^{12} kelvin, and the energy of the particles and photons was only 100 MeV. This was not enough to give rise to the lightest of the hadrons, the pions. The pions from the earlier period had disintegrated, and new ones were not able to form. This means that at the time when the Universe was 10^{-4} s old all the mesons had disappeared from it. This brought to an end the hadron era, since the pions are not only the lightest of the mesons, but also the lightest of the hadrons. There were never again strong interactions in the Universe as active as those in the hadron era, which lasted one ten-thousandth of a second.

• *The lepton era.* When the energy of particles and photons was falling from 100 MeV to 1 MeV, leptons were very numerous. The temperature was high enough for the intensive production of electrons, positrons, muons and neutrinos. The baryons (protons and neutrons) which had survived the hadron era were very rare compared with leptons and photons.

The lepton era began with the disintegration of the last hadrons – the pions – into muons and neutrettos. It ended within a few seconds at temperatures of 10^{10} kelvin, when the energy of photons fell to 1 MeV and the materialization of electrons and positrons ended. During this era the neutrinos and the neutrettos began their independent existence. Neutrinos and neutrettos from that time are called fossil. The whole of the Universe is saturated with a huge number of fossil neutrinos and neutrettos. This is the sea of neutrinos which has already been mentioned. But as yet our instruments are not sensitive enough to detect this sea of fossil neutrinos.

• *The photon or radiation era.* After the lepton era came the photon era. As soon as the temperature fell to 10^{10} kelvin, and the energy of the gamma photons to 1 MeV, only annihilation of positrons and electrons took place. New pairs of electrons and positrons could not come into being through materialization, since the photons did not have enough energy. But the annihilation of electrons and positrons continued until the pressure of radiation drove matter and antimatter apart. After the hadron and lepton eras a great many photons remained. At the end of the lepton era there were 2000 million times more photons than surviving protons and electrons. The most important component of the Universe after the lepton era was the photons – not only in numbers, but also in energy.

In order to make a comparison of the importance of the particle component of the Universe with its photon component, let us consider the magnitude of the energy density of each of them. This is the amount of energy in 1 cm^3 (0.06 cu in), or more exactly the average (mean) amount, supposing the matter in the Universe were evenly distributed throughout. If we add the energy of all photons in 1 cm^3 (0.06 cu in), we get the density of radiation energy, E_r. If we add together the rest energy of all particles in a cubic centimetre, we arrive at the mean energy of matter in the Universe, E_m.

With the expansion of the Universe, the energy density of both components – the photon and the particle – is falling. If the distances in the Universe increase twofold, the volume increases eightfold, and the particles and photons are spread over eight times the amount of space. Thus their number in 1 cm^3 falls eight times. In the course of expansion, however, photons behave quite differently from particles. While the rest energy of each particle remains unchanged with expansion, that of each photon falls. Photons decrease in frequency, as if they were growing tired. This results in the energy density of photons (E_r) falling more rapidly than that of particles (E_m). The prevalence of the photon component over the particle component (as far as energy density is concerned) decreased throughout the radiation period, until it disappeared altogether. At that moment the two components (E_r and E_m) were equal. The photon era ended, and with it the Big Bang itself. At that time the Universe was about 300,000 years old, and distances were a thousand times smaller than they are today (Figure 157).

The Big Bang did not last very long – only one thirty-thousandth of the present age of the Universe. But it can be said to have been the Universe's finest hour. Never again has its evolution been so rapid and dramatic as it was then, at the very start. At that time all events concerned free elementary particles – their transformations, appearance, disintegration and annihilation. Let us recall in what a short time (a few seconds) almost all the elementary particles disappeared: either through annihilation (transformation to gamma photons) or disintegration into the lightest baryons (protons) and the lightest charged leptons (electrons). This great simplification in the number and diversity of elementary particles is the main feature of the early period of the Universe. It took place in the expanding, cooling and rarefying universal primaeval matter, composed of free protons and electrons. The particles which we produce today as rare specimens in huge particle accelerators were very abundant at that time.

• *The stellar era.* After the Big Bang there followed the much longer era of matter, with the predominance of particles. Known as the stellar era, it has lasted from the end of the Big Bang (about 300,000 years) to the present. Evolution in the stellar era is not nearly as dramatic or as fast as it was in the beginning. Compared with the Big Bang, it seems very leisurely indeed. This is because of the low densities and low temperatures. We can look upon the Universe rather as a firework that has gone off. There are still some red sparks, ash and smoke. We are living on a piece of cooled-down ash, gazing at the slowly ageing stars, and thinking about the glory and the panache of the start of it all.

The birth of supergalaxies and galactic clusters

The evolutionary eras of the Universe are characterized by gradual prolongation – from the ten-thousandth of a second of the hadron era, to the stellar era which has already lasted 10,000 million years, and has not yet come to an end. The shorter the era, the more dramatic the events which took place in it.

In the third, the photon (radiation) era, the rapid expansion of the primaeval matter continued; this universe consisted mainly of photons, with the occasional free proton or electron making an appearance, exceptionally also alpha particles. (Let us not forget that there were 2000 million times more photons than protons and electrons.) In the course of expansion the volume increased and the density decreased. During the radiation era protons and electrons did not change, remaining free; only their speed decreased. At the start of the radiation era protons moved at a speed of about 10,000 km (6000 miles) per second, while at the end of it they had slowed down to a mere 10 km or so per second (about 6 miles per second). Photons were affected much more. Though their speed stayed the same, gamma photons gradually changed in the course of the radiation era into X-ray photons, ultraviolet photons and light photons. Towards the end of the radiation era, matter and photons had already cooled down to such an extent that each electron had united with a proton to form hydrogen atoms. At the same time one ultraviolet photon (or several light photons) was emitted. This meant the birth of the first atoms in the Universe.

With the arrival on the scene of the hydrogen atom the stellar, or particulate, era began. More exactly it was the era of protons and electrons. All the other particles and antiparticles, as we have seen in Figure 14, had long ago become extinct in the hadron and lepton eras. The protons and electrons could not become extinct, since this would be against the Law of the Conservation of Baryon Charge and of the Law of Conservation of Electrical Charge. All photons survived the radiation era, and towards the end of it they even increased in number through the arrival of many ultraviolet and light photons (in the production of hydrogen atoms from protons and electrons). But expansion cost the photons a great deal of their energy. From the original gamma photons (at the start of the radiation era) there gradually developed light photons, at the end of the radiation era.

The Universe began the stellar era as hydrogen gas mixed with a large number of light and ultraviolet photons. This hydrogen gas did not expand evenly in all parts of the Universe, and its density varied from place to place. It formed huge condensed clouds measuring many millions of light years across. The mass of these enormous clouds of hydrogen varied between hundreds of thousands and millions of times greater than that of our Galaxy today. Inside these clouds the hydrogen expanded more slowly than the sparse hydrogen in between them. The self-gravitation of individual clouds later formed them into supergalaxies and galactic clusters. Thus the largest structural unit of the Universe – the supergalaxy – is a result of the uneven distribution of hydrogen in the primaeval phase of the Universe.

The birth of galaxies

The giant hydrogen clouds – the embryos of the supergalaxies and galactic clusters – turned slowly. Inside them there were eddies similar to those which occur in water or air. They had a diameter of up to about 100,000 light years. They are called *protogalaxies,* meaning the embryos of galaxies. In spite of their immense size, such eddies of protogalaxies were only a tiny part of a supergalaxy, with a diameter smaller than one-thousandth of that of a supergalaxy. It

146

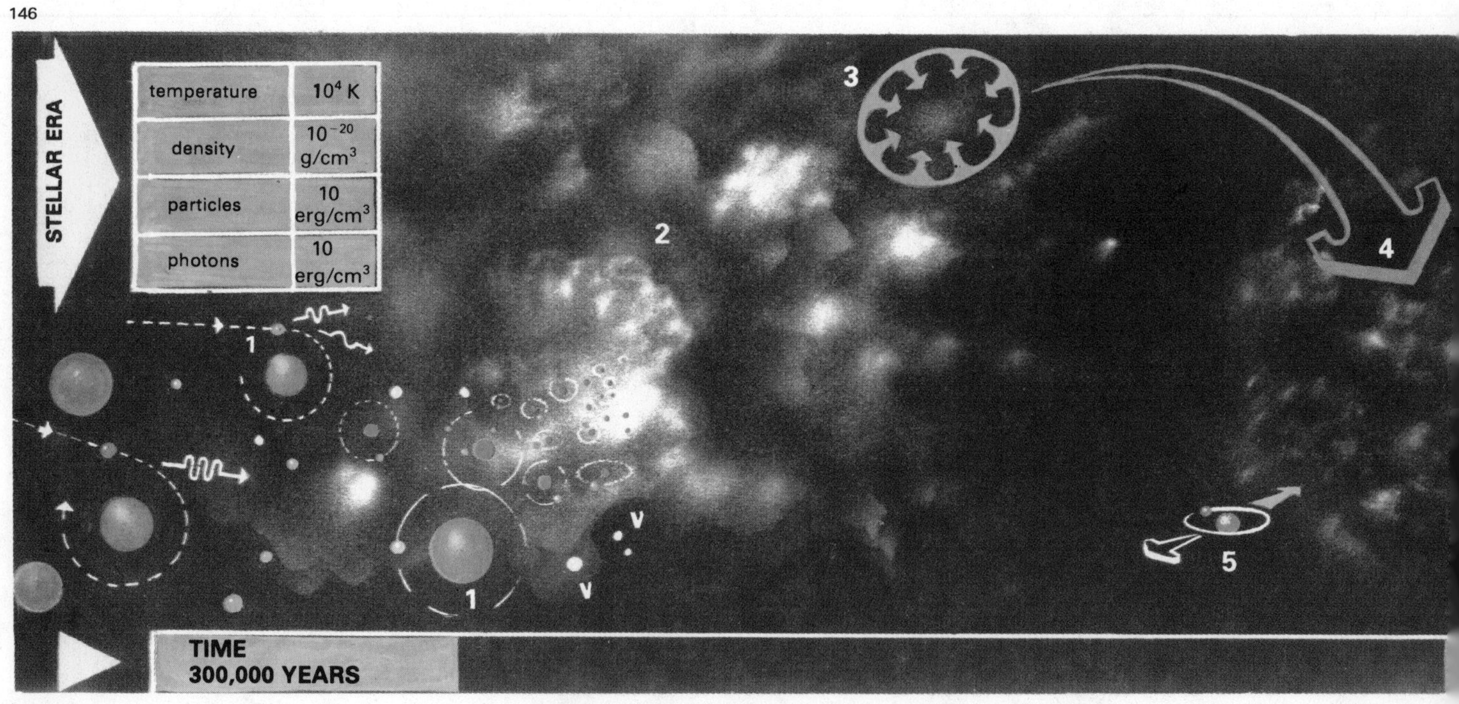

was from these eddies that gravitational force formed the systems of stars we now call galaxies. The shapes of some of today's galaxies are still reminiscent of a huge whirlpool (Figure 129).

Astronomical research shows that the speed of rotation of the eddy determined the shape of the resulting galaxy. Slow eddies gave rise to *elliptical galaxies,* while fast eddies produced *spiral galaxies* (Figures 137 and 147).

An eddy which moved very slowly was formed by gravitation into the shape of a slightly flattened sphere or ellipsoid. The size of such a regular giant hydrogen cloud could have been a few tens of thousands, or several hundred thousand, light years. We can well understand which hydrogen atoms became a part of the nascent elliptical galaxies (more precisely ellipsoidal galaxies) and which remained in the intergalactic space: if the binding gravitational energy of an atom on the perimeter was greater than its kinetic energy, it became a part of the nascent galaxy. This condition is called *Jeans' criterion.* According to it we can determine the dependence of the size and mass of the protogalaxy on the density and temperature of the hydrogen gas.

A protogalaxy which did not turn at all gave rise to a spherical galaxy. Flattened, elliptical galaxies arose from protogalaxies which turned slowly. The centrifugal force was small, and the gravitational force prevailed. The protogalaxy contracted and the density of its hydrogen increased. As soon as this density reached a certain value, concentrates with a mass of 0.06 $M_{\odot}$ to 100 $M_{\odot}$ began to separate and condense (the symbol $M_{\odot}$ stands for the mass of the Sun). Primaeval stars came into being, later to turn into stars (Figures 146–149).

The birth of all stars in spherical or slightly flattened protogalaxies took place almost simultaneously. The process lasted a relatively short time – around 100 million years. This means that in elliptical galaxies all the stars are of approximately the same age, and they are very old. In the elliptical galaxies all the hydrogen was exhausted right at the beginning, let us say in the first one-hundredth part of the life of the galaxy. In the remaining 99 hundredths of the galaxy's life no further stars were able to form, since there was nothing for them to form from. The amount of interstellar matter in elliptical galaxies is therefore tiny, or totally absent.

Spiral galaxies such as our own are composed of a very old halo component (a feature which they share with elliptical galaxies) and a young, flat component (in the spiral arms). Between the old halo component and the young, flat component in the arms there are several transitional components with various degrees of flatness, ages and speeds of rotation. Thus the structure of

146 The stellar era (also known as the matter era) began when the Universe was 300,000 years old and neutral hydrogen came into existence. The remaining free protons joined up with free electrons **(1).** The huge masses of hot hydrogen gas (10^4 kelvin) were saturated with light radiation **(5)**; they expanded and cooled **(2).** The large agglomerations expanded more slowly, and gravitation formed them into supergalaxies **(3,4).** In a nascent supergalaxy the embryos of galactic clusters formed **(6,7)** and in them condensed galaxies **(8,10).** In the galaxies gravitation formed stellar clusters at a tremendous rate, and in them stars **(9).** (Continued in Figure 147.)

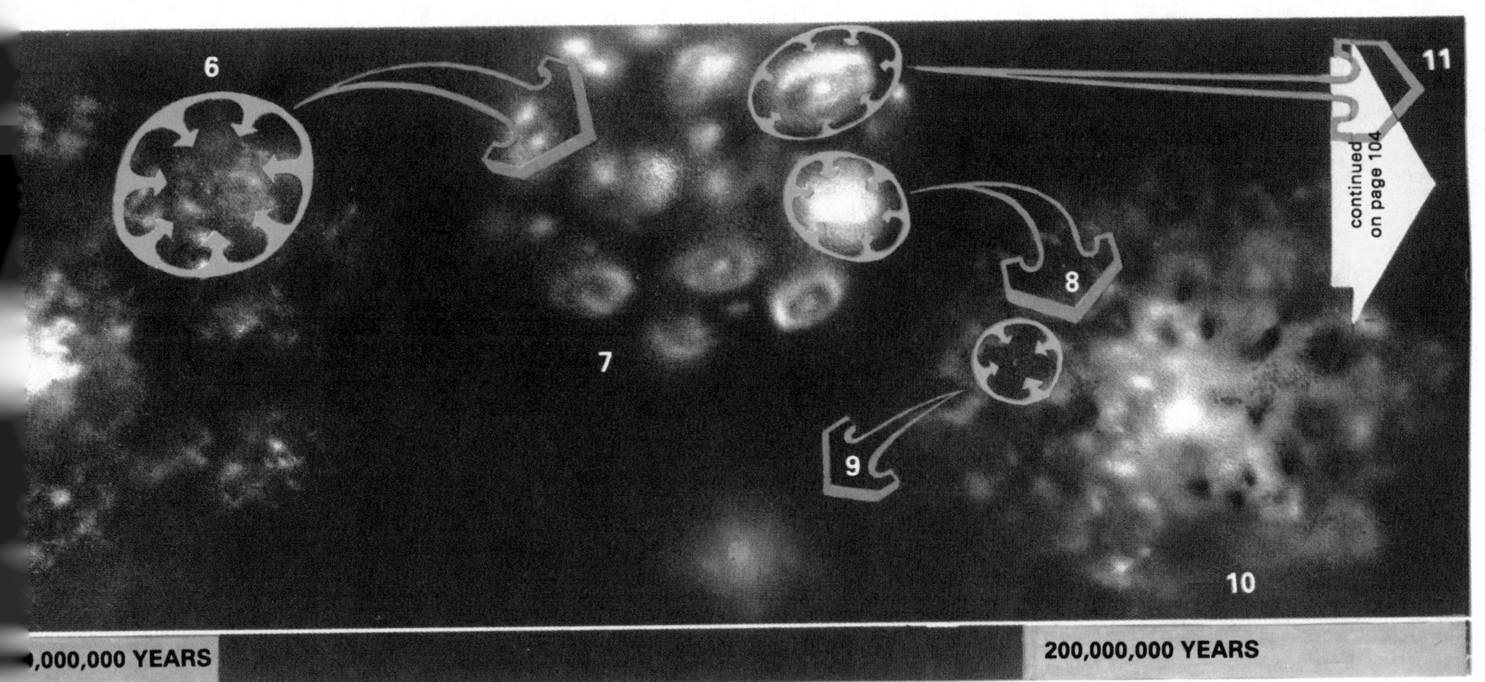

147 Gravitation formed galaxies from a large, rapidly rotating cloud **(11).** From the agglomeration first a halo **(12)** with spherical stellar clusters **(13)** formed. Two forces acted on the remaining gas in the halo – self-gravitation and centrifugal force **(14).** They compressed the gas from the whole galaxy into the thin layer of the disc **(15);** in this there formed, and are still forming, later generations of stars (called Population I; **16**). (Continued in Figure 151.)

spiral galaxies is more complex and more diverse than that of elliptical galaxies. In addition the spiral galaxies have a faster eddying motion than the elliptical ones. They did, after all, originate from the fastest eddies in the supergalaxy. For this reason both gravitation and centrifugal force played a role in the creation of spiral galaxies.

The cloud which gave rise to a spiral galaxy was large, and approximately spherical. It turned slowly at first. In the beginning there was no great difference between the protogalaxies from which the elliptical galaxies were formed, and those from which the spiral galaxies arose. So the stars in the earliest period in both our own and other spiral galaxies formed in a spherical space. They are called the halo component or Population II. They move around the centre of the spiral galaxy in elongated elliptical orbits, and they have the same spectra and the same age as the stars of elliptical galaxies. If all the interstellar hydrogen had escaped from our Galaxy after the first 100 million years (which is how long it took the halo component to form), further stars would have been unable to come into existence there, and it would have remained an elliptical galaxy. In such an event the Sun, its planets and the Milky Way would never have existed.

But the interstellar gas did not escape, so gravitation and rotation were able to continue the process of building our own and other spiral galaxies. As can be seen in Figure 147, the rotation of the interstellar gas which was left over from the formation of the halo component had to accelerate. The more it was attracted by gravity, the faster it had to move (according to the *Law of the Conservation of Angular Momentum*). Skaters are aware of this phenomenon: when a skater draws his or her arms in from the outstretched position to a position alongside the body, he or she starts to spin faster. Faster rotation means greater centrifugal force.

There were two forces acting on all interstellar gas: gravitation, which attracted it towards the centre of the Galaxy; and centrifugal force, which acted away from the axis of rotation. The resulting force pushes the interstellar gases towards the galactic plane (i.e. the plane which passes through the centre of the Galaxy and is perpendicular to the axis of rotation – see Figure 128, A and Q). At present the interstellar gas is compressed into a very thin layer spread out along the galactic plane. It is mainly concentrated in the spiral arms. It forms a part of the flat or disc component – also called Population I (Figures 128 and 131).

In each period of the flattening of the interstellar gas into an ever-thinner disc, stars came into being. We therefore find in the Galaxy and in other spiral galaxies stars from the oldest ones (about 10,000 million years old, such as those in the globular stellar

147

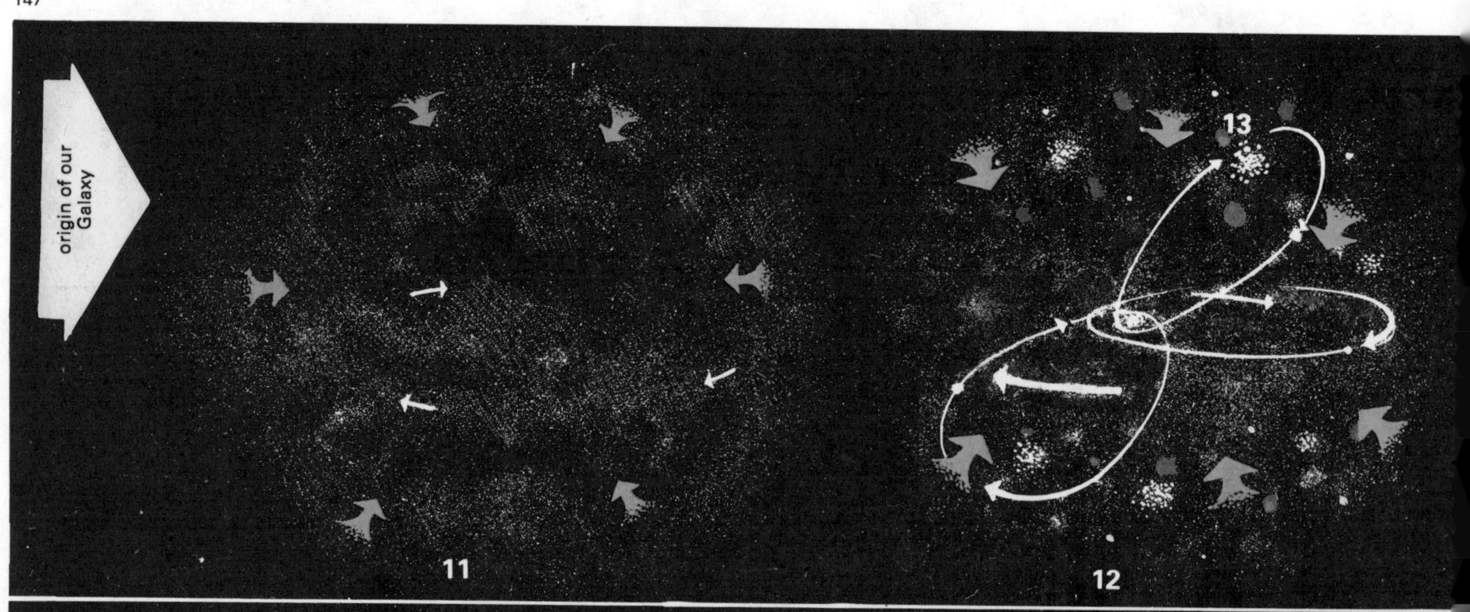

clusters) to stars which began their existence only recently (in the associations and open clusters). The younger the stars, the flatter the system in which they are arranged.

The youngest stars of all are still being formed from interstellar dust and gas distributed along the galactic plane. This gas is partly the remains of the ancient hydrogen of which the protogalaxy was made. But it also contains gas and dust thrown off by previous generations of stars. So, chemically, today's interstellar material is enriched with all the elements we know from Mendeleyev's periodic table. The oldest stars in the halo component of the Galaxy, on the other hand, formed from hydrogen. Thus the youngest stars, in contrast to the oldest ones, contain many heavy elements, such as metals. But this brings us to the subject of the origin and development of stars and the chemical elements in them.

The origin of stars

The fate of stars is in many ways similar to that of humans: they are born, transform energy, influence their environments, and then die. The material they arise from is called *interstellar matter,* spread out in the huge expanses between the stars. It consists of atoms (mainly of hydrogen, but also of heavier elements) and fine particles of dust. If interstellar matter is illuminated by nearby stars, we see clouds of it as *bright nebulae* (Figure 148). On the other hand, large clouds of interstellar matter rich in dust absorb the radiation of more distant stars, in which case we see them as *dark nebulae* (Figure 149). Sometimes very small dark nebulae can be seen against the background of luminous ones, in which case they are called *globules* (Figure 149). These dark clouds of interstellar dust and gas are about as big as *Oort's comet cloud* in our Solar System (Figure 114). They are the embryos of stars and planetary systems.

A globule contains a large amount of potential energy, which can be released and transformed into heat or radiation. This energy is gravitational and nuclear. Gravitation tries to pull the globule into the smallest possible space. By compressing it, it releases gravitational energy. The second form of potential energy in the globule is the nuclear energy of the hydrogen of which it is largely composed. Protons can join up through thermonuclear reactions into heavier nuclei, thus releasing their potential energy (see the valley of stability, Figure 63A).

A contracting globule is usually called a protostar. A protostar draws on its resources of gravitational energy, heating its interior until a thermonuclear reaction gets

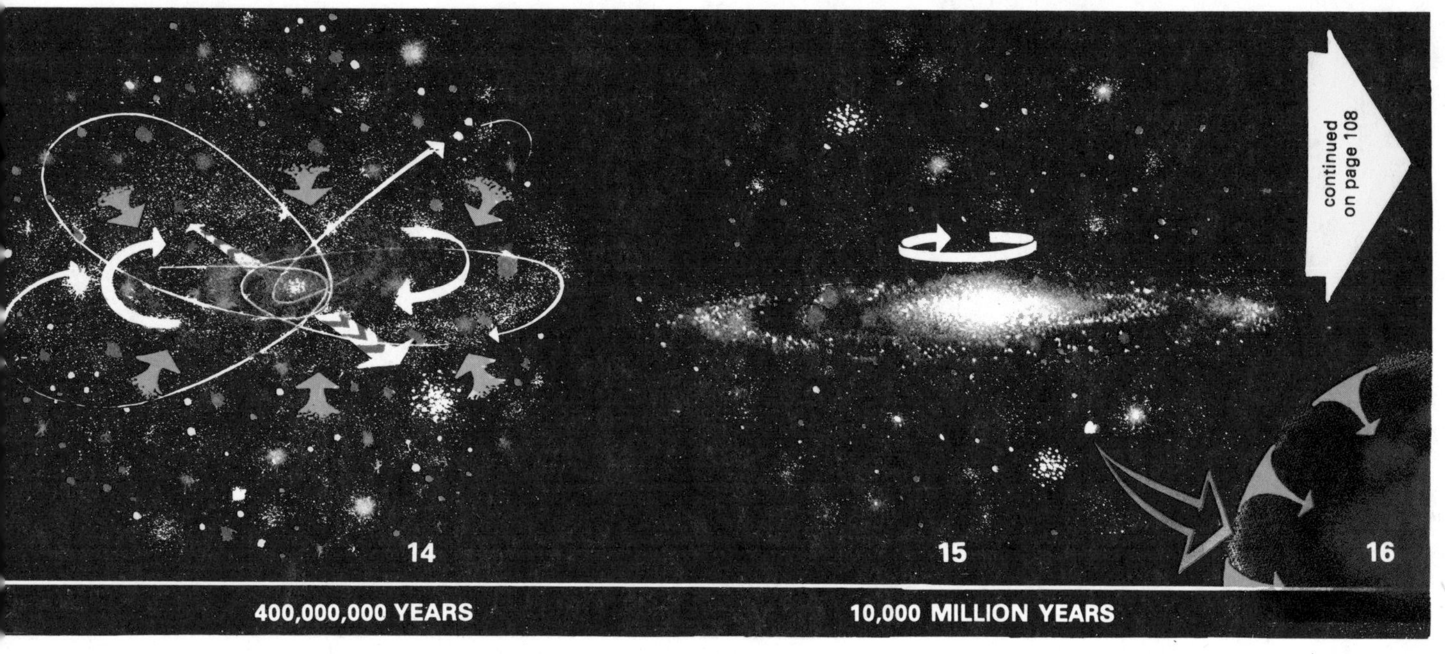

under way. The temperature in its core is in excess of 7 million kelvin, which is the point at which hydrogen starts to turn into helium. The contraction of the star stops at this point and the nuclear energy of the hydrogen starts to be released, taking over from the gravitational energy. Thus the protostar turns into a normal star with thermonuclear reactions.

Before the hydrogen starts to turn into helium, for a short time certain less important thermonuclear reactions take place in the star. Lithium and beryllium change, at temperatures of around 1 million kelvin, into helium, and at about 5 million kelvin the nuclei of boron turn into helium nuclei. These reactions are negligible as a source of energy, but they destroy lithium, beryllium

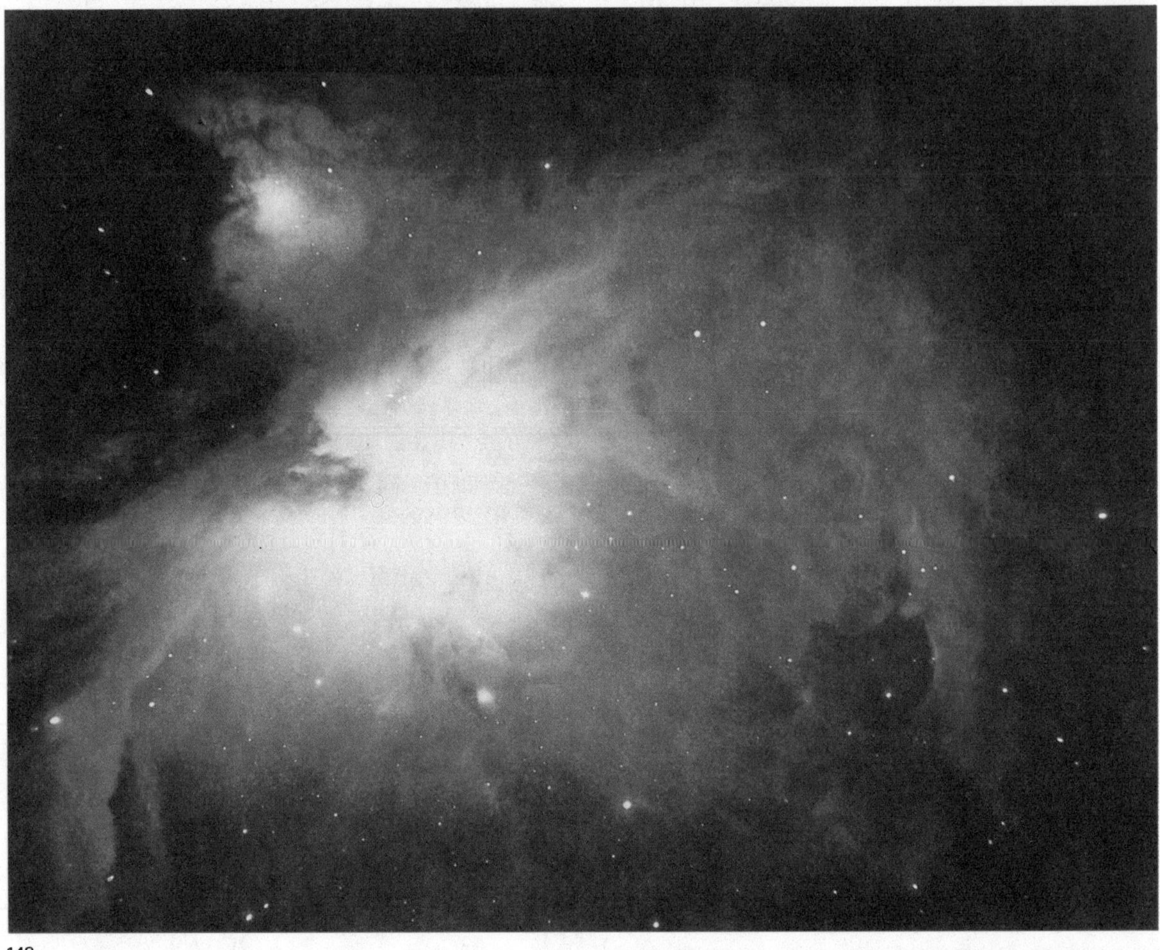

148

148 The Great Nebula in Orion consists of huge clouds of radiant hydrogen. It is about 1500 light years away. It can be seen with the naked eye in Orion's sword. There are many nascent stars in the nebula (which for the time being can be observed only in infra-red radiation).

149 Part of the nebula called Rosetta in the constellation Monoceros. A million years ago a group of very bright stars came into being in Rosetta, and these supply the nebula with heat and radiation. The dark patches in front of the nebula are the embryos of further stars and planetary systems (globules).

149

and boron. This is the reason there is so little of these light elements in the Universe.

Throughout its long life a star draws on its nuclear energy in such a way that its protons pass on down the valley of stability to lower and lower positions (Figure 63A). As they do so their rest energy is released, and their binding energy in the new atomic nucleus increases. The effort to lose its potential energy (which it acquired from the globule) determines the whole evolution of the star, and is the cause of all its transformations. After the nuclear energy has been exhausted, the gravitational energy again starts to be released, bringing the star's life to an end. All the heat and radiation of the star are thus contained in its embryo – the globule.

Globules vary in size (from 2 to 5 light months) and in mass. A star can only come into being from a globule which contains 10^{55} to 10^{59} nucleons, which represents a mass of 0.08 to 100 $M_{\odot}$. A globule with a mass greater than 100 times that of the Sun is not suitable for the formation of a star, since such a high radiation pressure forms in its interior that it is flung back into interstellar space. Too small a globule, on the other hand, has insufficient self-gravitation to get up to the temperatures required for the thermonuclear reactions to take place.

A globule which begins to contract and heat up by its own gravitation is called a protostar, until it reaches the main sequence in the Hertzsprung-Russell diagram (Figure 150). This diagram is useful for studying the evolution of stars. Its horizontal axis gives the surface temperature of the star, and its vertical axis its luminosity (output). Each star in the diagram is shown by a dot corresponding to its surface temperature and luminosity. But these two quantities alter in time, so that the dot of the star on the Hertzsprung-Russell diagram changes its position.

Figure 193 shows the evolution of a protostar from a globule to a normal star. Such a star transforms hydrogen into helium and its temperature and luminosity are represented by a point in the main sequence (Figure 150).

How long does it take a globule to turn into a normal star? That depends on its mass. A protostar with a mass of around 100 $M_{\odot}$ lives about 100,000 years; in other words, it contracts very rapidly. On the other hand,

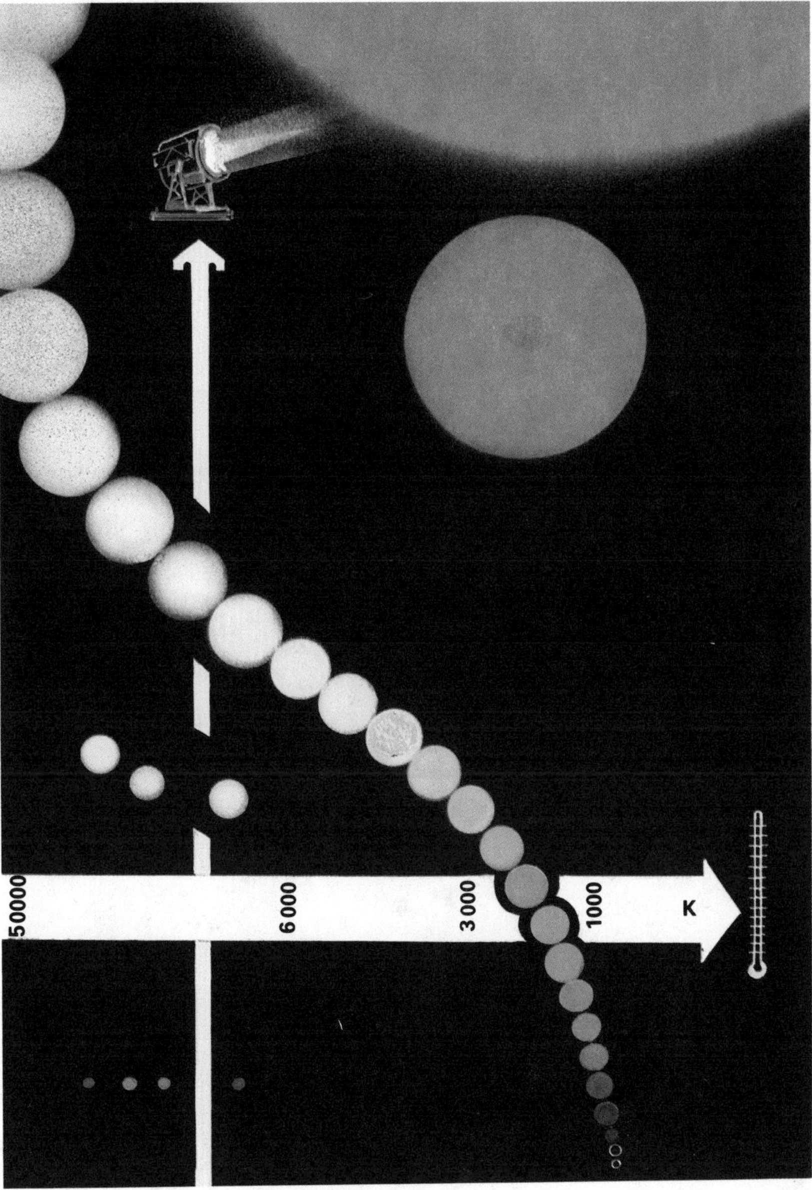

150

150 The Hertzsprung-Russell diagram. On the horizontal axis are the temperatures of stars (more precisely their surface temperatures). On the extreme right are low temperatures, and these increase towards the left. The stars on the right are dark red (cool); moving to the left their colours are in order: red, orange, yellow, white and, on the extreme left, blue. The vertical axis of the diagram shows the luminosity of the stars. Bottom left are neutron stars. At the top are giants and supergiants. Every actual star corresponds to a point in the diagram, according to its surface temperature and luminosity. The vast majority of stars are in what is called the main sequence, represented by the diagonal part of the diagram.

a very light protostar with a mass of about 0.08 $M_\odot$ takes some 100 million years to 'mature' into a star. Its mass is too small for the contraction process to take place any more quickly. During the period that man has been on Earth, only a handful of stars visible to the naked eye have appeared in the sky. Most of the visible stars were there before man came into existence.

151 The evolution of a star. Stars spend most of their lives in the main sequence, during which time hydrogen is converted to helium **(1)** in their core (white sphere **A**). When this process ends, gravitation compresses the star's core into a smaller volume and in doing so heats it up to 100 million kelvin **(B)**. At that temperature helium changes to carbon **(2)**. While the core contracts, the outer layers expand and cool: the star becomes a cool red giant.

The further development of stars is by the alternation of two processes **(C)**: the release of energy by means of nuclear reactions (blue rectangles) and by gravitational contraction (the temperature and density in the core increase). At 3500 million kelvin iron is formed.

The mass of a star also determines its position in the Hertzsprung-Russell diagram at the time it matures into a normal star. Very heavy stars have a very hot surface (about 30,000 kelvin), are blue, and have a luminosity almost one million times that of the Sun ($L_\odot$). They are shown in the upper left-hand corner of Figure 150. But very light stars have a surface cooler than 3000 kelvin; they glow red and radiate only one ten-thousandth of the energy of the Sun. These are shown in the bottom right-hand corner of the figure.

A star contains an enormous amount of hydrogen, from which it draws the energy necessary for its luminosity. This is also the reason why most stars remain in the main sequence of the diagram for a long time. The Sun will live off its hydrogen for a total of 15,000 million years. Less massive stars have a smaller luminosity and therefore live longer. Stars with greater mass than the Sun have a shorter life, since their higher luminosity means they consume their hydrogen much faster.

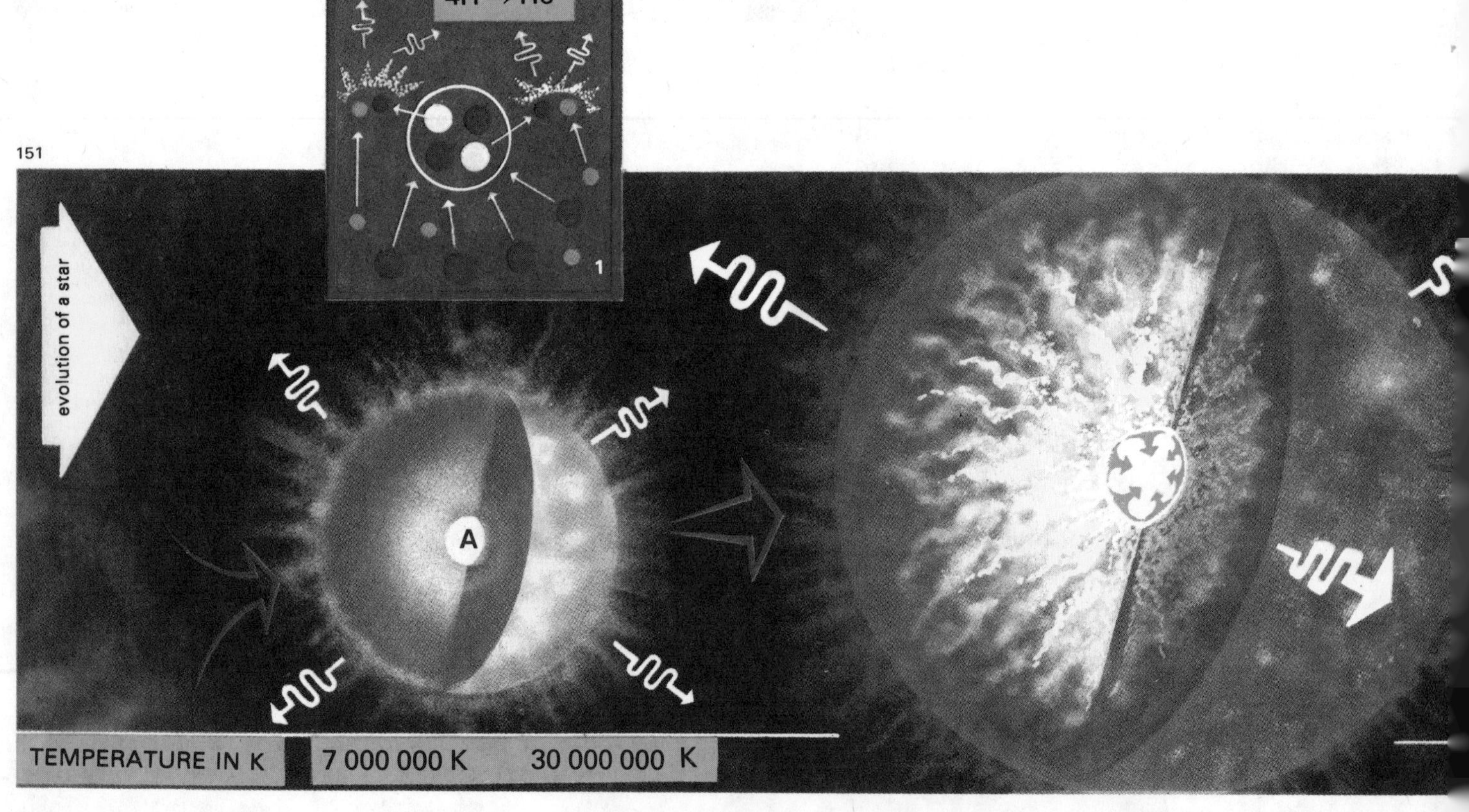

The conversion of hydrogen to helium takes place only in the deep interior of stars, that is in their central regions (the stellar cores, where there is one-eighth of the mass of the stars). Though the hydrogen supplies in the core of the star are huge, they are destined to run out one day, bringing to an end the longest and most peaceful period of the star's life.

The ageing of stars and the origin of chemical elements

In the transformation of hydrogen to helium, alpha particles (the nuclei of helium) are produced in the cores of stars. But after the transformation only three particles out of eight are left (one alpha particle and only two electrons, since two electrons annihilated with two positrons – Figure 163, bottom left). A large number of such reactions takes place in the core of a star – in the Sun, for instance, 10^{38} per second (see page 130). In each of them the number of particles is reduced from eight to three. This results in

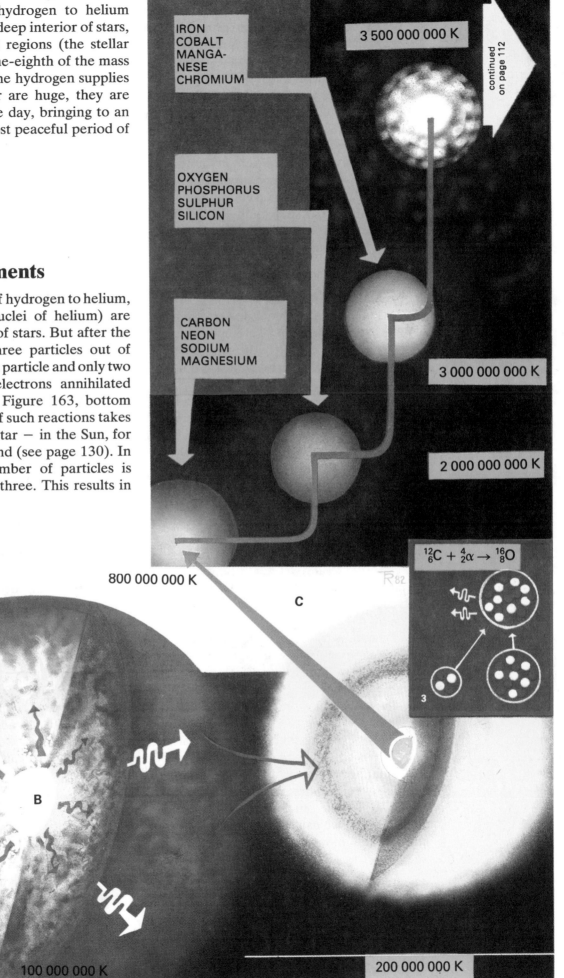

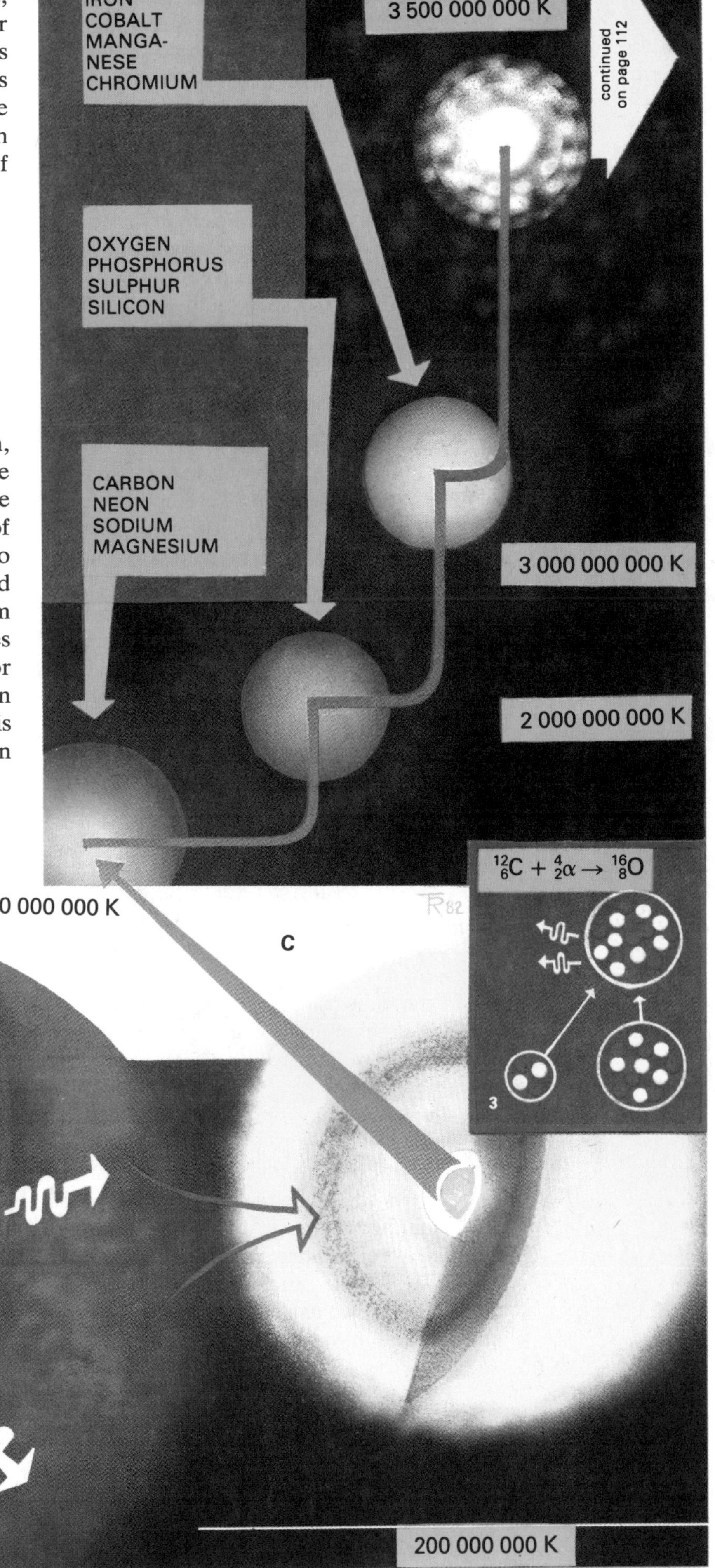

a fall in pressure in the core of the star, since the pressure is proportional to the number of particles. The pressure in the helium core of a star is therefore insufficient to support the gravitation of the upper layers. The outer envelope of the star thus compresses the helium core, which heats up, in the same way that the air in a bicycle pump gets hot. While the heat in the bicycle pump derives from the energy of our muscles, the heat in the core of a star comes from its gravitational energy. It is a part of the gravitational energy which the star was unable to use up in the period when it was still a protostar. After a very long period of transformation of hydrogen into helium, when the nuclear force was at work, gravitation once again becomes the star's energy source.

The hot core also heats up the layers of hydrogen surrounding it. At temperatures of over 7 million kelvin hydrogen starts to change into helium. At this stage the star has two sources of energy: gravitational contraction of the burnt-out helium core and the thermonuclear reactions in the layer around the nucleus.

A star with two energy sources increases its luminosity. While the nucleus contracts with gravitation, the burning of hydrogen progresses to the upper layers. This leads to an expansion of the external layers. The star's volume increases, and it becomes a giant. But in the course of expansion the surface of the star cools (and reddens), and the star becomes a red giant. In the Hertzsprung-Russell diagram the red giants are at the top right.

The heating up of helium in the core of a red giant continues until its temperature reaches 100 million kelvin. At this temperature the alpha particles collide so violently that they overcome each other's electrical repulsion and can approach to within 1 fermi (10^{-15} m) of one another. The huge nuclear force starts to act between the alpha particles, joining them in the more complex atomic nucleus. Three alpha particles then give rise to the nucleus of carbon, ${}^{12}_{6}C$. Symbolically, we can express this transformation of helium into carbon as follows:

$$3\,{}^{4}_{2}He \rightarrow {}^{12}_{6}C$$

As soon as the nucleons become a part of the carbon nucleus, part of their rest energy is again released – the carbon nucleus ${}^{12}_{6}C$ lies on a lower level in the valley of stability than the alpha particles (${}^{4}_{2}He$). The energy released in the transformation of helium to carbon is the source of the luminosity of red giants.

Before we consider the further fate of ageing stars, let us take a look at the reaction of the transformation of helium to carbon. Every atom of carbon which is to be found here on Earth or anywhere in the Universe had its origin in the core of a red giant at temperatures around 100 million kelvin. Carbon atoms form the basis of every living organism on Earth, since they are able to join together in long chains and to form complex organic molecules. The atoms of carbon, of which the whole biosphere is made, came into existence long ago when neither the Sun and its system, nor even the globule they derived from existed. At that time carbon atoms were being created from helium in the giant stars. That was more than 7000 million years ago. Subsequently, the carbon atoms escaped from the stars into interstellar space. There they became mixed with the interstellar matter from which the globule which gave birth to the Solar System later formed. Thus the carbon atoms travelled from the ancient red giants to our own planet, where they were incorporated in plants, and from them in the form of food into our own bodies. Without the red giants which existed 7000 million years ago there would be no carbon on Earth, and so no living organism. Thus from the astronomical point of view the red giants are among our distant ancestors.

But the life of a star does not end when it becomes a red giant, transforming helium to carbon in its interior. From the energy point of view the transformation of helium to carbon is not as efficient as that which transformed hydrogen to helium. The burning of helium therefore occurs over a short period only. The star again begins to draw on its resources of gravitational energy. Its core, in which most of the alpha particles have transformed into carbon nuclei, shrinks and heats up. The remaining alpha particles react with the carbon nuclei to form oxygen. The formation of oxygen nuclei is symbolically expressed as follows:

$${}^{12}_{6}C + {}^{4}_{2}He \rightarrow {}^{16}_{8}O + \text{gamma photon}$$

This exhausts most of the helium, so that the reaction at the next highest temperature, where the neon nucleus is formed, takes place less often than the formation of carbon and oxygen:

$$^{16}_{8}O + ^{4}_{2}He \rightarrow ^{20}_{10}Ne + \text{gamma photon}$$

We have already said that the birth, life and ageing of a star are a result of its persistent attempt to lose its energy. If the temperature in the core of the star is high enough, thermonuclear reactions take place. These continue until the element from the previous period is exhausted. Then the star draws on its supplies of gravitational energy: it contracts and heats its core until further thermonuclear reactions take place. Thus the star hurls into the space surrounding it huge amounts of radiant energy, which it obtains ('squeezes out') from its rest energy. Here the nuclear and gravitational forces alternate. The central temperature of the star rises to ever higher values (Figure 151). The alternating competition of nuclear and gravitational energy gets more and more dramatic. It is as though there was a huge wager as to which of them was to be the last, bringing the star's life to an end. In the case of light and very light stars with a small mass the gravitational force is small, and cannot manage to heat the star's core up to a further stage (Figure 151, upper right-hand corner), where the next thermonuclear reaction would take place. Very light and light stars end up as *infra-red* or *white degenerate dwarfs* (Figure 153).

The gravitation of heavy and very heavy stars, on the other hand, is a match for the nuclear force right to the end, that is until the thermonuclear reactions are exhausted. This happens at temperatures around 3500 million kelvin, when there are large amounts of both iron and elements close to it (in Mendeleyev's table) in the core of the star. This is understandable, since these elements lie right at the bottom of the valley of stability (Figure 63A), so that the nuclear forces cannot release any energy at all from them. A heavy or very heavy star in this condition is left completely at the mercy of gravitation, and *gravitational collapse* occurs (Figure 152).

But let us return to the temperatures of 100 to 200 million kelvin where the atoms of carbon, oxygen and neon are formed. The temperature of the core does not rise to 3500 million kelvin all at once, nor do these three elements suddenly change to iron. The increase in temperature is gradual. When the temperature in the core of the star reaches around 800 million kelvin, carbon starts to burn, and neon, sodium and magnesium appear:

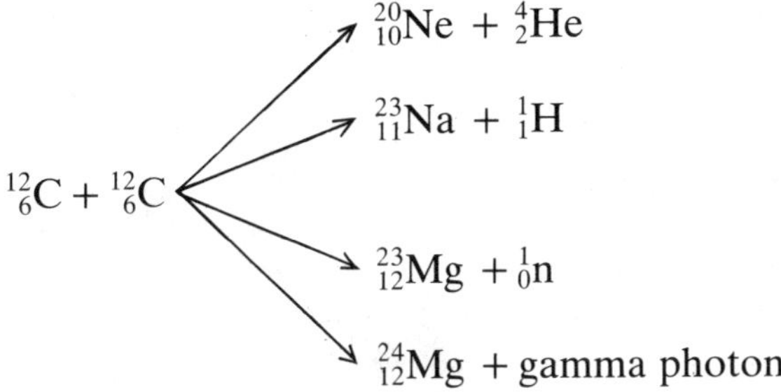

It is only here, at temperatures little short of 1000 million kelvin, that the life of light stars such as our Sun finishes.

Heavy and very heavy stars have sufficient gravitational force to compress the core of the star further and raise it to higher temperatures. Then all manner of nuclear reactions take place, in which further elements are formed. So, for instance, at a temperature of 2000 million kelvin oxygen is transformed to silicon (Si), phosphorus (P) and sulphur (S):

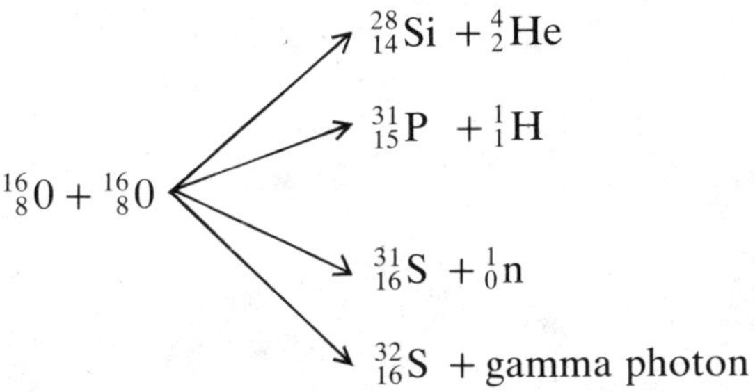

One of the processes which occurs in ageing stars is the burning of carbon and oxygen, in the course of which free neutrons $^{1}_{0}n$ arise. The free neutrons can easily penetrate into the nuclei of heavier elements, since they have no electrical charge. They are then trapped in those nuclei, which become one nucleon heavier. This process, called the *slow capture of neutrons,* gives rise to many elements. But there are not too many neutrons present, so that a nucleus has time enough to disintegrate through beta disintegration before it catches another neutron. In this way nuclei which may have 60 to 210 nucleons arise in ageing stars. There are many atoms (such as those of technecium, mercury, barium, rare elements, etc.) which can be explained only through the slow capture of neutrons in ageing stars.

A process similar to the building up of nuclei by the capture of neutrons takes place very rapidly when heavy stars die – in the explosion of supernovae (Figure 152). But there is a difference between the slow capture of neutrons in an ageing star and that in a star which dies in the explosion of a supernova. In the latter case a temperature of up to 200,000 million kelvin is reached (in the expanding envelope), some atomic nuclei are smashed, and in the enormous flow of neutrons new nuclei appear. In the envelope of a supernova an atomic nucleus acquires a large number of neutrons in quick succession, before beta disintegration can take place. Such nuclei, rich in neutrons, find themselves on the right-hand (neutron) slope of the valley of stability (Figure 63A), from where they cascade rapidly downwards (mainly by emitting electrons, i.e. by beta disintegration) to the bottom of the valley of stability. This is the difference between the fast and the slow capture of neutrons. In the case of the latter the nucleus has plenty of time to emit an electron before it captures a further neutron. Thus in the slow capture of neutrons the resulting nuclei remain low down in the valley of stability.

But by speaking about the capture of neutrons in supernovae we have got ahead of events. Heavy and very heavy stars end their evolution when, at temperatures of around 3500 million kelvin, they form cores composed of atoms of iron. The nuclear energy in such a stellar core is completely exhausted, and further evolution is in the hands of gravitation alone. The gravitational collapse of the core occurs, and the star dies.

152 A supernova – the death of a heavy star. When the hydrogen in the core of a heavy star has changed into iron **(7)** the gravitational collapse of the star's core ensues. The iron core is compressed very quickly by its own gravitation. A huge amount of heat is thus released. The heat escapes in the form of neutrinos **(8).** After a while, the envelope **(9)** collapses on the compressed core – the neutron star. As the precipitating envelope strikes the neutron star, a huge temperature is produced (about

152

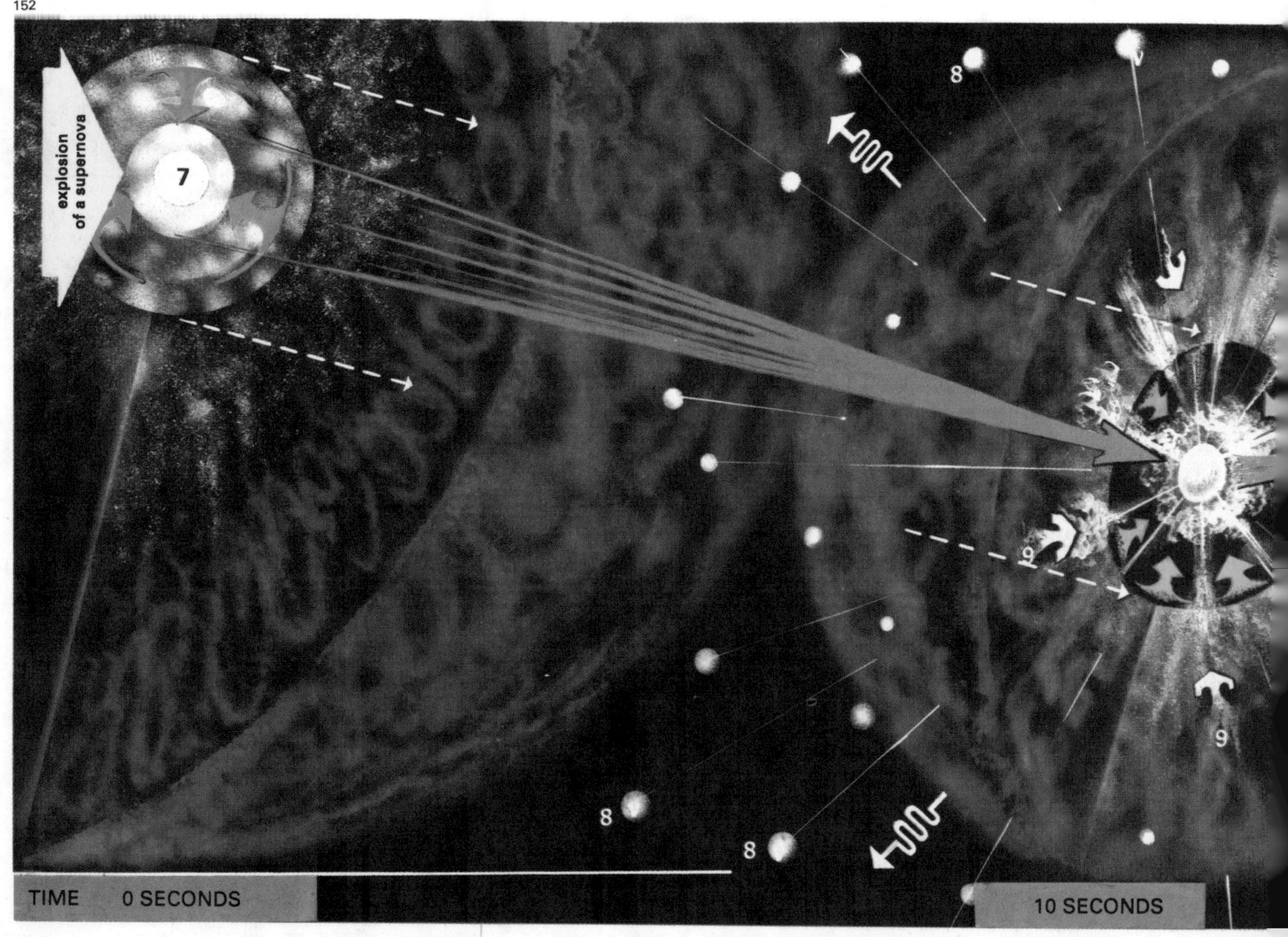

The death of stars

We have seen that the evolution of a star means a gradual rise in the temperatures inside it. It begins as a globule, at a temperature of about 5 kelvin, and can progress as far as an iron stellar core, with a temperature of 3500 million kelvin. The fate of a star is predetermined even before gravitation begins to shrink the globule into a protostar. According to the mass of the globule we can depict the evolution and demise of stars, as in Figure 153.

• *Infra-red and black dwarfs.* If the mass of a star is less than 0.08 $M_{\odot}$, the temperature in its core does not reach the point at which hydrogen burns. A star with a mass of 0.06 $M_{\odot}$, for instance, is heated by gravity to only 2.5 million kelvin, which is not enough to allow hydrogen to turn into helium. Stars such as this live off gravitation alone. Their radiation is mainly infra-red. When the contraction of the star by gravitation ends (the whole star is degenerate), it is left without an energy source. It gradually cools down to become a black dwarf.

• *White dwarfs and planetary nebulae.* If a globule has a mass of 0.08 $M_{\odot}$ to 4 $M_{\odot}$, it becomes a light star. The Sun is one of these. At the end of its evolution, the temperature in the core of a white dwarf may reach several hundred million kelvin. This means that all the thermonuclear reactions we have mentioned cannot take place in it. The heavier stars in this group (from 1.4 to 4 $M_{\odot}$) lose most of their plasma during their lives, especially towards the end of them, by hurling it out into interstellar space. The last of the plasma to be shed forms a planetary nebula (Figure 154).

A *red giant* is made up of a very dense degenerate core the size of the Earth, together with a large plasma envelope of low density. The plasma envelope expands and escapes from the degenerate core. It can be

200,000 million kelvin). Less stable nuclei of atoms (formed previously) are split and many protons and neutrons released. The free neutrons and protons easily penetrate the heavy nuclei. Thus all the elements heavier than iron (the right part of the valley of stability – Figure 63A) are formed in the expanding envelope **(12).** The envelope expands at the rate of many thousands of kilometres per second, taking with it into interstellar space all the nuclei **(10, 11).** (Continued in Figure 194.)

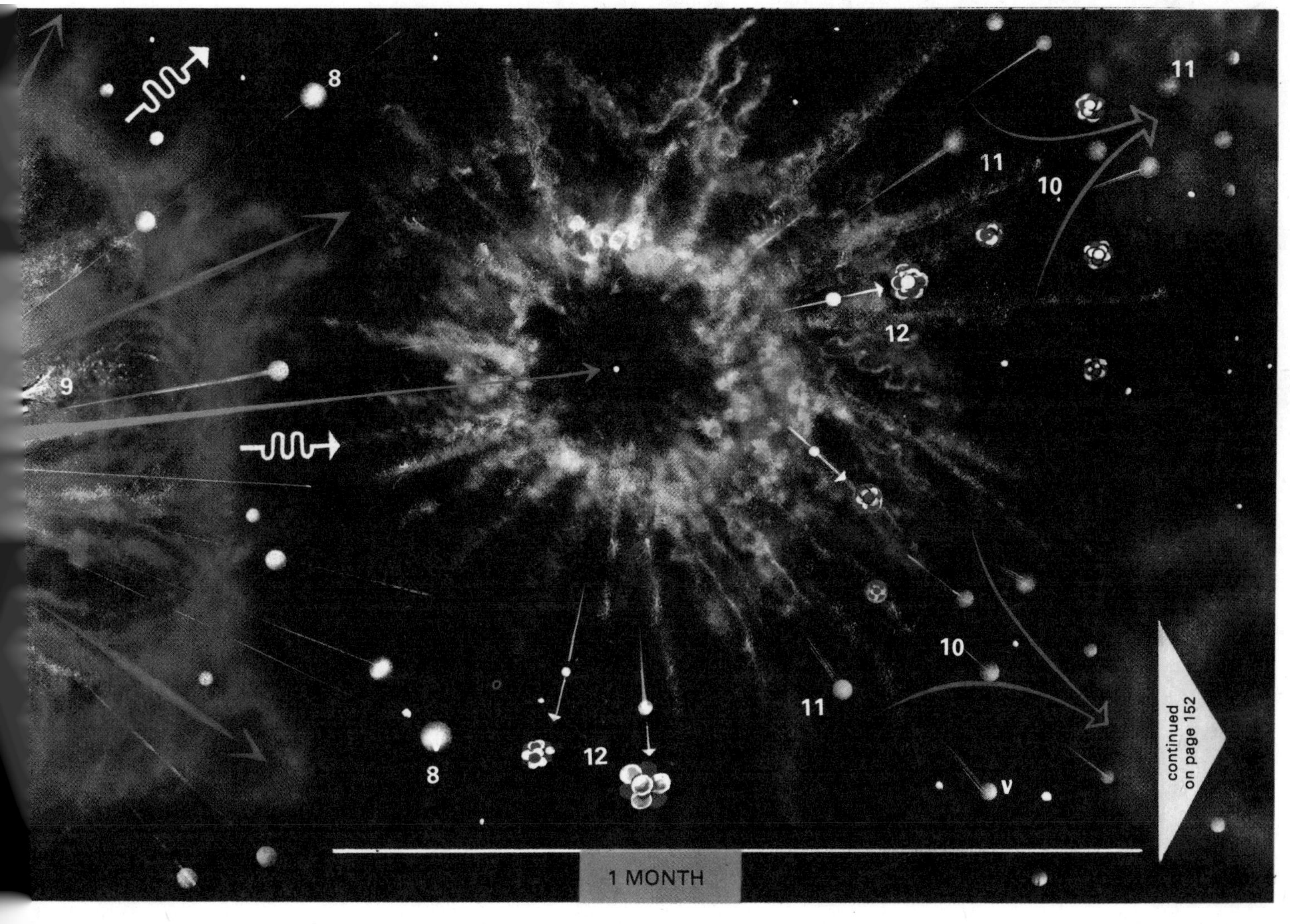

continued on page 152

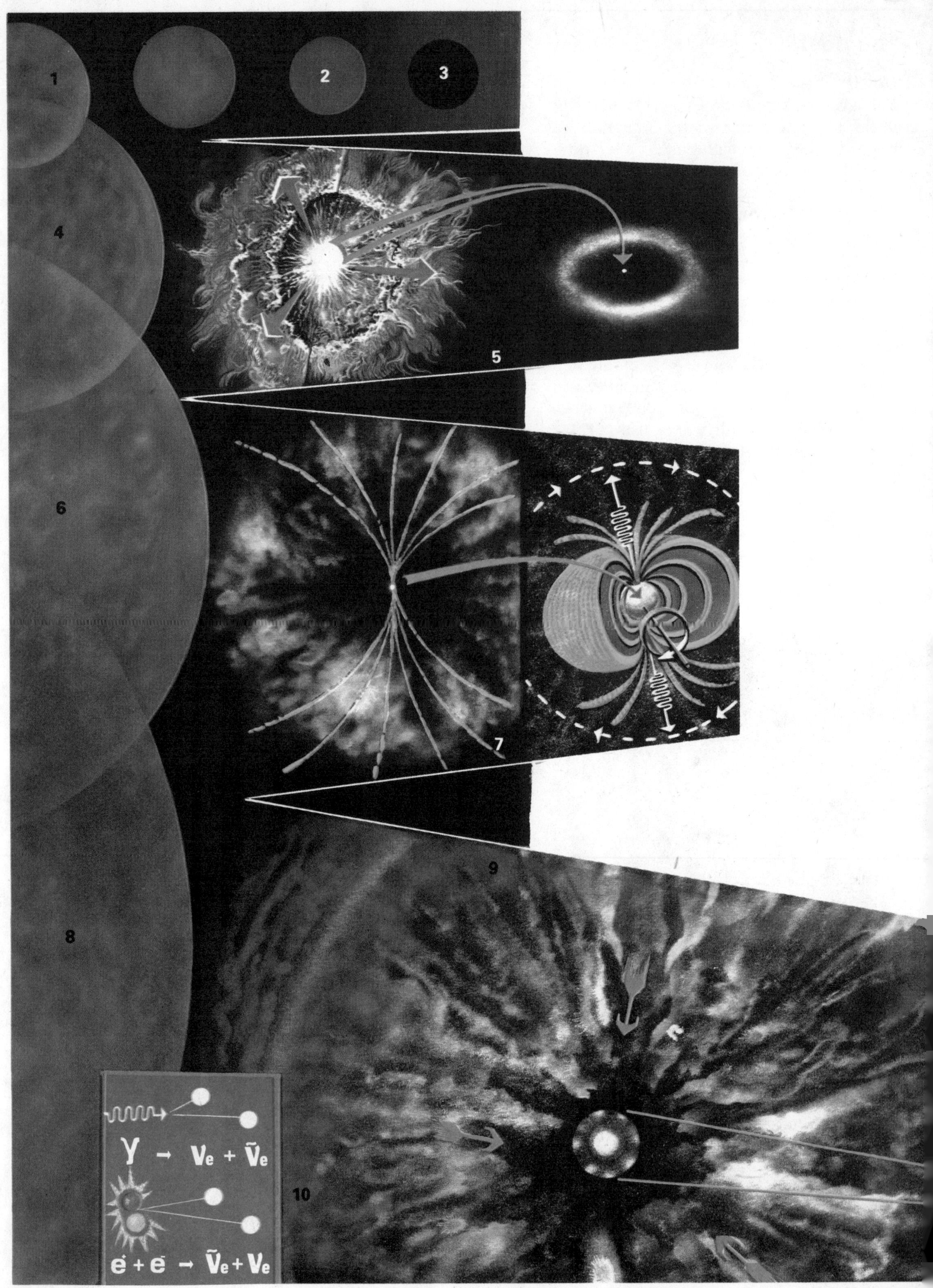
1
2
3
4
5
6
7
8
9
10
$\gamma \rightarrow \nu_e + \tilde{\nu}_e$
$e^+ + e^- \rightarrow \tilde{\nu}_e + \nu_e$

seen as a planetary nebula, in the centre of which a small, very dense, degenerate star remains – a *white dwarf* (Figures 153 and 154). In the course of several tens of thousands of years the planetary nebula disperses in interstellar space. The white dwarf has no source of energy of its own – neither nuclear nor gravitational. It radiates the heat which remains from the past (when it was a red giant). As it emits its heat, it slowly cools down, becoming a yellow dwarf, then a red dwarf and finally, after thousands of millions of years, a black dwarf. That is how the Sun, too, will end its days.

• *Neutron stars and supernovae.* A globule with a mass of from 4 $M_\odot$ to 8 $M_\odot$ develops into a massive star, the temperatures in whose core reach over 3000 million kelvin. After exhausting its nuclear reactions the star contracts violently under the force of its own gravitation into a ball some 30 km (18 miles) across. The density of this collapsed core is thus enormous ($10^{15}-10^{17}$ kg/m^3 or $10^{12}-10^{14}$ g/cm^3). At such an immense density matter can only exist as neutrons, all protons in the nuclei having fused with electrons and changed into neutrons (Figure 64). In the course of its gravitational collapse the core of the star pulled with it the magnetic lines of force. The number of these remained the same, and as they were compressed into the very small surface of the neutron star, the intensity of the magnetic field grew tremendously.

153 The death of stars. Stars end their life differently according to the mass they had. *The lightest red dwarfs* **(1)** have insufficient self-gravitation to heat up their cores to the temperature required for the thermonuclear reaction. They exist using gravitational energy only, slowly cooling down until they change into an infra-red dwarf **(2)** and then a cool black sphere **(3).** *Light stars* **(4**, such as the Sun) change into red giants with a degenerate core after exhausting their hydrogen. Their envelope recedes (as a planetary nebula – **5**), and the exposed degenerate core becomes a white dwarf. *Heavy stars* **(6)** end their lives quickly and dramatically. After their hydrogen has burned out, the temperature in their cores rises to as much as 3500 million kelvin, heavier elements up to iron being formed in the process. The core of the star collapses under its own gravitation ('gravitational collapse'), and a neutron star (or pulsar) arises **(7),** while the outer layers fly off, with the release of a huge amount of energy (supernova). *Very heavy stars* **(8)** have a similar evolution to heavy ones **(6),** but their gravitational collapse leads to black holes **(11).** The immense heat is transformed into neutrinos (rectangle **10** bottom left) which escape easily and thus cool the collapsing core.

154

154 The ring nebula in Lyra. In its centre is a white dwarf – the exposed nucleus of what was once a red giant. The expanding nebula is the outer envelope of the red giant. The Sun will meet a similar fate.

153

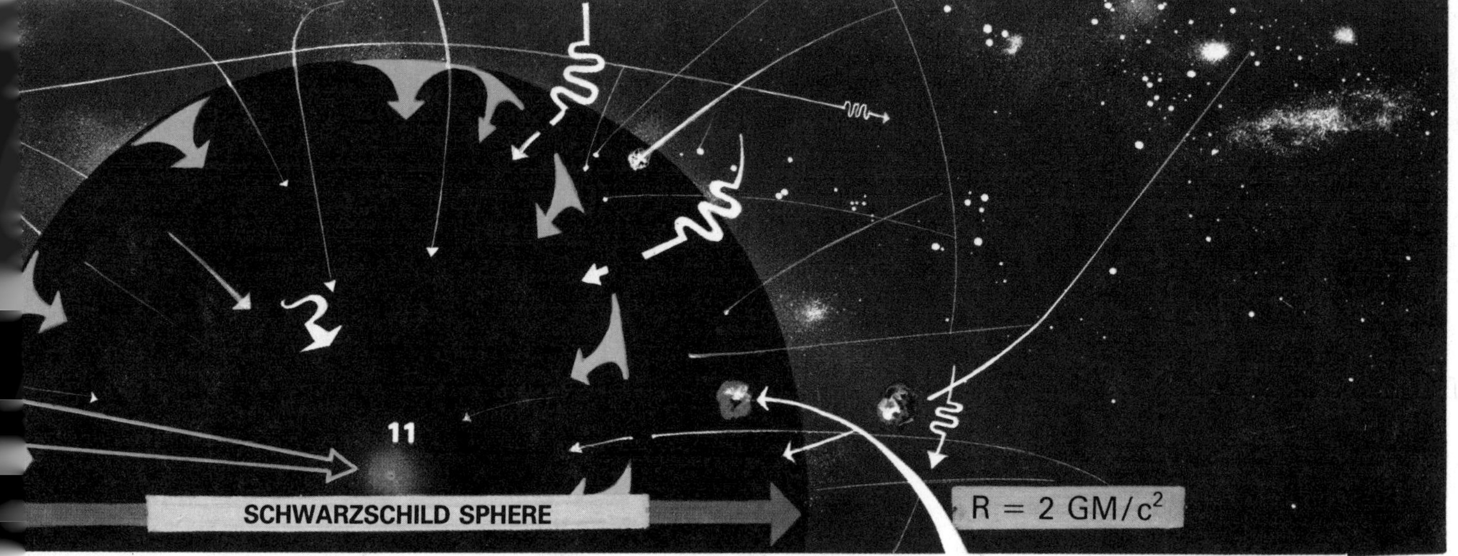

155

In collapsing, the neutron star began to spin very rapidly – here we may recall the Law of the Conservation of Angular Momentum, i.e. the example of the skater with arms outstretched. The magnetic field of the neutron star carries many electrons with it; these radiate whenever they move towards us. The radiation of a neutron star (mainly radio waves) is reminiscent of the flashing light of a rotating beacon. It thus pulses, which is why these stars are also known as *pulsars.* According to detailed research carried out by Australian astronomers, there are more than a million pulsars in the Galaxy.

So far we have considered only the fates of the core of the star, which collapses into a neutron star (pulsar). The envelope layers are left without support, thousands of kilometres above it. But this situation does not last long. The strong gravitation of the neutron star sucks the envelope layers like an enormous and extremely fast-moving waterfall, down to the surface of the neutron star which formed less than a minute before in the centre (Figure 153 number 7). When the fast-moving plasma of the envelope, which is rich in hydrogen, strikes the rigid neutron star, it heats up a great deal, so that various thermonuclear reactions take place, instantaneously. It is really a huge 'hydrogen bomb', which flings the plasma into the surrounding interstellar space. This explosion, called a supernova, is so huge that the remains of the expanding envelope can be observed around the neutron star for millenia (Figure 156). The best-known example

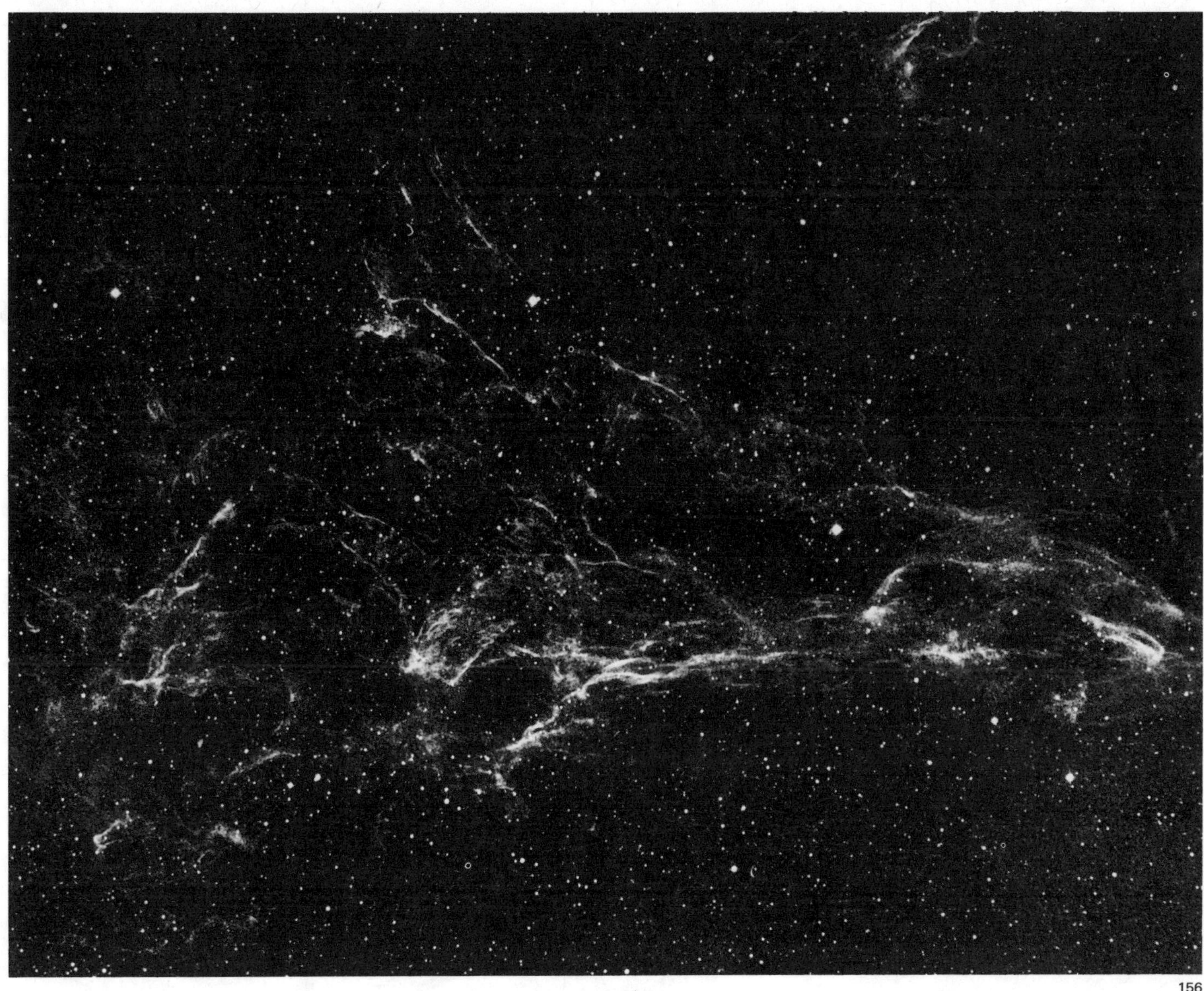

156

155 The Crab nebula – the remnants of a supernova in the year 1054. The pulsar (neutron star) in its centre rotates 30 times a second. Its huge kinetic energy is converted into the luminosity of the nebula. It is still expanding rapidly, with a present diameter of more than 10 light years. It is 6000 light years away.

156 The remnants of a supernova which exploded tens of thousands of years ago in the constellation Cygnus. It is called the Network nebula. It is expanding at the rate of several hundred kilometres per second. Its collisions with interstellar gas cause it to shine. It contains many heavy atoms formed during the supernova explosion.

is the supernova in Taurus, which exploded in 1054. Its neutron star pulses not only in radio waves, but also in infra-red, visible light, and gamma radiation. The expanding plasma is what we see as the Crab nebula (Figure 155).

Black holes

We have seen that the gravitational forces compress dying stars in proportion to their mass. We might therefore expect the most massive group of stars ($8\,M_\odot - 100\,M_\odot$) to be affected most by gravitation. This is, in fact, the case – very light infra-red dwarfs have a diameter around a 100,000 kilometres, white dwarfs around 10,000 kilometres, neutron stars about 20 to 30 kilometres, and black holes, the remains of the heaviest stars, only a few kilometres. To be more precise, this diameter of a few kilometres is not that of the small but immensely dense stellar corpse, whose mass is over $8\ M_\odot$, but of the *Schwarzschild sphere* in whose centre it lies. The space and time around this remnant are so distorted and enclosed by the immense gravitation that not a single particle or photon can escape. This space and time are quite different from the rest of the Universe. The Schwarzschild sphere represents their surface, and thus separates the mysterious relic from the Universe outside.

The core of a massive red giant or supergiant reaches a temperature of around 3500 million kelvin towards the end of its life. It contains iron and other nuclei from the bottom of the valley of stability. This iron inferno at the heart of the star brings to an end its thermonuclear evolution. Iron and the neighbouring elements at the bottom of the valley of stability (Figure 63A) are unable to provide any energy at all for thermonuclear reactions. The star is entirely in the hands of its own gravitational force, which acts most strongly in its core.

In the hot core of the massive star all sorts of processes are in progress at the end of its thermonuclear evolution. Those which are important for further events are the ones which release huge flows of neutrinos and antineutrinos. For instance, the equivalent of materialization:

$$\gamma \rightarrow \nu_e + \tilde{\nu}_e$$

or annihilation:

$$e^+ + e^- \rightarrow \nu_e + \tilde{\nu}_e$$

There are huge numbers of energy-rich gamma photons in the stellar core at temperatures of 3500 million kelvin. These may give their energy direct to neutrinos and antineutrinos, or may first materialize into an electron/positron pair and then transfer the energy to a neutrino and an antineutrino. There are other ways in which energy can become 'embodied' in neutrinos at high temperatures. But the most important is the ability of neutrinos and antineutrinos to pass easily through even the thick layers of the stellar envelope over the core. The energy in the core takes the form of neutrinos and thus easily escapes into surrounding interstellar space. This would take photons thousands of years to achieve.

The core cools rapidly, since its heat is carried away by the huge efflux of neutrinos and antineutrinos. On cooling, the pressure in the core decreases, and there is no longer anything to prevent the gravitational forces from compressing the core of the star into a Schwarzschild sphere. The envelope layers of the supergiant soon follow the core into this sphere, and nothing is left for our optical instruments to observe.

In the centre of the Schwarzschild sphere are the immensely dense remains of the supergiant, and no one knows what happened to the elementary particles of which it was once made. We know only that the gravitational force remains, acting over a distance of several light years through surrounding space. Like some huge, invisible spider, the giant's corpse draws into the Schwarzschild sphere interstellar gas, dust, comets, everything – just as if they were all falling into a dark hole – a hole that gets bigger as more objects fall into it, and is never filled in. The radius R of the Schwarzschild sphere is greater, the greater the mass M inside it, since

$$R = 2G\,M/c^2$$

where c is the speed of light and G the gravitational constant. The Sun would form a Schwarzschild sphere with a radius of around 3 km (1.8 miles). That means that if it were compressed into a ball with a radius of 3 km (1.8 miles) or smaller, we should not see it – there would be only a black sphere in its stead. Since neither a single photon nor a particle can escape from it (so strong is its gravitational force), we could never observe its interior.

While photons cannot escape from a Schwarzschild sphere, they are easily drawn into one. This means that rays falling on the sphere are entirely absorbed. But this is a property of all black bodies. Because of these two properties the remains of very massive stars are called *black holes.*

It is a strange destiny – the greatest giants of the Universe, visible from distant galaxies, eventually leave nothing at all behind: not a trace to enable us to observe the most dramatic processes in the whole Universe. The stellar giants are quite simply squeezed out of the visible Universe.

The birth and death of the Universe

In Chapter 2 we spoke of the expansion of the Universe, due to cease in 30,000 million years. After that the gravitational contraction of the whole Universe will take place, ending after a further 40,000 million years with the Big Collapse. Let us recall that the expansion of the Universe started about 10,000 million years ago with the Big Bang. The whole immense pulse of the Universe from the Big Bang to the Big Collapse will thus have lasted about 80,000 million years. This is indicated by certain observations on light and radio wavelengths. It must, however, be said that our observations are still imperfect, and the measurement of immense expanses of time and space imprecise. There are some important data on the Universe, such as its average density, which we do not yet know. Thus certain hypotheses, such as those contained in this chapter, must be regarded as possibilities only. It is probable that in the near future we shall obtain new data on the Universe, showing some present theories and hypotheses to be incorrect, while confirming others.

At the start of this chapter we deduced what went on in the first few moments of the Universe. In doing so we carefully avoided the question of where the tremendously hot matter of the nascent Universe came from and what caused its expansion. We spoke of the hadron era, lasting one ten-thousandth of a second, but we did not say what came before it. Similarly, we mentioned the Big Collapse, but said nothing of what was to follow. These are of course very important questions. The answer to them can be formulated as *the cosmology of matter and antimatter,* or *the cosmology of the pulsating Universe.*

There are in the Universe – or so the hypothesis goes – equal amounts of matter and antimatter. This means that every proton has a corresponding antiproton, every electron a corresponding positron, and so on. The separation of matter and antimatter into different parts of the Universe took place right at the beginning. Figure 157 shows what will happen in the Big Collapse. Matter and antimatter will be compressed by gravitational force into the state of very hot, dense plasma. On the contact surfaces between the regions of matter and antimatter intensive annihilation will take place. The resulting gamma photons will grow in energy and number to such an extent that they bring the collapse to a halt, flinging matter and antimatter apart before they have the chance to annihilate completely.

The end of the Big Collapse is also the beginning of the Big Bang. It is the zero point on the graph of distances in Figure 157. If this idea is correct, we should be looking for a region of antimatter in the Universe by, for instance, seeking a source of large streams of antineutrinos. While ordinary stars emit neutrinos from their cores (Figure 163 number 9), antistars will emit antineutrinos. But we do not as yet have equipment for this, so for the moment we do not know whether the Universe really is divided into a part made of koinomatter (i.e. ordinary matter) and one made of antimatter.

If the supposition is correct, then the energy for the expansion of the Universe would come from the annihilation of matter and antimatter, i.e. the most efficient process of the release of energy (Figures 55 and 56). Even the rotation of supergalaxies, galactic clusters and galaxies themselves would contain energy released long ago (10,000 million years ago) by annihilation on a huge scale. The orbiting of the planets around the Sun, the rotation of the Sun and other stars, and the movement of the Moon around the Earth all have kinetic energy which is very old indeed, conserved from the start of our Universe.

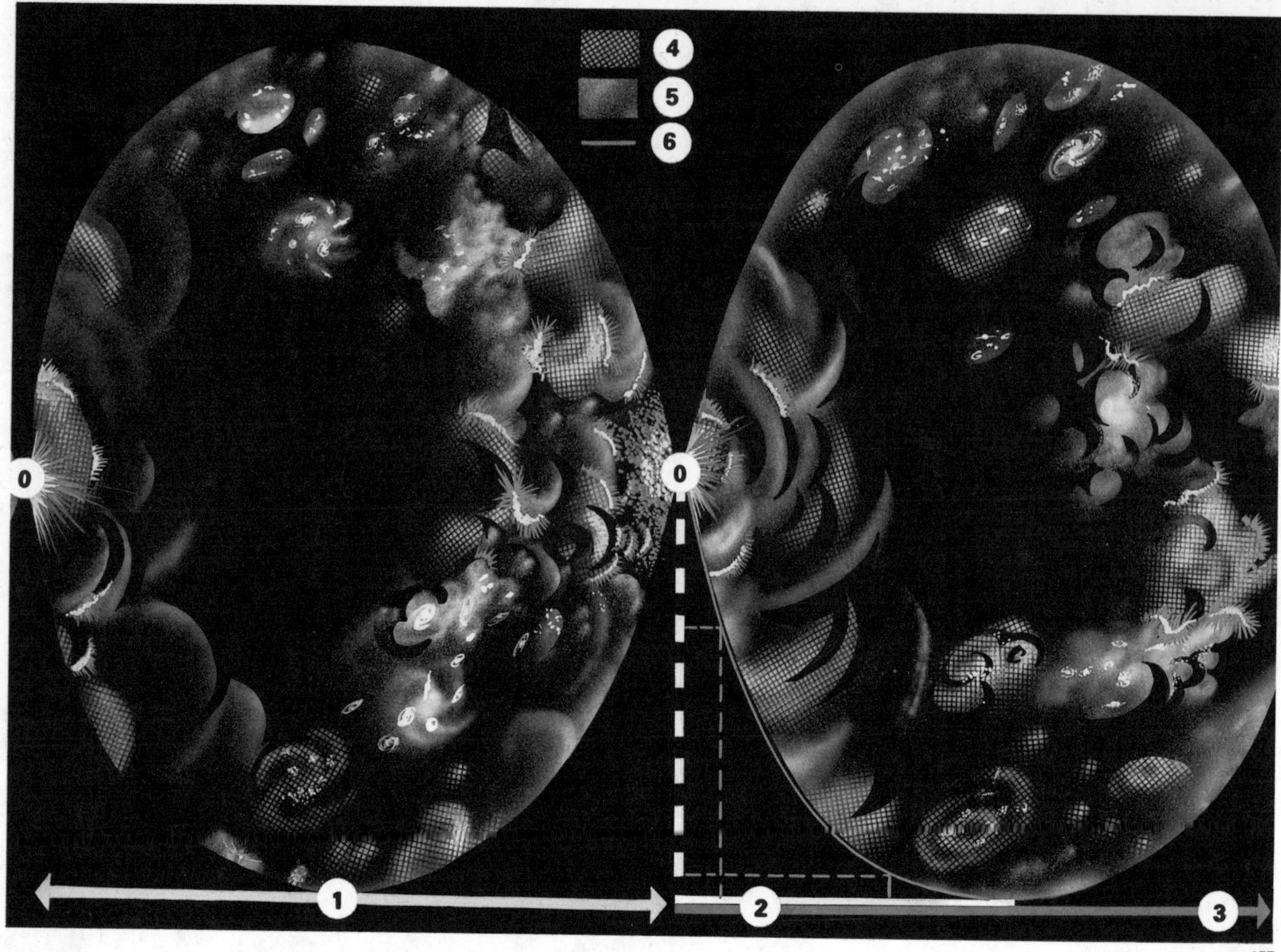

157 At present (vertical red dashed line) the Universe is expanding **(2).** The expansion began with the Big Bang ('zero' in the middle), about 10,000 million years ago (white dashed line). In the left half of the picture is the Universe before the Big Bang. The gravitational collapse before the Big Bang (just before the 'zero') is called the Big Collapse. It brought the Universe to an immense temperature and density; part of the matter **(5)** annihilated with antimatter **(4)** at their contact surfaces **(6).** In annihilation huge pressures built up, not only stopping the Big Collapse, but turning the movement of the hurtling masses in the opposite direction: the Big Bang occurred **(0).** A similar Big Collapse will come at the end of our Universe, in about 70,000 million years, giving rise to another Big Bang **(3)** and another Universe.

When we speak of our Universe, we mean the pulse of the Universe in which we are living (right half of Figure 157). If the hypothesis of the cosmology of matter and antimatter is correct, this means that following the Big Collapse of the last Universe, the Big Bang of our Universe took place (the zero in the middle of the figure), and the Big Collapse of our Universe 70,000 million years from now will herald the Big Bang of the next. The driving energy of these huge explosions at the start of the pulse (the Big Bangs) is released by the annihilation of matter and antimatter. The driving energy of the Big Collapse is gravitational energy.

Man has been forced to accept the unpleasant certainty that he will never know anything about the previous pulse or pulses. All systems built by the evolution of the Universe in the previous pulse are completely destroyed in the immense fire of the Big Collapse – broken down into elementary particles. Not a single atomic nucleus survives into the next pulse, and all the photons from the last pulse are transformed by annihilation and materialization. There is nothing which can exist to give the tiniest clue to what went on in the previous Universe.

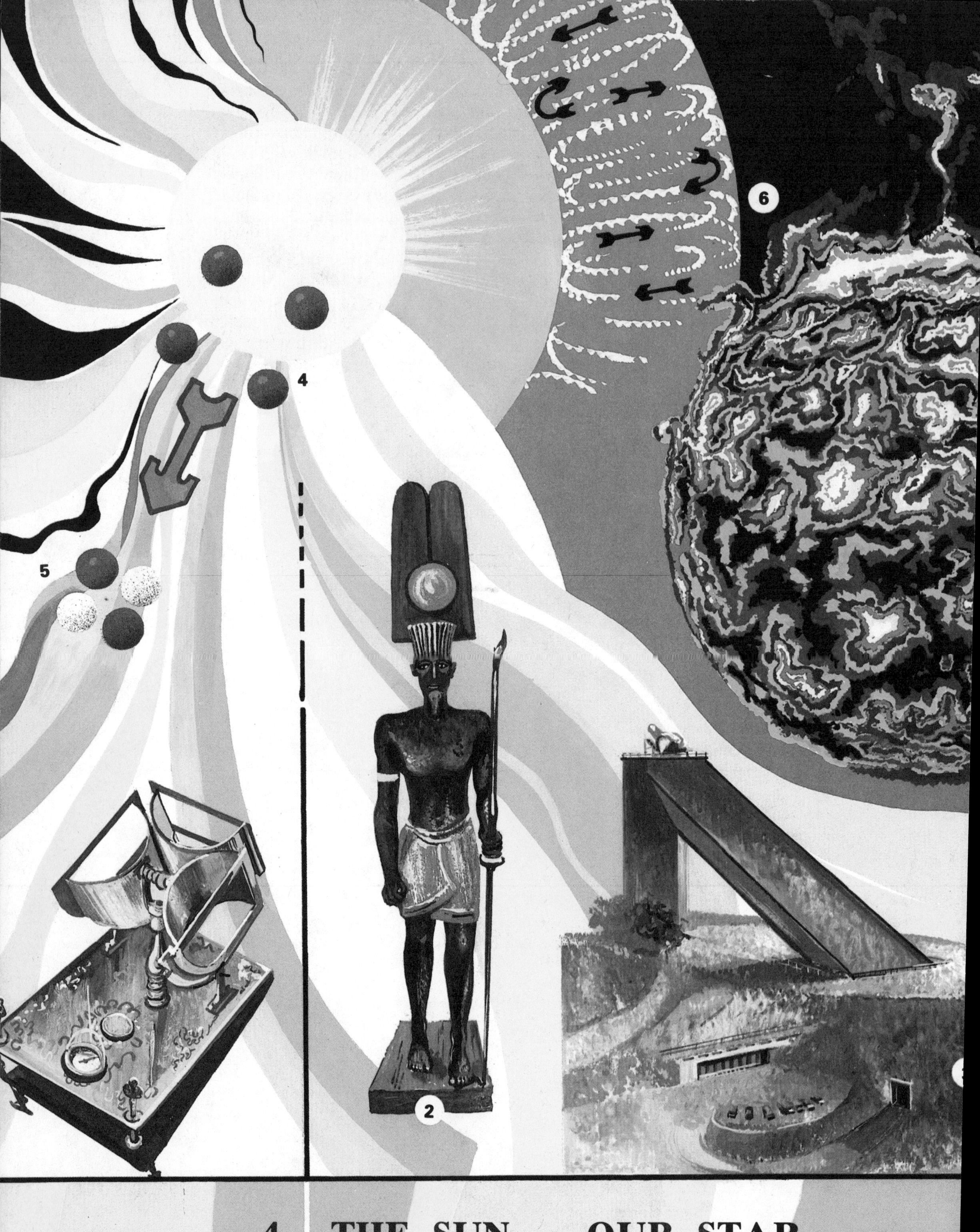

4. THE SUN – OUR STAR

What is the Sun?

Let us for a moment suppose that we live in ancient Egypt, in the valley of the Nile. Our sovereign is the pharaoh Akhnaton (whose name means 'sheen of the sun disc'), who composed a splendid hymn in honour of the Sun god; this is an excerpt from it:

'Thou shineth beautifully on the sky's horizon, O living Sun, who lived at the start of all things. When thou riseth on the eastern horizon, thou filleth each land with beauty. Thou art great, beautiful, shining, soaring high above all lands. Thy rays embrace the world to the very end of all thou hast created. Thy rays nourish all the fields; so long as thou shalt shine, they shall live and grow for thee... Thou createth the year's weather, in order to endow with life all thou hast created...'

The Indians, Mesopotamians, Greeks, Incas, Aztecs and many other nations also considered the Sun a god; they built temples to it, carved statues, sang hymns, worshipped it with prayer and dance and offered various sacrifices to it – sometimes human hearts.

Instead of Sun temples we now build solar observatories and make telescopes and other instruments to learn about the Sun. We observe the Sun from the surface of the Earth, from mines deep in the ground (using instruments called neutrino telescopes), from artificial satellites and from spaceships. In this way we have learnt much about the Sun: about its structure, and what takes place both on its surface and deep inside its hot core.

The Sun is, in effect, a huge, very hot ball, in the centre of which energy is released from hydrogen. In the process the hydrogen changes to helium. The energy which is released then penetrates to the surface and from there it is radiated into the frozen space of the Universe. Only a tiny proportion of the solar radiation is captured by the planet Earth. On the surface of the Earth this radiation is mainly absorbed and changed into heat. It also causes the movement of the winds, the sea currents and the flow of rivers. By photosynthesis the solar energy enters the biosphere and keeps everything alive.

The Sun (in Greek *helios*) is the source of all energy on Earth. The old civilizations were of course wrong to think it a god, but they were right in thinking that the Sun is essential to all life.

The distance from the Earth to the Sun

The Sun is 150 million km (93,200,000 miles) away from us. This distance is called an astronomical unit. It is used for measuring the distances in the planetary system. A rocket flying at 1000 km (620 miles) per hour would take 17 years to cover an astronomical unit. To man, it is an enormous distance, but from the astronomical point of view it is tiny. After all, it is 250,000 times further from Earth to the next nearest star (Proxima Centauri).

There are various ways of measuring the distance to the Sun. Using radar is one of them; a huge transmitter is employed, and the time it takes the signal to get to the Sun and back again is measured. It is in fact 16 minutes and 40 seconds, or in other words 1000 seconds. Since a radar signal travels at the speed of light (300,000 km/186,400 miles per second), it covers a total distance of 300 million km (186,400,000 miles). Thus the distance to the Sun is half that, or 150 million km (93,200,000 miles).

Exact measurements show that the distance to the Sun changes in the course of the year, however, from 147 million km (91,300,000 miles) to 152 million km (94,500,000 miles). The Earth has an elliptical orbit, in which the Sun is at one of the foci. The ellipse of the Earth's orbit is, however, very similar to a circle. The astronomical unit is in fact the average distance between the Sun and the Earth, which is exactly 149,597,906 km (92,900,300 miles), but the approximate figure mentioned above is usually used.

158 The Sun is the star of our Galaxy **(1)**. For us terrestrials it is the most important star of all, since it gives energy to all living things. That was why the ancient nations worshipped it as a deity **(2)**. Today we observe and study it through sophisticated instruments **(3)**. In the interior of the Sun protons **(4)** change into alpha particles **(5)**. The energy which is released in the process is emitted as light, infra-red and ultraviolet radiation, partly also as radio and X-ray radiation **(6)**. The Sun is a perfect thermonuclear reactor, a perfectly clean and inexhaustible source of a huge flow of radiant energy.

Measuring the Sun

The size of the Sun may be measured from a dark room into which its rays enter through a small hole. You take a sheet of stiff white paper and hold it perpendicular to the Sun's rays. The Sun's image appears on the paper. The further away from the hole the paper is

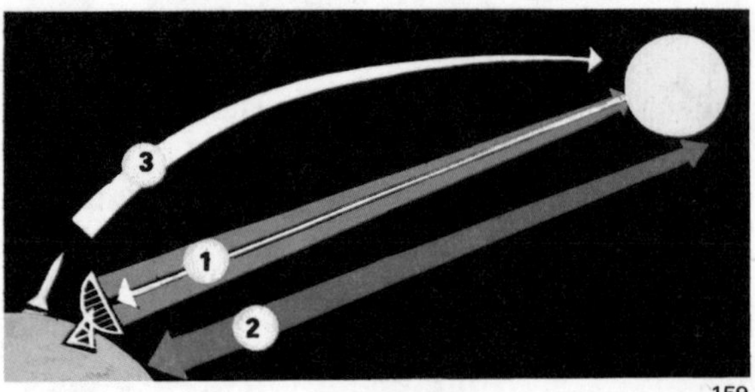

159 The distance to the Sun can be measured by means of radar **(1).** It takes the signal 500 seconds to get there, and the same time to get back after being reflected **(2).** This time corresponds to a one-way distance of 150 million km (93,200,000 miles). It would take a rocket 17 years to get to the Sun **(3).**

held, the greater the size of the image. For instance, at a distance of 107 cm the diameter of the image will be 1 cm; at a distance of 214 cm it will be 2 cm, and so on. The image of the Sun will always be 107 times smaller than the distance between the paper and the hole. The diameter of the real Sun will also be 107 times smaller than the distance of the Sun from the Earth (Figure 160).

The distance of the Sun is 150 million km (93,200,000 miles) so that the diameter of the Sun is 107 times smaller. By division we arrive at a figure for the diameter of 1,400,000 km (871,000 miles).

If we were to place spheres the size of the Earth next to each other, it would take 109 of them to equal the diameter of the Sun. If the Earth were in the centre of the Sun, the Moon would orbit about half way to its surface.

The volume of the Sun is more than a million times larger than the volume of our planet. There are giant stars with volumes many million times larger than that of our Sun. On the other hand, neutron stars, or pulsars, are about 100 million times smaller than the Earth.

The mass of the Sun

The Earth orbits the Sun once a year on a nearly circular path. A year lasts about 31 million seconds, and in that time the Earth covers a distance 6.28 ($= 2\pi$) times that from the Sun to the Earth. We can easily calculate from this that the Earth moves at the rate of 30 km (20 miles) per second.

Every body moving in a circular orbit is subject to centrifugal force, whose magnitude can easily be calculated – on a body of mass m moving along a circle of radius R at a speed of v, the centrifugal force is mv^2/R. The Earth is also subject to a centrifugal force. If it were not attracted by the Sun's gravitational force, it would fly off into space. According to Newton's well-known law (page 34), a force of GMm/R^2 acts between the Sun and the Earth (G is the gravitation constant of 6.67×10^{-11} newtons $\times$ m^2/kg^2, M is the mass of the Sun which we want to know, m is the mass of the Earth, and R is the distance between the Sun and the Earth). If only the gravitational force acted on the Earth and there was no centrifugal force, in a quarter of a year the Earth would fall into the Sun. But the centrifugal force has the same value as the gravitational attraction between the Sun and the Earth. This can be expressed as follows:

$$mv^2/R = GMm/R^2 \quad \text{or} \quad M = v^2R/G$$

The velocity v is known (30 km/20 miles per second or 30,000 m/s), as are the distance R and the gravitation constant G. From these we can calculate the mass of the Sun: $M = 10^{30}$ kg. Astronomers call this $M_\odot$. A circle with a dot is the ancient symbol used to denote the Sun. The more exact value is 1.99×10^{30} kg. The Sun's mass is the sum of the masses of all its nucleons; their number is 10^{57}. Electrons contribute much less, since their mass is around 2000 times less than that of the nucleons.

The mass of the Sun is 760 times greater than that of all its planets put together. It is 330,000 times greater than the mass of the Earth. We know of stars in the Universe whose mass is up to one hundred times greater than that of the Sun. On the other hand, there are also stars many times lighter. Our Sun is just an average sort of star.

The chemistry of the Sun

The Sun is composed of 92 per cent of hydrogen atoms, rather less than 8 per cent of helium atoms, and less than 1 per cent of the other elements. Thus for every 100,000 atoms of hydrogen there are 8500 atoms of helium, 66 atoms of oxygen, 33 atoms of carbon, 9 atoms of nitrogen, 8 atoms of neon, 4 atoms of iron, 3 atoms of silicon, 3 atoms of magnesium, 2 atoms of sulphur, and so on. All the other elements occur in still smaller numbers.

There is no element on the Sun that we do not know on Earth. At one time scientists thought that helium was one. It was first discovered in the spectrum of the Sun (Figure 161), which is how it came to be called helium, meaning 'Sun element'. But it was later also discovered on Earth, though in much smaller quantities.

There are many dark lines in the solar spectrum, each of which corresponds to atoms of a particular element. The darkness of each of these lines indicates the amount of the element which is present there. A study of the spectra of other stars has revealed that their chemical composition is much the same as that of the Sun. The same goes for the chemical composition of the whole Universe as we know it so far.

We should also mention here that the composition of the Earth was also like that of the Sun in the beginning. But right from the start of the life of our planet hydrogen and helium, the most abundant elements, escaped, leaving behind the heavier elements. Thus the Earth represents a sort of cosmic impurity, consisting as it does of the elements which in fact pollute the most abundant cosmic materials, hydrogen and helium. On Earth only a small part of the original hydrogen remained, in combination with oxygen in the form of water. Since helium was unable to combine with any of the other elements (it is an inert gas), most of it escaped into space. But because of its greater gravitational attraction and greater distance from the Sun, Jupiter managed to retain all its hydrogen and helium. The chemical composition of the Sun and of Jupiter is thus the same, since both these bodies, like the whole of the Solar System, came into being from the same maternal cloud (see page 152).

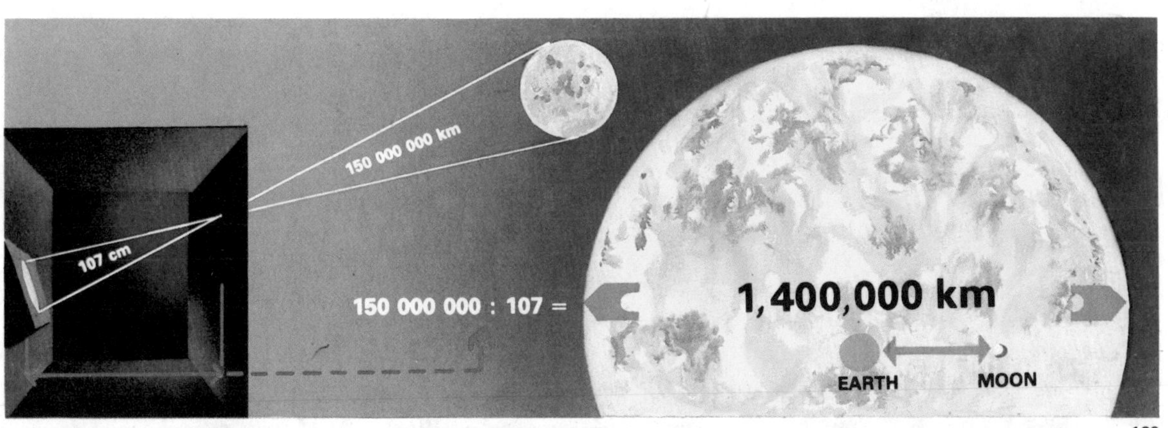

160 The diameter of the image of the Sun is 1/107th of the distance between the screen and the hole. The real diameter must therefore be 1/107th of the distance between the Sun and the Earth, in other words 1,400,000 km.

The temperature of the Sun

The temperature of the surface of the Sun is determined using the solar spectrum. The intensity of the radiation in individual colours of the spectrum corresponds to a temperature of 6000 kelvin. This is the temperature of the solar surface, the photosphere. In the higher layers of the Sun's atmosphere – the chromosphere and the corona – the temperature is higher. In the corona it is around 2 million kelvin. Above large flares the temperature may, for short periods, reach about 50 million kelvin. The corona above flares emits much radio and X-ray radiation.

Though the interior of the Sun does not emit a single ray, we know how to calculate the temperature at any point in it. These

161 The spectrum of the Sun in colour. The long band of the spectrum is divided into sections which are shown one below the other, in order to fit the photographic plate. Every dark line in the spectrum corresponds to a certain element. The spectra of the Sun, the stars and other bodies are important sources of information on the Universe.

calculations show that the deeper we go beneath the surface, the higher the temperature of the plasma. The temperature rises from 6000 kelvin in the photosphere to 13 million kelvin in the centre of the Sun.

Let us look at it another way. The higher the temperature of any matter, the more rapidly its particles move. In the photosphere, for instance, the protons and hydrogen atoms move at a speed of around 7 km (4 miles) per second, and the light electrons at about 300 km (186 miles) per second. In the hot corona and in the core the protons move at speeds in the order of 350 km (217 miles) per second, and the electrons about 15,000 km (9320 miles) per second.

The lowest temperature on the Sun is in the sunspots. In large sunspots the temperature is lower than 4000 kelvin.

The surrounding white photosphere with a temperature of 6000 kelvin emits about five times as much radiation per m^2 as the spots. That is the reason the spots seem dark, even black.

Every body which falls on to the Sun is very soon reduced to its individual atoms, and then the electrons are pulled away from these atoms. On the Sun matter can exist only in the form of plasma.

The structure of the Sun

The properties of the solar plasma – density, temperature, pressure and chemical composition – depend on the distance from the centre. The central part of the Sun, for instance, called the core, has a density nine times that of lead. On the other hand, the density of the highest part of the Sun, the corona, is one billion times less than that of the air we breathe. The difference in density between the corona and the core is so great that 1 cm^3 (0.06 cu in) of the core contains the same amount of matter (100 g/3.5 oz) as 1000 km^3 (240 cu miles) of the corona.

There are also huge differences in temperature (Figure 162). In the Sun's core the temperature is extremely high (13 million kelvin). The temperature of the visible surface is 6000 kelvin. This is much lower than that in the core – 2000 times lower. On the other hand, compared with the temperature of the interstellar space around the Solar

161

System (around 3–5 kelvin), the temperature of the Sun's surface is high (about 2000 times higher). Nature abhors large temperature differences, and tries to even them out. There is a huge difference in temperature between the core of the Sun and the cold interstellar space, which nature is trying to remove. For this reason there is an enormous flow of energy from the core to the surface, and thence to the chilly depths of space.

Nature's effort to transfer heat from a hot body to its cold surroundings is called the *Second Law of Thermodynamics,* also known as the *Law of the Growth of Entropy*. Let us imagine a small amount of heat, 1 calorie (very rapidly moving nuclei and electrons in a microscopic volume – about one-hundredth of a millimetre cube) in the core of the Sun. If we divide this heat by the temperature, we get the entropy, which is 1 cal per 13 million kelvin. After about one million years this energy makes its way to the Sun's surface, where the temperature is only 6000 kelvin. The entropy there is 1 cal per 6000 kelvin, i.e. more than 2000 times higher than in the core. Thus in the transfer of the energy from the core to the surface the entropy is increased about 2000 times. The entropy is further increased – also about 2000 times, when the radiation is emitted from the solar surface into cosmic space. The law of the growth of entropy explains why the Sun and the stars shine.

The outer part of the Sun (or any of the other stars) is called the *atmosphere*; and from it photons escape directly into the cosmic space. Unlike the interior, the atmosphere is visible. The border between the visible atmosphere and the invisible interior is formed by a thin layer called the photosphere (Figures 163 and 166). Compared with the radius of the Sun it is very thin (around 250 km/155 miles). It radiates almost all the solar light, however, and thus seems to us to be the Sun's surface.

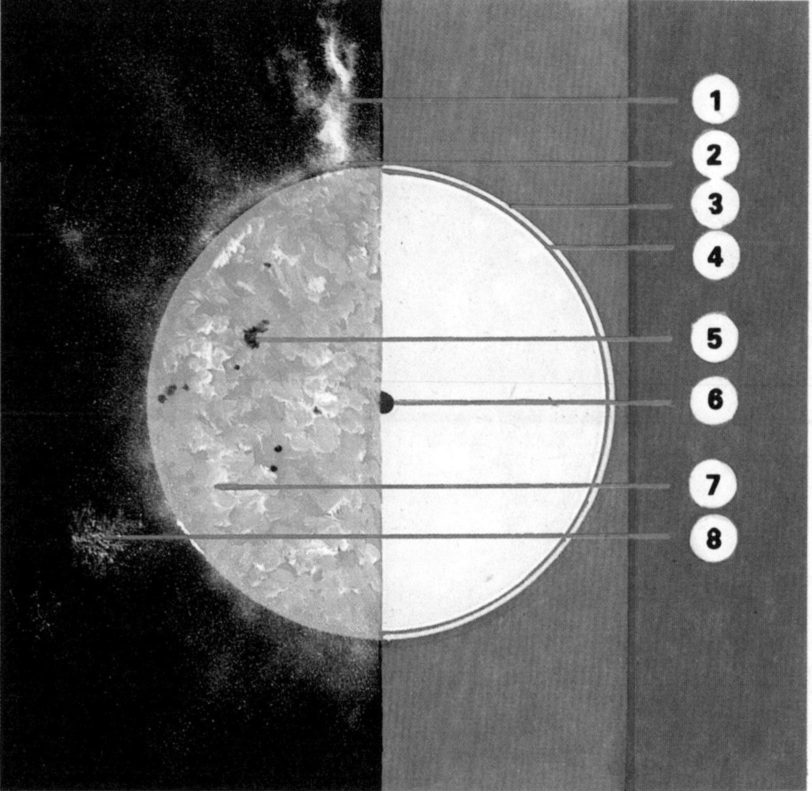

162

162 Different parts of the Sun have different temperatures. The corona **(8)** is at 2 million kelvin, while the region of coronal condensation **(1)** over a flare **(2)** may reach as much as 50 million kelvin. The temperature in the chromosphere **(3, 7)** is around 10,000 kelvin, that in the photosphere **(4)** 6000 kelvin, and in sunspots **(5)** only 4000 kelvin. The temperature rises with increasing depth to 13 million kelvin in the centre of the Sun **(6)**.

The interior of the Sun is divided into three parts (Figure 163 numbers 1, 3 and 4):

- The *core* which is about the size of the Earth, is where hydrogen changes to helium, releasing huge amount of energy.
- The extensive layer surrounding the core to a distance of 650,000 km (403,600 miles) from it. The energy from the core is propa-

9
9
9
7
12
10
11
6
5
4
13
9
14
4
P
e⁻
17
3
P
e⁻
8
9
e⁺
e⁻
α
P
P
P
P
2
1

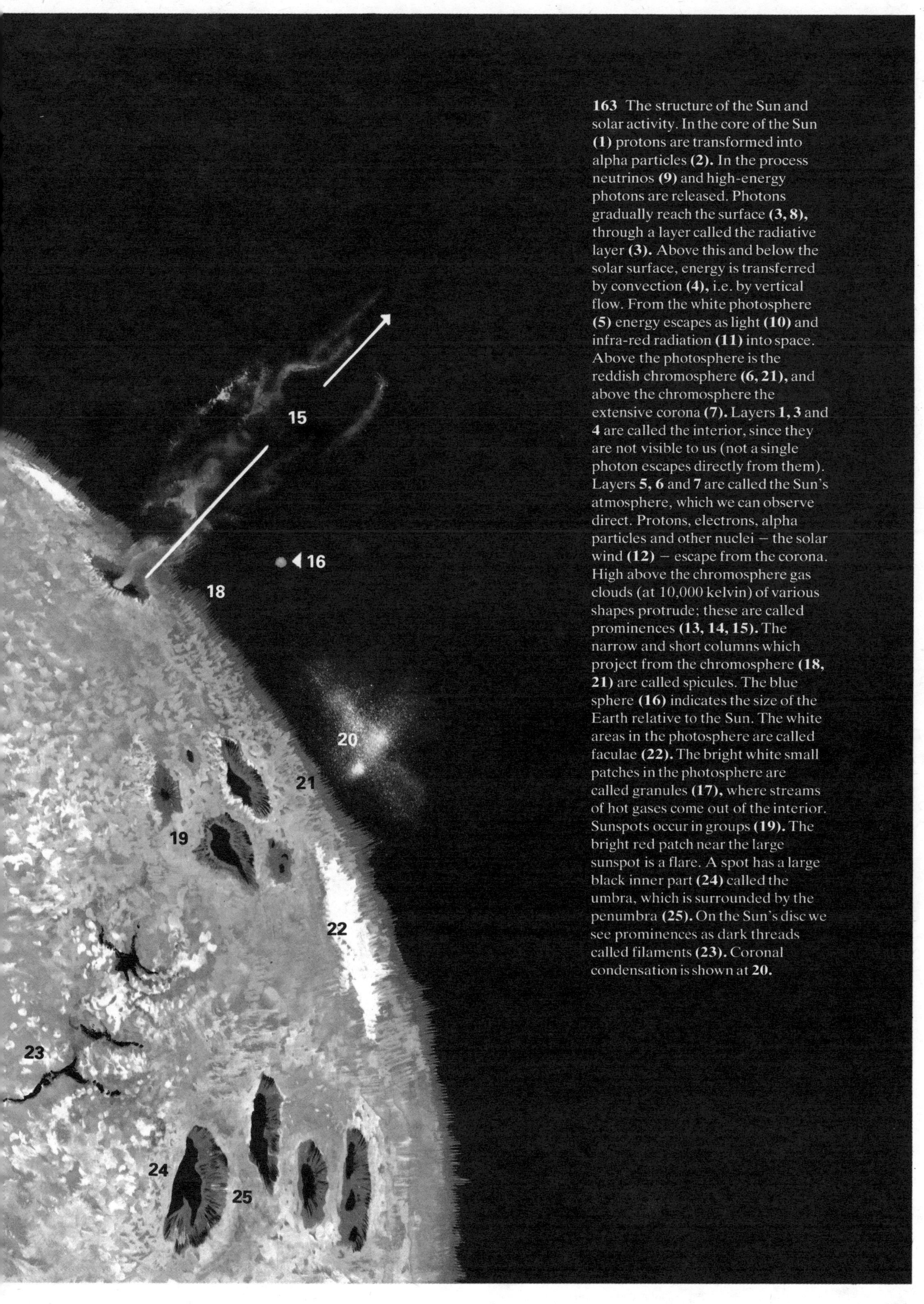

163 The structure of the Sun and solar activity. In the core of the Sun **(1)** protons are transformed into alpha particles **(2).** In the process neutrinos **(9)** and high-energy photons are released. Photons gradually reach the surface **(3, 8),** through a layer called the radiative layer **(3).** Above this and below the solar surface, energy is transferred by convection **(4),** i.e. by vertical flow. From the white photosphere **(5)** energy escapes as light **(10)** and infra-red radiation **(11)** into space. Above the photosphere is the reddish chromosphere **(6, 21),** and above the chromosphere the extensive corona **(7).** Layers **1, 3** and **4** are called the interior, since they are not visible to us (not a single photon escapes directly from them). Layers **5, 6** and **7** are called the Sun's atmosphere, which we can observe direct. Protons, electrons, alpha particles and other nuclei – the solar wind **(12)** – escape from the corona. High above the chromosphere gas clouds (at 10,000 kelvin) of various shapes protrude; these are called prominences **(13, 14, 15).** The narrow and short columns which project from the chromosphere **(18, 21)** are called spicules. The blue sphere **(16)** indicates the size of the Earth relative to the Sun. The white areas in the photosphere are called faculae **(22).** The bright white small patches in the photosphere are called granules **(17),** where streams of hot gases come out of the interior. Sunspots occur in groups **(19).** The bright red patch near the large sunspot is a flare. A spot has a large black inner part **(24)** called the umbra, which is surrounded by the penumbra **(25).** On the Sun's disc we see prominences as dark threads called filaments **(23).** Coronal condensation is shown at **20.**

gated through this layer in the form of radiation (photons), and it is therefore called the *radiative layer*.

• Above the radiative layer, up to the surface, extends the *convective layer*. It is about 50,000 km (31,000 miles) deep. In it energy is transferred to the surface through convection currents (marked by black arrows in rectangle 4 in Figure 163). The hot clouds rising through the convective layer reach the photosphere, where their motion stops. We can easily observe them as resembling bright grains of rice (they are called granules). They are as large as Britain or Texas (Figure 167).

The atmosphere of the Sun is also made up of three layers:

• The *photosphere* (Greek meaning light sphere; Figures 163 and 166), a thin, white layer about 250 km (155 miles) thick enveloping the Sun. It is the coldest, densest and lowest part of the solar atmosphere. (Below the photosphere lies the interior.)

• The thin, reddish *chromosphere,* reaching to a height of about 10,000 km (6200 miles) above the photosphere (Figures 163 numbers 6 and 21, and 172).

• The very extensive, extremely rarefied and hot *corona,* which stretches for millions of kilometres above the surface of the Sun (Figures 163 number 7, 174 and 192). The chromosphere, and especially the corona, can be seen with the naked eye for only a short time during a total eclipse of the Sun (Figure 190).

The atmosphere is very extensive, and takes up much more space than the interior of the Sun. But the interior contains 10,000 million times more matter than the atmosphere. This means that we are able to observe only a very small part of the Sun's matter. Almost all the solar matter is contained in the invisible interior. Using modern computers, however, we can determine the density, temperature, pressure and chemical composition at any point inside the Sun. The properties of the solar plasma are much simpler than those of solid and plastic minerals such as those of which the Earth's interior is made. We therefore know more about the interior of the Sun than we do about the interior of our own planet.

The transformation of hydrogen to helium

The most important process in the Sun is the transformation of hydrogen to helium. It is the source of all the Sun's energy.

In the core of the Sun the density and temperature are high; there are therefore very frequent violent collisions between electrons, protons and other nuclei. Many of the proton collisions are so violent that in spite of electrical repulsion the colliding protons approach to within their own diameter (1 fermi) of each other. At such close distances the attraction of the nuclear force can act (Figure 33), and the protons fuse together. This fusion of protons in the nucleus of helium is shown in Figure 163 number 2 (in its lower left corner). The square shows a very small space (a cube with sides measuring 0.02 nm) in the central region of the Sun. This small space contains only four protons (red circles) and four electrons (blue dots). The four protons gradually fuse in the helium nucleus, with two protons changing into neutrons (white circles), two positive charges being released as positrons (red dots) and two small neutral particles – neutrinos (white dots) – being formed. Both positrons e^+ change into photons of gamma radiation on meeting electrons e^- (annihilation). The rest energy of the atom of helium is smaller than the rest energy of the four atoms of hydrogen. This difference in energy has changed into gamma photons and neutrinos. The total energy of all the gamma photons arising and the two neutrinos is 28 MeV.

A huge number of such transformations takes place inside the Sun (10^{38} every second). In the process about 567 million tonnes of hydrogen is changed into helium. The resulting helium weighs only 562.8 million tonnes, i.e. 4.2 million tonnes (4.2×10^9 kg) less. And it is this loss of mass each second that changes into solar radiation. It is an energy of:

$$4.2 \times 10^9 \text{ kg} \times 9 \times 10^{16} \text{ m}^2/\text{s}^2 = 3.8 \times 10^{26} \text{ joules}$$

This energy is radiated by the Sun every second. The energy per second is called the output, and 1 joule per second is 1 watt. Thus the output of the Sun is 3.8×10^{26} W. This output is in the form of radiation.

It is called *solar luminosity* and denoted by the capital letter L with the ancient symbol of the Sun, i.e. $\odot$. Hence:

$$L_{\odot} = 3.8 \times 10^{26} \text{ watts}$$

The age of the Sun

With the aid of helium in the solar core we can calculate how many years have passed since the Sun was born. By the birth of the Sun we mean the moment when the hydrogen in the core began to change into helium. At the moment of its birth the composition of the Sun was the same in all its parts, in the atmosphere as well as in the core. With the exception of its core, the chemical composition of the Sun remains constant. The older the Sun becomes the more helium accumulates in the core. This accrues at the rate of 562.8 million tonnes per second. By calculating the amount of helium present in the Sun it is possible to estimate how long it has been accumulating. This will give an approximation to the age of the Sun. Astrophysicists are able, with the help of complex computer programs, to determine the density, temperature, pressure and amount of helium in each layer below the photosphere within the Sun, including the core.

An estimated 89 quadrillion (89×10^{24}) tonnes of hydrogen have been transformed into helium. Assuming a conversion rate of 563 million tonnes per second, then the Sun is approximately 1.58×10^{17} seconds old – 5000 million years.

The Sun is thus 'in its prime'. There are stars one hundred to one thousand times younger (such as the Pleiades at 50 million years of age). But there are also older stars (such as some in the globular clusters which are around 10,000 million years old).

Up to the present moment the Sun has exhausted about one-third of the nuclear fuel – hydrogen – in its core. Thus the transformation of hydrogen to helium in the Sun will continue for another 10,000 million years or so. But even then the Sun will not be extinguished. In fact it will shine even more brightly; the helium in its nucleus will start to turn into carbon, the carbon to oxygen, etc. (Figure 151). The Sun will continue to shine for another few thousand million years. We certainly have no reason to fear that it will 'go out' in the near future, or die in some explosion. If mankind is threatened by a catastrophe, it will not come from the Sun.

How much energy does the Sun release?

We have already mentioned the value of solar luminosity $L_{\odot}$. How can it be determined? A simple answer to this question is depicted in Figure 164. At noon on a clear day we light a powerful bulb whose luminosity we know. We designate it with the letter l. We then close our eyes and look alternately at the bulb and the Sun. If the bulb seems brighter than the Sun, we move away from it. If the Sun is brighter, we move closer to the bulb. As soon as the Sun seems to us to be the same brightness as the bulb through our closed eyelids, we measure the distance from the bulb to our eyes. This distance (r metres) depends on the luminosity of the bulb. The distance of the Sun is 150,000 million metres (page 123). We already know that the luminosity of the Sun is denoted $L_{\odot}$. The brightness of any source decreases with the square of the distance. The simple relation:

$$L_{\odot}/l = (150{,}000{,}000{,}000/r)^2 \text{ or:}$$
$$L_{\odot} = 2.25 \times 10^{22}\, l/r^2$$

tells us how much greater is the luminosity of the Sun compared with that of the bulb. It is, of course, only a very rough estimate.

In order to make an accurate determination of the amount of radiation emitted by the Sun every second, we must first measure what is called the *solar constant.* This is the amount of solar radiation which falls on 1 m^2 per second at a distance from the Sun equal to that of the Earth and perpendicular to the solar rays (Figure 165).

Many precise measurement have set the solar constant at 1353 Wm^{-2}. This is an average value, since the distance between the Earth and the Sun changes in the course of the year. The Earth's orbit around the Sun is elliptical, so that more radiation falls on it in winter (on January 1, for instance, 1438 Wm^{-2}) and less in summer (on July 1, for instance, only 1345 Wm^{-2}). Here summer and winter refer to the northern hemisphere, and the one-metre square area is situated above the Earth's atmosphere. The atmosphere absorbs or reflects a considerable proportion of the solar radiation (Figure

274). The solar constant must be therefore measured outside the Earth's atmosphere, with satellite-borne instruments.

Now we can calculate the luminosity of the Sun exactly. Let us imagine a large sphere, with the Sun in the centre and with a radius equal to the distance between the Sun and the Earth. On every square metre of the large sphere an energy of 1353 watts (the solar constant) falls. Since we can calculate from a known radius the surface of a sphere, we can easily ascertain that 3.8×10^{26} watts passes through the entire sphere. This is the output of the Sun, or the solar luminosity, $L_{\odot}$. It is a huge output, yet there exist stars with outputs a million times greater. Beside one of them the Sun would appear quite faint and insignificant.

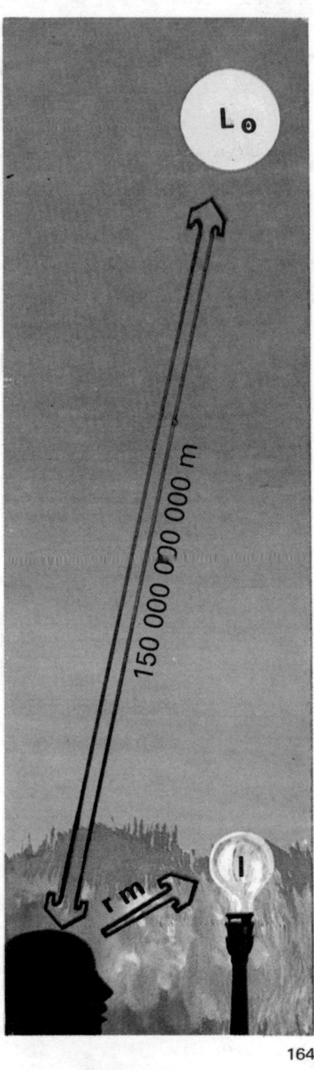

164

164 An estimate of the Sun's luminosity. The distance between the bulb and the eye is changed until it and the Sun seem equally bright through closed eyelids. This distance (in metres) is called r. Then the luminosity of the Sun is

$$L_{\odot} = \left(\frac{1.5 \times 10^{11}}{r}\right)^2 l$$

where l is the luminosity of the bulb.

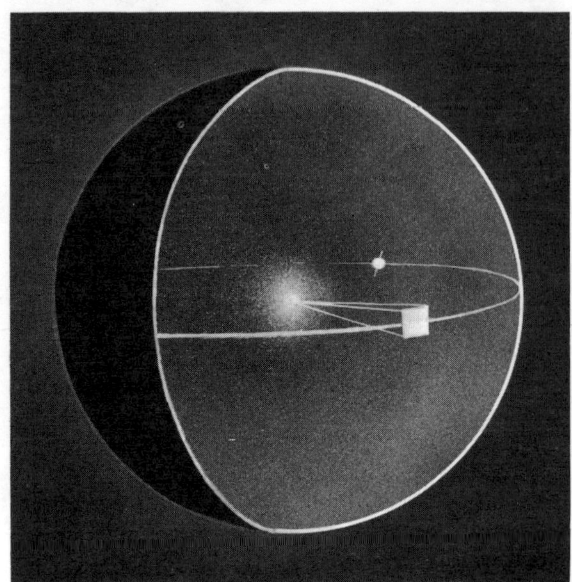

165

165 The solar constant is the solar radiation which passes every second through an area of 1 m^2 lying on the Earth's orbit and perpendicular to the Sun's rays. It is 1353 W/m^2. Since we know the radius of a notional sphere around the Sun (1.5×10^{11} m – see Figure 164), we can easily calculate the solar luminosity $L_{\odot}$, i.e. the total radiation passing through the whole sphere.

Journey of the photons

The energy released in the transformation of protons into alpha particles is mainly in the form of gamma photons. Soon after their origin gamma photons gradually change into X-ray photons. Their journey to the surface of the Sun and into cold interstellar space is not as straightforward as that of the neutrinos (Figure 163 numbers 8 and 9).

Though photons move at the same speed as neutrinos, after a few decimetres they are absorbed by an electron and immediately emitted again in another direction. The path taken by photons in the interior of the Sun is very long and tortuous.

When photons are absorbed and again emitted, sometimes two photons appear instead of one (but with less energy). Thus in the course of their wanderings inside the Sun the photons gradually 'crumble'. In this way the energy of the original gamma photon is distributed among several hundred thousand light photons, which leave the solar surface.

But it takes about 2 million years for the 'crumbs' of the wandering photons to reach the surface of the Sun. As we have seen, the neutrino gets from the interior of the Sun to its surface in 2 seconds flat (Figure 163 number 9).

Is it not interesting that the light of the Sun which we are now seeing and being warmed by was released from the Sun's core in the days of the first men on Earth? In the time it took the photons to make their way from the core of the Sun to its surface, primitive man has developed into the educated and intelligent being he is today.

If the journey of the photons through the interior of the Sun is a difficult one, it becomes almost impossible as they near the surface. In the layers below the visible surface of the Sun the hydrogen is mostly in the neutral state, which means that almost every electron is attached to one proton. The layer of neutral hydrogen reaches to a depth of about 50,000 km (31,000 miles) beneath the Sun's surface.

The electrons of the hydrogen atoms absorb photons avidly. The photons are therefore virtually unable to continue their journey to the surface. The transfer of energy through the hydrogen layer cannot take place by means of radiation. The Sun therefore chooses another, much more effective means of energy transfer – at the bottom of the hydrogen layer (about 50,000 km/31,000 miles below the Sun's surface), it sets in motion clouds of hot plasma. These reach as far as the visible surface of the Sun, so that we can see them as light (Figure 167).

Then the photons escape direct from the visible surface into the surrounding cosmic space.

The photosphere – the surface of the Sun

Through thin clouds in the Earth's atmosphere we see the Sun as a shining white ball. In fact what we see is the photosphere, of which we have already spoken (Figure 163 number 5). We know that the photosphere is the lowest layer of the Sun's atmosphere. Its thickness is only 250 km (155 miles). From it light escapes directly into space. Hidden below the photosphere is the convective layer – the top layer of the invisible part of the Sun's interior (Figure 163 number 4).

In the convective layer, clouds of hot plasma rise in to the photosphere. They are about 1000 km (620 miles) across. The speed at which they rise upwards to the surface (the photosphere) is $\frac{1}{2}$ km (0.3 miles) per second. Since they have a temperature of more than 6000 kelvin they carry a great many photons with them. Such a form of photon transfer with hot plasma clouds is called convection. These clouds do not stop until they reach the photosphere, and from there the photons can easily escape into space, since the atmosphere over the photosphere is very rarefied.

Viewed through a telescope the hot clouds look like white grains of rice. They are known as granules. In emitting photons granules lose energy, cool down, and become heavier. The heavier plasma falls back into the convective layer below the surface of the

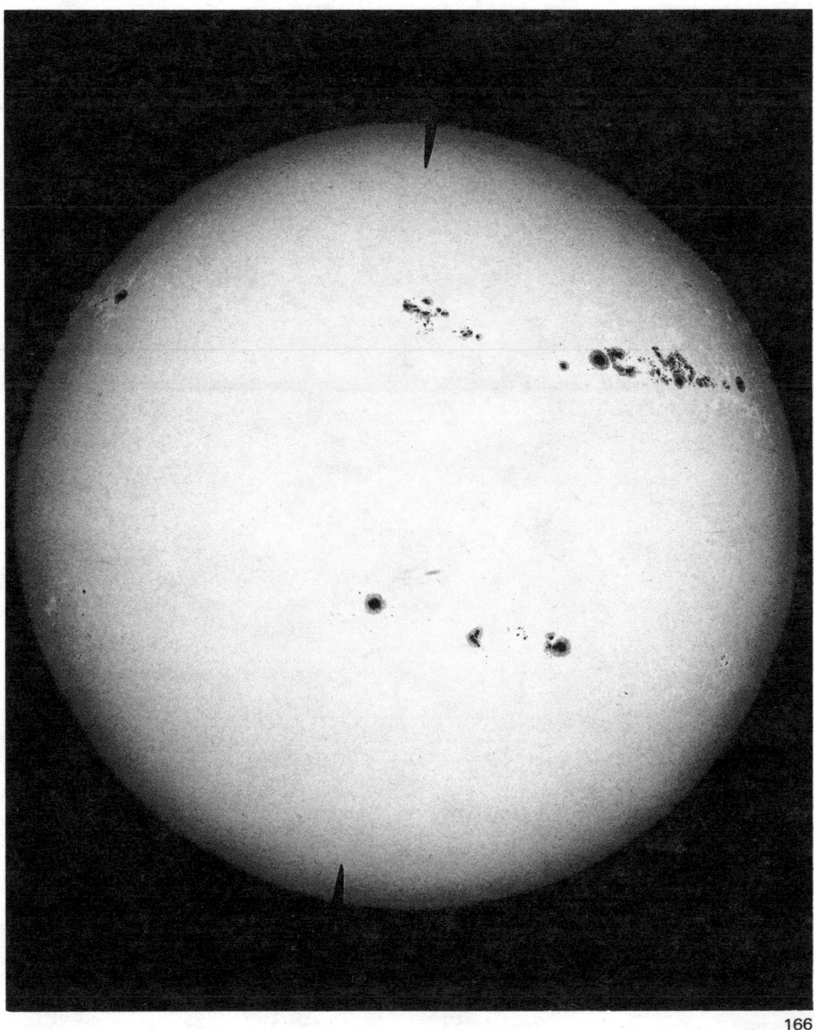

166

166 The solar photosphere with large groups of sunspots (photographed on February 20, 1956). In the case of the large spots you can see a black umbra surrounded by a grey penumbra. The bright areas at the edges are faculae (the two black pointers – produced in the telescope – are used to help locate the sunspots).

Sun, and is visible to us as a dark space in between the granules (Figure 167).

From the photosphere the Sun's radiation escapes directly into space. Almost all the solar energy (3.8×10^{26} watts) is emitted from the photosphere as light and thermal radiation. Only a negligible part of the Sun's luminosity is emitted from the upper, very rarefied layers of the atmosphere.

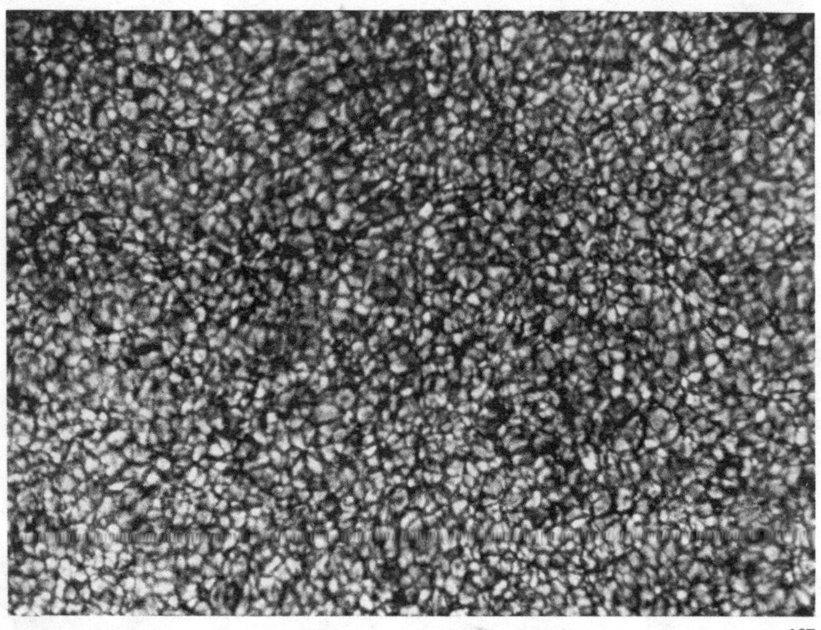

167

167 Solar granulation. We should see the surface of the Sun like this if we could approach within 250,000 km (155,000 miles). The white areas – granules – irradiate their heat within minutes, and on cooling fall back to the interior of the Sun (black spaces between granules). A granule may be observed for five to eight minutes.

Sunspots – huge magnets

From time to time black spots appear in the brilliance of the photosphere. They are called *sunspots.* Amid the shining photosphere they look dark, because they are relatively cold – less than 4000 kelvin, while the surrounding photosphere is at a temperature of 6000 kelvin (Figures 166 and 169).

The Sun is thus not the 'pure unstained' sphere the ancient philosophers took it for. The Italian astronomer Galileo Galilei was the first to observe sunspots through the telescope. Though he had only a simple instrument he observed how the spots appeared, grew in size, changed their shape and appearance, and within a matter of days or weeks disappeared again. He also noticed that all the spots move from the eastern limb of the Sun to the western limb – a fact caused by the rotation of the Sun around its own axis (Figure 168).

The sunspots may be small, perhaps about the size of France. Spots as small as that are called pores. Large sunspots may be several times the size of the surface of the Earth. These large spots have two clearly defined areas: the central, black part called the *umbra,* and around it the *penumbra,* forming the transition between the umbra and the photosphere. The penumbra is composed of fine, bright fibres, pointing in the direction from the photosphere to the umbra (Figure 169).

If we want to learn about a heavenly body, we photograph its spectrum. The spectrum of a sunspot has some spectral lines split into two components. This splitting is called the *Zeeman effect.* The distance between the two components increases in size with the intensity of the sunspot's magnetic field. We can thus determine from the Zeeman effect the intensity of the magnetic field of the spot. The resulting values vary from 0.1 tesla to 0.3 tesla (1000 to 3000 gauss). The magnetic fields in the sunspots are many thousand times stronger than that of our Earth.

The magnetic field of sunspots is due to the huge electric currents which flow around them. Positive ions move in one direction, negative electrons in the other. The dotted green lines in Figure 170 show the magnetic lines of force. Their number per m^2 (10.76 sq ft) indicates the intensity of the magnetic field. The straight line which is tangential to the force lines at a certain point shows the direction of the magnetic force at that point.

The intensity and direction of the magnetic field in a sunspot below the photosphere cannot be measured directly, since not a single ray escapes. But we conclude that the field is very entangled (Figure 170 left). The clouds of hot plasma which rise to the surface cannot, however, move through the entangled lines of force, and avoid them. (The motion of rising hot clouds near sunspots is marked by black arrows in the figure.) For this reason the spots receive less energy than the surrounding photosphere, which is why they are dark and cool.

Sunspots tend to appear in groups (Figure 169). The photosphere around groups of sunspots is warmer and brighter than that further away from them. This bright area of the photosphere is called a *facula.* Faculae are clearly visible if the group is near the limb of the Sun (Figure 166). The chromosphere around groups of sunspots is also brighter than that further away from them (Figure 172). This bright area is called a *plage.* There are other phenomena accompanying groups of sunspots, such as *flare, prominence, coronal condensation,* etc. All these, including the sunspots, the faculae and the plages, are called *solar activity.* A group of spots together with all the phenomena of solar activity is called a *centre of solar activity* or an *active region.*

Groups of sunspots

A sunspot comes into being in the photosphere as a small black spot in a place where a strong magnetic field (at least 0.1 tesla or 1000 gauss) arises.

At first it has no penumbra, and is called a pore, with a diameter of 1000–3000 km (620–1860 miles).

If the magnetic field increases in intensity and size, the sunspot also grows, surrounding itself with a penumbra, and other sunspots appear in its vicinity. A large group may contain 50 or more sunspots. In some sunspots the magnetic lines of force emerge, and these are said to have northern polarity. They represent the north pole of a magnet, designated N. In other sunspots the magnetic lines of force submerge into the sunspot, which then has southern polarity – S. Figure 170 shows the polarity of only two sunspots of the group. Note that the magnetic lines of force link the northern polarity with the southern. If we could look beneath the surface of the Sun we should find that the lines of force continue from southern polarity to northern.This means that they form closed loops, only the upper part of which is in the atmosphere. The lines of force have neither beginning nor end, but are closed (Figure 170, left).

Individual groups of sunspots differ in the number of spots and their size. There are many amateur astronomers and observatories in the world which help to keep a daily record of sunspots, observing and drawing or photographing them.

The number of sunspot groups, and individual sunspots in them, changes from day to day. New spots appear, while others disappear. The Sun rotates, and at the eastern limb sunspots which came into being on the far side of the Sun come into view. On the western limb, on the other hand, groups of sunspots disappear from the visible hemisphere.

The Swiss astronomer Rudolf Wolf introduced the index (designated R), linking the total number of sunspot groups g, and the total number of sunspots (designated S), irrespective of their grouping. The number R is called the *relative number,* or *Wolf's number.* It is calculated by multiplying the number of groups g by 10, and adding the total number of all sunspots, S. It means:

$$R = 10g + S$$

Scientists studying the influence of the Sun on Earth and life on it, frequently use the relative number R. They compare the value of R from day to day (sometimes also the average for a month, or even a year) with various terrestrial phenomena. The relative number has been recorded by astronomers regularly since the 18th century (Figure 171).

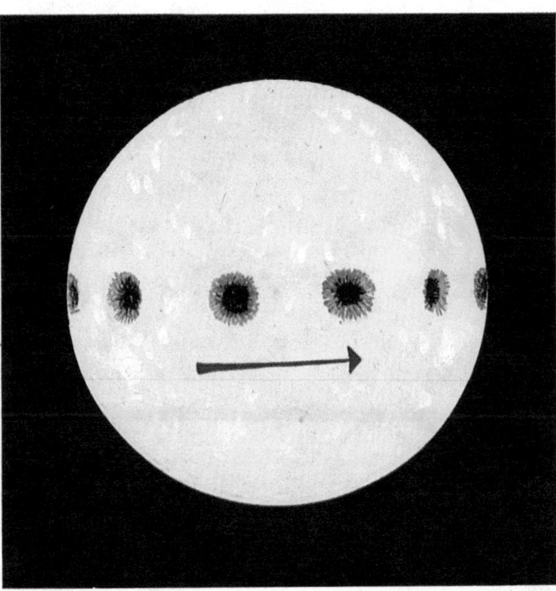

168

168 Sunspots move from the eastern edge of the Sun (left) to the western (right). It takes 14 days for them to make the crossing and is a consequence of the rotation of the Sun.

The curve of sunspots

Because of the daily changes in the number of sunspot groups and of sunspots, the relative number R also changes.

Since the relative number over a period of several years would form a series difficult to comprehend, we use the mean monthly numbers.

If one is interested in changes in sunspots over a period of centuries, one uses the annual mean number. It is calculated in the same way as the monthly number; the daily numbers are added together and divided by the number of days in the year. Figure 171 shows the annual relative numbers from the year 1750. The resulting curve is called the *sunspot curve* or the relative numbers curve.

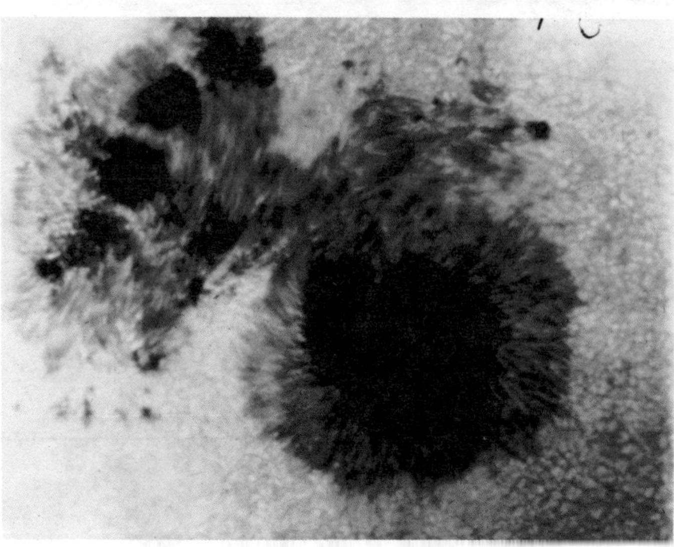

169 A group of sunspots. Their temperature is 4000 kelvin, so they seem dark compared with the surrounding photosphere, which is at a temperature of 6000 kelvin.

The lowest points on the sunspot curve are called the *minima of solar activity*. In 1856, for instance, the annual relative number was only 4.3, and in 1867 it was 7.3, etc. These years are known as the years of solar minimum.

The peaks in the sunspot curve are called *maxima of solar activity*. In 1778, for instance, the annual relative number was 154, in 1917 it was 104, etc. These years are called years of solar maximum.

Similarly, we can depict the changes in other solar phenomena (flocules, faculae, prominences, flares, etc.). We arrive at curves similar to that of relative numbers. The curve of relative numbers is therefore sometimes called the *solar activity curve*.

The solar cycle

The part of the sunspot curve between two neighbouring minima is called the solar cycle or *solar activity cycle* (Figure 171). Between 1856 and 1867, for instance, the solar cycle lasted about 11 years. Indeed, the solar cycle is sometimes known as the 11-year cycle. But, as the sunspot curve shows, it may last anything from seven to 16 years – it is not entirely regular.

Sunspots and other solar phenomena which occur in their vicinity affect the Earth. Thus the solar cycle is reflected by events here on Earth. These include the aurora, the formation of the growth rings of some trees, the weather, the incidence of certain diseases, the number of traffic accidents, and so on (Figure 399).

The chromosphere

The movement of clouds of hot plasma in the convective layer is several times faster than the fiercest gale known on Earth. There is thus a lot of noise in the convective zone and in the photosphere. All noise is simply a form of energy. The noise in the photosphere is propagated upwards, into the upper layers of the solar atmosphere. There this energy is absorbed and converted to heat, which warms the chromosphere and the corona.

The chromosphere is about 10,000 km (6200 miles) thick, and consists of very rarefied plasma (Figures 163 number 6, and 172). Its temperature reaches as much as 20,000 kelvin. It shines much less than the photosphere, and only in a few colours (spectral lines). The brightest is the red radiation given off by the most abundant element, hydrogen. This is known as H-alpha-line radiation. We can thus observe the chromosphere only by means of special telescopes (called chromospheric telescopes) which 'see' H-alpha-line radiation only (Figure 172).

There are constant geysers from the chromosphere called spicules. These are columns of plasma around 1000 km (620 miles) across. They move at some 20 km (12 miles) per second upwards to a height of 10,000 km (6200 miles) above the chromosphere and disappear after about 10 minutes. There are over one million spicules on the whole Sun at any moment (Figure 163).

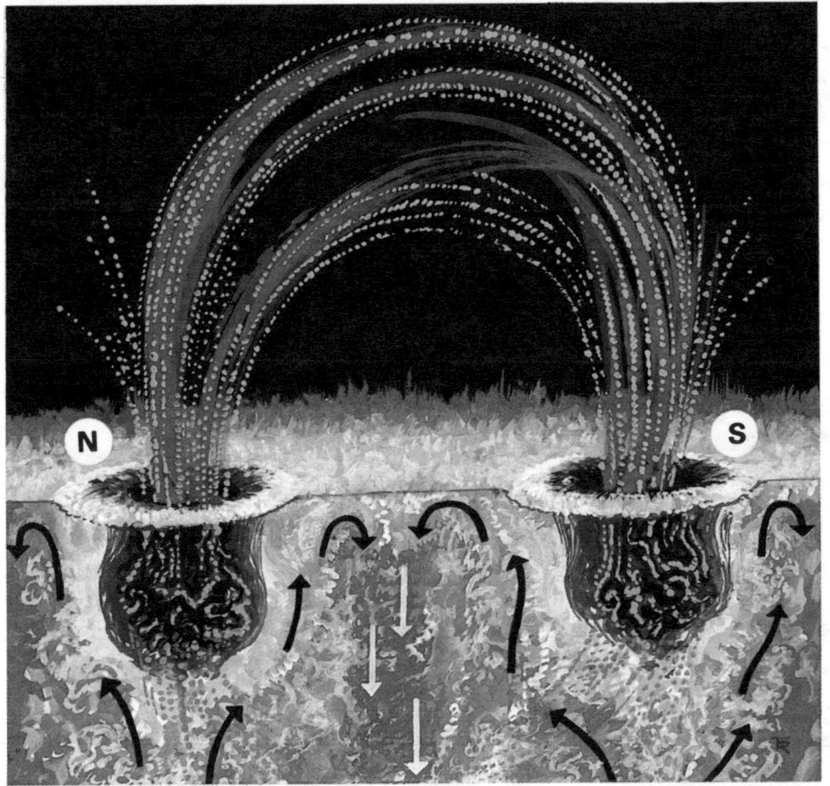

170

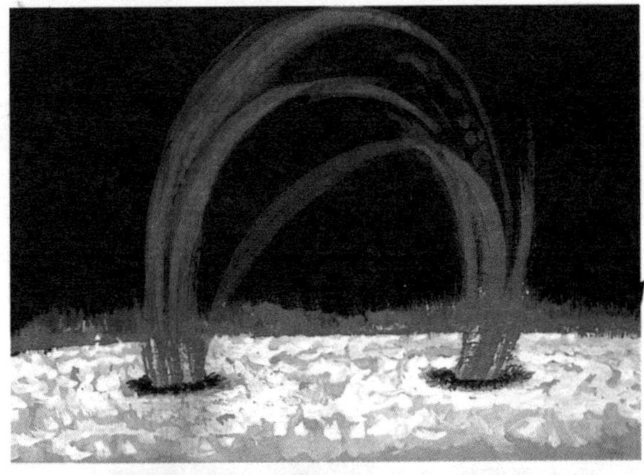

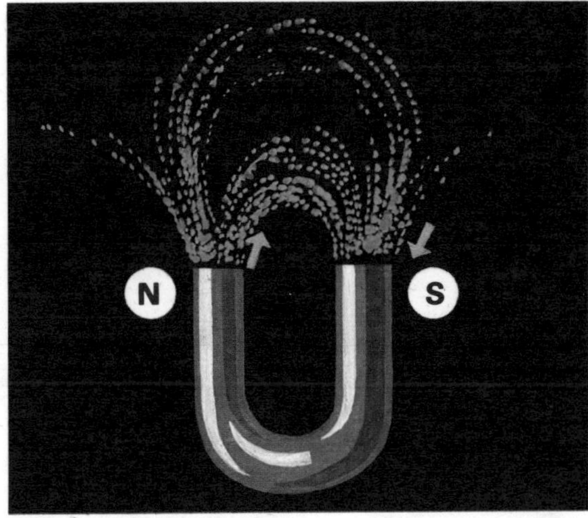

The corona

170 The area around a magnet (bottom right) is surrounded by magnetic lines of force (dotted green lines). Sunspots are huge magnets, along whose lines of force radiant hot gas moves, forming prominences (top right). The lines of force are closed below the photosphere (figure on left).

The highest part of the Sun's atmosphere, the corona, is visible with the naked eye only during a total solar eclipse, when the Moon covers the photosphere and the chromosphere (Figure 190). Then the solar corona appears in a dark sky as a large white ring around the Moon.

Other than during eclipses, astronomers observe the corona using radio telescopes on the surface of the Earth or with X-ray telescopes in spaceships or satellites. X-ray radiation and radio waves are emitted only by the hot corona, with a temperature of from 2 to 50 million kelvin. The photosphere is too cool to emit either X-rays or radio waves. We can therefore observe the X-ray and radio corona even over the disc (Figure 173). In the visible light spectrum the corona can be observed outside eclipses only at the limb of the solar disc. Very delicate instruments called coronographs and coronometers are used for the purpose (Figures 174, 190 and 192).

The corona is the largest part of the solar atmosphere. It extends for several hundred thousand kilometres over the Sun's surface. It consists of very rarefied plasma, heated from below by energy arriving in the form of noise from the photosphere. The air in a matchbox contains the same number of particles as 1 km^3 (0.24 cu miles) of the corona. Its temperature is 1–2 million kelvin.

171

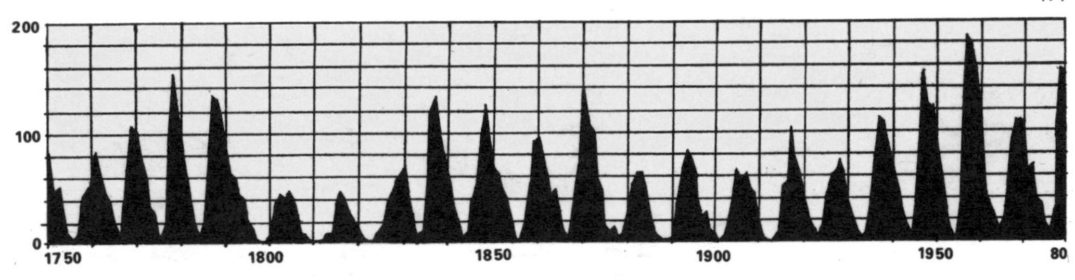

171 A sunspot curve or graph of relative numbers. The horizontal axis gives the year, the vertical one the mean (average) relative numbers per annum.

Over the sunspots it is even higher. Exceptionally, and for short periods, the temperature of the corona over the sunspots may reach 50 million kelvin, which is much hotter than that in the core of the Sun (i.e. 13 million kelvin). But the corona is too rarefied for thermonuclear reactions to take place. The close collisions of protons are extremely rare events.

100,000 km (62,000 miles) over a group of sunspots (Figures 163 number 14, and 176). Loop prominences occur during large flares and last for several hours.

The largest of all are the quiescent prominences (Figures 163 number 13, and 178), which are often up to 200,000 km (124,000 miles) long, or even more. They are irregular in shape and slow-moving. After several

172

172 A small part of the chromosphere close to the edge of the Sun. Its height is about 10,000 km (6200 miles) above the photosphere (it is seen over the light border at the top). The long, dark filaments are cooler gas flowing along the lines of force into sunspots. The photograph was taken in the red light H_α of hydrogen.

Prominences

Observed in the hot and highly rarefied corona are clouds of cooler plasma (10,000 kelvin). They are known as prominences. Prominences contain mostly atoms of hydrogen, so they emit red H-alpha-line radiation. The plasma of prominences is similar to that of the chromosphere; prominences are in fact protrusions of the chromosphere into the corona.

Prominences can be observed with the naked eye only during a total solar eclipse. If you want to observe them at other times you must use an instrument called a prominence telescope. This makes an artificial eclipse using a metal disc.

Close to sunspots, prominences are clear-cut; the strong magnetic field of the sunspots gives them a regular shape. One of the finest sights in astronomy is the regular, loop-shaped prominence which arches up to

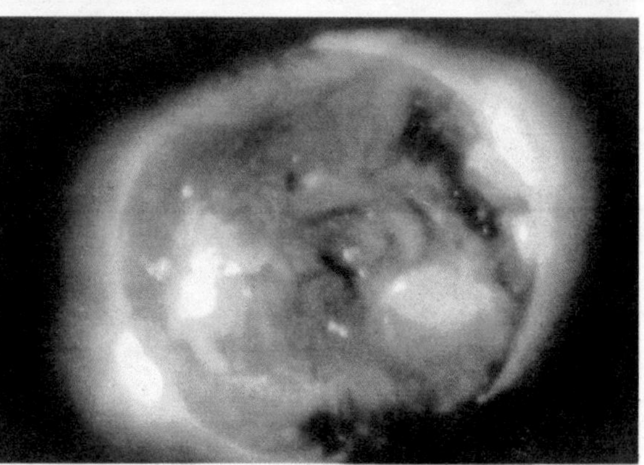

173

173 The Sun, photographed in X-rays. Only the corona shines above groups of spots, where there is increased density of gas and temperature.

weeks a quiescent prominence suddenly expands, rises, and sometimes escapes from the Sun. These dying prominences are called eruptive prominences (Figure 179). In them, thousands of millions of tonnes of plasma are thrown out in a short time and with velocity up to 1000 km (620 miles) per second. This is the most dramatic spectacle to be seen in the Solar System.

vatories throughout the world observe, measure and photograph flares. We should like to know how they occur, and to learn more about the effect they have on living creatures and on the Earth in general.

Flares occur close to sunspots, whose magnetic fields contain energy, part of which is transformed in a flare. In the course of a few minutes up to 10^{26} joules of magnetic

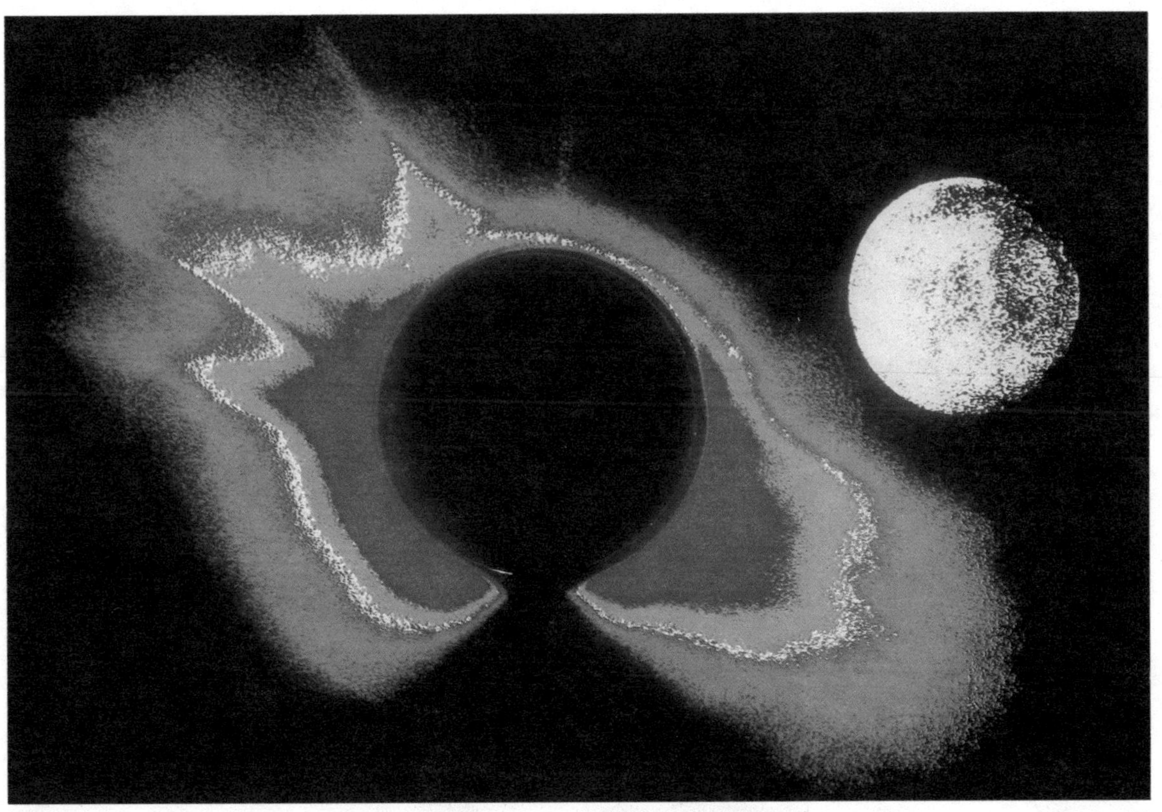

174 The solar corona photographed from a satellite just before an eclipse. The Moon is very close to the Sun. It is new, and we would not see it from the Earth with the naked eye. The faint luminescence of the new moon is the reflection of the light of the Earth (called earthshine). The colours of the corona correspond to its brightness (they are produced artificially). The corona itself is white.

174

Flares

A flare is a sudden brightening of the chromosphere (Figure 177), the corona, and sometimes also the photosphere. Flares occur over groups of sunspots or in their vicinity. Roughly speaking, we can say that the larger the sunspot, the more flares occur in it. In the biggest sunspot groups, 10 or more flares will occur each day. Small flares have an area about the size of Europe, while large ones cover a part of the chromosphere bigger than the surface of the Earth. In the corona they reach a height of up to 100,000 km (62,000 miles). Small flares last a matter of minutes, while large ones may last an hour or more. Several satellites and many obser-

energy can be converted to heat, photons, the kinetic energy of particles, the corpuscular radiation of the solar wind, and other energy forms. (The solar wind is dealt with in the following section.) The energy released in a flare escapes as huge streams of radio, ultraviolet, X-ray and sometimes even gamma photons. A few per cent of the energy released in a flare is carried away by the streams of corpuscular radiation, whose destructive force will be discussed below. A large flare may throw out up to 10^{40} particles (protons, electrons, and the nuclei of various atoms). Their total mass may be many thousands of millions of tonnes. These particles cause a large increase in the solar wind – a phenomenon called an interplanetary storm. Against the solar wind and interplanetary storms life on Earth is protected by its atmosphere (Figure 386).

Ultraviolet, X-ray and gamma photons are also very dangerous to all living organisms. Fortunately, the Earth is surrounded by its atmosphere, which absorbs these harmful photons. If it were not for the magnetosphere and the atmosphere around our planet, solar flares would long ago have destroyed all life on Earth.

The energy of a single large flare – emitted as photons and particles – is about 10 times larger than the energy consumption of all of humanity since the first fire was lit by man.

The solar wind

The corona can be observed to a distance of several million kilometres from the surface of the Sun. The transition between the corona

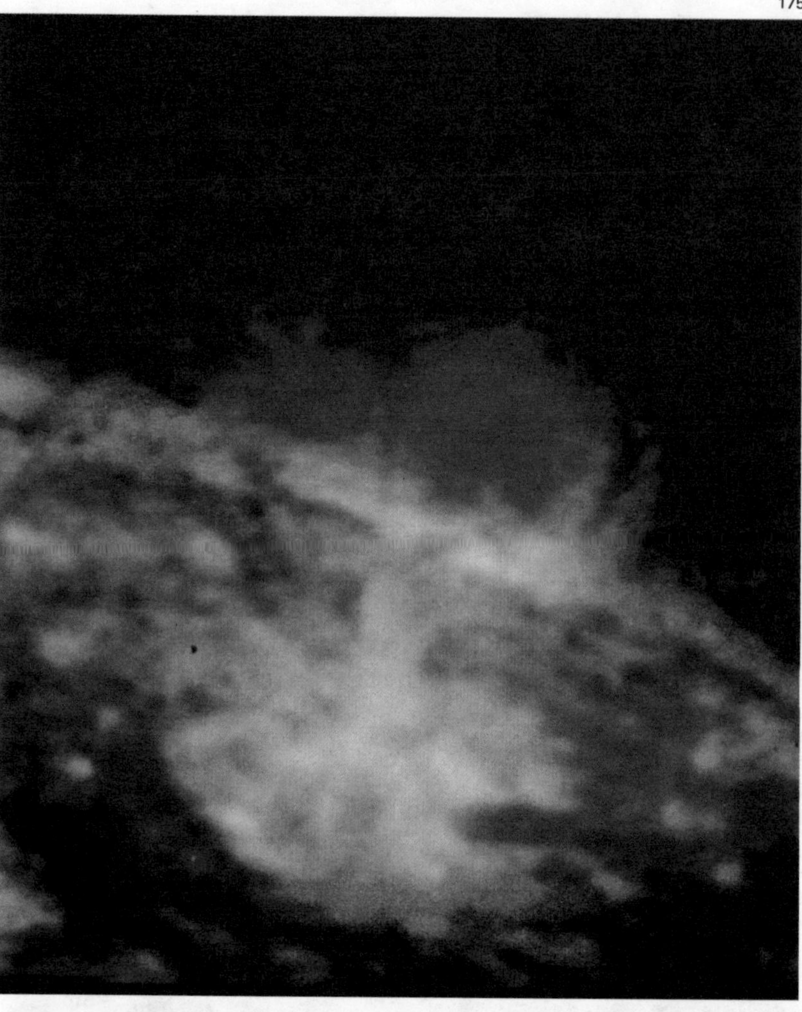

175 Two active regions close to the edge of the Sun. This photograph was taken in ultraviolet radiation on November 23, 1973 by an instrument on board Skylab (i.e. outside the Earth's atmosphere). The blue colour corresponds to a temperature around 600,000 kelvin, which is the transitional layer between the chromosphere and the corona. The green colour corresponds to a temperature of about 10,000 kelvin, which is the temperature of the chromosphere and prominences, while red is a temperature of 2,200,000 kelvin, corresponding to the hot corona.

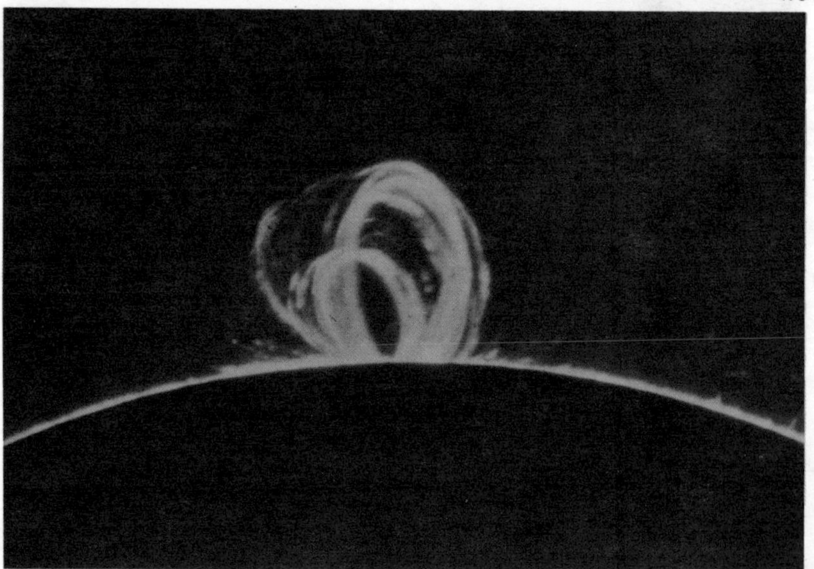

176 Loop prominences in red hydrogen light (in the H-alpha line). The photosphere is covered by a disc in order to make the picture clearer. Loop prominences are formed by magnetic force lines in which plasma is caught up. They occur particularly after major flares.

and interplanetary space is indiscernible. No one can say for certain where one ends and the other begins. The very rarefied, hot plasma of the upper part of the corona is constantly expanding into space and escaping from the Sun. Interplanetary probes measure a constant flow of particles, especially protons, alpha particles and electrons. This stream moves in the direction away from the Sun, and is called the solar wind (Figure 163 number 12). The particles in the solar wind move 500 times faster than the bullet from a gun. They are very dangerous to all living organisms, but the Earth's magnetosphere acts as a shield against them (Figure 386).

The solar wind has nothing in common with the wind on the surface of the Earth; it 'blows' at 300–600 km (186–372 miles) per second, and after large flares at up to 1000 km (620 miles) per second. Terrestrial winds are several thousand times slower. The wildest gales on Earth have speeds up to 200 m (656 ft) per second, making them 5000 times slower than the solar wind at its fastest, when it is known as an interplanetary storm. The density of the wind on Earth is 1.2 kg per m^3, or about 10^{27} nucleons per m^3, which is 100 trillion (10^{20}) times more than that of the solar wind (a few million nucleons per m^3). The only feature the two winds have in common is their kinetic energy, which is of solar origin.

The solar wind blows from the Sun in all directions. In this way about 1 million tonnes of hot plasma escape from the solar corona every second. Calculations indicate that it probably reaches well beyond the bounds of the Solar System – two to three times further

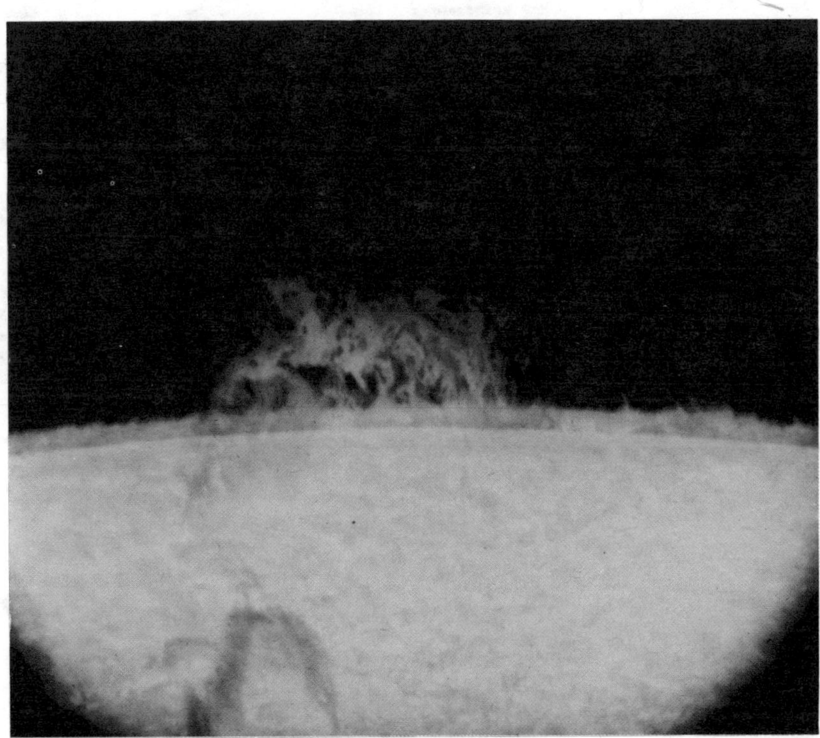

178

178 A small section of the solar limb photographed in hydrogen light (H-alpha line). Over the limb (photosphere) is an irregular band (chromosphere). Above the chromosphere is a quiescent prominence. Seen against the disc the prominence seems dark, and is called a filament. The whole of the Sun in this photograph would measure about 1 m (3.3 ft) across.

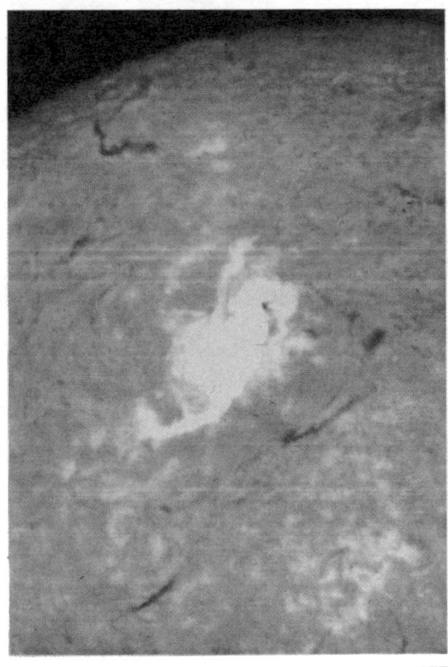

177

177 A large solar flare.

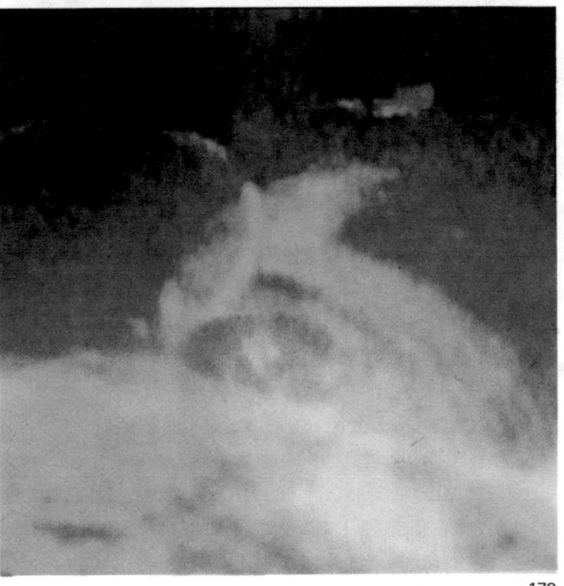

179

179 Prominence on the western limb of the Sun. The photograph is in ultraviolet radiation. The blue colour shows the plasma of chromosphere and prominence (10,000 kelvin), while the surrounding hot corona (1 million kelvin) is red. The height of the prominence is 130,000 km (80,700 miles).

180 Six selected regions of the solar spectrum. In the lines of helium (second strip), calcium (fifth strip) and hydrogen (first, third, fourth and sixth strips) there is strikingly strong emission. It originates from a solar flare and after 20–40 minutes or so it weakens and disappears. A detailed study of such emission lines reveals the temperature, density and other properties of the flare.

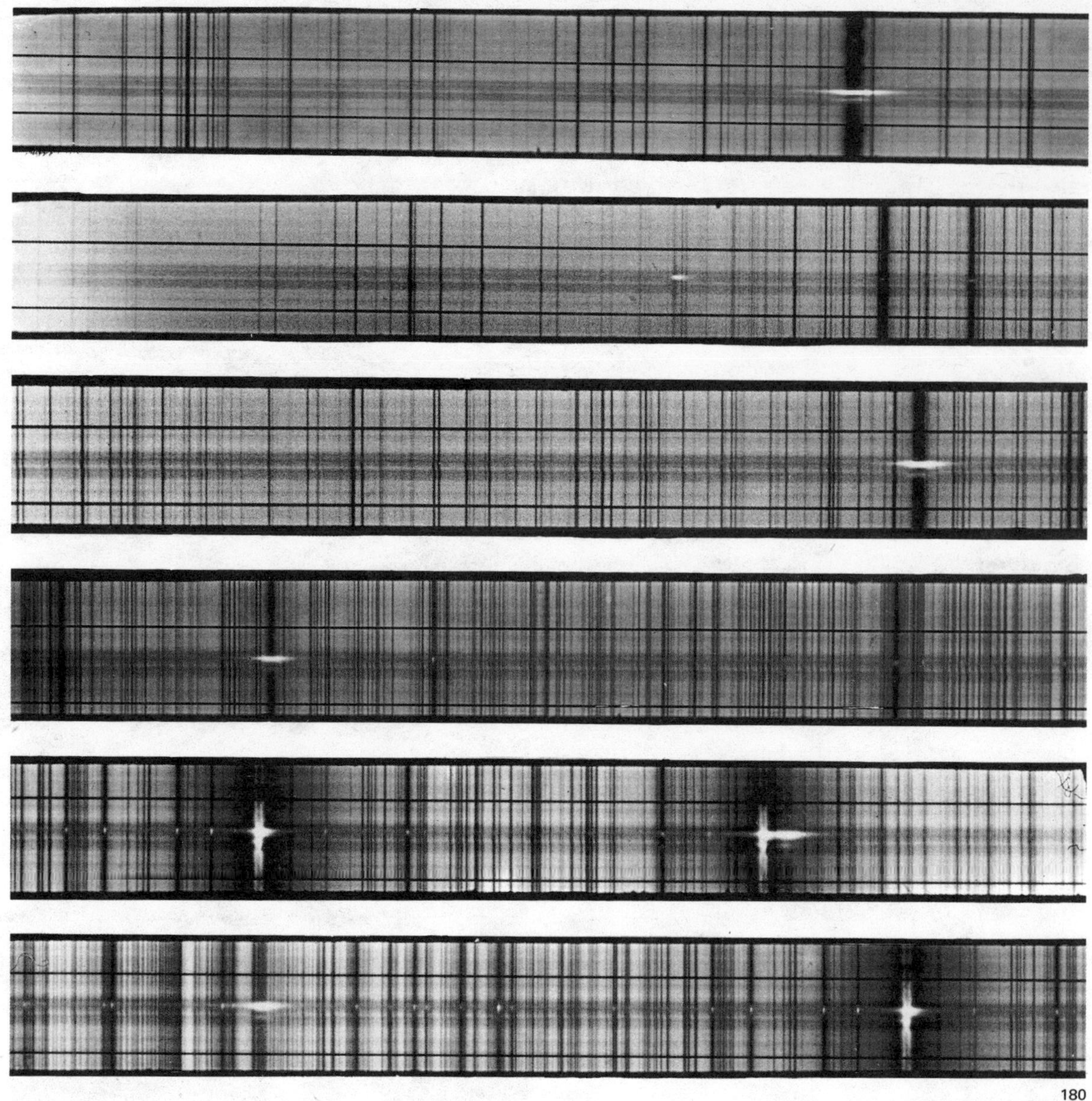

180

than Pluto. We would expect to find the solar wind at distances of 15,000 million km (9300 million miles) from the Sun – which is a distance one hundred times that between the Earth and the Sun. The extensive space around the Sun in which the solar wind blows is called the *heliosphere*. It is roughly spherical, with the Sun and its planets in the middle (Figures 114 number 2, and 185 number 2). The Sun and its gravitational force rule the entire heliosphere, not allowing the particles of the solar wind to leave it and escape into distant space between the stars.

Solar neutrinos

So far we have spoken only of photon and corpuscular radiation. Both of these carry away from the Sun energy released in the thermonuclear reactions in its core. For the sake of completeness we must mention a third type of radiation, which is quite different from the two above. It is called *neutrino radiation*. While photons and particles are emitted from the Sun's atmosphere, neutrinos are born in the solar core, i.e. close to the very centre of the Sun.

In the Sun's nucleus 4×10^{38} protons a second are changed into helium nuclei. This gives rise to 2×10^{38} electron neutrinos per second (page 130). As soon as the neutrinos appear, they move in a straight line outwards from the Sun's nucleus (Figure 163 number 9). In their journey through the Sun the neutrinos encounter a huge number of protons, electrons, alpha particles and the nuclei of other elements. But none of these particles has any effect on the neutrinos, nor do they absorb them. They easily pass through the Sun into cosmic space in $2\frac{1}{3}$ seconds.

Only one solar neutrino in 2000 million falls on the Earth (a total of 10^{29} neutrinos a second). This means that 700 billion neutrinos a second fall on 1 m^2 (10.76 sq ft) perpendicular to the Sun's rays ($7 \times 10^{14}\ s^{-1}m^{-2}$). They bring with them direct information on the hot core of the Sun. If our eyes were sensitive to neutrinos, we should see a 'neutrino Sun', or the core of the Sun, which is much smaller than the visible Sun (the photosphere) – Figure 163 number 1.

Unfortunately, neutrinos pass unnoticed through the retinas of our eyes, without passing any information on to our brains. In fact they pass quite unaffected through the whole of the Earth. We should thus be able to observe the neutrino Sun even at night, since the Earth is transparent to neutrinos. How, then, are we to catch neutrinos, since our instruments are made of the protons, neutrons and electrons to which they are so indifferent?

Some neutrinos can be caught, very rarely, in the atomic nucleus. For this purpose scientists have made a 'neutrino telescope', situated deep inside a mine. It is in fact a huge tank (the size of a swimming-pool) filled with a chemical called perchlorethylene. The perchlorethylene molecule (C_2Cl_4) consists of two atoms of carbon and four of chlorine. It is in the nuclei of chlorine that neutrinos are, very rarely, caught, and the chlorine nucleus changes to an argon nucleus. Thus:

$$\nu_e + {}^{37}_{17}Cl \longrightarrow {}^{37}_{18}Ar + e^-$$

where ν_e is the solar neutrino, Cl the nucleus of chlorine with 37 nucleons (17 of them protons), and Ar is the argon nucleus with 37 nucleons (18 of them protons). In the chlorine nucleus one of the neutrons captured a neutrino ν_e and changed into a proton:

$$\nu_e + n \longrightarrow p + e^-$$

The remaining nucleons were not affected.

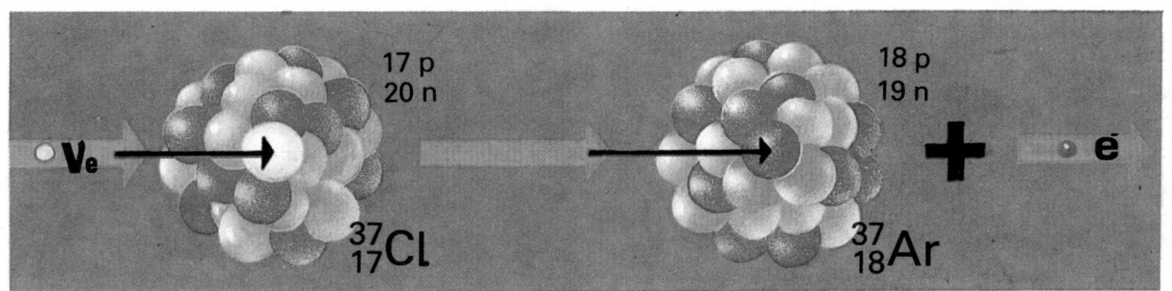

181 By the capture of a neutrino ν_e in one neutron of the chlorine nucleus the neutron is changed into a proton and the nucleus of chlorine into a radioactive argon nucleus.

The capture of a neutrino just described is represented in Figure 181.

The actual measurement consists of obtaining a few resulting argon atoms (${}^{37}_{18}Ar$) from around 400,000 litres (88,000 imperial gallons) of perchlorethylene. After 'fishing them out' they must be counted. The resulting argon nuclei are radioactive, and this makes it possible to identify them and count them.

The vast majority of solar neutrinos cannot be captured in this way at all, since the neutrino telescope described here is only sensitive to neutrinos with a high energy, of which there are very few. Nor are the results obtained by this method entirely reliable; but improved neutrino telescopes which should be sensitive even to low-energy neutrinos are being developed.

The calm Sun and the stormy Sun

The energy released in the core is transported to the photosphere, from where it escapes into space. It is carried away by three types of radiation: neutrino, photon and corpuscular. Neutrinos are measured deep below the ground. Photons are studied either in observatories on the surface of the Earth, or in satellites above the Earth's atmosphere. Satellites are used to study those photons which are absorbed by the atmosphere, i.e.

X-ray, ultraviolet, gamma and some infra-red photons and radio photons with very low frequencies. We have already dealt with radiation of the neutrino and corpuscular (solar wind) types (see pages 140–143). Here we shall take a look at photon radiation. The photons carry away much more energy from the Sun than the neutrinos and particles put together. The photon radiation of the Sun can be divided into two components: the *basic (constant)* and *the supplementary (variable)*.

• The *basic photon component* does not change. Its photons are radiated from the photosphere, and it contains the vast majority of solar energy. The chromosphere and the corona contribute only a minor amount to the basic component. Their contribution is

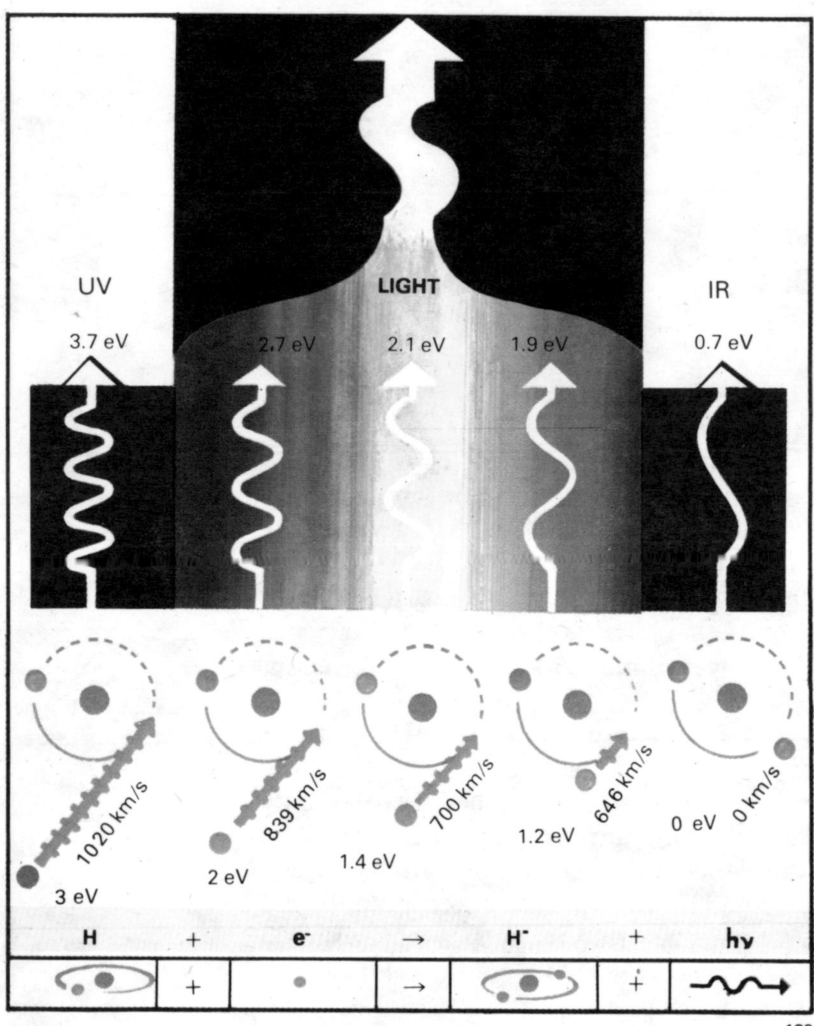

182

182 A photon is emitted from the photosphere when an electron is captured by a neutral atom of hydrogen. In this way a hydrogen atom with two electrons arises. The faster the electron was moving when captured, the greater the kinetic energy it gives to the photon. An electron which moves slowly emits an infra-red (IR) or red photon. Fast electrons emit blue or ultraviolet (UV) light.

183

183 The constant (basic) component of solar radiation (infra-red, light) is emitted by the granules of the photosphere. The variable component, on the other hand (radio and X-ray radiation), is emitted from the hot and rarefied corona.

shown in Figure 183, from which it is clear that most of the radiation is emitted from the photosphere (light and infra-red photons). Most of the solar photons are born in granules. We might say figuratively that their parents are electrons and atoms of hydrogen (Figure 182). By linking a hydrogen atom with an electron, a negatively charged atom (or negative ion) of hydrogen arises. In the process the energy of the electron is released as a photon. This is in fact the last emission of the photon and its parting from the Sun. Prior to that the photon had been emitted, absorbed and re-emitted countless times in the interior of the Sun.

When an electron is captured by a hydrogen atom in the photosphere, it irradiates all its energy – binding and kinetic. An electron which is not in motion emits only binding energy (0.75 eV), like an infra-red photon.

That sort of photon does not have enough energy to be visible to us, but we can feel it on our skin in the form of heat.

An electron which is in motion has, in addition, kinetic energy ($\frac{1}{2}mv^2$, where m is the mass of the electron and v its velocity). The faster an electron moves, the greater its kinetic energy. When it is then captured by a hydrogen atom, it emits in one photon both its binding (0.75 eV) and its kinetic energy. The faster the electron was moving, the greater the energy of the photon. The solar photon thus obtained its energy from a free electron in the photosphere. The electron passed its kinetic and binding energies on to the photon emitted during its attachment to the hydrogen atom. This process is called *recombination,* and it can be written as follows:

$$H + e^- \longrightarrow H^- + h\nu$$

where H is the neutral atom of hydrogen, H^- is the negative hydrogen ion, $h\nu$ is the photon emitted, h is Planck's constant, and ν is the frequency of the photon. The product $h\nu$ is thus the energy carried off by the photon.

A free electron can have various amounts of energy. If it is at rest, the photon $h\nu$ which is emitted carries off only the binding energy (0.75 eV). But electrons in the photosphere are rarely at rest. They move at different speeds, and thus have various amounts of kinetic energy, so that the photons which arise have different energies, and together make up white light. The photosphere emits light photons of all colours, and also infra-red (by recombination of slow electrons) and ultraviolet (by recombination of the fastest electrons).

That is how solar photons, which bring us energy – and with it heat, light and movement – are born. Without them we should not be here. The birth of photons of the basic component is described in Figure 182.

• The *variable component* of the solar radiation consists of X-ray (sometimes also gamma), ultraviolet and radio photons. Figure 184 shows how their intensity can change with time. The photons of the variable component are emitted from the upper layers of the solar atmosphere – the chromosphere and the corona. The variable component is weak compared with the basic component, and depends entirely on solar activity, especially on sunspots and flares. The more tumultuous the solar activity, the more the variable component intensifies. At solar maxima the variable component is much greater than at solar minima.

As an energy source, the basic component is absolutely essential to life on Earth. It remains constant, as may be seen in Figure 184 number 2. The variable component (Figure 184 numbers 1 and 3) does not bring us much energy, and it does so irregularly. It is not essential to life – on the contrary, it can be harmful to the health of living creatures (Figure 399).

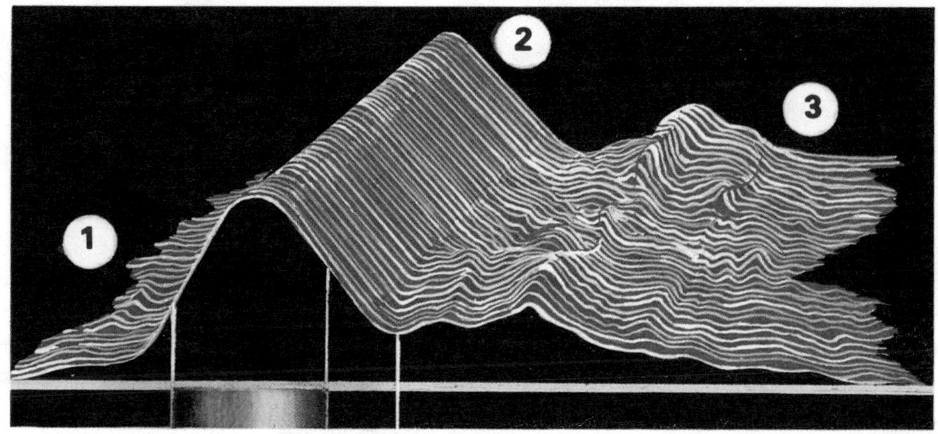

184

184 Each white curve (solar spectrum) shows the dependence of intensity on frequency. The figure shows a series of many solar spectra at hourly intervals. The part containing light, ultraviolet and infra-red radiation contains most energy and is constant **(2).** The radio **(3)** and X-ray **(1)** parts of the spectrum, on the other hand, change with time. At maximum solar activity both parts **(1)** and **(3)** are elevated and variable. During minimum solar activity they are very low and inactive.

Destiny of the Sun's rays

Within a few seconds of their birth in the photosphere, photons have passed through

185 Light travels for five hours across interplanetary space to reach Pluto's orbit **(1).** About 15 hours after leaving the Sun the photons meet the particles of the solar wind in the heliosphere **(2).** Then for several months the photons pass through Oort's Comet Cloud **(3).** About two years after they left the Sun, the photons meet comets in interstellar space **(4)** which may belong to a neighbouring star.

the extremely thin corona into the Solar System. Here only a very few of them are trapped by the various cosmic bodies (Figure 185). Just one photon in several million hits a planet or one of its moons, an asteroid, meteor or lump of iron, a comet or a speck of fine dust, all in the close vicinity of the Sun.

After about five hours' breathless flight (covering around 300,000 km/186,000 miles per second) the photons reach the planet Pluto with their numbers scarcely diminished. But this is not the end of the Solar System, which extends many thousand times that distance. The outer region is the heliosphere, where the photons meet the protons, alpha particles and electrons of the solar wind and fine dust expelled from interplanetary space. About 15 hours after leaving the Sun, the photons reach the edge of the heliosphere, where the solar wind blows itself out.

For another year and a half the photons may meet the farthest-flung members of the Solar System, the snowballs of the Oort's Cloud. That is about half way to the nearest star, Proxima Centauri. By now the photons are leaving the realm of the Sun; here solar gravity ends. So after a year and a half of flight, they reach the vast area of interstellar space, where the 'emptiness' is almost total. There is just the occasional atom of hydrogen or helium, and even fewer specks of dust. In some places extensive dust clouds can be observed. The grains are extremely small (about one-tenth of a micron), and are made of carbon and silicates. Some are coated with ice. A large number of photons end their lives in these dust clouds, their energy being converted to heat as they strike the grains. Photons which are not absorbed in a dust cloud escape the Galaxy after several thousand years (Figure 128). They then spend millions of years flying through the huge expanses of intergalactic space. On reaching another galaxy they are absorbed by either a dust cloud or one of the billions of stars.

The farthest visible galaxy (photographed by the huge Mt Palomar telescope in California) is in the constellation Hercules, 5800 million light years away. The first photons from the Sun will take still 800 million years to get there. (They left the Sun 5000 million years ago – so they have already covered the distance of 5000 million light years). The most distant galaxies detected by radio waves are about 10,000 million light years away. It will be about another 5000 million years before solar photons get there. Our Sun doesn't exist yet for those galaxies.

185

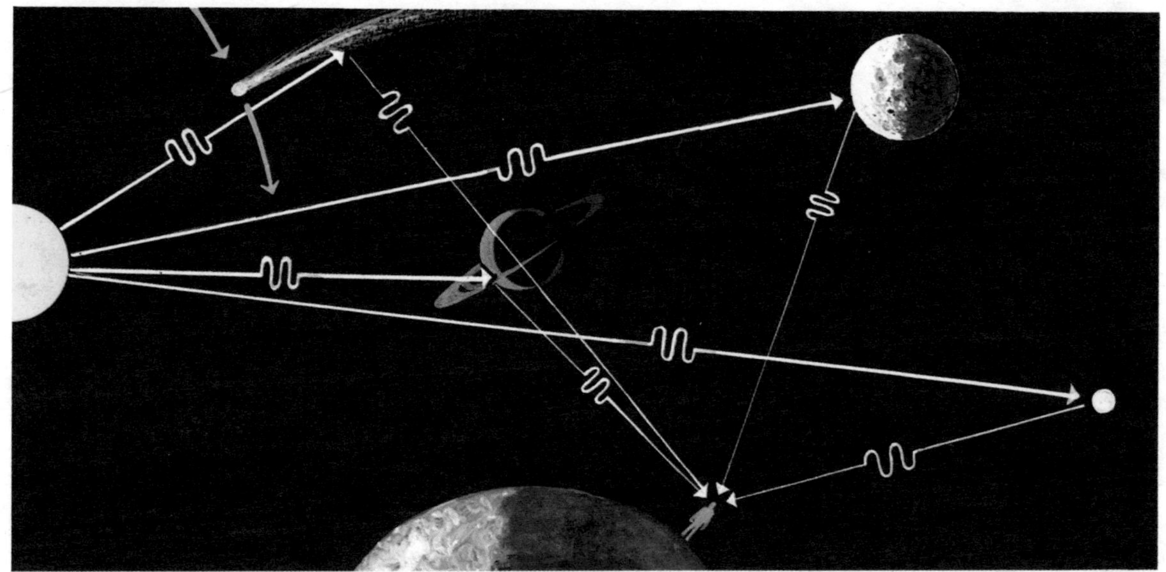

186 The light of planets, comets and satellites is reflected sunlight. Without the Sun these cool bodies would not be visible.

Solar radiation in the planetary system

Only one photon in several million emitted by the Sun remains in the Solar System. The rest escape into the Universe. Thus the Sun gives only a tiny proportion of its energy to its own 'family'. Nearly all the solar radiation ends up in the freezing space between the stars. But for us even that tiny part of the energy which is captured in the Solar System is immensely great and vitally important.

The bodies of the planetary system, large and small, react in different ways to solar radiation. Some absorb it (changing it into heat and thus warming up), while others reflect it back again. It is due to this reflected part of the solar radiation that we are able to see and learn about the individual members of the Solar System (Figure 186).

The brightest planet in our skies is Venus – our closest neighbour. Other planets which are easily visible with the naked eye are Mars, Jupiter and Saturn (Figure 108). The

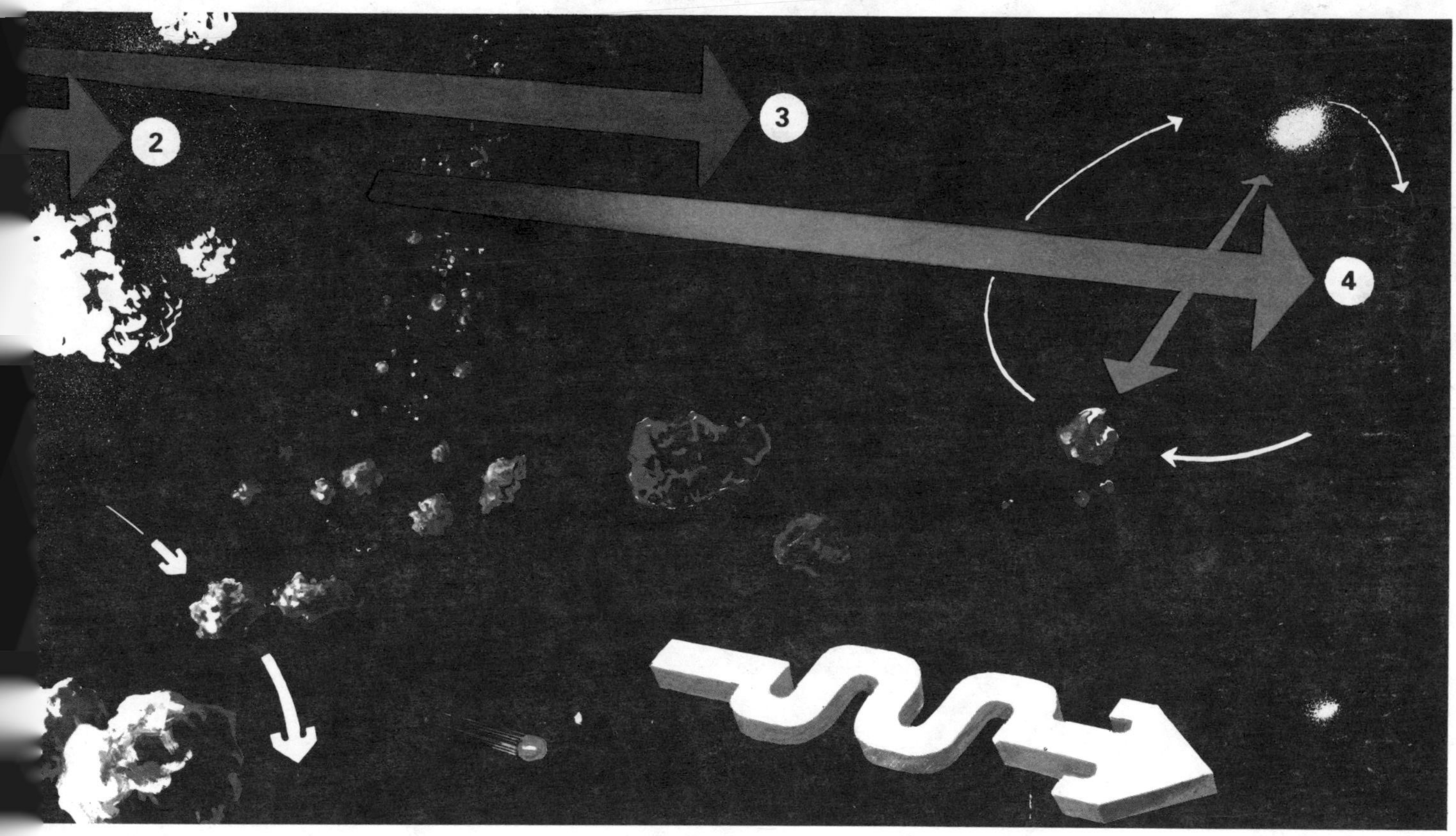

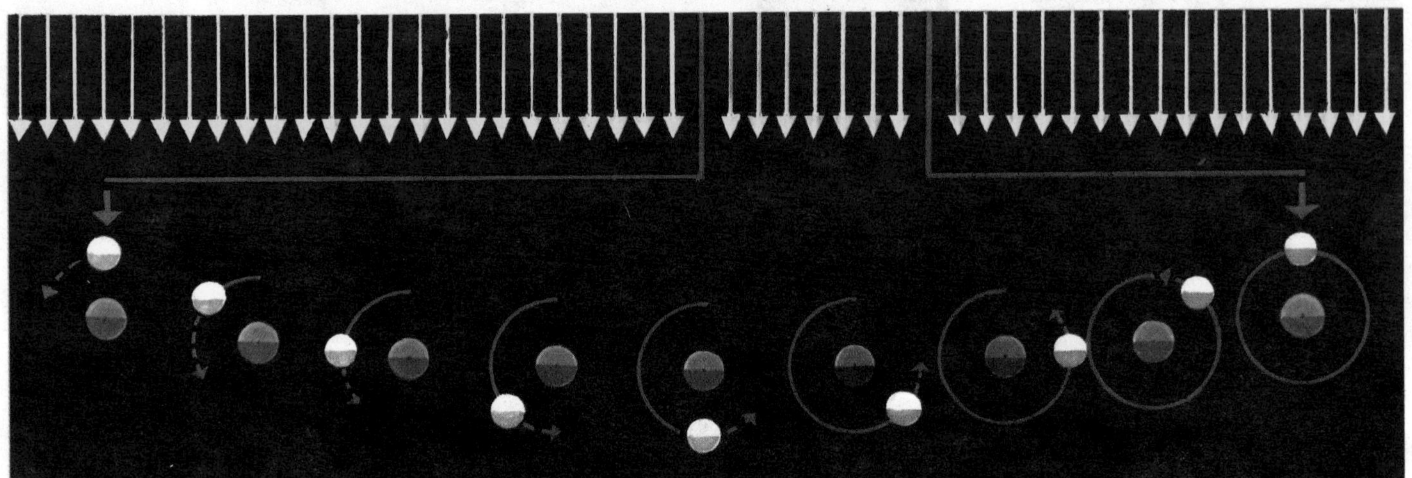

187 The Earth is accompanied on its orbit around the Sun by the Moon (top and middle figures). The orbit of the Moon around the Earth is shown by the red arrow. The lunar month is the time taken for the Moon to complete a cycle of phases (bottom figure). It lasts 29.5 days.

planets look like stars, but their positions in the heavens change. However, stars are many millions of times farther away than planets. The Sun and most of the stars have roughly one million times the volume and mass of planets, and surface temperatures of many thousands kelvin. The Sun and other stars are self-luminous bodies. Planets, on the other hand, receive light from the Sun,

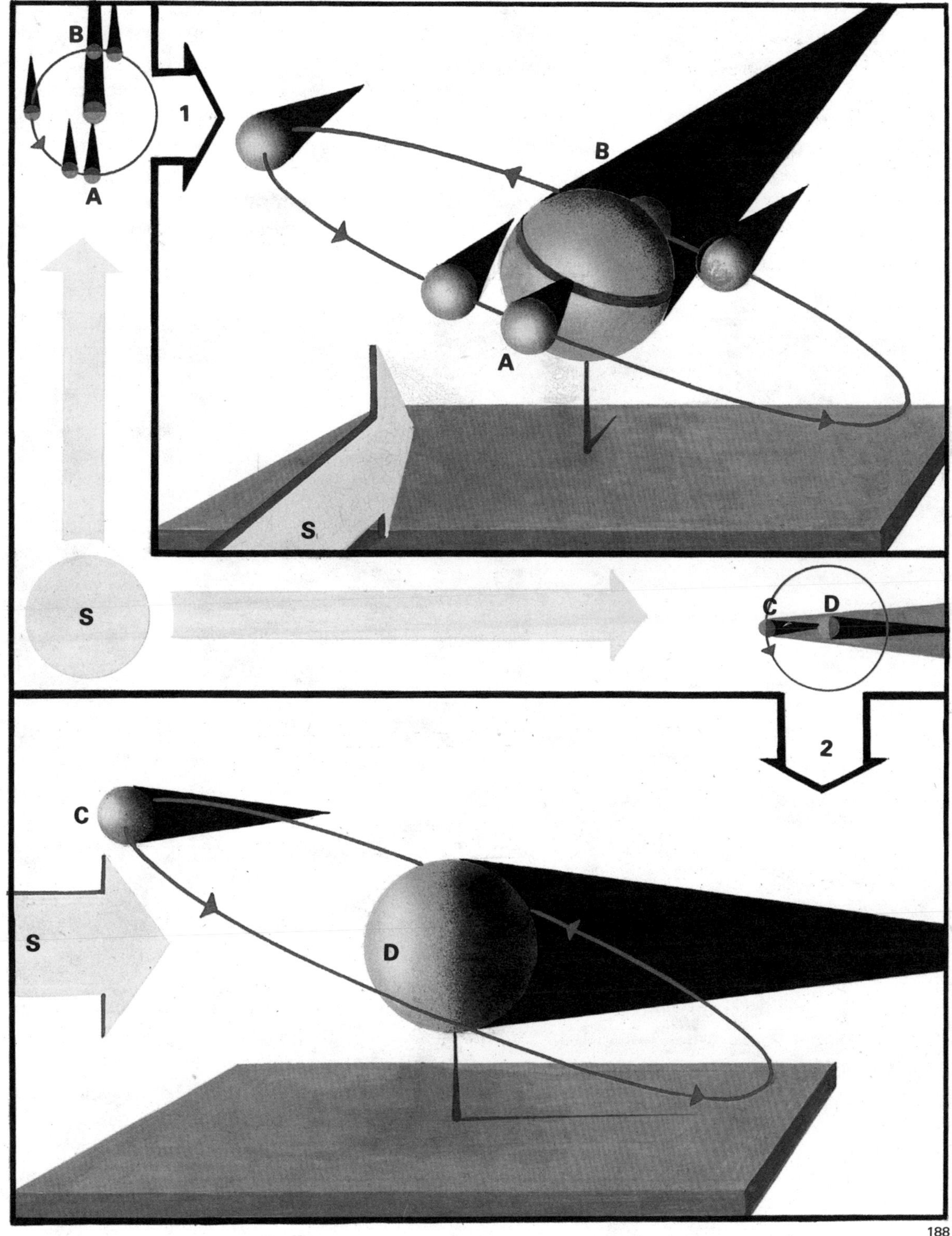

188 Eclipses of the Moon and the Sun. If the Moon **(A)** passes between the Sun **(S)** and the Earth **(B),** its shadow falls on the Earth. In the region where the Moon's shadow falls a solar eclipse is observed. But the Moon's orbit is slanted with respect to the Earth's, so that the Moon's shadow **(C)** at the new moon often passes above the Earth **(D)** or below it. There is then no solar eclipse on the Earth. If the Moon gets into the shadow of the Earth, a lunar eclipse is observed. The lunar eclipse is seen from the whole nightside hemisphere of the Earth. However, the total solar eclipse may be seen from a narrow band of the dayside hemisphere only.

which is reflected from their surfaces (Figure 186). Planets are too small and cool to emit light of their own.

Planets have satellites which orbit them. With the naked eye we can see only our own, the Moon. The Moon arose from the same parent cloud as the planetary system and the Sun. It is a twin of our Earth. Solar radiation illuminates only one part of the Moon. As the Moon orbits the Earth we see all, part, or none, of this illuminated half. At the full moon, it is all illuminated, in the first or last quarter it is half illuminated, and at the new moon, none is illuminated. A full moon occurs when the Moon is on the opposite side of the Earth to the Sun (Figure 187). That is why we can see it only at night. It is easy to deduce from the figure that the new moon is visible only in the daytime (i.e. when we are on the illuminated half of the Earth), the first quarter is visible from midday to midnight, and so on.

The orbit of the Moon round the Earth is inclined towards the plane of the ecliptic (upper part of Figure 187 and lower part of Figure 188). If it lay in the plane of the ecliptic, the Moon would be in the Earth's

seen only partly. The umbra of the Earth is divided in two by the plane of the ecliptic, and stretches for 1.5 million km (nearly 1 million miles), i.e. much farther than the Moon. If a full moon is close to the ecliptic, it must be in the shadow of the Earth, and a lunar eclipse occurs (Figure 188).

The conical shadow stretching out behind the Moon is the same length as the distance

189 A total solar eclipse can be observed only in the narrow band through which the Moon's shadow passes. This varies from one eclipse to another (top 1977, bottom 1979).

190 During a total solar eclipse, the Moon covers the glaring photosphere, making the extensive faint white corona clearly visible. It consists of the light of the photosphere diffused by the electrons of the hot coronal plasma. This photograph was taken by the astronomers of Kiev University during the solar eclipse of June 30, 1954.

shadow at each full Moon, so that we should see an eclipse of the Moon at each full moon. At each new moon, on the other hand, the Moon's shadow would fall on the surface of the Earth, and there would be an eclipse of the Sun.

Behind every body in the planetary system there are long *umbras* and *penumbras* (Figure 188). From the umbra the Sun cannot be seen, whereas from the penumbra it can be

from the Earth to the Moon, so that its apex touches the surface of the Earth (Figure 188 at A). At the places where the Moon's shadow falls on the Earth there is a total solar eclipse; at this point the new moon must be in the plane of the ecliptic. From places in the penumbra of the Moon only a partial solar eclipse can be observed.

A lunar eclipse can be seen from the whole of the dark side of the Earth. A total solar

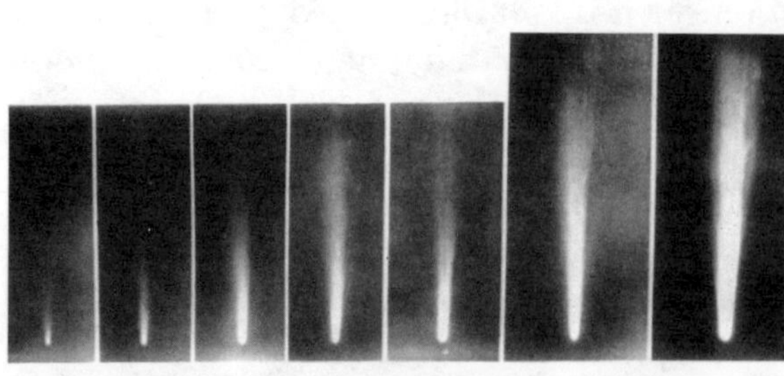

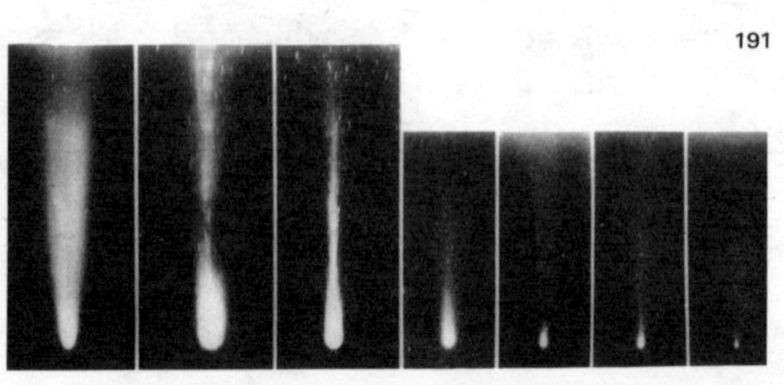

191 Halley's Comet photographed on April 26 to June 11, 1910. The head and tail grow larger as a comet approaches the Sun. Then the comet's brightness again fades as it recedes from the Sun. Halley's Comet orbits the Sun once every 76 years. Its last visit (at the end of 1985 to the beginning of 1986) was a disappointment for many: when brightest it was in the day sky and could not be seen. For the first time in history scientists were able to send their instruments into the comet's head and tail to make observations.

eclipse can be seen only from a narrow belt (approximately 160 km/100 miles wide) on the illuminated side of the Earth along which the lunar shadow moves (Figure 189).

At the time of a new moon, an observer on the Moon's surface would see our planet in full, i.e. he or she would see the entire illuminated half. The 'full' Earth illuminates the dark part of the Moon so brightly that we can see its features from the Earth – this is called earthshine.

Figuratively speaking, the old Moon is said to be in the arms of the new. The earthshine is thus the light of the Sun reflected from the Earth and falling on the Moon, whence it is reflected back to the Earth. Sometimes the earthshine is so bright that lunar mountains and craters can clearly be seen.

The satellites of other planets (Mars, Jupiter, Saturn, Uranus, Neptune and Pluto) are observed and photographed using telescopes. The planets and their satellites also have long, conical shadows behind them, and the equivalent of solar and lunar eclipses can be observed. These can be seen particularly well in the case of the brightest satellites of Jupiter (Figure 87). The passage of a satellite's shadow across Jupiter's disc (the equivalent of a solar eclipse) and the passage of a satellite into the planet's shadow (the equivalent of a lunar eclipse) can be observed. It is also possible to see the satellite disappearing behind the planet's disc (a phenomenon called an occultation).

One of the beautiful manifestations of the solar radiation in the planetary system are comets. From time to time a comet visible to the naked eye appears. When it gets as close to the Sun as the planet Mars, a large head (the coma) and a long tail of gas and dust appear (Figures 191 and 192).

The grains of dust which a comet trails behind itself in interplanetary space subsequently orbit the Sun on their own. There are very many such specks – grains, pebbles and stones orbiting the Sun. These are called meteoroids (Figures 111 and 112).

They make up a large disc formation spread along the plane of the ecliptic (Figure 112). This formation can easily be seen with the naked eye, in the evening in spring and in autumn just before sunrise (in the northern hemisphere). At these times the ecliptic is almost perpendicular to the horizon. The light of the Sun, diffused and reflected by the countless meteoroids, looks like a light cone along the ecliptic. This phenomenon, called the zodiacal light, was described on page 64.

Due to the light of the Sun, we can see the bodies of the planetary system. The solar light reflected from the surface of the Earth makes it possible to see our planet from space (Figures 384, 385 and 389) and also enables us to see all the things on Earth around us.

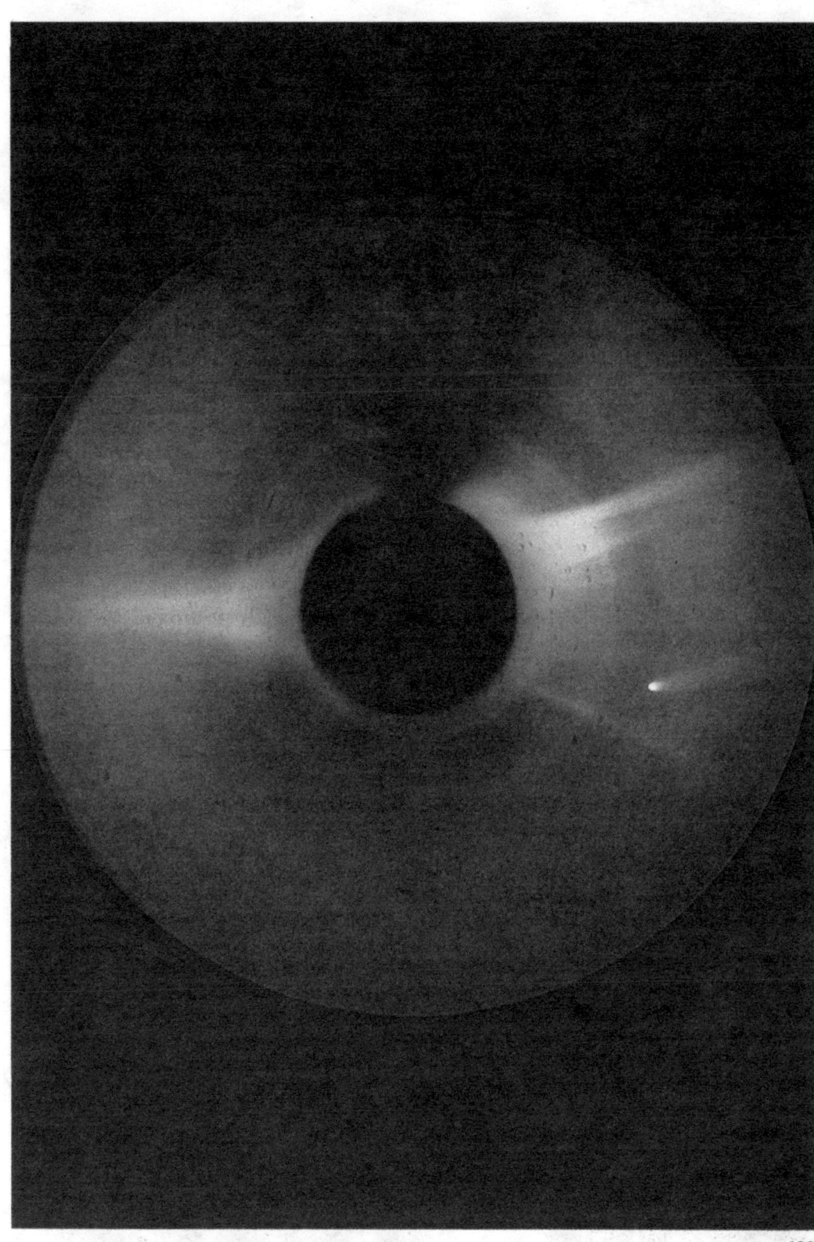

192

192 Kohoutek's Comet close to the Sun. It provided a fine spectacle in 1973–1974. The photograph was taken with a coronograph, which artificially shades the photosphere, making the corona clearly visible in the form of long white streamers.

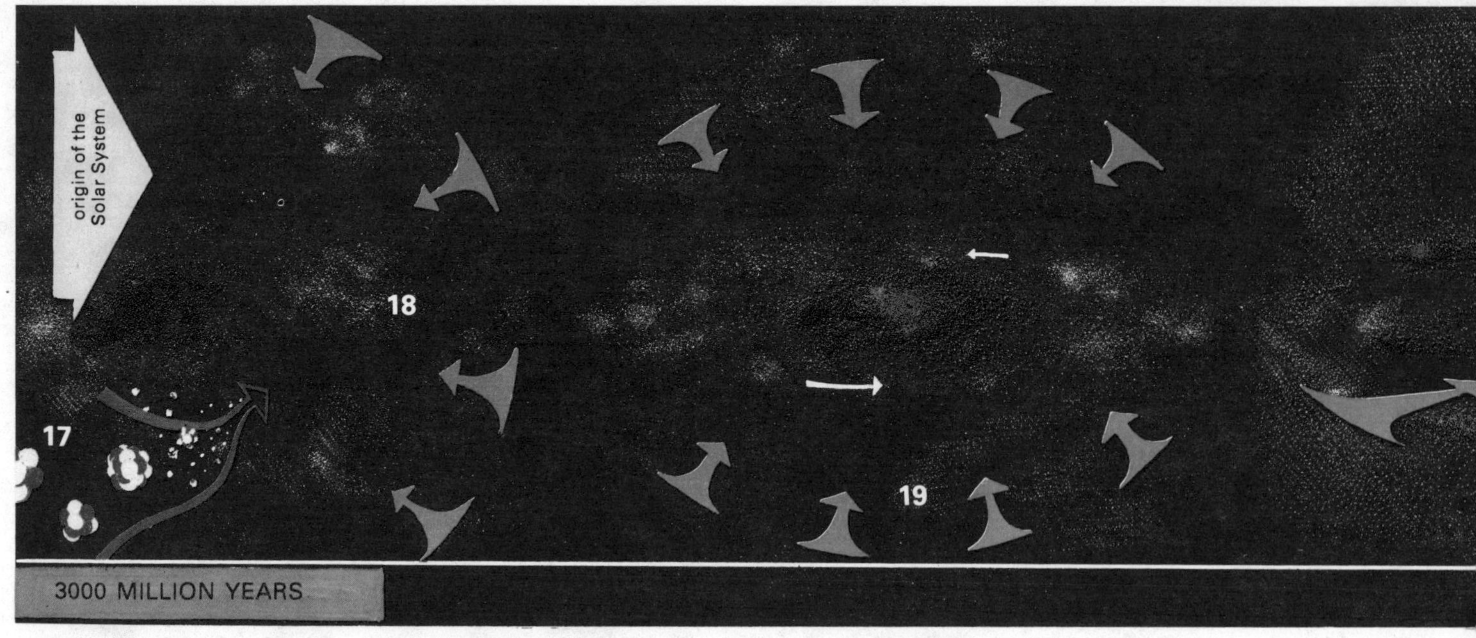

193 The curving arrow in this Hertzsprung-Russell diagram shows how the surface temperature of the Protosun rose, and how its luminosity changed. As soon as the Protosun reached the main sequence (the diagonal row of stars), it became the Sun. Similarly all stars developed from globules (bottom right-hand corner).

The origin of the Sun and the Earth

Five thousand million years ago there was still neither a Sun nor an Earth. Instead there was a huge cloud of very cold and rarefied gas and dust. Photons had to travel for half a year to get from one side of it to the other. This giant cloud of dust and gas is called the *parent nebula* (Figure 194), or the *parent globule*.

Five thousand million years ago that parent nebula began slowly to shrink as a result of its own gravitational force. The grains of dust and molecules of gas were attracted to each other by gravitation.

On shrinking, the nebula grew denser and warmer. The gravitational force drew on the rest energy of the particles which formed the parent nebula. The energy released partly changed to heat and was partly emitted as infra-red photons.

The density and temperature increased most of all in the middle of the nebula. There the Sun came into being: a huge, dark-red ball which gradually contracted and heated up. As the temperature in the centre of the nascent Sun (the Protosun) reached 13 million kelvin, the Sun came of age. It became like it is today.

What important changes took place at the time for a hot, contracting ball of gas to become an adult star? When the temperature in the core of the Sun reached 7 million kelvin, hydrogen gradually started to change into helium. But gravitation continued to shrink and heat the nucleus, up to a temperature of 13 million kelvin. At that temperature the pressure in the core grew to such an extent that it matched gravitation (the weight of the upper layers) and prevented further contraction. The dot representing the surface temperature and luminosity of the Sun in the Hertzsprung-Russell diagram moved into the main sequence (Figure 193). Since then the Sun has released all its energy only by the transformation of hydrogen to helium, and has not needed its self-gravitation (which had heated its core from freezing to an inferno of

193

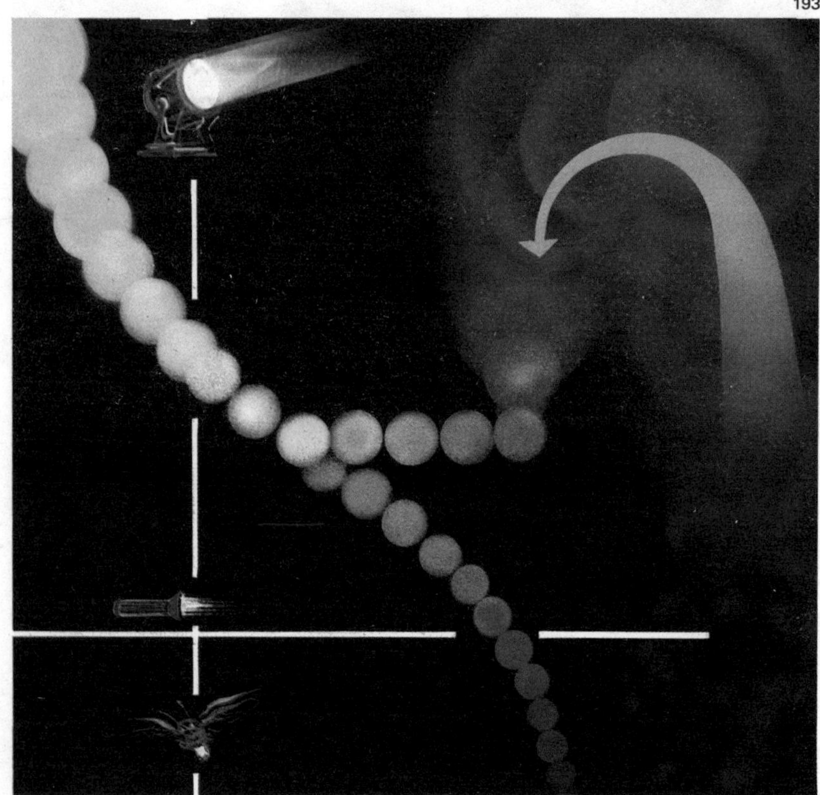

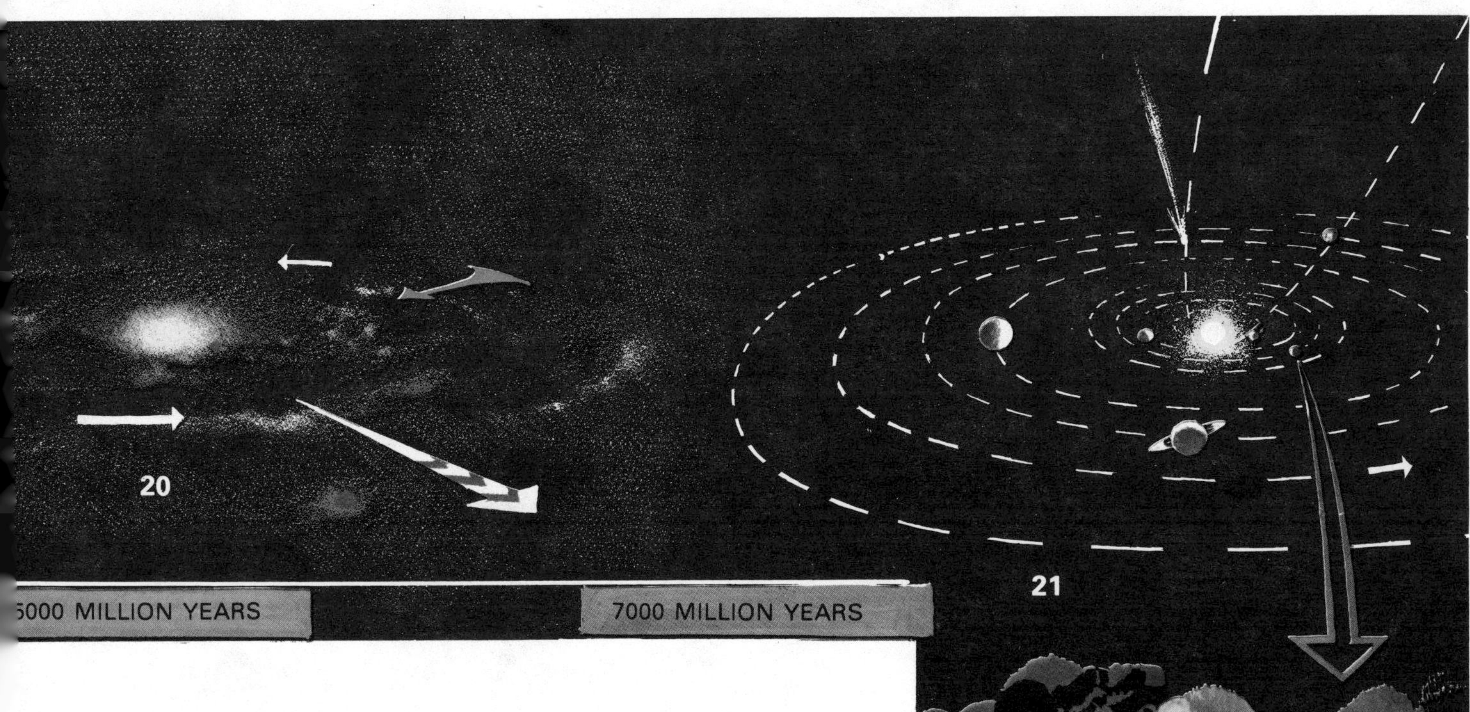

13 million kelvin). Gravitation was replaced for a long time to come by nuclear forces. While gravitation took only around 10 million years to form an adult Sun from the parent cloud, the nuclear forces have been releasing energy from the Sun for 5000 million years, and will continue to do so for another 10,000 million years.

The remains of the parent nebula formed the planets around their shining Sun. The large shapeless lumps of nebula circling the young Sun (Figure 194 number 20) were moulded by their self-gravitation into spherical shapes – planets (Figure 194 number 21). The Earth, Venus, Mars and the other planets came into being. The Moon, the Earth's little 'sister', and the satellites of other planets, were also born that way.

The Sun is 30,000 light years from the centre of the Galaxy. It orbits around it along an approximately circular path. The Sun and all the bodies of its system are moving at about 230 km (143 miles) per second in the direction of the constellation Cygnus (Swan). Once every 250 million years the Sun orbits the centre of the Galaxy; it has made the journey 20 times during its existence.

We have seen that the Sun is the elder and bigger 'sister' of the Earth; both arose from the same parent cloud. The Earth is about one million times smaller than the Sun around which it orbits (Figure 187). The Sun holds it by its gravitational pull, not allowing it to roam off into the frosty darkness of space.

194 When our Galaxy was around 3000 million years old, there were many supernova explosions (Figure 152). Thus numerous atoms of all elements were formed, and flung out into interstellar space **(17).** The new atoms (for example carbon, oxygen, silicon, iron, uranium, etc.) also enriched the globule **(18)** which was later to give birth to the Sun and its planetary system **(19–21).** This means that the atoms of which the Earth is built (and all living things on it) are around 7000 million years old **(22).** For instance, the atom of iron in the molecule of haemoglobin of red blood cells was formed in a supernova which exploded 2000 million years before the Sun and planets were born. Nothing remains today from that supernova, but we are its descendants.

194

195 The final aim of the geologist's work is to get to know our planet to such an extent that we can maintan 'natural' conditions and at the same time exploit the Earth's resources without being destructive.

5. MAN STUDIES THE EARTH

Geology
– a young science

The modern history of man began when he found out how to use stone and, as time went on, other mineral substances such as iron, copper and tin. The names of prehistoric epochs, such as the Palaeolithic and the Neolithic, are derived from the Greek word *lithos,* meaning stone. Soon after humans learned to make use of stone, they began actively to look for it and quarry it. The coast of western Europe, formed from limestone sediments, became one of the first locations where men deliberately went underground in search of mineral resources.

But an acquaintance with minerals is merely the first step in getting to know the Earth. You may know something about the activity of volcanoes, and perhaps also about how rivers can wash away weathered rocks, and you may have seen huge cliffs, storm clouds, avalanches, or pounding surf, perhaps fossilized animals, too. All these form familiar parts of the pattern of the Earth, and belong to such branches of science as meteorology, vulcanology, geophysics or geochemistry, to name just a few.

We often speak of mineralogy, petrography, geochemistry or geophysics as if they were quite separate; you hear about the atmosphere, the hydrosphere and the lithosphere. These divisions are necessary, but they should not hide the fact that the Earth sciences are all part of the same family. There is, for instance, a constant exchange of substances between the atmosphere and the hydrosphere, so that these two cannot be studied separately, nor can you draw a clear dividing line between, say, petrography and mineralogy. They are closely connected. Petrography examines rocks, while mineralogy deals with minerals. But rocks are composed of minerals, so every petrographer has to be a good mineralogist. And if we want to understand the workings of the Earth, both in relation to the surface world and from the point of view of what goes on inside it, we have to look at the various sciences and research methods both separately and as a whole, as a system. Let us begin by relating the sciences that have the Greek word *geos* – Earth – in their names.

A knowledge of the geological structure of our planet calls for a sound understanding of geography. Thus the ideas different nations have of its shape, size and position vary. It was the Greek men of learning who deduced

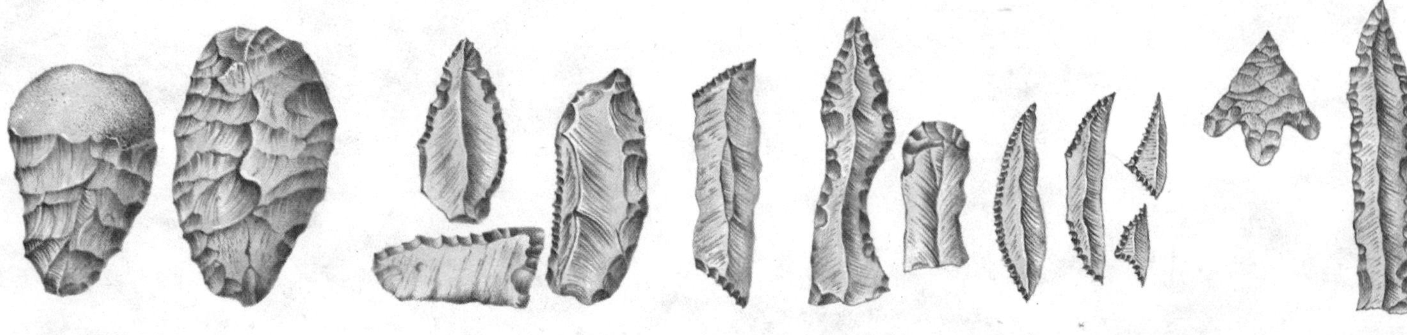

196 Flint was one of the first materials sought by man; he was performing his first 'geological experiments'. The manufacture of flint tools developed from the Palaeolithic to the more modern Neolithic, as illustrated by this series of tools. Later on stone became inadequate, and men started to seek new materials – copper, bronze and iron.

197 The mining of mineral raw materials as an organized activity is something men have engaged in since the age of classical Greece (which is where this 6th-century BC picture comes from).

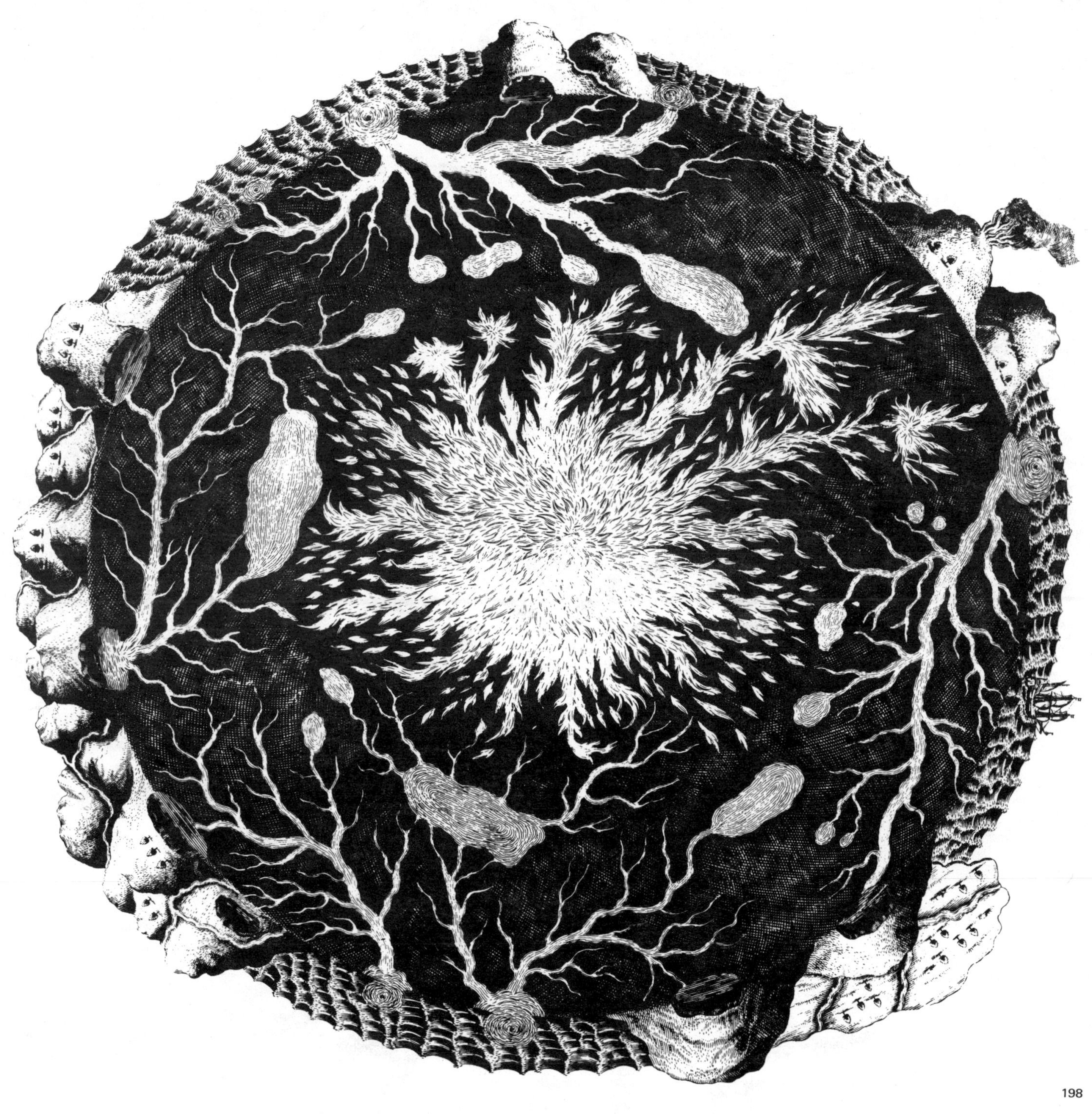

198

that the Earth was round, and in ancient Greece and other centres of civilization, particularly the Arab countries, information began to accumulate on our world and its minerals and rocks, and on the processes taking place on its surface and in its atmosphere. This laid the foundations of the different Earth sciences we know today.

Knowledge of the fundamental constants of the Earth – its mass, volume, density and shape, and the layout of its continents and oceans – was well established during the 18th and 19th centuries. Geology played an important part in the industrial revolution, which brought a huge new demand for minerals, so that experts were needed who had studied the structure and composition of the Earth, and understood its materials.

198 Aristotle's teaching dominated not only ancient times, but also the early Middle Ages and the start of the modern era. The idea of the eternal struggle between fire and water is to be found in an engraving for the book of Athanasius Kircher of the 17th century.

What is geology, and what is Earth science?

Laymen usually think the only way of investigating the Earth, of revealing its insides, is by drilling into it. But drilling is a very expensive, complicated, technically limited and time consuming business. What is more, it is obvious that even the deepest of bores, perhaps 12 km (7 miles) deep, are insignificant in comparison with the diameter of the Earth. In geological research and surveying, drilling to a depth of a few tens, hundreds or even thousands of metres is usually just the finishing touch to investigation using cheaper and more readily available means. Before considering the question of how man investigates the Earth by means of drilling, we shall therefore take a look at what usually precedes such drilling, and at the other ways in which we obtain information about our planet.

Research into the interior of the Earth presents an immense array of problems. One of the first, though it may seem paradoxical, is ascertaining the position of the Earth in the Universe. Naturally, the geologist consults astronomers and astrophysicists in order to understand where the Earth is, what fate awaits it in the Universe, whether it is an ordinary planet or an exceptional one, and so on. The structure of the Earth as a body cannot, of course, be studied without a knowledge of other bodies, so that comparative

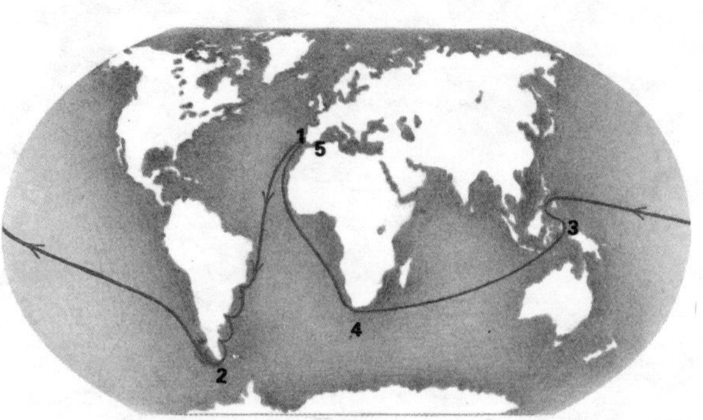

199 A century of discovery and learning reached its climax in 1519–22 with Magellan's circumnavigation of the globe (completed by Elcano after Magellan's death, April 27, 1521, in the Philippines).
1 – San Lucar, September 20, 1519;
2 – The Magellan Straits, October 25, 1520;
3 – The Philippines, March 16, 1521;
4 – The Cape of Good Hope, May 18, 1522;
5 – San Lucar, September 6, 1522.

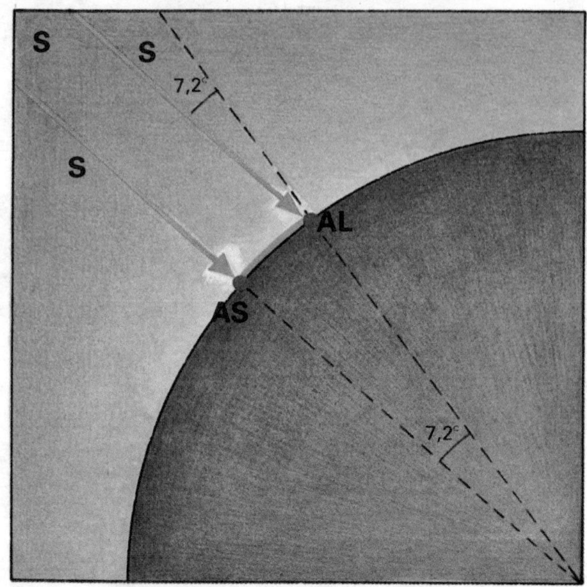

201 A simple geometric problem, whose solution allowed Eratosthenes to measure the circumference (and thus also the radius) of the globe. If you know the distance between two points on one meridian (in this case Alexandria **(AL)** and Aswan **(AS)**), as well as the angle between the Sun's rays **(S)** and the perpendicular in both places on the same day, the resulting triangle gives the dimensions of the Earth.

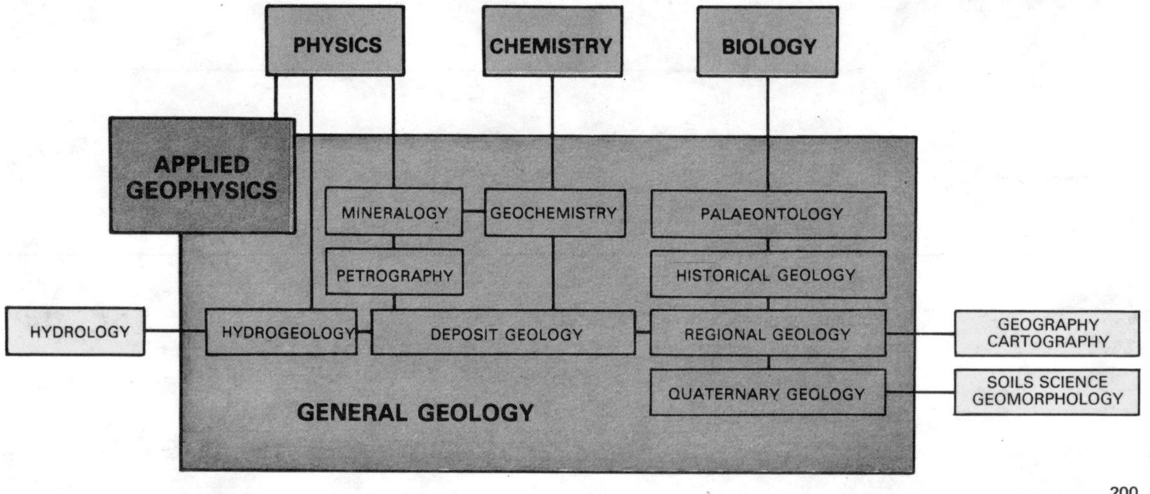

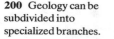

200 Geology can be subdivided into specialized branches.

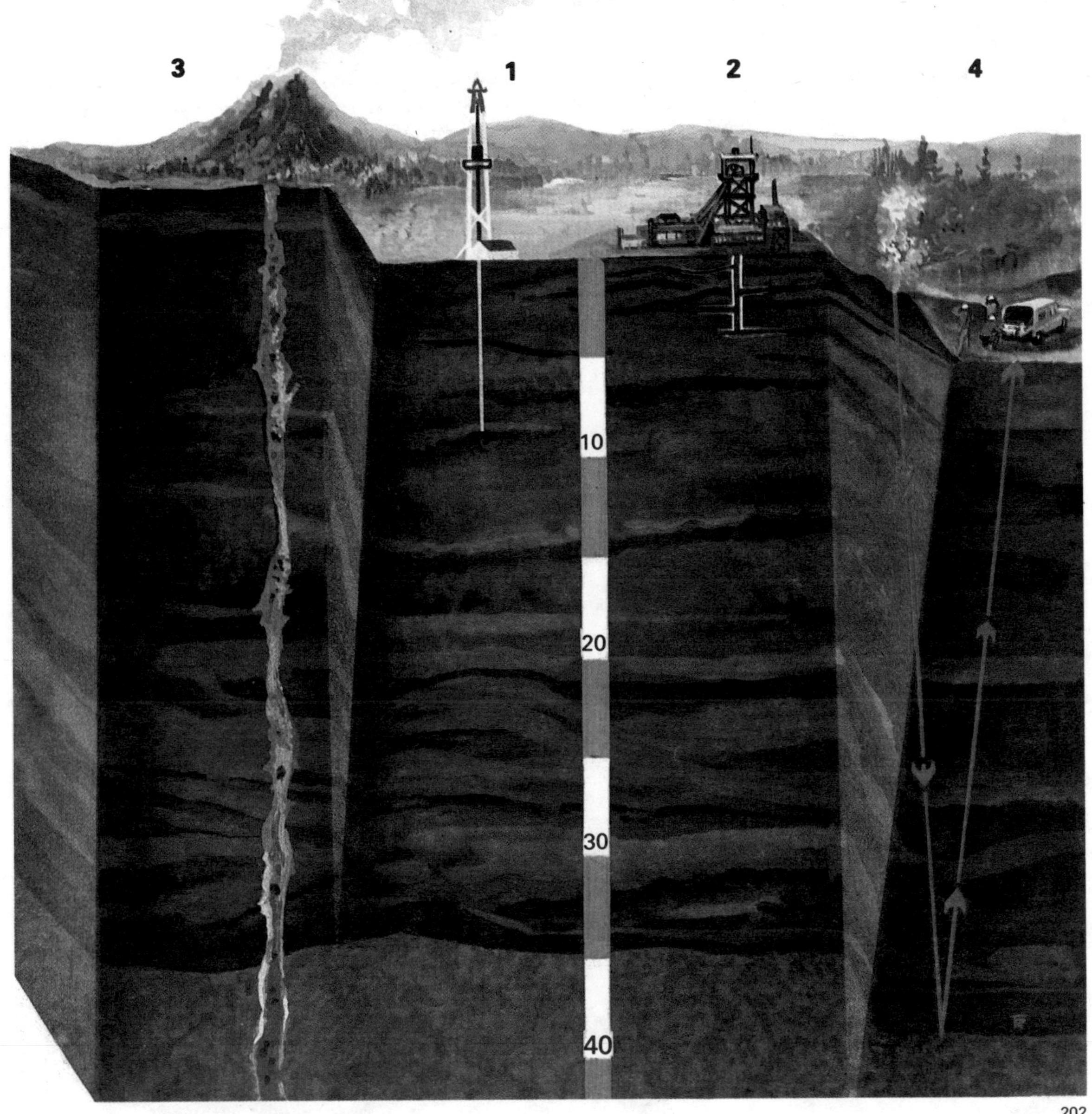

202 Direct investigation of the interior of the Earth by means of bores **(1)** or mines **(2)** is limited to the very uppermost part of the Earth's crust. Indirect evidence is offered by igneous rocks **(3),** which carry fragments of rock from the lower crust and the upper mantle to the surface. Seismic sounding **(4)** is one of the most successful methods used today.

planetology, though a very young discipline, exists to determine the structure and organization of other planets. This sort of research is based on a thorough knowledge of the physical properties of bodies. Mass, heat flow and other factors are measured, and tests are made on the elastic properties of rocks and minerals, including those which man has never actually seen, since they are hidden hundreds or even thousands of kilometres beneath the Earth's surface. The relationship between Earth sciences and physics is a very close one, and the basic knowledge of the composition of the Earth's crust, its internal organization, the pressures and temperatures inside it, and its magnetism are derived from physics or, more exactly, from geophysics. But if we wish to study terrestrial minerals we are obliged to move into the sphere of chemistry, of cosmochemistry, and if we confine ourselves to the Earth, then we must study geochemistry. The science of the uppermost parts of the Earth, the top layers of soil and rocks, is also not only a geological matter – weathering takes place with the participation of the hydrosphere and the atmosphere, and organisms take part in it as well. The Earth sciences include meteorology and hydrology, and in cases where they try to decipher questions of life in the past, they turn to a branch of biology called palaeontology. Biology, on the other hand, relies on geology to provide evidence of the evolution of individual forms of life.

There are few unexplored areas of the Earth's surface left today, so that geologists would find it hard to make a living if they confined their attention to these. But mankind requires raw materials for its industries and sources of energy, and a planet on which it can live. All manner of geological professions are called for to achieve these goals – chemical geologists, physical geologists, palaeontologist-geologists and planetologist-geologists.

The final goal of the geologist is to learn enough about the Earth so that we are able to make use of our planet without upsetting the balance of nature, and to provide the raw materials for human civilization. Problems such as finding new deposits of ore or suitable sites for oil drilling call for cooperation between many different specialists in branches of geology, from map-makers to mineralogists, and from petrographers to mining engineers and economists. Learning about the Earth is also a matter for international cooperation, since the geological boundaries (those between individual geological formations) never correspond to national frontiers. Therefore concerted cooperation between all those who study the Earth is essential.

The concentration of efforts on a study of the seabed, and projects for investigating the lithosphere (the Earth's crust) are supported by many different countries and by international organizations, and in the past have produced much information which is of use to mankind as a whole.

203 The aerial photograph permits the geologist to recognize immediately tectonic features such as the large fault dividing a mountain chain from a plain. This picture shows an area of the San Andreas fault in California.

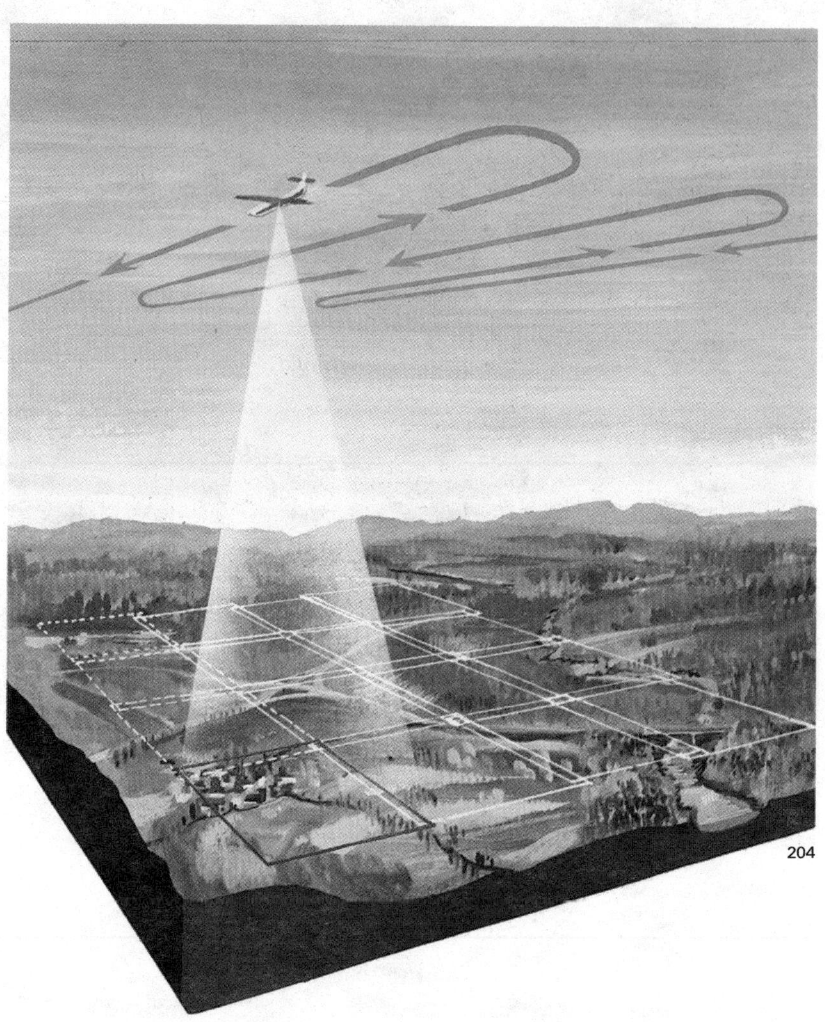

204 Over the last couple of decades aerial photography has become an important source of information not only for cartographers, but also for geologists. Here we see the principle; it is important for shots to overlap adequately.

The geological map

Although geologists can easily ascertain that the sands of the seashore are mainly silicates, or that in those parts of the Earth where caves occur they are formed from limestone, it is a less straightforward matter to discover and identify some of the other minerals on the Earth's surface and below it. This is, and always has been, the main task of geological map-makers. All other geological activity and work is based on this knowledge. It may be a matter of finding rocks for roadmaking, clays for ceramics or deposits of ore.

Geologists are also called in to help determine the best sites for dam walls, for instance. The geologist must know the type of rock which occurs in a particular place; that is where his work begins. It would be foolhardy to build a dam in a place where there was a large amount of clay, or harder rocks but with cracks in them. Similarly, we should never look for salt in granite, nor oil in basalt. Experience shows that salt or oil are found in complexes of sedimentary rocks of marine origin. Some of these criteria are very clear, and quite obvious to the geologist. Every spot on Earth has its own long geological history, which is encoded in the rocks of the area. For the geologist, understanding the 'language' of the rocks is of paramount importance. The organization and composition of the rocks can reveal their past, their neighbours' identity, and associated phenomena; geology can then predict and seek deposits of mineral raw materials.

205

207

206

205 Aerial view of cliffs with atolls and lagoons.

206 Satellite shot of the Caribbean coast of Venezuela. In dry areas there are many sand dunes. The water is coloured by fine material which has apparently been blown from the mainland.

207 The dead bodies of reef-building corals form the foundation of a coral island.

Right at the start of geological work comes the making of maps. But even before that there must be the topographical basis – a topographical map. Aerial or satellite photographs can be used instead. For the geologist they have a number of advantages. We shall see, in the chapter on cosmic research, the revolutionary changes brought by satellite photography of the Earth's surface. Aerial or satellite photographs have the advantage over topographical maps that geological differences are visible on them: the differences in colour or tone show differences in the mineral composition of the baserock. These differences are not so readily apparent from terrestrial observation or from topographical maps.

The geologist does not need a great many tools or instruments to begin preparing a geological map. It is sufficient for him to have a geological hammer, a notebook, pencil, coloured crayons, a bag for samples, a magnifying glass and a geological compass.

208 **208** Geological map illustrating the distribution of rock types in the terrain. The geologist chooses his route across country in such a way as to take in all rock outcrops and to cross the borders between the rock types several times. The resulting effect is a sort of coloured mosaic. On the left-hand side of the picture is a representation of the actual territory, and on the right-hand side the resulting geological map. The geologist's route is shown in white, and the places where he found the rocks are shown by coloured dots, and on the map by numbered crosses also.

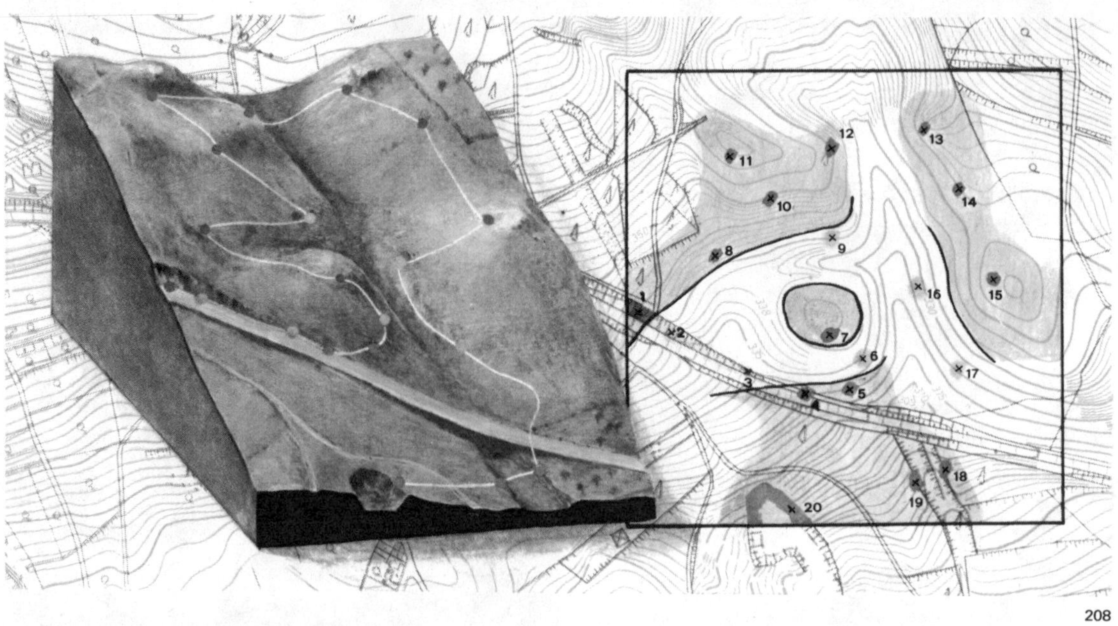

208

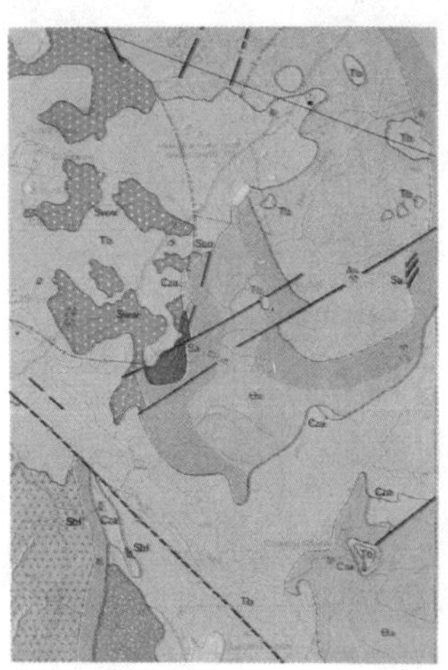

209

209 Large-scale geological maps are used for special purposes, for example when seeking mineral deposits or to show the area where major engineering works such as a dam are to be sited.

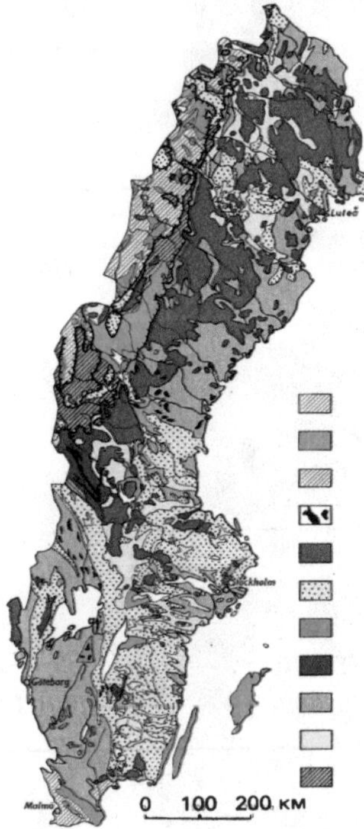

210

210 Geological maps showing a large area have scales of more than 1:200,000. They usually show only the main geological units. (Greatly reduced.)

He can carry all these things with him as he sets out into the countryside. He tries to note on his map – using, of course, recognizable symbols and colours – all the observations of the rocks he finds. He chooses his points of investigation at sufficiently close intervals to gain an idea of the boundaries of individual rocks. He has to cover every corner of the territory he is studying in order to find what sort of rock occurs beneath the layer of soil. In places where he is sure of the occurrence of a rock, he makes a coloured point or line corresponding to that rock according to an agreed code. Granites and volcanic rocks are usually marked in shades of red, gneisses in browns, etc. Dots then become lines, lines become areas, and the areas with the same colour on the map indicate the occurrence of the same rock. The geologist collects samples of the rocks he finds in order to study them more closely in the laboratory. He measures all the characteristic features, such as the system of breaks and cracks, and the orientation of the structural units – the minerals. So, for instance, he has to include on his map the direction and inclination of the sedimentary strata or the orientation of volcanic rocks, which help to decide the direction in which they run.

After months of work in the field and then the laboratory, the geological map becomes a fully fledged document. The scale, amount of detail and emphasis of one aspect or another of the map depend on the purpose for which it is intended. The sites of dams, for instance, must be mapped in great detail. All breaks and cracks in the territory must be shown, as must all porous and impervious rocks. Neglect of some characteristic feature might jeopardize the stability of the dam. Another type of map is the general geological map. This must give the geologist his bearings or inform the economist of the possible sites of deposits of raw materials, which rocks might provide a source of water, for example, or where gravel for the building of a new road might occur.

211 Today's solid and brittle rocks come from those parts of the Earth's crust where the temperatures and pressures achieved conditions under which molten rocks behaved like plastic material. The white areas represent cavities into which granite magma intruded deep below the surface.

211

212

213

212, 213 Andesites are among the most common rocks of active geological regions, the points of collision of lithospheric plates. But a rock which undergoes many transformations plays a major role in the composition of the future, growing continents. Under the petrographic microscope a seemingly uninteresting lump of rock becomes a mosaic of diverse colours. (The largest coloured patch is the mineral pyroxene.)

214, 215 The continents and their crust have their basis in rocks which underwent a lively geological evolution in the distant past. They crystallized under conditions of high pressures and temperatures, deep beneath the surface. The parallel arrangement of their parts are visible even with the naked eye, but are more striking than ever under the microscope.

214

215

The geological microcosm

If we take a fresh piece of rock, say a piece of granite, we find that in some places it has more mica in it, in other places less. We see that the rock is not uniformly of one type of mineral: it has components. Sometimes we can make out mica, or the milky colour of feldspar, with the naked eye, while on other occasions we need the help of a magnifying glass; often even that is not enough, and we have to use a microscope.

It was the invention of the microscope which marked the beginning of the true study of rocks, called petrography or petrology. Of course, anyone who has looked at an onion skin, an ascarid or a protozoon through a microscope will find the appearance of a piece of rock rather disappointing. It is opaque, which is why a special technique for observing rocks under the microscope was developed in the middle of the last century; the preparation of the sample for observation is very important. The rock must be ground to such a thin sheet as to be transparent. If granite, gneiss, limestone or any other type of rock is ground down to sheets three or four-hundredths of a millimetre thick, they are wonderfully translucent, and can be studied under a microscope. This method was invented in the last century by an English geologist, Henry Clifton Sorby. He and his successors also developed methods of distinguishing different sorts of mineral relatively easily under the microscope. They set out with the assumption that crystalline materials deflect the plane of polarized light which is passed through them. If a light polarizer and analyser are placed in a petrographic microscope, the thin sheet of rock under the microscope looks almost like the mosaic of a kaleidoscope (Figures 213 and 215).

216 Small wrinkles, visible only under the microscope, also betray the complexity of the particles in the Earth's crust. This is green schist.

216

218

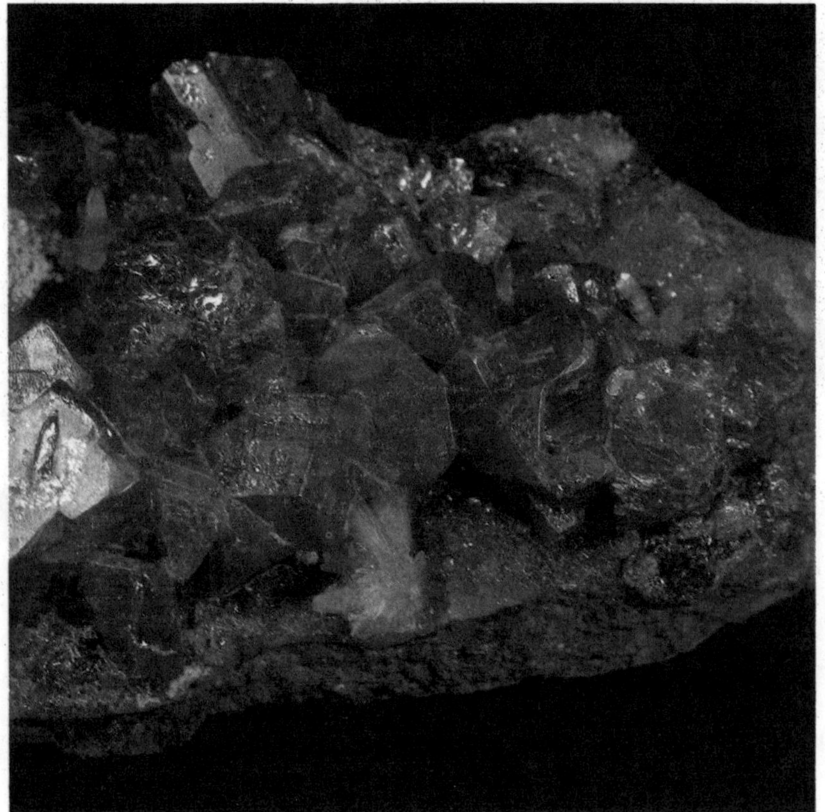

217

217 Minerals containing silver are not only beautifully coloured, but like the silver compounds used in photography are also photosensitive. These crystals of realgar are rare.

218 Galena is one of the minerals which are easy to recognize. The crystal shapes, indicating cubic symmetry, the leaden grey shine and the clear cleavage into cubes make it quite clear that this is galena.

It is clear from the figures that almost every rock is a collection of minerals; so ordinary granite is made up of the main components, i.e. silicon, feldspar and mica, plus minerals which usually make up less than 1 per cent of the rock (such as magnetite and apatite). These are called accessory. Gabbro contains mainly feldspar, pyroxene or amphibole. Only exceptionally does one find rocks made up of a single mineral. An example of this rarity is limestone, consisting of the mineral calcite.

Thus petrographers examine rocks under the microscope, determining their *modal* (volume) composition, which is the proportion of the main component minerals. According to the way in which the components are arranged, their ratios, and their size, the rock gets a name, and often also an adjective. Since the size and shape of the components, and also the ratios between them and their quality reflect processes which formed the rock, the classification of rocks is not made just for its own sake. It gives a clue as to where in the Earth's crust the rock was formed, and what conditions were like in the place where it crystallized or was sedimented. This applies to volcanic and sedimentary rocks, as well as to metamorphic ones.

Petrography yields much useful information. Determination, for instance, of the porosity of a rock may show its capacity to absorb water, or even oil or gas. In other cases the presence of a small quantity of cassiterite – tin dioxide – makes granite into tin ore. Often a very small change in the mineral composition (such as the presence of sulphide in the rock) indicates that ore may be present, or that the rock is unsuitable for roadbuilding.

It is not only petrographers who require a good knowledge of mineralogy. A sound knowledge of mineralogy is also necessary in many branches of industry. For instance in the ceramics industry the expert must first determine the mineral composition of the clay, and then recommend the means of processing it accordingly.

The extent of the science of mineralogy is considerable. Today's mineralogist investigates the properties of minerals from the chemical and physical viewpoints, and looks at communities of minerals. Physical properties such as hardness, colour, lustre, density,

219

220

219 Though agate is a compound of very simple silicon dioxide, SiO_2, the variability of this semi-precious stone illustrates the diversity of the mineral kingdom.

220 Crystals of sulphur are a mineralogical curiosity, while most sulphur deposits are processed by the chemical industry. Colour, softness and ease of grooving are characteristic of sulphur.

221

221 Calcium carbonate (calcite) is often found in lodes alongside useful ores of copper, lead or zinc. Its crystal shape is also typical. Apart from a characteristic cleavage, it can be identified by adding a few drops of hydrochloric acid when, like other carbonates, it releases bubbles of carbon dioxide.

streaking, etc. are very variable and not always conclusive on their own. Mineralogists also study the crystalline shapes of their minerals. The Danish scientist Nils Stensen defined one of the basic rules of mineralogical crystallography, according to which the angles between the faces of the crystals of the same mineral are always the same. This means that a quartz from Brazil has exactly

In order to find out about the internal structure of minerals one needs more than just a microscope or a goniometer (an instrument used for measuring the angles between the faces of crystals). One requires much more sophisticated equipment and methods. A study of the arrangement of the structural components of minerals is closely allied to the physics of solids. A major role

222 Agate – another example of the beauty of semi-precious stones.

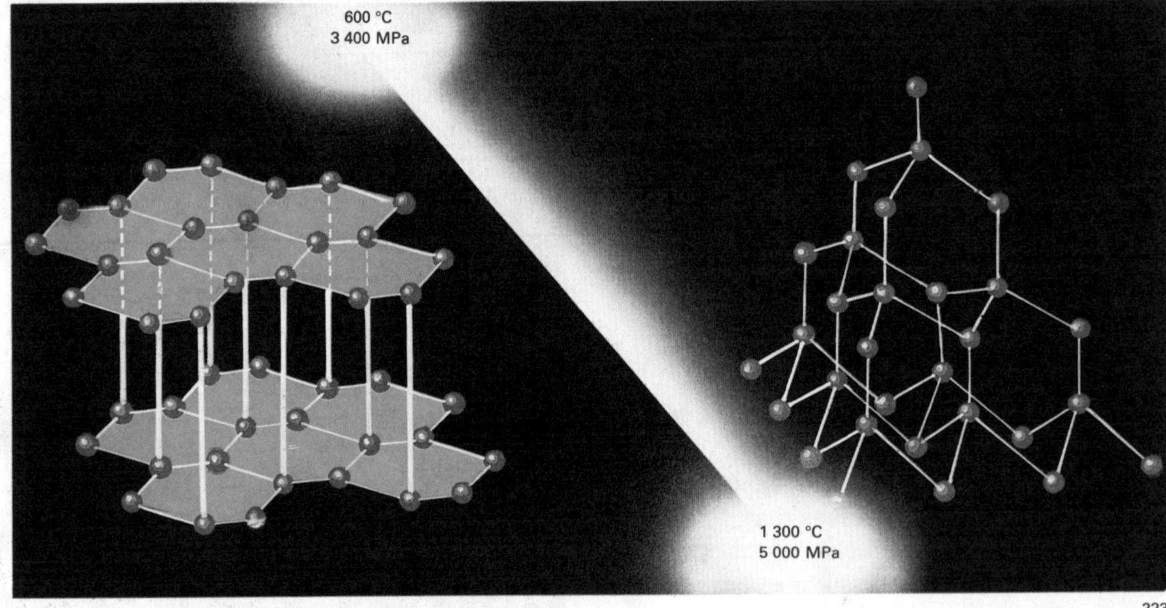

223 Temperature and pressure determine whether carbon, one of the commonest substances on the Earth's surface, crystallizes into graphite (left) or diamond (right). The diagram also shows that at high temperatures and very high pressures diamonds can be made from graphite. The process is already used.

the same angles as one from Siberia. The outward appearance of the crystal is a reflection of its inner structure – of molecules and atoms.

has been played here by X-ray technology. If crystals are X-rayed their structure can be seen in the resulting picture, and the distances between the structural components can be calculated. A knowledge of the structure of a mineral is the first step along the path to its identification. The mineralogist must add to this the results of the analysis of the exact chemical composition, and only then is the determination of the mineral complete. Thus a mineral is a chemical compound with a certain organization of its structural components. The differences which the organization of these components can bring about are shown by the example of graphite and diamond. Chemical analysis shows both of them to be pure carbon. The difference lies only in the organization of the structural components. Figure 223 shows the difference in the crystalline structure. The appearance of the two modifications of carbon is also very different – you would scarcely find a woman prepared to wear a graphite ring!

The study of the internal structure of a mineral has progressed to the stage where modern instruments are capable of determining the internal structure of the mineral, and at the same time observing an object under an optical microscope and ascertaining its chemical composition. The means used to study cosmic material – meteorites and moon rock – requiring a minimum amount of the material, are now quite routine in mineralogy.

The minerals which make up a rock and which can scarcely be seen with the naked eye are chemically analysed in sophisticated electron beam analysers, allowing each – often chemically very variable – component to be characterized much more fully than was possible 20 years ago. In parallel with the changing views on the structure of the Earth, now seen as a system of moving plates, the geological microcosm has also undergone many transformations.

The geologist needs chemistry and physics

There are over 90 elements involved in the composition of the Earth. But when we take a look at the chemical composition of rocks on the surface or in the top 400 km (250 miles) of the Earth, we find that only 10 of those elements are present in any large amount. For the Earth's crust they are as follows: oxygen (60.5 per cent), silicon (20.4 per cent), aluminium (6.2 per cent), hydrogen (2.8 per cent), sodium (2.5 per cent), calcium (1.9 per cent), iron (1.9 per cent), magnesium (1.8 per cent), potassium (1.4 per cent) and titanium (0.2 per cent). Then come elements such as carbon, phosphorus, manganese, sulphur and fluorine. So far we have mentioned only 15 elements. The remaining 75 make up a mere 0.19 per cent. They include elements as rare as irridium, platinum, tantalum and niobium. This list shows the rarity of mineral deposits, which may contain only 2.5 per cent of metal in the ore, and the coincidence that was required for such a concentration of metals to occur on Earth.

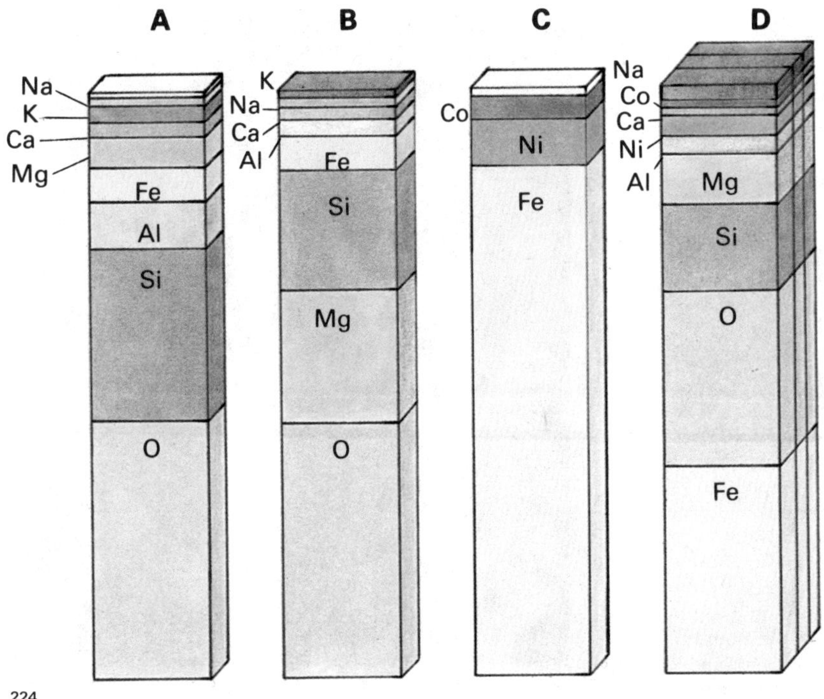

224 The various layers of the Earth differ in composition. Their role in the structure of the Earth is shown here. **A** – crust, **B** – mantle, **C** – core, **D** – the Earth as a whole.

The composition of the Earth is studied by geochemists, who examine the composition of rocks, air and water, and compare the composition of rocks and soils. They try to understand how, where and why elements are concentrated or dispersed, and how they are cycled in nature. This activity also has its purpose. The Earth is like a living organism: elements change places, are digested and released. To understand the sulphur cycle, for instance, is to understand just a little about the metabolism of the Earth. And an understanding of that metabolism is an important prerequisite for some major questions regarding the survival of human civilization.

Another branch of geochemistry is the endeavour to find out about the composition of the Earth's interior. Practical geochemistry is quite different again, attempting to find mineral deposits by purely chemical methods. Increased contents of elements are sought in the soil, the sediments of rivers, even in the leaves of trees, since these may indicate usable deposits below the soil horizon. And since these 'high levels' are in fact very low, perhaps one-tenth or one-thousandth of 1 per cent, geochemists need a very complex research base, with laboratories and instruments.

When we spoke of minerals, we said that their basic properties are the structural arrangement of their components and their chemical composition. The same goes for the Earth; if we wish to describe it and define it as a cosmic body, we must know its internal structure, which is possible only on the basis of its physical properties and chemical composition.

The most basic property of any body is its mass. Every mass corresponds to a certain gravity – the gravitational acceleration, which can be calculated on the basis of Newton's well-known laws. Mass and gravitation are properties which are inseparable. Of the other physical properties which can be used to characterize the Earth as a body, the main ones are geomagnetism, geo-electricity, heat flow and the elastic properties of the Earth's rocks. The physics of the Earth is interesting in that the laws which govern the physical principles of the Earth as a whole are the same as those which govern all of its minor formations.

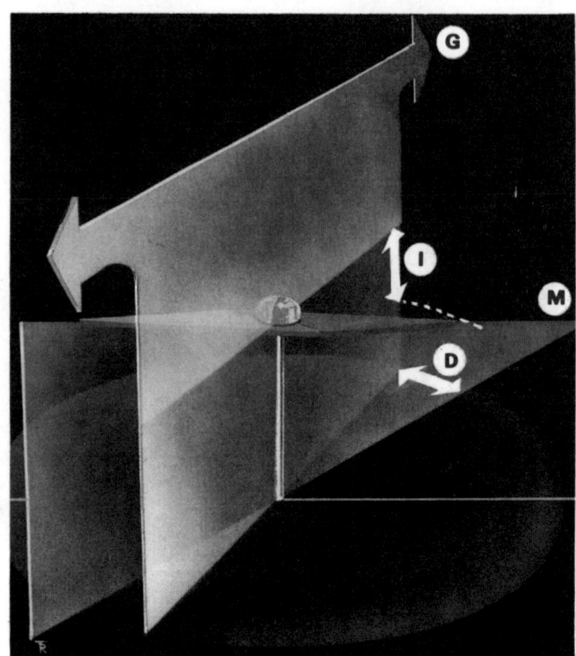

225

225 Most scientists suppose the source of magnetization to be the Earth's core, as shown in the picture. Here you can also see that the axis of the Earth's rotation (blue) and the magnetic axis of the Earth (red) are not the same. For this reason a magnetic needle does not point towards the geographical north **(G)** but the magnetic north **(M)**. The difference is called the declination **(D)**. The needle of a compass has a second deviation, since it points slightly downwards; this is known as inclination **(I)**. Declination and inclination are said to be the horizontal and vertical components of the magnetic vector. (Picture on the right is an enlarged section marked in the left-hand picture by the red circle with an arrow.)

226

226 The principle of gravimetry is very simple. At points where there is an exceptionally large amount of matter, gravitational pull is greater. Both the weights in the picture are of equal mass, but the one on the left is affected by the greater specific gravity (greater gravitational pull) of the rocks beneath the surface.

Let us begin with gravity. The measurement of gravity (Figure 226) is used to determine the shape of the terrestrial globe (it is potato-shaped), and also to find the boundary between two masses of rock with different densities. Or take the magnetic properties: the Earth as a whole acts like a huge magnet, and the geophysicist studies the changes in that magnet in the course of days or months, and must thus know the magnetic properties of rocks and their components, minerals.

Earthquakes cause immense damage every year, so it is little wonder that so much attention is paid to them. Men have known of their existence since time immemorial. They try to predict them and avoid them, and in recent times there have even been attempts to control them. The actual prediction of an earthquake is an extremely difficult task, but since it is of immense economic and social importance, from the beginning of this century there has existed an extensive network of stations which continuously record earthquakes.

227 Most geophysical instruments are constructed on a very simple principle. A large, high-inertia mass – the weight – remains still, while the Earth's surface and the seismograph attached to it shake. This gives a seismographic reading. Such seismographs have to be installed in sets, to record earthquakes coming from different directions.

The instruments used to record and measure earthquakes are simple. They are called *seismographs*. The principle of their operation is the inertia of matter. Let us imagine a vessel, say a saucepan, in which a heavy stone is placed. If the pan is shaken from side to side, the stone rattles against the sides of the pan. This is the principle upon which a seismograph works. The outer shell of the instrument (our saucepan) is fixed to the ground, and shakes with it. The stone is replaced by a more massive object. It has a rest inertia, and if it is suspended or otherwise placed so that the shell can move independent of it, and a recording device is attached, we have a seismograph. In order to detect the direction from which the shock waves of the earthquake come, the body of the seismograph is adapted to movement in one direction only. In order to record seismic waves in several directions at once, a number of seismographs are placed in the same location.

Geologists are interested in earthquakes for other reasons also. They have learnt their significance for understanding the structure of the Earth's interior. If it were not for earthquakes, we should know much less about the crust and the core of our planet. We have also learnt much about our neighbour the Moon thanks to seismographs. From the rate of the propagation, the course, reflection and refraction of seismic waves, an idea has been formed of the onion-like

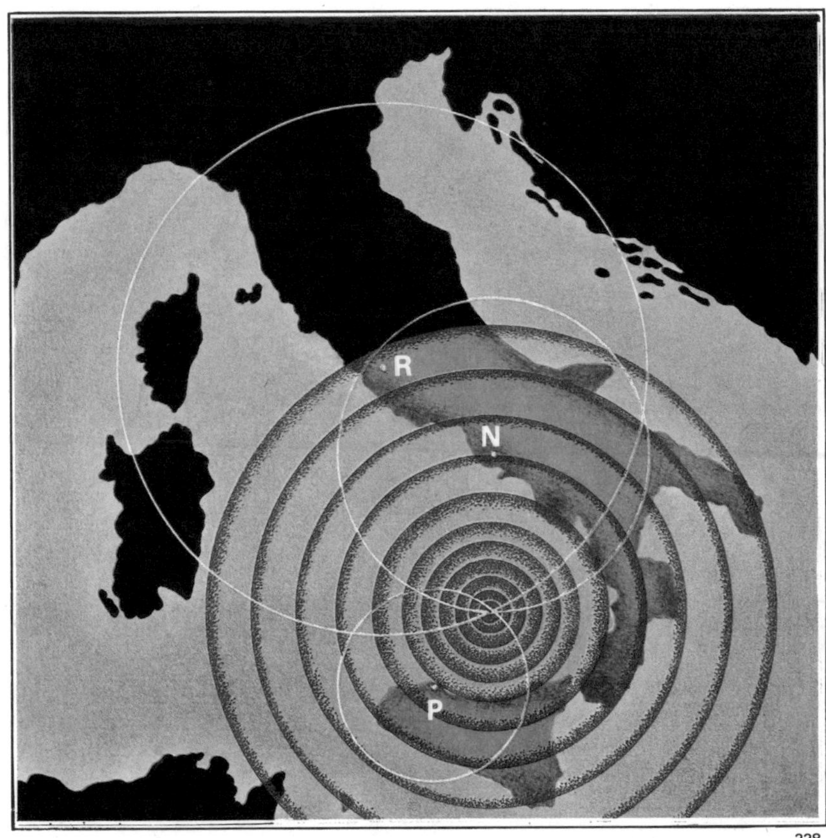

228 Finding the epicentre of an earthquake is a relatively simple geometric task provided the readings from three different seismometric stations are available (for example **P** – Palermo, **R** – Rome and **N** – Naples) and precise times are recorded. The more stations which take part in the determination of the epicentre, the more accurate it is.

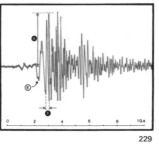

229 The arrival of seismic waves and their recording on a seismograph. Value **A** is the amplitude – the size of the waves – while **T** is the duration of one oscillation. **E** indicates the time of the first wave's arrival.

structure of the Earth. It seems incredible, but an earthquake and its waves pass right through the whole Earth, from their origin in, say, Tahiti, to the other side of the globe. But they are not always recorded successfully. Seismic waves are bent at the boundary between two media, according to exactly the same laws as the refraction of light: they are reflected, decelerate or accelerate according to the medium in which they are moving. There are several types of seismic waves, differing in the way individual particles oscillate in the course of the wave. The most important ones are those known as longitudinal and transverse. These are propagated through the body of the Earth. Surface seismic waves are of less importance.

230 We owe our knowledge of the Earth's metabolism to the enormous development in research into the ocean floor in the 1960s. One of the first tasks was the mapping of the morphology of the seabed, with indications of the deep-sea trenches and mid-oceanic ridges. Thus it was ships which played a major role in determining the origin of the land areas of the Earth and the history of the continents.

There are important differences between longitudinal and transverse waves and on them, geophysicists have based a theory of the Earth's structure. In a solid body the longitudinal waves are propagated more rapidly than the transverse ones. In volcanic and some metamorphic rocks on the surface of the Earth their speed is around 6 km (3.7 miles) per second. The transverse waves are propagated more slowly, and in addition have the property of not being propagated through matter in the liquid state. And since they are not transmitted through the entire body of the Earth, they actually form one of the proofs that inside, though at a great depth (about 2900 km/1800 miles below the surface), the Earth is liquid – its ferro-nickel core is molten. This is not a new idea. It originated at the start of this century, at the same time as it was discovered that the Earth has only a thin crust, and that beneath it there is a layer called the *mantle*. The boundary between the crust, which is only some 35 km (22 miles) thick, and the mantle was first discussed in 1908 by a Yugoslavian scientist, Andrij Mohorovičić, following one of the disastrous earthquakes in Skopje. This interface is therefore known as the *Mohorovičić discontinuity,* or the *Moho,* and is at a depth of between 10 and 75 km (6 and 47 miles), though the average is 35 km (22 miles).

If we wished to compare the Earth with some other object we know well, this might be an apple into whose centre we have placed a rotten plum. The skin of the apple is the Earth's crust, the flesh of the apple is the upper mantle, the rotten plum is the liquid core, and the stone of the plum is the inner core, which is probably solid. It is interesting to consider that man, who makes artificial satellites and flies into space, is helpless when it comes to exploring the inside of his own planet. If you take a pin (in your imagination only, of course) and cut it off with a pair of pliers as close as you can to the head, however hard you try a small part of the shank will be left. If you then push the stub of the pin into an apple, it will just about manage to get through the skin, which is about as far as man's drilling can get into the Earth.

Apart from the rate of propagation of seismic waves, however, further physical properties of the Earth can be studied, and the same, basically simple but accurate instruments are used. Thus, for instance, measurement of the Earth's magnetic field is based on the fact that the Earth's magnetism has two components – one horizontal and one vertical. Geologists and geophysicists use the Earth's magnetism to obtain much information, for example to determine the age of rocks. Rocks retain in their structure traces of the components of the magnetic field of the time when they came into being.

231

231 Lowering a boring apparatus to a depth of several kilometres is a difficult operation. The ship must be kept in one place, and when bits are changed drilling must recommence in the same spot. Modern electronics help here.

Some minerals, especially iron ores, are themselves highly magnetic, so that deposits of mineral raw materials containing such substances can be sought using magnetometers.

Geophysicists even make their own 'earthquakes' to discover the structure of the uppermost layers of the Earth's crust. They place charges a short distance below the surface, locate small seismographs at strategic points, and wire the whole thing up. Then they set off the explosives, thus causing a small 'earthquake', and use the shape of the waves to construct a cross-section of part of the Earth's crust.

The temperature of the Earth, or the heat flow escaping from the ground, is a physical value which indicates what the interior is like. It also shows the dispersion or concentration of radioactive elements in the ground, which emit heat on decaying. This is the reason why in parts of the crust or mantle with high concentrations of uranium and thorium there is more heat escaping than in places with low concentrations. High heat concentrations are found in volcanic regions, but also in regions where volcanic activity existed until recently (in geological terms). In these places heat is also probably of radioactive origin. But it comes from a great depth, and arrives at these sites by flowing through the matter of the mantle.

Man studies the sea

Most of the evidence about the composition, evolution and metabolism of the Earth comes from the seabed. Twenty years ago we should have still thought it nonsense to study the 'organism' of the Earth on the seabed. Today, it is a matter of course. Research into the ocean and its bed, or into the edges of the ocean, has shown that the Earth is a dynamically living planet, on whose surface almost all points, including the continents and the ocean bed, are in motion.

Enormous progress has been made in oceanographic techniques since the Second World War. New instruments which are technologically advanced, but simple in principle, have been developed to measure the temperature of the seabed and the heat flow from it.

The sputnik era helped to make detailed maps of the surface currents of the ocean.

The depth of the seabed has for half a century been measured not by lead weights on thin lines, but by echolocators. Seismic techniques of registering the epicentres of earthquakes have also helped solve some of the mysteries of the seabed and are also used to find oil deposits on continental shelves. In many places where there was previously an empty sea there are now anchored huge

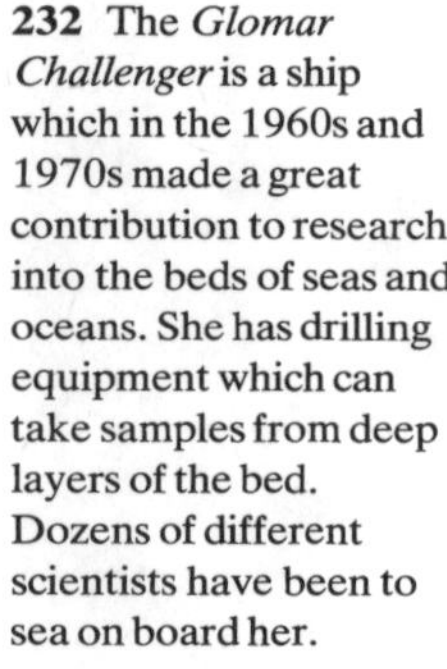

232 The *Glomar Challenger* is a ship which in the 1960s and 1970s made a great contribution to research into the beds of seas and oceans. She has drilling equipment which can take samples from deep layers of the bed. Dozens of different scientists have been to sea on board her.

233 The fact that there are not many sedimentary rocks in deep-sea areas came as a surprise to many scientists in the 1960s. Only when photographs were taken did it become clear that the seabed is covered with a layer of lava. The lava pillows in this photograph clearly show that the lava poured out on to the ocean floor.

234 One of the first craft used for marine research – the submarine *Turtle* from the Scripps Oceanographic Institute.

232

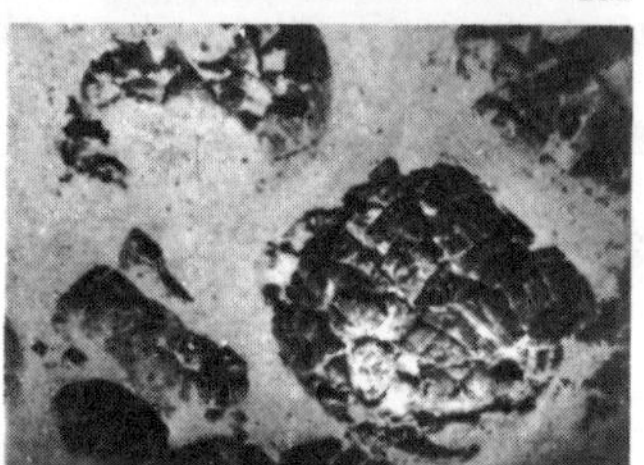

233

234

drilling platforms which bore down to the continental shelf and allow seabed pipelines to carry ashore that highly prized black liquid, crude oil. Dozens of research vessels from the major countries of the world ride the ocean waves.

Apart from classical methods of water sampling, special nets are used to take plankton samples for research into the top layers of water.

There are already many submarines used for scientific research, some of which, like the *Trieste,* have been to the bottom of marine trenches at depths of tens of kilometres. Research vessels have discovered fresh lava on the seabed, and springs of mineralized water, with useful minerals precipitating around them. Living organisms have been discovered, not only in the vicinity of sulphurous springs (which were always thought quite unsuitable for life) but also at depths of several thousand metres. Even photographing the seabed is not a simple matter: one needs only recall that all light is absorbed at a depth of 100 m (328 ft). But we have nonetheless been able to take a look at the bottom of the ocean basin.

For decades geologists were convinced that the main rocks of the seabed were sedimentary. But modern oceanographic research has shown that the reverse is true. Much of the rock on the seabed is volcanic basalt rock. Samples are obtained not only by dragging and scraping, but also by means of modern boring equipment.

The atmosphere – a part of the Earth

We can easily distinguish the atmosphere from the hydrosphere. After all, one is the Earth's gas envelope, the other its water envelope. But we shall try to show that this distinction is too artificial; there is a lively exchange between the two. The atmosphere and the hydrosphere, and even the top layer of the Earth's crust, form a single, dynamic system, with exchange of materials and of energy.

Let us consider water. Water vapour is carried into the atmosphere; it condenses and falls to form lakes and rivers, or enters the Earth's crust as free water, for example in the pores of rocks or as water chemically bound in minerals. The methods of researching into the atmosphere, the hydrosphere and the lithosphere are basically different. In the first case it is a gas that is being studied, in the second case a liquid, in the third case a solid. But one should not forget that all three of these envelopes arose from long-term geological activity.

The atmosphere and the hydrosphere both

originated through the escape of matter from the Earth's mantle. The Earth's crust is also the result of the differentiation of material from the mantle. Today's composition and structure of these media are the result of long interaction between them. But the atmosphere has another very important neighbour of its own – space. Certain gases, such as hydrogen, escape from the atmosphere into space, because the Earth has not sufficient gravitation to keep hydrogen back. As a result of cosmic rays, layers of the atmosphere are formed which exclude further cosmic rays from the surface of the Earth. Research into the upper layers of the atmosphere has also taken great strides since man entered the space age. Satellite technology is routine today. The satellite pictures of the weather situation we see daily on our television screens have become so familiar that it is hard for us to recall how, a few years ago, the weather forecast was based solely on data from earthbound meteorological stations regarding air pressure, temperature, wind speed, and so on. This does not, of course, mean that the techniques of investigating the atmosphere which have been used for decades have become entirely obsolete with the flood of new methods (including balloons, satellites, rockets and aircraft flying at great heights). Air pressure and temperature, humidity, chemical composition, wind direction and speed are still the most important factors in studying the atmosphere.

A new and important, though unfortunate, branch of the study of the atmosphere is the monitoring of its pollution. Increased industrialization, burning of low-quality fuels such as high-sulphur coal, and other emissions fill the atmosphere with many chemical compounds which are detrimental to the vegetation and inhabitants of this planet. Sulphur, in the form of sulphur dioxide, makes rainwater quite acid, so that what falls from the skies is in fact weak sulphuric acid. This makes rivers and lakes acid, too, killing certain forms of life and setting in motion a long cycle of changes which lead to a loss of equilibrium in nature. Modern research into the atmosphere concentrates on its chemical composition, with a view to eliminating harmful components from it. These include many other substances besides sulphur, such as carbon monoxide, lead, mercury, etc., which attract the attention of geochemists. Here the Earth sciences serve the interests of human health and the conservation and modification of the environment.

The Earth from space

In 1957, when the first sputnik went into orbit, many people, including geologists, had no idea of the revolutionary changes which were shortly to be brought about in the field of research into the Earth by the coming of satellites, rovers and automatic space stations. It is still too early to form a judgement about the space age in the history of knowledge of the Earth, since it is only just beginning, so we shall confine ourselves to just two areas where enormous strides have been taken.

The first is that of research into the Earth itself, photographing its surface, establishing

235 The origin of the Earth's atmosphere is attributed to the degasification of the interior of the Earth in the course of geological eras. This illustration shows what the degasification of the Earth's interior during the present-day volcanic process is like, and the role played in the transformations of volcanic gases by solar radiation and the biosphere (green plants).

236 The illustration shows the regions of the atmosphere and the main means of research into it: **1** – helicopters and jet-propelled aircraft, **2** – radiosondes, **3** – rocket-powered aircraft, **4** – meteorological rockets, **5** – communication satellites. Today there are certain 'milestones' in space research: **6** – Yuri Gagarin's flight, **7** – Aleksey Leonov's space walk, **8** – the release of the *Apollo* lunar module. Many space phenomena actually have their origins and causes in the atmosphere – shooting stars are burning meteorites. Most of them burn up in the mesosphere **(9).** A typical feature of the ionosphere is the aurora **(10).**

distances and determining precisely the physical properties of the planet.

The second area is research into other planets. Because our Earth is still a very lively and dynamic body, the early stages of its development have been obscured by more recent processes. Information from a study of the surfaces of other planets or moon rock, and the analogies and peculiarities they reveal, show that development elsewhere has not been so dynamic, and a study of these other planets' surfaces can help us to understand the evolution and history of the earliest stages in the development of Earth.

Let us start with the Earth itself. The most basic property of matter is attraction. Gravitational attraction causes bodies to fall to the ground, and those to which the required speed has been imparted, to orbit the Earth. Since the circular or elliptical path of satellites round the Earth is affected by gravitation, an analysis of the flightpaths of satellites can be used to prove that the attraction of the Earth varies from place to place, and that it is a heterogeneous and varied body. In some places it is 'heavier' than in others, which means that gravitational values vary.

Thus the same object would fall more quickly in one place and more slowly in another. These differences can easily be measured with great accuracy from the trajectories of satellites orbiting the Earth (or another planetary body). Similar gravitational 'hills and vales', i.e. differences in the amount of gravitational acceleration, have been found on the Moon and on other planetary bodies.

What is the point of research on the shape of the Earth? The measurement of gravity and gravitational acceleration is one of the most effective methods of finding mineral deposits. Let us take the example of oil. Oil-bearing rocks have a relatively low density, and consequently lower gravitational pull, so that they may appear as areas of diminished gravitational field. On the other hand rocks containing the sulphides of heavy metals, such as iron, nickel and copper, have greater density and greater gravitational pull, and appear on the gravitation map as sizeable anomalies.

But let us consider the area of human effort and research where satellite geology (research from space or space orbit) has brought direct and unusual views. We have already

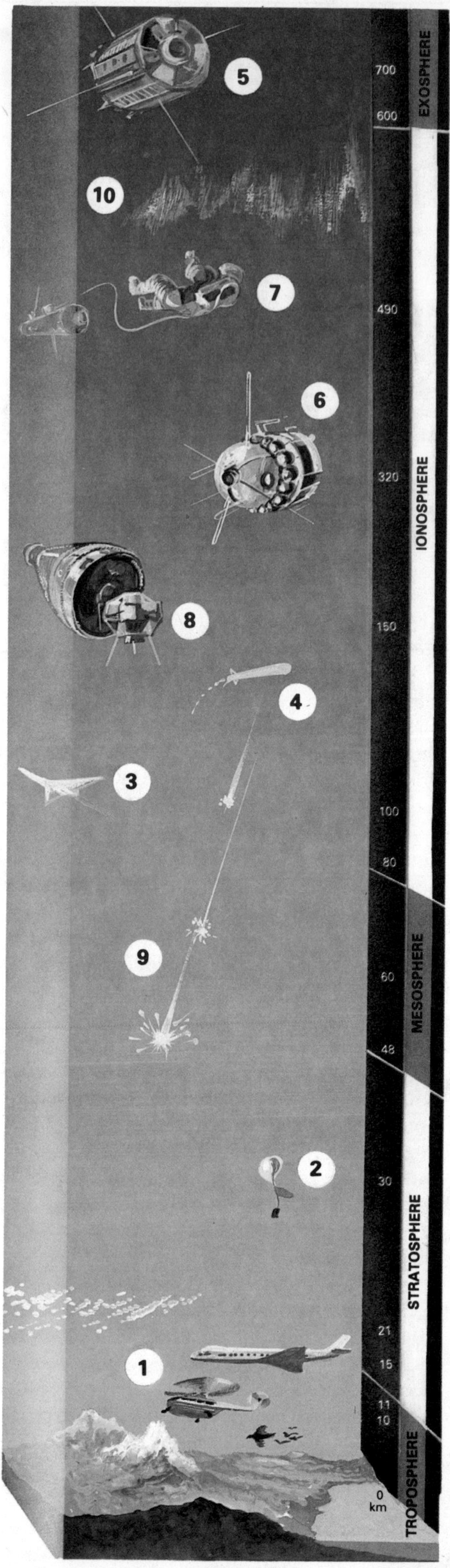

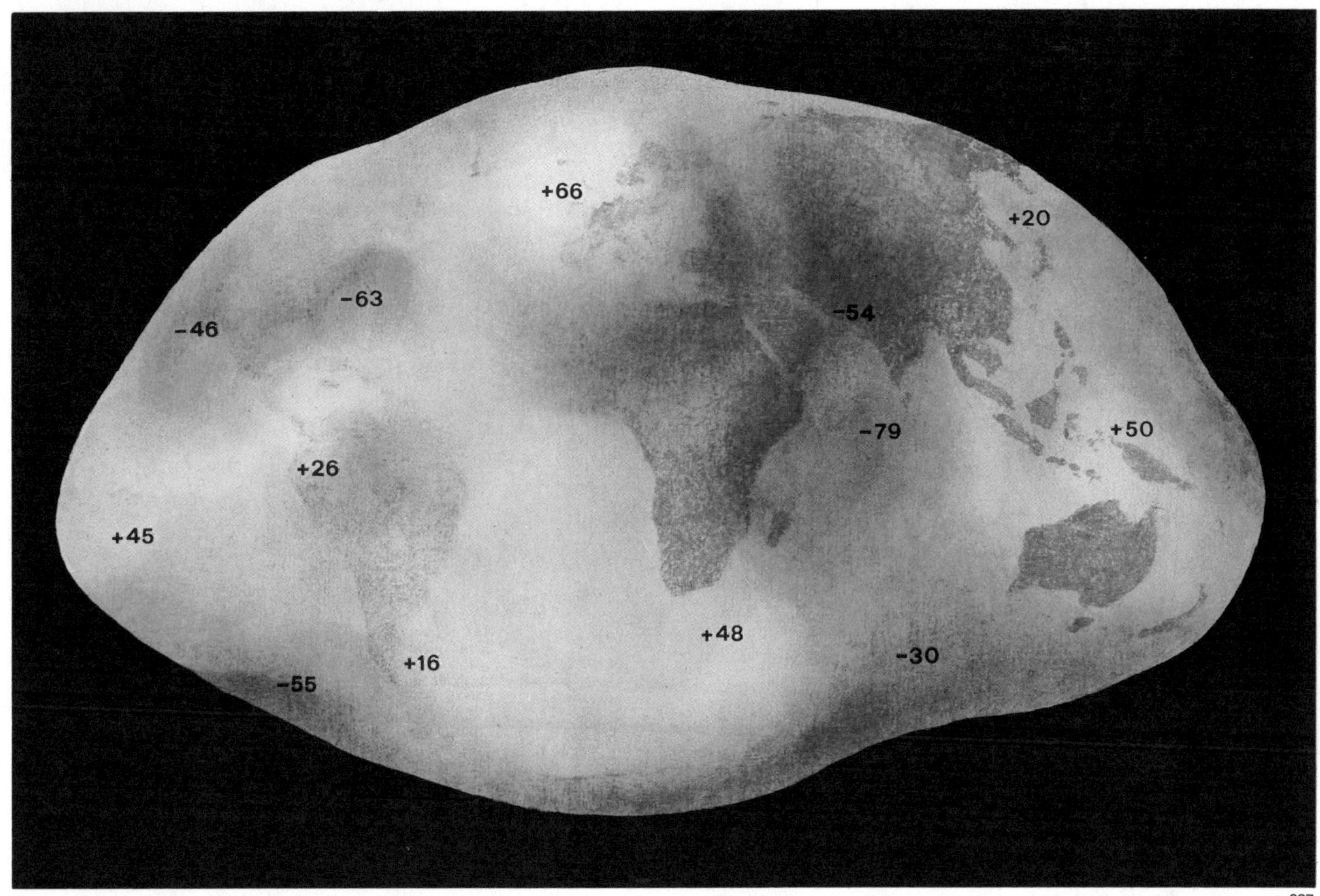

spoken of how this form of research allows man to achieve the required distance away. You know how you have to step back from some paintings (especially big ones) in order to get an overall view. It is this distance which allows the observer to get a correct picture of what he is seeing. So from heights of 270 to 1000 km (about 170 to 620 miles), from which satellite pictures have been taken, man for the first time was able to take that essential 'step back' to see the Alps, the Himalayas and the coral reefs bordering the Australian coast. In Antarctica we saw mountain ranges we didn't even know existed, and got an insight into areas which are not easily accessible, such as the Sahara and the Saudi Arabian peninsula.

Cosmic research has also made it possible to view the Earth in areas of 'vision' of which the human eye is not capable. We know that the part of the spectrum of electromagnetic waves which we see is very narrow. It is the *visible part of the spectrum.* Infra-red radiation emitted by the Earth can be registered on a special film, so that we can distinguish even small differences in the temperature of the planet's surface. And infra-red satellite pictures, i.e. from the region with wavelengths below those to which the eye is sensitive, have revealed details of which we never dreamed. Not only is it possible using shots such as these to distinguish clearly between hot and cold areas (such as completely cooled and not yet fully cooled lava streams), but even the temperature of surface rocks and their humidity can be determined.

Satellites for research into the Earth – and there have been many of them – have provided such a large number of photographs that the present number of expert analysts is too small for a detailed analysis to have been made of all of them. Satellites mass-produce these shots taken at various wavelengths. Most of them are in the visible spectrum, but often records at lower or higher wavelengths are used. Shots are taken singly, and only later are they combined and assembled. The result is multicoloured photographs which are both pleasing in

237 The blue potato characterizes the shape of the Earth. The Earth is naturally spherical – the artist has exaggerated its potato shape here. The figures show in metres how its real shape – a geoid – diverges from a rotational ellipsoid.

appearance and interesting – they are real works of art. For the geologist they are yet another source of information. Many previously unknown structures, faults, folds and arches, which are not perceptible from the Earth, are revealed. Space photography is becoming the basis of good geological maps, showing the main features of regions, the interfaces of the different compositions, colours, moisture contents, etc. of rocks.

the satellite which takes the pictures of a given region passes over at various time intervals. The reason is that the height of the Sun over the horizon, or morning frost, may emphasize details which are not otherwise visible from space. Modern geological research is now unimaginable without space shots.

The second area in which satellites and space stations have meant great advances in

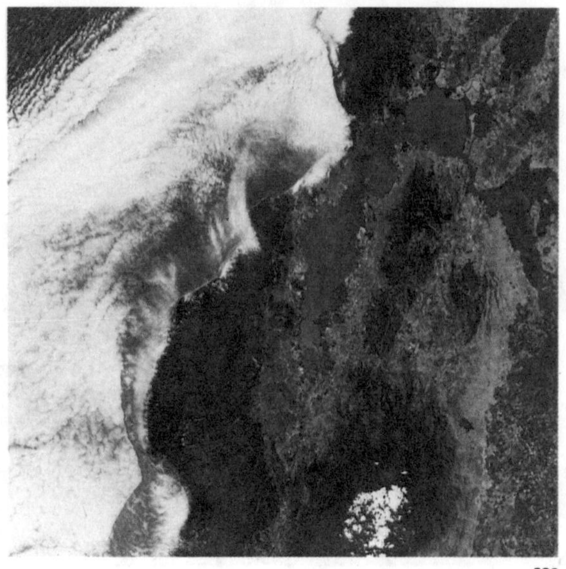

238

239

238 A view of the San Francisco area in the west of the United States. Much of the sea (left) is covered by cloud: the sky above the land is clear. Densely populated areas can be seen around the bay. Mountain ridges and their structure are also clearly discernible. (*Skylab* – NASA, 1973)

239 In this black-and-white shot the cameras on board *Skylab* photographed part of Sicily, with the volcano Etna in the top centre. The smoke trace of the volcano and numerous parasitic craters on its slopes can clearly be seen.

Hydrogeologists and hydrologists who are concerned with the provision of fresh water supplies for home and industry gain information from space shots not only about rocks, but also about areas of snow and ice, and on the amounts of melting snow and water in rivers. Cartographers can use satellite photographs to make both general and detailed maps, and they can have them ready in a matter of days, compared to the months or even years of work which went into traditionally produced maps.

In the field of environmental control, satellite shots offer information on affected areas, showing contaminated regions, protective belts of water, landscapes polluted or devastated by mining activity or disease, and symptoms of a 'sick' vegetation cover. For the agriculturalist and the economist pictures of wide areas allow estimates of yields, fertility, crop maturity or the extent of pest attack to be made without having to take to the field at all.

For the geologist it is also important that

Earth sciences is in the field of research into the Universe around us. Now you might suppose that a knowledge of conditions on the Moon or on Mars would not be relevant to the Earth. But imagine for a moment that you had no other people around you: living in isolation you would have nothing with which to compare yourself, no standard by which to judge your own qualities. An intimate knowledge of the Earth also calls for some sort of comparison. We know, for instance, from lunar research, that the Moon has been geologically passive for a long time now, that there are no active volcanoes there, no atmosphere or hydrosphere, and for a long period of its history it has done no more than passively accept the blows of meteorites. We can even ascertain from Moon rock that the amount of matter falling on to its surface was many times greater in the past than it is today. The further back we go into the history of the Moon, the higher the rate of meteorite strikes becomes. This means that the Earth, too, must have been subjected to this cosmic bombardment, and that the present surface of our planet is different from that at the start of its history. Since the earliest stages of development are preserved on the Moon, we can speculate on what the Earth was like at the dawn of its history.

Experiments in geology

Geologists studying the Earth investigate its matter. If we were to reduce their work to its basics, we should inevitably arrive at physics and elementary chemistry. For a complex of rocks consists simply of aggregations of minerals – minerals of chemical compounds with a certain definable structure of basic components. But to describe some geological processes in terms of the simplest laws of physics and theorems of chemistry is as yet beyond the powers of man. The reason is the immense breadth and interrelation of natural events. Take volcanic activity as an example: there are many factors involved – the temperature, pressure and composition of the rocks in the place where it occurs, the route to the surface of the resulting magma, the assimilation and crystallization of components, etc. If we were to describe these processes using fundamental physical theorems, we should only succeed in confusing the reader completely.

Apart from observations, geology has recently been assisted by experiments – a method which has brought major successes in other branches of the natural sciences. Most of what the geologist spends his time doing could be compared to reading records or decoding existing processes. It is in fact the work of a detective.

Geologists try to get to the truth concerning the interior parts of the Earth in the laboratory, too. The validity of studying geological events by using laboratory models was once in doubt, but in the last 20 years experiments in geology have found their place. They are very complex, and require a great deal of care. Experts work with pressures such as those which exist at depths of hundreds, even thousands of kilometres beneath the Earth's surface, and with extremely high temperatures, at which most known metals and silicate rocks (such as basalt, phonolite or granite) are molten. Care is again called for in interpreting and explaining the results. The experiments are carried out with amounts of material about the size of a pea, but the apparatus used is as big as industrial presses which take up at least as much space as your living-room.

Experiments have shown that the ideas of those geologists who are concerned with earthquakes and the state of the Earth's interior are basically correct. The uppermost

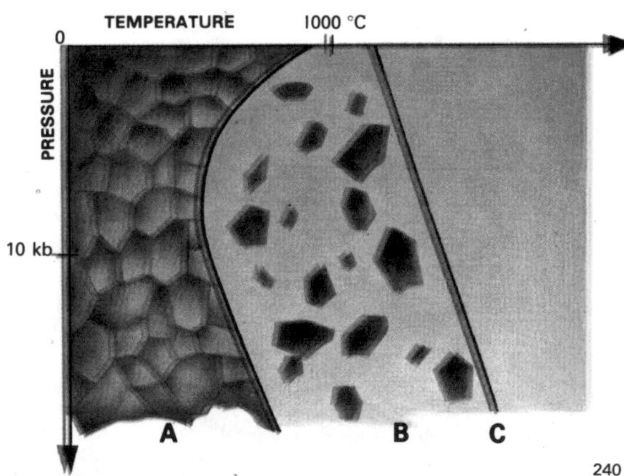

240 With increasing temperature and pressure, the melting and crystallization points of rocks alter. The curve dividing totally molten rock from crystals **(C)** is called the liquidus, that which characterizes the complete crystallization of rocks the solidus. The melting and solidification points depend on the content of volatile crust components in the magma, and also vary with the chemical composition of the rock. The area **A** represents solid rock, the area **B** the zone of partial melting.

part of the Earth is solid, with only occasional 'pockets' or zones of molten matter at depths of 80 to 350 km (50 to 217 miles) – they are actually only partly molten – and that it is only at the edge of the mantle that fully molten material occurs; the outer core is for the most part liquid. At present scientists are discussing whether or not there is another, solid core inside the outer one (as in the example of the rotten plum and its stone). There is geophysical evidence to support the idea, but it has yet to find support in the laboratory. Such high temperatures and pressures have not yet been produced for any length of time. Laboratory evidence of this solid inner core and molten outer core will probably not be forthcoming for some time.

Among the simplest experiments to simulate natural processes is the evaporation of seawater, and perhaps one of the most complex is the study of lavas at high temperatures and pressures. Computers have proved to be very useful in geological experimentation. They are used to model the limiting situations in the propagation of seismic waves, the statistical analysis of large quantities of data, and even to calculate the probability of the occurrence of mineral deposits. But man's best laboratory is still nature herself.

241 Our mother planet is a very well organized mixture of cosmic material, metals, stones and gases. Current knowledge of the surface composition and internal structure of the bodies of the Solar System is derived mainly from precise measurements taken by unmanned probes, exploiting our experience with research into the structure and composition of the Earth.

6. THE EARTH AMONG THE PLANETS

Mother Earth

The Earth is flattened at the poles, and in addition has many 'bumps' of various shapes and sizes. The polar flattening is just over 0.3 per cent. The polar radius of the Earth is 6356.755 km (3947.54 miles), and its equatorial radius is 6378.16 km (3960.83 miles). The equatorial circumference is 40,073 km (24,885 miles).

As we have seen, the Earth's mass can be calculated on the basis of Newton's gravitational laws. It is 5.976×10^{27} grams. Its density, which is a measure of mass per unit volume, is 5.5 g per cm^3.

This data, combined with a knowledge of the physical properties of rocks on the Earth's surface, allow one to deduce that the matter in the interior of the Earth is denser than that on the surface. A study of the propagation rate of seismic waves has shown that the Earth has a crust which is around 35 km (22 miles) thick beneath the continents, and an average of 10 km (6 miles) thick under the oceans. Below the crust is the mantle. At a depth of 2900 km (1800 miles) beneath the surface is the interface between the core and the mantle.

Over the layer which makes up the crust is a discontinuous layer of water. It is known as the hydrosphere, and it includes all the oceans, the rivers, the lakes and the water in the pores of rocks. The Earth's gaseous envelope is the atmosphere, composed of nitrogen, oxygen and other gases.

The layout of the continental and marine sections of the Earth's crust is asymmetrical, the basic outline of the continents and sea shallows forming the boundary between them. There is more ocean than land. The continents bear the traces of stages in the development of the Earth 3700 million years

242

242 The Earth – the planet and home of men, seen through the window of a space ship on its way to the Moon *(Apollo 10)*.

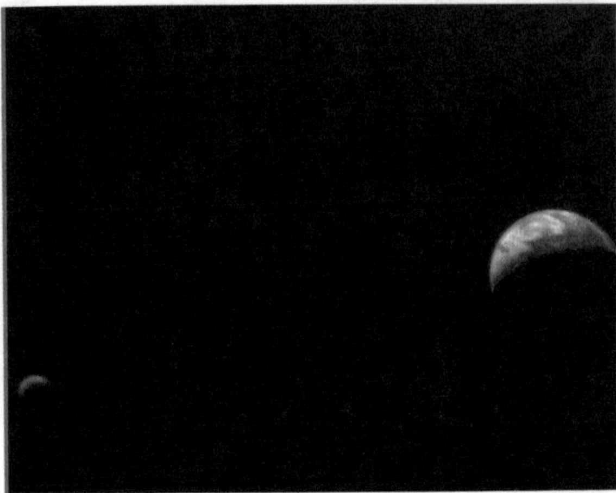

243

243 The Moon and the Earth – a pair of cosmic bodies. The home of mankind as seen by the *Voyager* probe at a distance of 11.5 million km (7.1 million miles).

ago, whereas the oceanic crust is much younger. But the Earth is 4600 million years old. The oldest regions of the ocean which have been found are only 200 million years old. The oceans are, geologically speaking, very young, and the crust beneath them is in motion. This movement begins in the mid-oceanic ridges and ends in the region of the island arcs and continental shelves. The whole Earth is in constant dynamic motion. The main driving forces are its internal heat, and the movement of the Earth along its orbit around the Sun. The most intensive energy source for the evolution of the planet's surface is, however, solar radiation.

What is the composition of planets?

The Earth has its 'brothers and sisters' in the Solar System. Very probably other stars, like the Sun, also have planetary systems.

The planets nearer the Sun are called *terrestrial,* because they resemble the Earth in size, density and composition. They comprise Mercury, Venus, Earth and Mars. The more distant planets are called *giant,* or *Jovian,* after Jupiter. They differ from the terrestrial planets, but like the latter have many things in common with each other. While the nearer, small planets are made of rock, the more distant giants are gaseous. We know much more about the closer planets, though it is no easy matter to study them. Even an astronomer as great as Copernicus complained that he had not had a good view of Mercury during his whole life, since it is visible only for a short time just before sunrise, and then only just above the horizon. Naturally, this limits the possibilities for observing it.

244

244 The first human footprint in the Moon dust not only symbolizes the second half of the 20th century – the start of the conquest of space – but also provides us with information. It shows the mechanical properties of the uppermost layer of the Moon.

Our knowledge of the planets of the Solar System is much greater today than it was 20 years ago. Research into planets is one of the most attractive subjects of scientific investigation today – despite, or perhaps because of, there being much we do not know about them. Due to intensive research the information we do have is rapidly growing. Thanks to modern space probes like Voyager, Mariner and Venera we are daily learning much that is new, confirming our ideas in some cases, in others completely overturning our impressions of our nearest neighbours, setting new questions for scientists to seek the answers.

Present knowledge on the composition of the surface and internal structure of planetary bodies is derived from precise measurements from automatic space stations, and the application of our experience with research into the structure and composition of the

Earth. But what are all these measurements, carried out or received from a distance, compared to a real piece of rock brought back from the Moon, for instance, which can be weighed, measured and pondered over – we can even think up new ways of getting information out of it.

But it is not only pieces of the Moon which are available here on Earth. We have something which can tell us more even than the surface of the Moon; something which has come to the end of its development. There are witnesses from space that can tell us a great deal, though at present we have only succeeded in unravelling part of their mystery. Those vital objects are meteorites. It is as though someone had written a book about the evolution of our system, and was hurling single pages down to the Earth's surface at random. And as soon as we manage to put them in the right order and learn to read the book...

The sight of a shooting 'star' has a different effect on different people. Some suppose that if they make a wish at that moment, it will be fulfilled. But the yearnings of the geologists and mineralogists who watch the fiery path of the shooting star seldom come true. For they are rather ambitious: their wish is that the shooting star's fall will last as long as possible, that the 'star' will not burn up but will reach the Earth's surface in the form of a meteorite, and that the meteorite will be found and taken to a laboratory for investigation.

Most meteorites which enter the atmosphere are small pieces of interplanetary matter, weighing hardly a gram. For the most part they are either deflected or burn up in the upper layers of the atmosphere. Their density is very low – 0.25 g/cm^3. Only larger and heavier bodies, provided they strike the atmosphere at the correct angle, actually reach the Earth's surface. But even then there is a problem: most of them end up in the sea. If one of them does get into the hands of a scientist, then it is a minor miracle, and must be fully exploited. For it offers untold information on what the world around it was made of, what the environment of the Earth is like, maybe even about the intensity and composition of the cosmic radiation which acted on it. Careful research and sophisticated equipment reveal where the meteorite originated, when it became separated from its maternal body, whether it was part of the surface or even of the internal structure of some planet, even how old that planet was. Even more meticulous analysis shows whether the meteorite contains components of material from a part of the Universe other than the Solar System.

Meteorites, or in other words meteoroids which have fallen to Earth, are classified according to their composition. The fundamental facts of physics and chemistry mean that meteorites contain the same elements as the Earth, the Sun and other bodies. The basic components which make up a meteorite are well known – minerals occurring on Earth, the silicates, and alloys of iron and nickel, together with sulphides. Metallic meteorites, with a predominance of iron and nickel, are the most popular, and the most impressive in the glass cases of museums. But they make up only 5.7 per cent of all the meteorites, which fall on to our planet, while stony meteorites, i.e. those which are mainly composed of silicate minerals, make up 92 per cent. The remainder are 'crosses' between the two main types.

In the early history of science it was not known that meteorites come from space. But in the 18th and 19th centuries much convincing evidence of the fact was assembled. Then a new problem came to the fore, one which led to long discussions between scientists. From which part of the Universe did they actually arise? Exact observation and measurement of their fiery paths, together with the calculation of their velocity, showed that meteorites belong to the Solar System, in which they move for a long period. The paths of meteorites before they fall show that before coming to Earth they move along circular or eccentrically elliptical orbits in the Solar System.

It has been possible to photograph, from several places at once, the fall of a large meteorite. The site of its impact was precisely calculated, and subsequently a successful search was made for its remains, near the town of Příbram in Czechoslovakia. This experiment, and two similar ones in North America (Innesfree and Lost City), proved that meteorites come from the Solar System and that before falling most of them were moving in the asteroid (minor planet) belt, which orbits between Mars and Jupiter.

Many scientists thought that meteorites might come from the Moon, from where they

245 The classification of meteorites is based on the ratio of silicate, sulphide and metallic content. Achondrites are most reminiscent of terrestrial or lunar basalts. Chondrites, on the other hand, represent a material which has undergone relatively few chemical changes from the time of its origin, and are considered part of the original Solar System. Stony metal and metal meteorites consist of strongly differentiated (layered) material which originated somewhere inside planetary bodies which today are in a disintegrated state.

were thrown off by the impact of another cosmic body. At first this was quite an acceptable explanation, but calculations later showed it to be unlikely for such a body to reach the Earth's surface. What is more, study of the Moon rocks brought to Earth by astronauts and remote-controlled probes showed that the rocks of the Moon differ from those of meteorites.

Stony meteorites are called *chondrites,* because they contain a peculiar formation, a sort of small nucleus called a *chondrule,* which is a spherical, crystalline formation, dispersed in the form of fine-grained fundamental matter. Chondrules are made of quite ordinary minerals known here on Earth, such as olivine and pyroxene. More than a century's work on chondrules, mainly investigating their origin, shows them to be the remains of molten matter which crystallized out. Chondrules are small formations, and some of them are scarcely visible to the naked eye in a fragment of a meteorite. They were discovered more than 100 years ago by the English scientist Henry Clifton Sorby, and are clearly visible in a meteorite section or in a petrographic thin section.

Other minerals contained by the chondrules of meteorites are feldspars, which are known from terrestrial rocks. It is interesting that pieces of cosmic material from distant planets should have a similar mineralogical composition to rocks here on Earth. There are really very few minerals in meteorites which geologists do not know from their own 'back-yard'. Apart from silicates, including olivine, pyroxene and feldspar, chondrite meteorites usually contain fragments and grains of metal and pieces of sulphides. In this respect they differ from terrestrial rocks. The combination of sulphides and silicates is very unusual in most rocks, and the presence of metal in terrestrial rocks is quite exceptional. The chemical analysis of meteorites

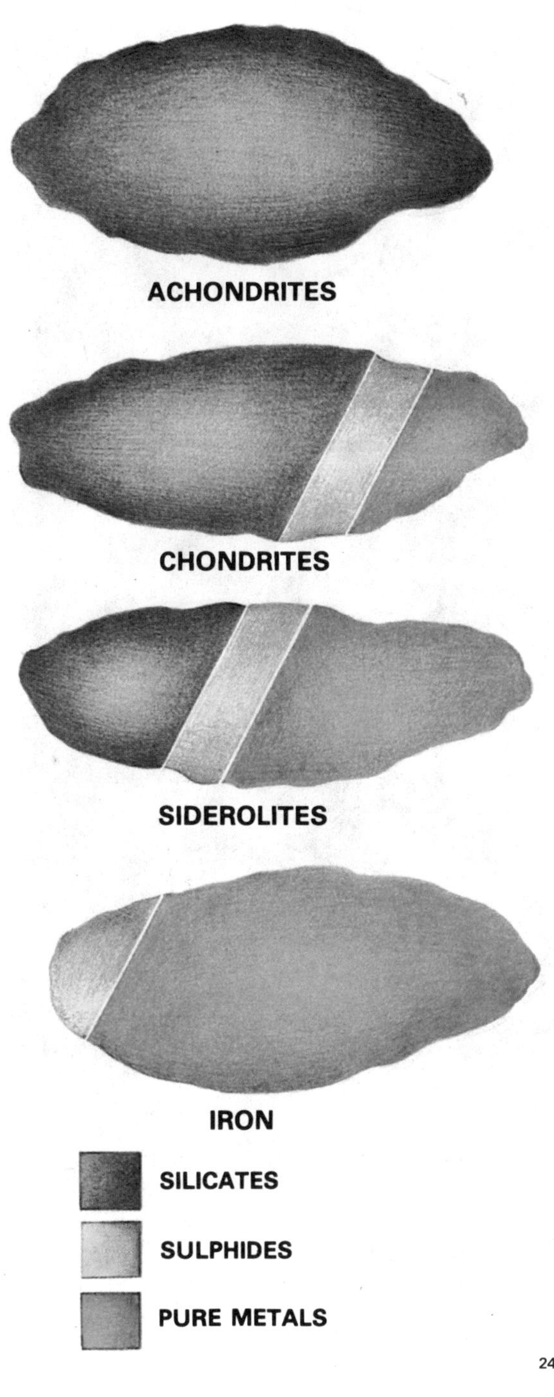

245

246 Planets orbit with elliptical paths. **M** – Mars, **E** – Earth, **J** – Jupiter, **SA** – Saturn. But smaller bodies, too, such as asteroids, approach the Earth on eccentric, highly elliptical orbits. **A** – the main belt, **AD** – Adonis, **IC** – Icarus,

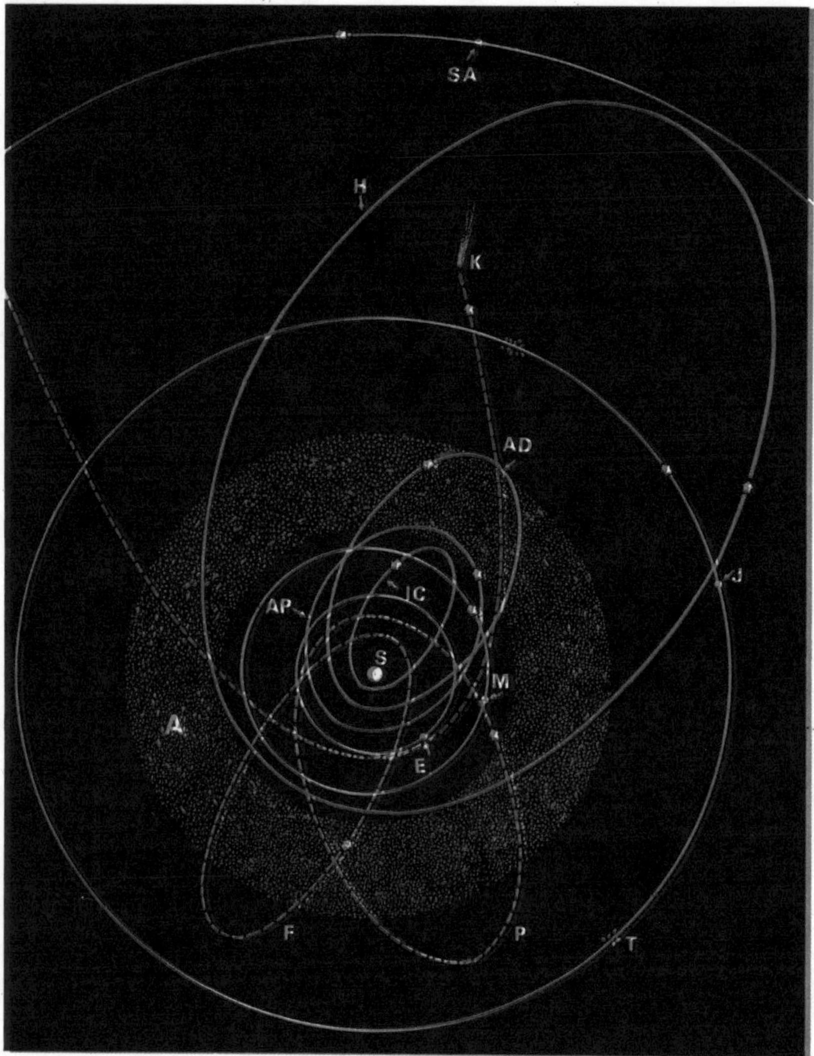

AP – Apollo, **H** – Hildago, **T** – the Trojans. The paths of the Příbram **(P)** and Lost City **(F)** meteorites, whose flight was photographed and which were later found, indicate the eccentricity of meteorite paths and their origin in the asteroid belt between Mars and Jupiter. Comets also follow highly eccentric paths **(K)**.

emphasizes their extraordinary nature, but at the same time shows their similarity to Earth rocks.

Nor are the other terrestrial planets, whose chemical composition shows them to have had a history similar to that of the Earth, unrelated in composition to meteorites. The more closely the chemistry of chondrite meteorites is studied, the more striking is the resemblance to the composition of the Sun. We must of course discount hydrogen and helium, after which we see a similarity in the proportions of the main elements and the content of trace elements. If we then consider the fact that the Sun is a very average star, we may assume meteorites are composed of some sort of commonplace cosmic material. Some chondrites even contain carbon, water and a certain amount of other fluids, which prove that they have not gone through any transformation from the time of their origin. If they had become molten, like volcanic rocks, for example, certain of the elements would have separated out: metals would have parted from silicates and sulphides, and water and carbon would have been lost. Scientists therefore think that chondrites are the primitive building blocks of the Solar System. Matter similar to that found in meteorites is probably the material of which planets of the terrestrial type and others are fundamentally composed.

The world of meteorites is an extremely interesting and mysterious one. In itself the presence of carbonaceous material in carbonaceous chondrites is a great mystery, but at the same time a challenge to scientists, offering as it does a possible answer to questions of the origin of life. For if the carbonaceous compounds of meteorites are subjected to detailed chemical analysis, it turns out that they contain amino-acids. Biologists suppose these to be the basic building-blocks of organic life. Thus the embryos of life would seem to have been cruising around in space for a long time, and occasionally some of them happen to fall, quite by accident, on to the Earth.

Meteorites – and we are still speaking of ordinary, primitive meteorites which have not undergone many changes since they came into existence – have other surprises in store for us. Scientists studying the isotope composition of their components have found that some of them contain components which are not from the Solar System at all, but are older than that. Before the Solar System came into existence there existed in the Universe material of other, older systems and stars. Very detailed research has revealed the remains of such material in meteorites. Thus the study of meteorites brings a quite new view of the evolution of the planets, the Solar System and the Universe in general.

Among the meteorites which have undergone a transformation and are the opposite of the primitive chondrites, are those called *achondrites.* As the name suggests, they lack those typical spherical formations, the chondrules. They are also different in having no iron or sulphides. They are reminiscent of volcanic rocks – basalts – and are almost indistinguishable from the basalts which occur on the Moon. They are in fact lava from a planet or minor planet which we do not as yet know, but whose chemical and isotopic composition show that it is a body from our Solar System.

We might also explain the origin of metal (iron) meteorites as the transformation of the original material. If we melt a chondrite – i.e. a very primitive, unchanged meteorite – we separate, as we would in a furnace, the metal and the silicates. The latter form an achondrite – a basalt meteorite, and we are left with the metal.

So scientists studying meteorites have created, on the pages of scientific literature, hypothetical planets. These planets, 'built up' on the basis of the chemical and mineralogical study of meteorites, have a nucleus made of material similar to that of the iron meteorites, and a surface of achondrite or carbonaceous material, but they mainly consist of ordinary chondrites. This idea is a very rough and ready one, but it does show us that the link between the work of astronomers (who study the composition of stars) and the investigations of geologists (who study the composition of the Earth) is meteorites.

Previously special glassy materials which were thought to fall to Earth from space were also ranked among the meteorites. Collectively they are called *tectites,* and they have further designations according to the place where they were found. So, for instance, in Europe there are moldavites from Czechoslovakia, on Java there are javaites, and in Indo-China there are indochinites. Similar tectites can be found in the Libyan desert and in Texas. They have beautiful shapes, a highly modelled surface, and a rich variety of colours from bright green to brown and even black, and may be transparent or quite opaque. At present they are very popular in the jewellery trade, since each of them is in fact a unique 'original'.

The shapes of tectites show them to be melted and solidified pieces of rock which have passed through the atmosphere. For a long time scientists were puzzled by these chunks of glass, and tried all sorts of tests on them. Some thought they came from the Moon, whence they were wrenched by the impact of large meteorites. For in such

247 The surface of the Moon as seen by the *Apollo 11* astronauts. The material filling the large, circular craters is basalt rocks, while the 'continents', the older part of the Moon, are formed of feldspar – anorthosites.

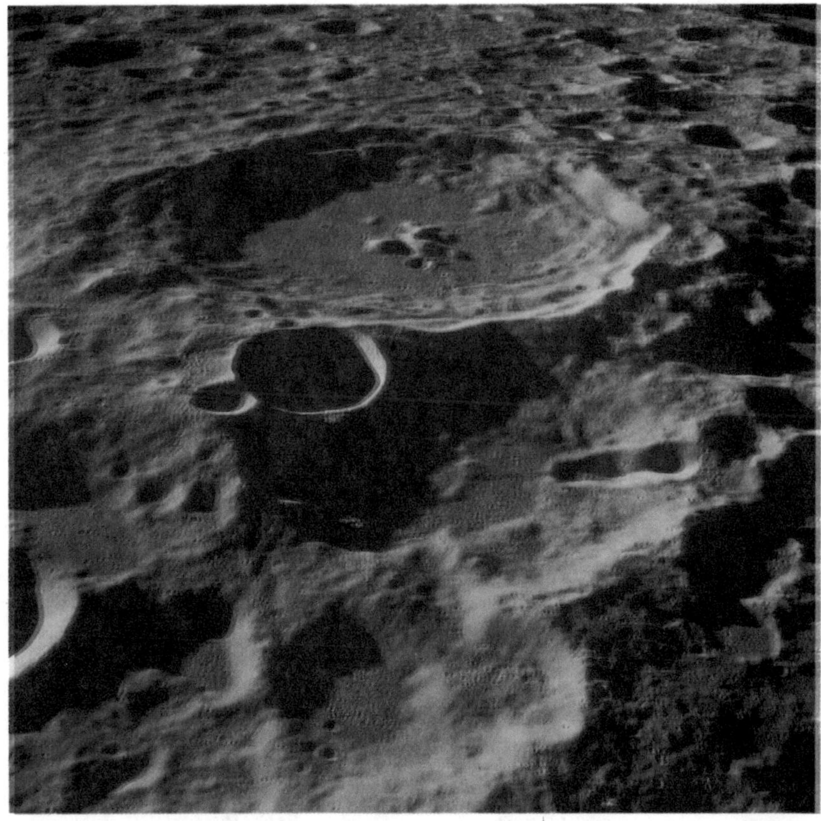

247

248 It was only after man's flight to the Moon that the relative scale of the age of the lunar surface could be established. The number of craters per unit area can, using this graph, be converted to real age. It is clear from the graph that in the course of geological eras the probability of collisions between planetary bodies and meteorites declines.

249 A study of the age of the surface of planets would seem to be a very difficult matter, but in fact the relative age of two areas can be determined from the number of craters per unit area. The figure illustrates the 'ageing' of a planetary surface and shows that even small meteorites, whose craters are not visible at first sight, obscure original features. One often speaks of erosion by cosmic particles.

a collision the kinetic energy of the meteorite is transformed into heat, and the rock at the point of impact is melted. But a comparison of tectites with Moon rocks shows that they could not have come from there. Comparison with terrestrial rocks, however, shows tectites to have a similar chemical composition to these. So in the end the two views were combined in the 'impact' hypothesis, accepted by the vast majority of scientific circles. Tectites – moldavites, javaites and all other varieties – are the product of the collision of large meteorites with the planet Earth. We are unlikely to live to see such an event, since they occur around once every million years.

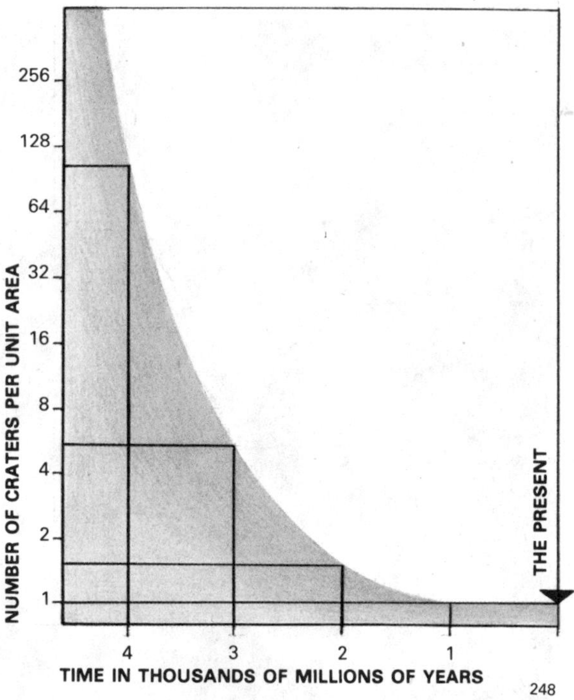

248

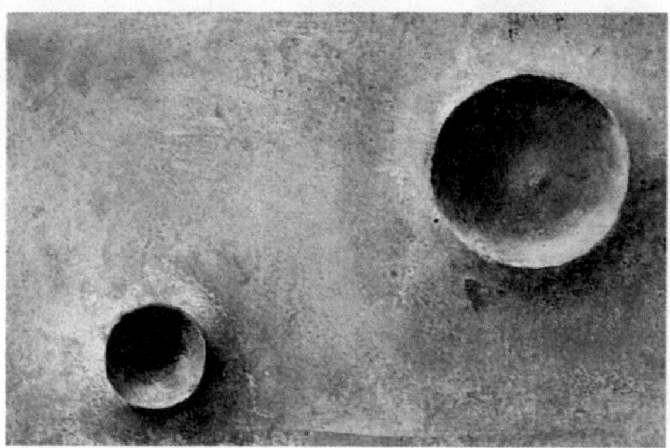

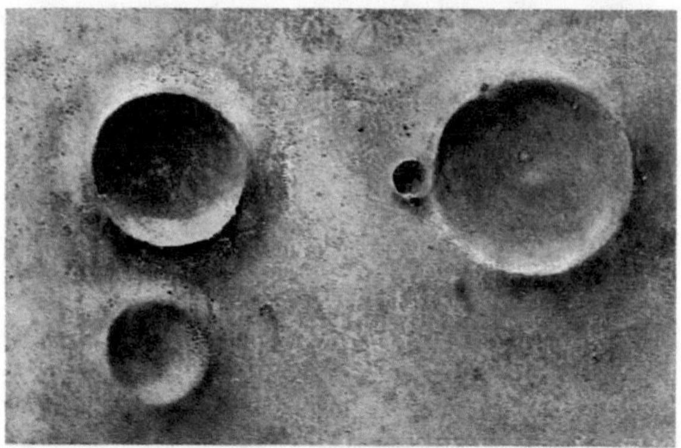

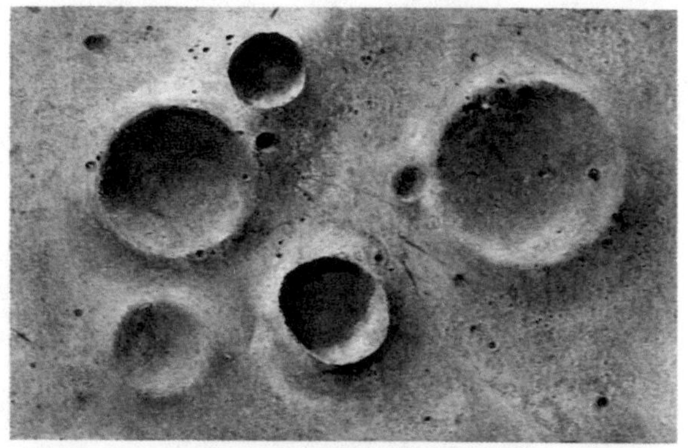

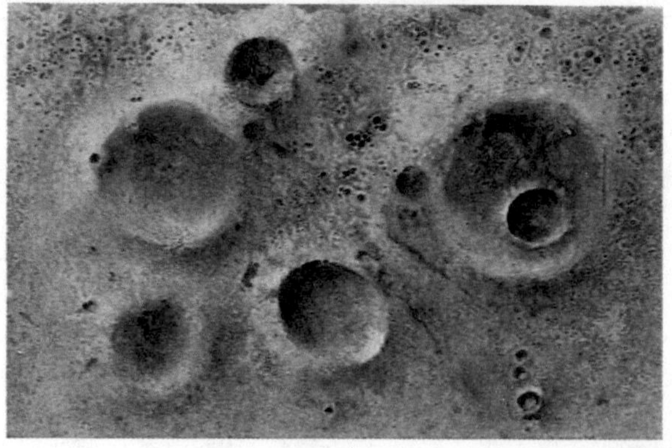

249

The planetary system

Since time immemorial, man has observed the stars and planets. We can see this from the ancient observatories in the centres of civilization which existed in Britain, Mexico and South America. At first it was only possible to observe the nearer heavenly bodies – the Moon, Mercury, Venus, Mars, Jupiter and Saturn – while the rest, those further away, remained hidden to the naked eye, to be discovered only with the aid of telescopes in the 18th, 19th and 20th centuries.

Our nearest neighbour, the Moon, is the most striking body in the night sky. At present it is under close surveillance both by man and by the satellites which now orbit it. Until recently, even the Moon was a very

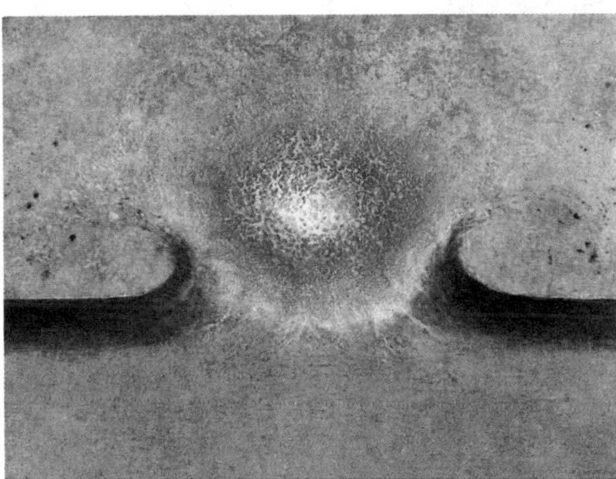

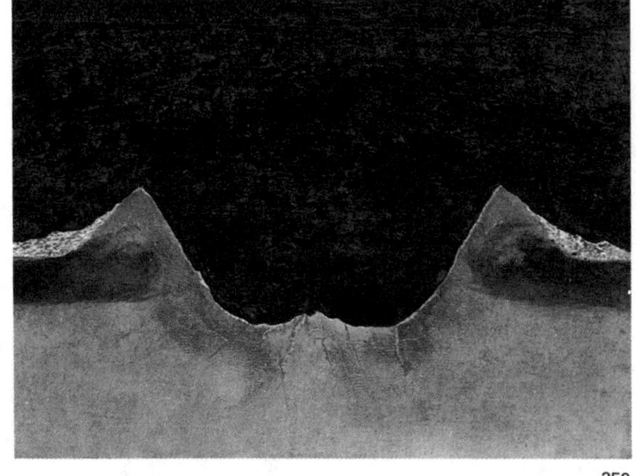

250

250 The occurrence of an impact crater on the surface of a planetary body is an event which had a great effect on the living conditions there. It is even supposed that such a major collision may have been the chief factor in the extinction of the dinosaurs on Earth. Cosmic bodies lose their kinetic energy in a fraction of a second, and it is partly converted into work, in forming the crater, and especially into heat. This brings about the evaporation of the falling body and a melting of rocks.

mysterious body. But modern research has revealed what was on the dark, hitherto unobserved, side of the Moon. It was the detailed Moon surface research programme of the 1970s that began a series of investigations into the inner planets – Mercury, Mars, Venus – and some of the moons of Jupiter and Saturn.

Even a glance at the Moon shows that there are two distinct parts – dark areas, improperly called seas (they have no connection with water), and lighter sections of 'land'. The latter areas make up 80 per cent of the lunar landscape, the seas the remaining 20 per cent. The two types of material differ in their chemical and mineralogical composition. Only the light type of 'land' rock occurs on the far side of the Moon. The surface is covered with large and small craters; crater formation was the most important feature of the evolution of the lunar surface. The number of craters in a given area is a good guide to the age of that area – the older it is, the more craters it has.

251

251 Impacts, the collision of planetary bodies with smaller cosmic bodies (meteorites) are a basic feature of the shaping of the surface of those planets which have no protective envelope in the form of an atmosphere, and where the internal resources of the planet are incapable of regenerating the surface.

The most important milestone in the history of planetary research has been the landing of men on the Moon. (It seems paradoxical – you might say the Moon is a moon, not a planet). Men brought back more than 400 kg (880 lb) of samples from the Moon, set up seismographs on its surface, measured the amount of heat it gives off, its magnetic field, and the magnitude of its gravitational pull at various points. What is the result of all this? The Moon, like other planets, has an onion-like structure, and the rocks on its surface are different from those in the interior. In the early stages of development, about 3000–4000 million years ago, volcanoes were very active there. Lava poured from these, filling the great depressions which had been created by meteorite collisions, and these became the dark seas, or *maria* (the plural of *mare,* the Latin word for a sea).

The Moon is a body which underwent a quite independent development, like the Earth and the other planets. Samples of the lunar surface, rocks, and even soil, have

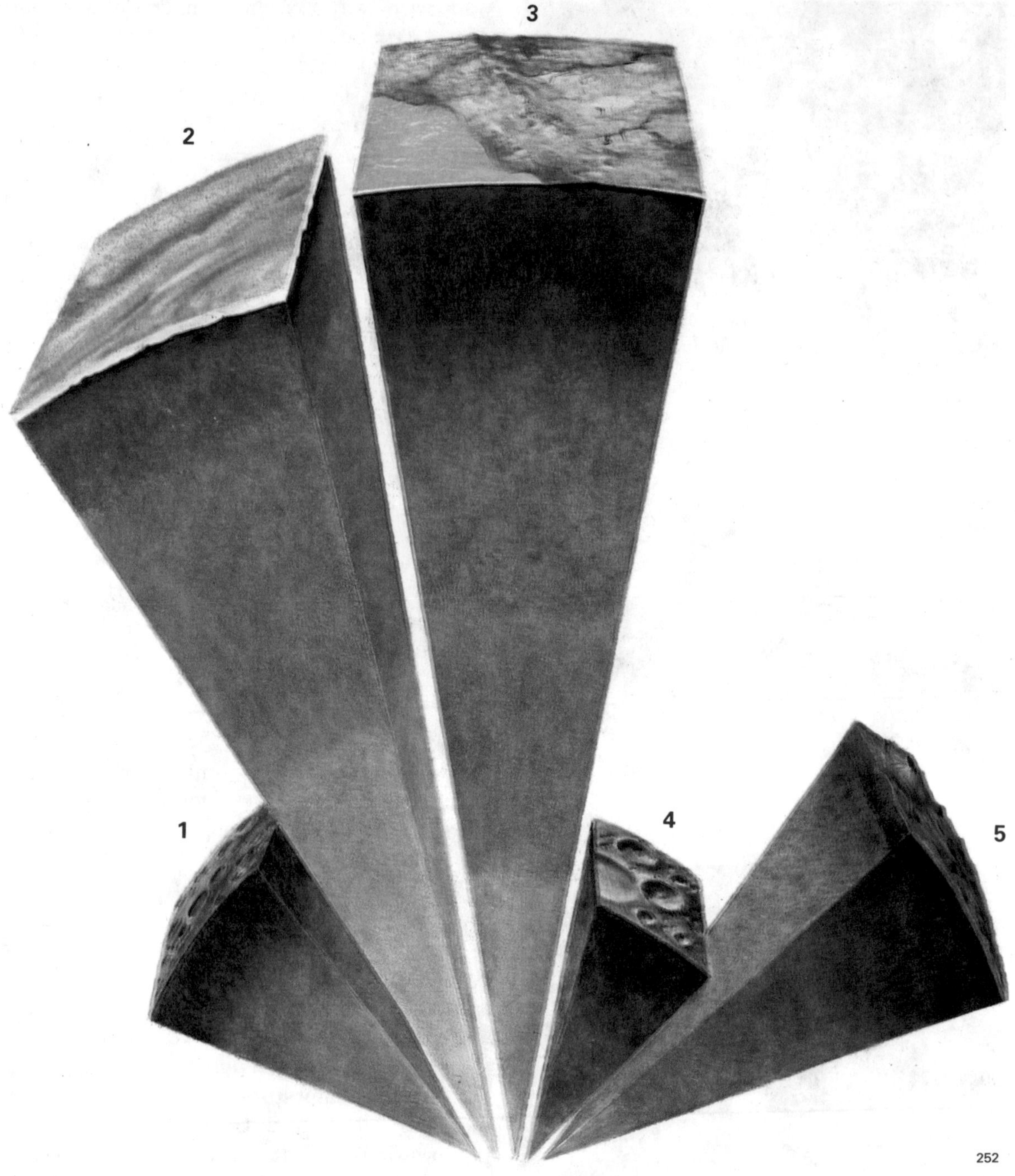

252 Planetologists and geologists are able, after almost two decades of intensive research, to offer such cross-sections of the terrestrial type of planets. Mercury **(1)**, Venus **(2)**, the Earth **(3)**, the Moon **(4)**, Mars **(5)**.

allowed man to explain some enigmatic features of meteorites. Thanks to the fact that samples of the Moon's surface were brought to Earth, the information from automatic probes can now be interpreted more fully. In addition, man has learnt to appreciate his own planet Earth all the more, with its strength and dynamic development. For the Moon has no water, and never did have; because of its low mass, it has no atmosphere, and all activity (in the sense of internal geological forces, such as volcanic activity) ceased there more than 2000 million years ago. The Earth, on the contrary, is flourishing, thanks to the presence of water – and volcanic activity, too, since this created not only our gaseous envelope, but also all the water on the surface.

The information obtained on the lunar surface can be used to help solve the mysteries of the early history of the formation of the Earth's surface and, thanks to experience of the Moon, to explain craters on the Earth, caused by the impact of large meteorites.

The Moon has many relatives in the Solar System. Mercury, for instance (the planet nearest the Sun), is quite indistinguishable from it in some photographs. Its surface is also covered with innumerable craters of various depths and sizes, but a comparison of the physical properties shows that there are major differences. While the density of the Moon is about 3.3 g/cm^3, that of Mercury is high – 5.4 g/cm^3. This means that Mercury probably has a metal core of considerable size. This is also indicated by the strong magnetic field which it has, and the Moon does not have. And though the surface of the Moon and that of Mercury are very similar, their internal structure is certainly different. The surface features of Mercury show that apart from cratering, a major role was played in the formation of the surface by contraction and tidal forces. The volcanic activity of the surface, what little there was, was much less than on the Moon.

Mars is another planet on which men have carried out a great deal of research. The Mariner and Viking probes showed that the evolution of the surface of this enigmatic body – the red planet – has been much more complex than that of Mercury or of the Moon. This seems to be connected with the fact that Mars is a rather larger body, and a little further away from the Sun, so that it has retained at least a part of its history – a content of fluids. Mars, too, is a differentiated (layered) body; it has an onion-like structure like the Moon, Mercury or the Earth, and volcanic activity has played, and is playing, an important role in the formation of its surface. Physical measurements show Mars to have a nucleus formed of high-density matter and a crust made of low-density matter. Morphologically, it is one of the most varied bodies in the Solar System. Apart from its geological curiosities, among them polar caps made of real ice, areas of red and orange colour occur on its surface. The discovery of volcanoes and riverbeds also suggested a resemblance to the Earth. That was why there was for a long time hope

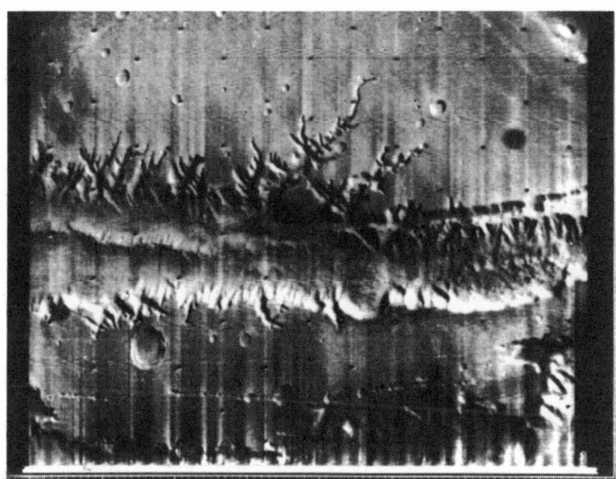

253

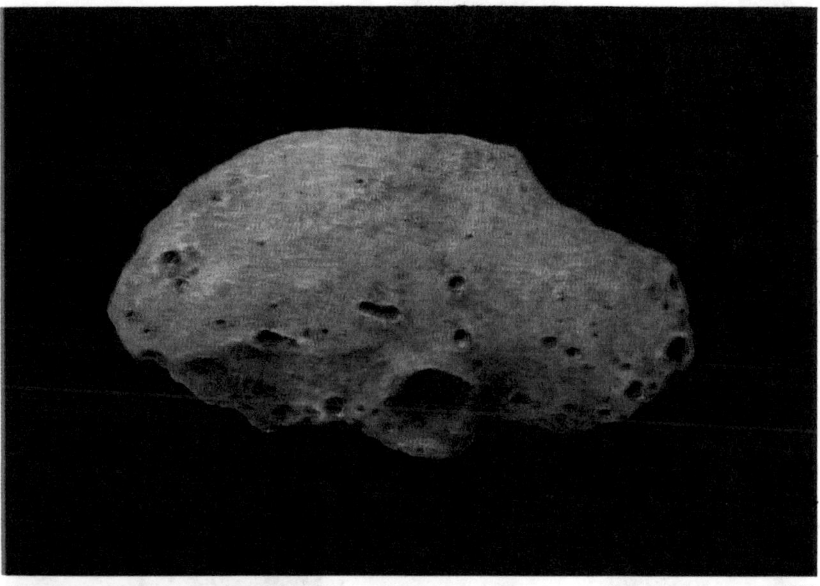

254

255

253 A view of part of a huge valley on Mars – Coprates, which is in the area called Tithonius Lacus. In comparison the Grand Canyon in Arizona would be 'baby brother'. The length of the valley, its depth, and the system of inlets have no equivalent on the Earth. Though everything points to the 'canals' having been formed by water, there is no trace of water on Mars today.

254 The body orbiting Mars was once considered to be an artificial satellite launched by intelligent inhabitants of the planet. But it turned out to be the uninteresting and even unshapely Phobos, covered in impact craters.

255 The planet Venus, often referred to but just as frequently denied as the sister planet of the Earth, is constantly veiled by rapidly rotating clouds. The surface temperature is around 450°C (842°F), and the atmosphere unbreathable for humans. This shot was taken by the US probe *Mariner 10*.

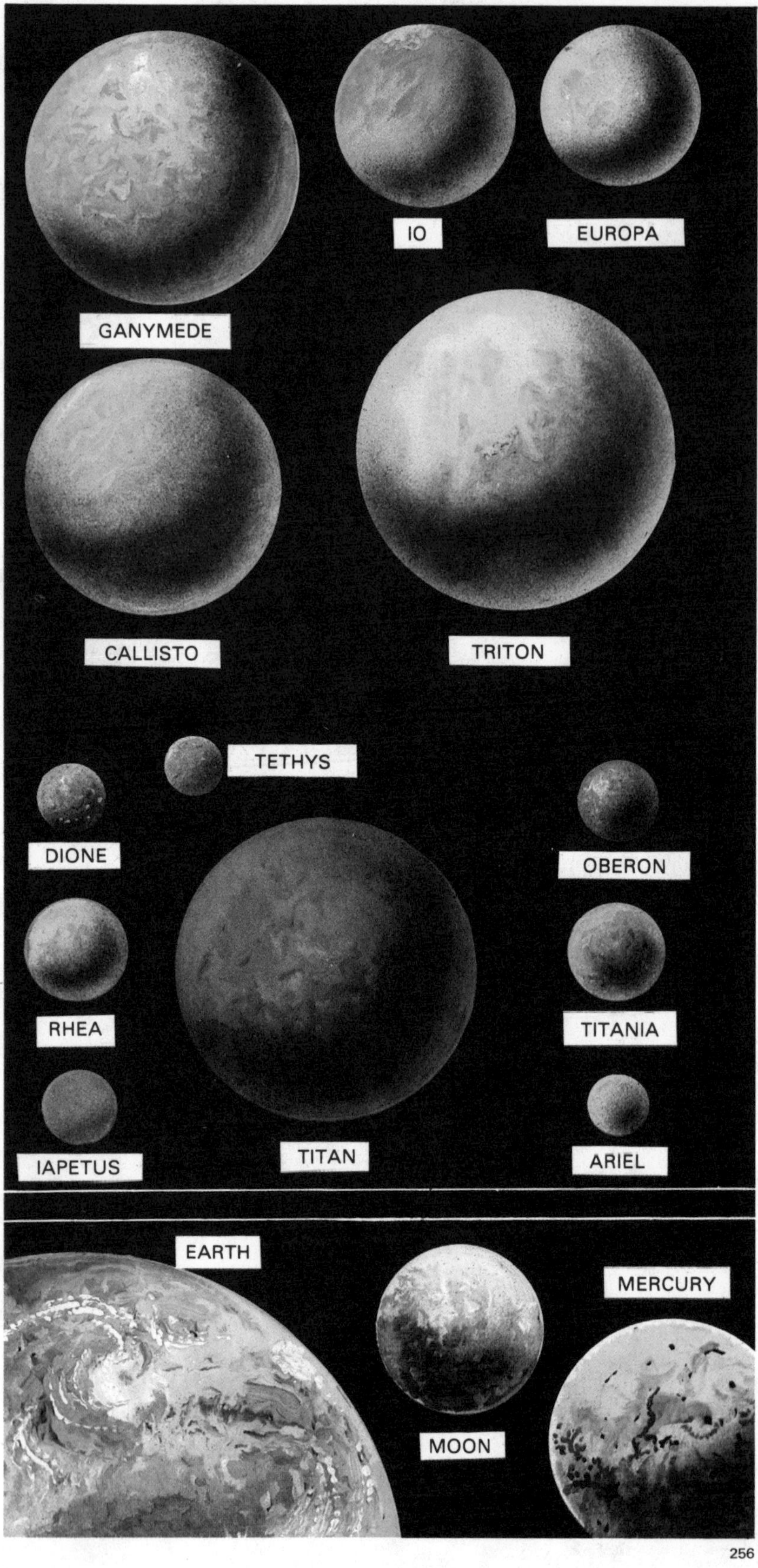

256 The large moons of the Solar System compared with the size of Earth, Mercury and the Moon, which is 3476 km (2158 miles) in diameter.

of finding life on Mars. But no evidence has been found for this. Mars seems to be an extinguished planet, which has lost both its atmosphere and the water it once had.

The riverbeds are an indication of this. On the surface there are signs of the windstorms which (in spite of the rarefied atmosphere it has today, mainly composed of carbon dioxide) still afflict the planet. Volcanic peaks, which are unmatched for size in the whole of the Solar System, occur in the region of large faults. The differences in elevation between mountains and valleys are enormous. The distance from the bottom of the deepest valley to the top of the highest mountain is 20–25 km (12–16 miles).

Two moons, Phobos and Deimos, orbit Mars; in fact they are no more than lumps of rock, riddled with craters and only 13 km (8 miles) and 10 km (6 miles) long, respectively, along their longest sides. They are probably captive asteroids, though until about 20 years ago some scientists thought they might be artificial satellites sent up by intelligent Martians.

The most enigmatic and apparently the least hospitable of the inner planets is Venus, often called the Earth's sister-planet. Its dimensions, mass and density are indeed very similar to those of our own planet. But conditions on the surface are a good deal less pleasant than those on Earth. We humans would be less than comfortable, with a temperature of around 450°C (842°F) and an atmospheric pressure about one hundred times greater than that on Earth. Nor is the composition of the atmosphere exactly to our taste – it contains not only carbon dioxide, but also sulphuric acid. Soft-landed probes, though designed to resist such conditions, usually break down in a matter of minutes. The surface of Venus is, like that of other planets, heavily cratered, and radar research has shown that the huge round depressions like the Mare Imbrium on the Moon or Caloris on Mercury are characteristic of Venus, too. But Venus remains the least well studied of the terrestrial type planets. At present radar instruments have managed to characterize further topographical features: tall mountains similar to the Himalayas (the Venusian range is called the Maxwell Montes) or quite unique features such as the plateaux Ishtar Terra and Aphrodite Terra.

These are similar to the rift valleys and peaks of Earth. The surface rocks, whose chemical composition we know from a single locality, are also similar in many respects to terrestrial rocks.

Let us examine the planets of the outer Solar System. The planet farthest from the Sun is Pluto, which is similar in size to the inner planets, but moves very rapidly, and has a highly elliptical orbit. It is not a giant planet compared with Jupiter, Neptune or Saturn, and at first sight does not seem to belong in that group at all. The next planet in towards the Sun is Neptune. It is somewhat larger, and perhaps more interesting than Pluto, if only because it shines with its own light (it radiates more light and heat than it gets from the Sun). It has a long orbit around the Sun, and takes 165 times longer to make the journey than does the Earth. On Neptune, and also on Uranus, another of our neighbours which is very similar to it, conditions are very unhospitable. It is very cold (−200°C/−328°F), and the amount of energy received from the Sun is minimal. There are also smaller bodies – moons – orbiting Uranus and Neptune.

Men have only known of the existence of the last three planets for a comparatively short time, their discovery was regarded on each occasion as a great success for astronomy (Uranus 1781, Neptune 1846, Pluto 1930). Our knowledge of the composition and structure of these distant neighbours is very poor. We know their temperatures, something of the composition of their atmospheres (methane, ammonia, hydrogen, helium), and recently we have learnt a little about their moons. The rings around Uranus, made up of boulders, were not discovered until 1977. The latest satellites to be discovered were found by the space probe Voyager in 1986, and belong to Uranus. The possibilities of classical astronomy are almost exhausted today. There remains only the prospect of craft like Voyager bringing new data on the outer planets of the Solar System. The usefulness of such probes is shown by the results obtained from investigations of the closer planets, such as Jupiter or Saturn.

The second largest planet of the Solar System, Saturn, is surrounded by a number of rings and satellites. It is only in recent years that the mysterious, apparently solid rings

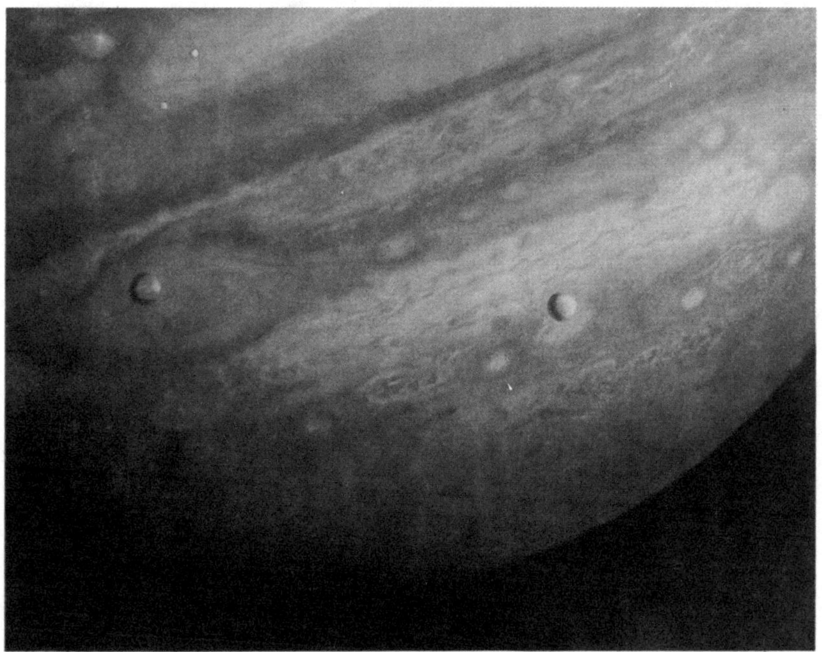

257

257 Jupiter, often compared to an unsuccessful sun, is the largest planet in the Solar System. Its moons, too, are desert-like. The US probe *Voyager* photographed Io and Europa. The surface of Jupiter, with a large red patch, can be seen in the background.

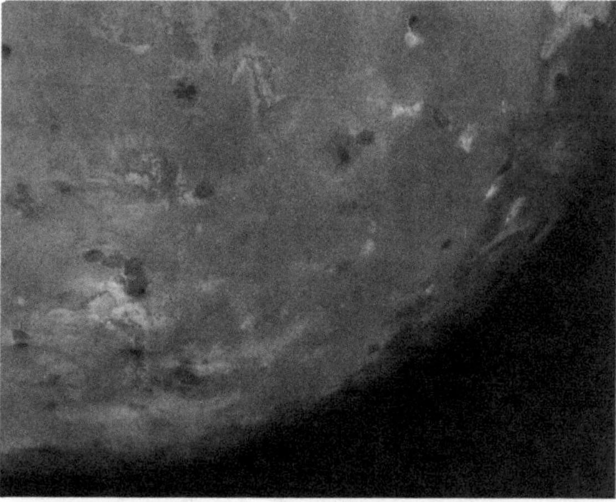

258

259

258, 259 Jupiter's moons are among the most interesting bodies studied in the 1970s and early 1980s. Though some of them had already been discovered by Galileo himself, the fact that they were 'planets', with their own evolutionary features, became apparent only after *Voyager*. The moon Io, for instance, was shown still to be volcanically active, while the surface of Callisto is similar to that of our own dead Moon.

have had their secrets revealed. There are not in fact five or six of them, but thousands, made up of small fragments of rock, perhaps a few metres across. Saturn is a planet with a low specific gravity, less than that of water. It is relatively cool, with a temperature of −170°C (−274°F), and its atmosphere consists of methane.

Jupiter, the largest of the Sun's companions, is about 320 times more massive than the Earth. It emits more heat into space than it receives from the Sun. It therefore sometimes seems to us terrestrials like a failed sun, in which the reactions required to produce great amounts of heat and light have never really started properly. Jupiter's atmosphere is opaque, with different 'climatic zones' visible, where the temperatures are different. There is a huge red spot caused by the convection of the atmosphere, which is visible from a great distance.

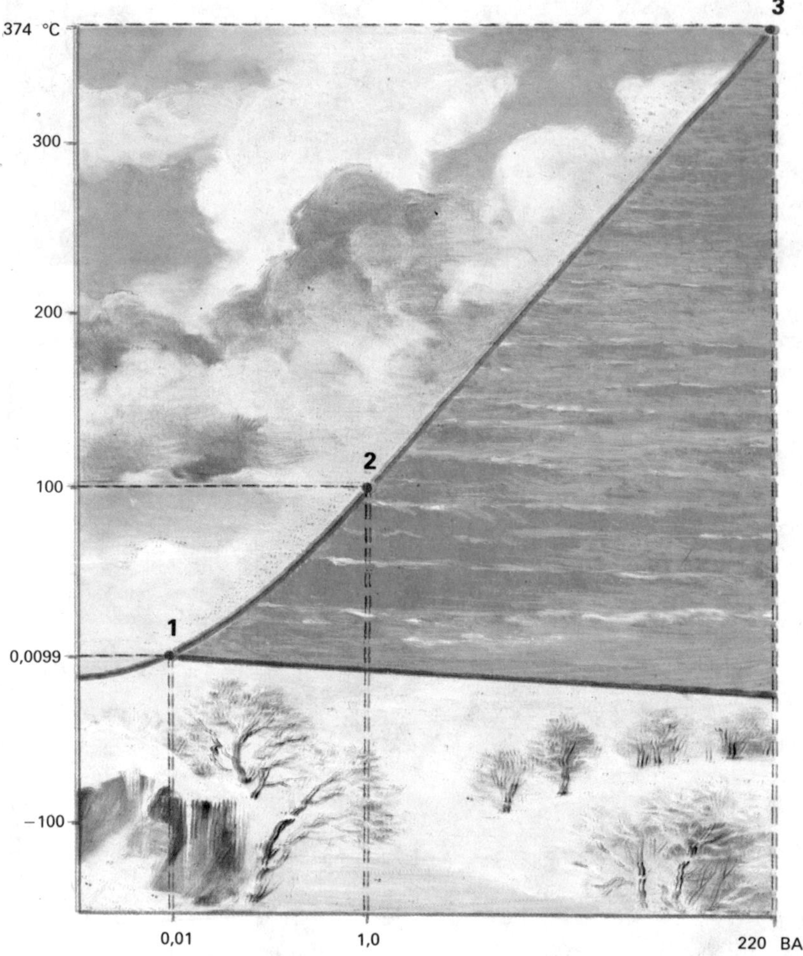

260 The graph shows the three states of water with the triple point where all three can exist at the same time **(1).** Under different conditions, however, there are never more than two states of water at once **(2),** or one only. Above the critical point **(3)** water does not exist as a compound.

The large size and small mass of the planet are the reasons why the fission reactions which take place in the Sun do not occur on Jupiter. Nevertheless, it makes its own light and heat. This phenomenon, previously thought to be exclusive to the Sun, is explained by the contraction of the planet. It has been calculated that a rate of contraction of 1 mm (0.04 in) in radius every year is enough to account for the energy produced by Jupiter. Like the Sun, it is mainly composed of hydrogen and helium, with methane and ammonia present in the atmosphere. The great intensity of the magnetic field is explained by the presence of a metallic core. Whether it is made of iron or metallic hydrogen, no one yet knows.

Jupiter's moons are interesting enough for geologists to spend a good deal of time in the near future studying them. They are made of solid matter – stones. On one of them, Io, explosions can even be seen, in which matter flies into the atmosphere to a considerable height, subsequently falling back to the surface. Analysis has shown the material to consist of compounds of sulphur. The rest of Jupiter's moons seem dead, with no signs of continuing activity having been observed as yet.

The most beautiful of the planets

We are now left with the last, but for us the most important of the Sun's companions, the blue, or maternal planet, Earth. In more prosaic terms, it is a very well organized and ordered mixture of cosmic material, metal, stone and gas. In its proportions, onion-like structure and chemical composition it is similar to the other planets. The main difference lies in its hydrosphere – the oceans, seas and rivers – and the presence of green vegetation and animals, including intelligent ones. Earth is ideally suited to life: water exists on it in all three states (Figure 260), its distance from the Sun is ideal for maintaining optimal temperatures, and its mass and dimensions allow it to retain its own atmosphere, composed of gases. As yet, we know of no similar planet. There are certainly none in the Solar System. We humans have not been able to look beyond that system, and it will be a long time before that is possible.

Most of the planets, including the Earth, orbit the Sun in a single plane, except for the outermost planet, Pluto, which is inclined away from that plane. They also orbit in the same direction, and most even have the same direction of rotation about their own axes. It corresponds to the direction in which they orbit the Sun. The only exceptions here are Venus and Uranus, which rotate in the opposite direction, and so are said to have *retrograde* orbits.

The planets and their moons also move in the plane of the ecliptic, and some of them are synchronized, i.e. they always display the same side to an observer. An observer on Jupiter would always see the same side of its moons. The fact that the majority of the planets orbit in the same direction and have the same direction of rotation is one of the proofs that they had the same origin in some sort of parent nebula, which also rotated. It is interesting that in the modern age of artificial satellites, spacecraft, radar, minicomputers and microprocessors, we retain the old hypothesis on the origin of the Solar System put forward by the German philosopher Immanuel Kant (1724–1804), whereby the planetary bodies arose from a contraction and agglomeration of the original nebula. Modern branches of science such as isotope geology and the mineralogy of meteorites, of which neither Kant nor the Frenchman Pierre Laplace (1749–1827, who developed the theory) had the slightest inkling, have offered new evidence for the justification of the idea, however.

Geologists and planetologists, who are concerned with how planetary bodies come into being, are agreed that cosmic matter wraps itself around, or adheres to, a gravitational centre. The more massive the body, the greater the attraction it holds. So a larger and larger body is gradually built up, which evolves its own independent course of development only after many thousands or even millions of years.

In the course of the long history of the Earth's evolution, its surface changed; its atmosphere also developed, and is doing so to this day. The configuration of its continents and seas alters, too. New mountain ranges come into being, while old ones disappear; where there was once an ocean, there is land today. Not even the Earth's magnetic poles have stayed put.

All these forces, changing and developing the surface and the interior of our planet, work in interrelation. There must be a sort of unison among them for evolution to move forwards, and not towards a global catastrophe. But minor catastrophes, such as collisions with cosmic bodies or climatic changes, are a part of that evolution. It is therefore important for man to know and understand the underlying laws. Rather like a doctor, he must learn about the metabolism of his planet, and not try to use the wrong medicines, let alone perform the wrong sort of surgical operations.

7. THE FACE OF THE EARTH

Continents and landmasses

The first thing an extra-terrestrial visitor would notice on approaching the blue planet is its colourfulness. The blue of the oceans is predominant, making up two-thirds of the total surface area of the globe. This caused the first spacemen to speak of the blue planet. The remainder of the Earth is brownish, but also green in places, and with the white of snow and ice the dominant colour at the poles. Most people have seen television pictures of the shining, blue Earth. The clouds that envelop some parts of the planet are also familiar images from our television screens, thanks to photographs sent back from meteorological satellites. Europe often appears as a white blob, in which the outlines of countries must be artificially superimposed.

Geologists and geographers differ as to where they think landmasses and continents start and finish, however. Though a geographical map shows the sea coast as the borderline, the geological boundary between sea and land lies elsewhere. The geologist considers many stretches of water not as part of the oceans and seas, but as part of the landmasses, which differ from the ocean in geological structure. So where the geographical atlas shows blue seas, geologists might still consider this to be part of the continent. For instance, the whole area between Norway and the northernmost tip of the British Isles is not, geologically, sea at all. For the geologist it is a part of the continent which is called the *continental shelf.* In that part of the world geologists behave as if they were on dry land, and this attitude pays dividends. They find oil and natural gas there, and many other raw materials characteristic of land; the continental shelf is a very good place to look for oil.

It is only when he reaches that part of the ocean where there are no longer continental rocks that the geologist considers himself at sea. And, even if we were to let all the water out of the ocean, or to freeze it, the geologist would still be able to tell from the composition of the rocks and the whole structure of the geological seabed whether he was on land or at sea.

In spite of these limitations and geological definitions, the oceans are much more extensive than the continents. Geologists' maps show Australia, Africa, North and South America and Antarctica – only Europe seems to be missing. It is drawn together with Asia, since there is no geological reason to divide the two. We shall see that the positions of the continents on the surface of the Earth were not the same in the geological past as they are today. Much has been written on the movement of continents *(continental drift).* Maps to illustrate it used to adorn textbooks as a sort of curiosity. But much evidence was found in the 1960s for the fact that the continental masses are in motion, along with their substratum. Europe and Africa are moving away from America, while India was once a part of Antarctica. Australia, in the course of hundreds of millions of years, has 'floated' towards Asia until the two are geologically touching. In the last 20 years many such pictures have been published.

Geologists study the movement indirectly using magnetic methods. They determine the magnetism of old rocks, observe the course of geological units from one continent to another, and compare the occurrence of characteristic rocks on two continents which are today separated by a deep ocean. So, for example, identical types of basalt are found in southern Africa, Australia and Antarctica. Since they are basalts of the same age, and have the same mineralogical and chemical composition, they were probably part of the same continent – Gondwana. This would mean that the basalts were connected, and that the present position of the continents is different from that in the past. In the course of geological eras the continents shifted in relation to each other and separated, but also collided with each other.

The theory of continental drift was put

261 The hydrosphere, an essential condition of life on the Earth, arose through volcanic activity. The binding of water, its evaporation and condensation in ice masses are all part of the thermal regulation of the Earth. The balance is a very delicate one, and could be upset by human rashness.

262 The continental shelf is that part of the sea which still has a continental structure. It is merely flooded. But the shelf region provides huge quantities of oil. The map shows all continental shelves shallower than 200 m (656 ft), i.e. seas in which petroleum deposits might be expected.

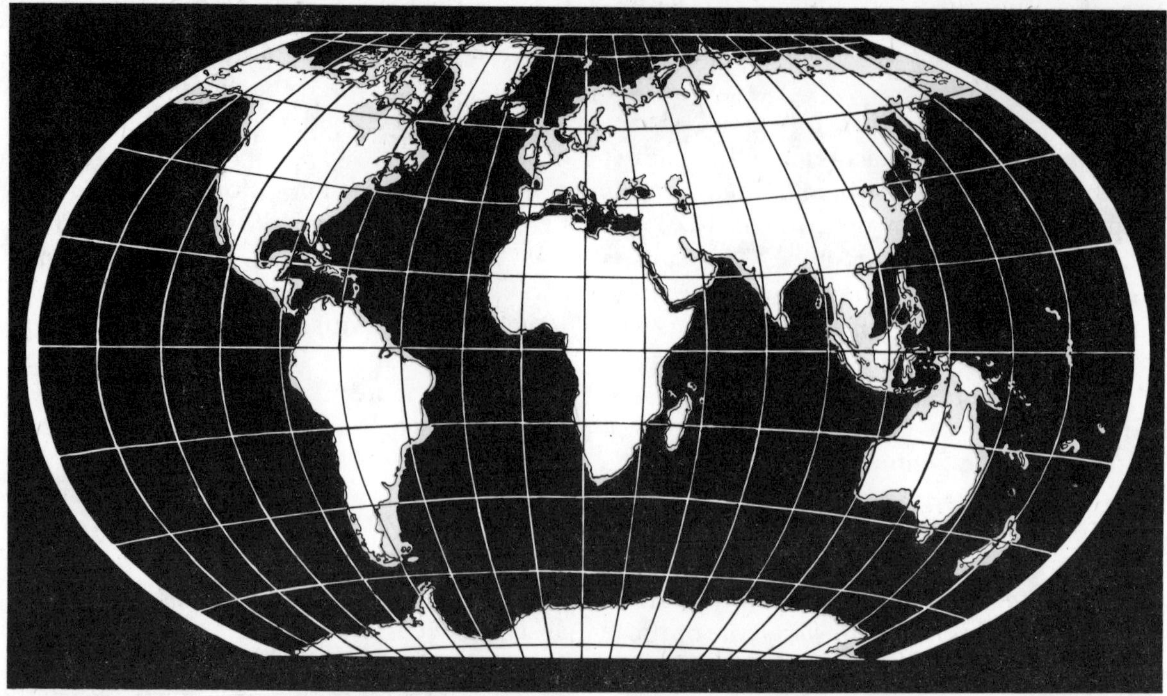

262

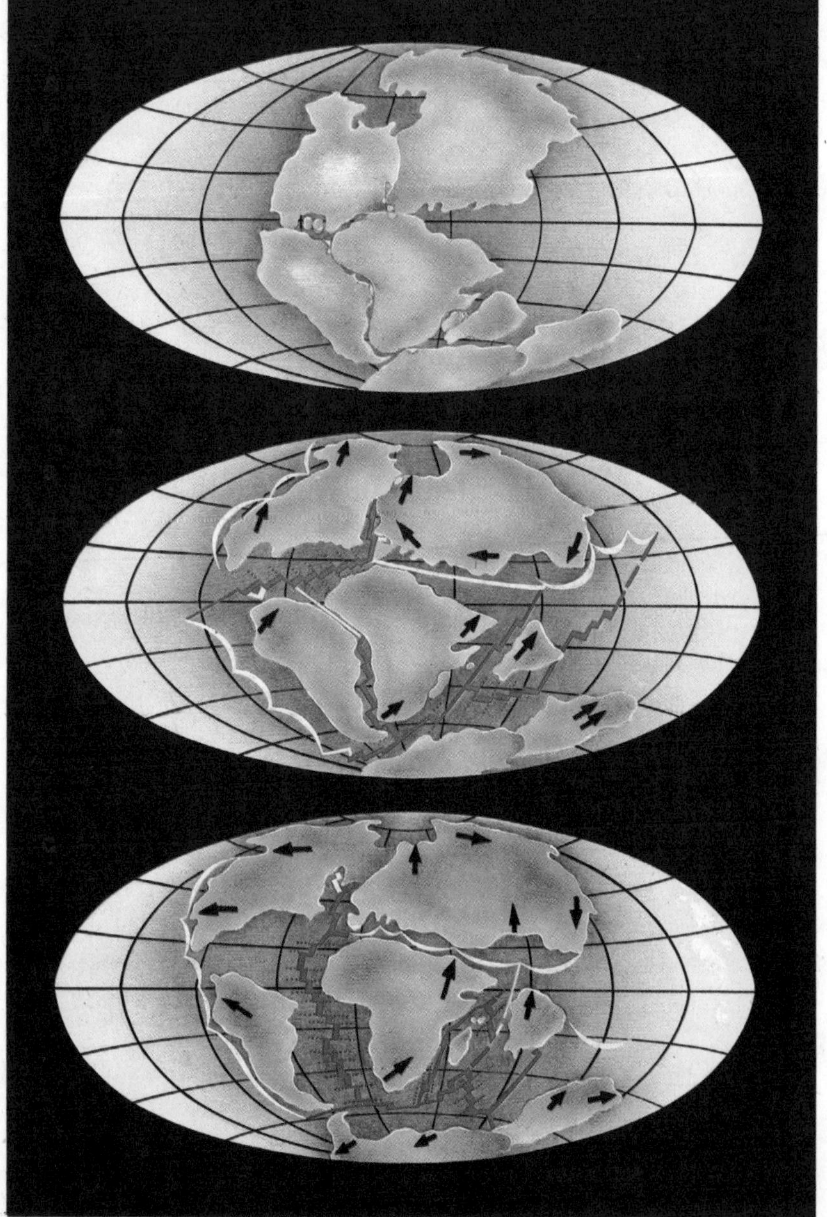

263

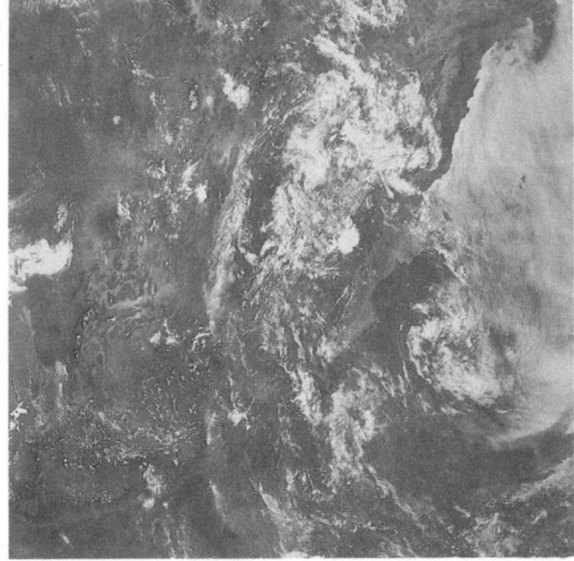

264

263 The theory of continental drift (of the early 20th century German scientist Alfred Wegener) was updated in the 1970s. The figures illustrate the idea of the separation of continents and the origin of areas of expansion (red) and subduction or collision (white). The oldest stage, the existence of a single landmass, Pangea, is shown at the top, the stage of Laurasia and Gondwana in the middle, and the origin of today's configuration of the continents at the bottom.

264 The Gulf of California, as seen by the *Apollo 11* astronauts, illustrates the possibilities of space photography. Not only meteorological features, but also the geological structure of the continent, are visible.

forward at the start of this century by the German scientist Alfred Wegener, but it was not until the 1970s that it was, in an updated and better supported form, generally accepted. In the Wegener theory the continents, i.e. 'floes' of the mainland crust some 35 km (22 miles) thick, moved along the plastic mantle of the Earth. Today's idea, supported by geophysical data, involves the movement of whole plates of the Earth, i.e. of crust and part of the mantle together. These blocks are some 200 km (124 miles) thick. Apart from continental plates there are also oceanic plates, which do not have continents on them. The plates slide under each other, or simply collide. The movement of these plates, which is caused by the internal forces of the Earth, is responsible for the great majority of geological processes. The internal forces of the Earth, meaning its heat, decide the layout of the landmasses and continents on the globe. It is strongly asymmetrical; it is even possible to divide the Earth into one hemisphere which is almost entirely water and one which is almost totally land. (To do so the poles must be placed in positions different from those they actually occupy, see Figure 266).

The fact that the continents are not symmetrically arranged is connected with continental drift, but such a layout is typical of the other terrestrial planets, Mars, Venus and Mercury. It can also be seen in the layout of the 'continents' on the Moon, or the positions of the great basins on Mars or Mercury. Let us recall once again that in the case of the Earth we are speaking of continents in the geological sense. But in the text below we shall consider the ratio of land and sea in the geographical sense, especially with a view to heights above sea-level and ocean depth. Even from that point of view we obtain interesting data for geological observation and comparison.

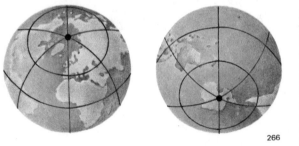

Let us look at Figure 267. The vertical axis shows the height above sea-level, the horizontal the amount of land or ocean at that altitude. Such a diagram is known as a *hypsographic curve.* If we want to know the area occupied by, let us say, land between the altitudes of 100 and 200 m (328 and 656 ft), we have to do some mental arithmetic. The diagram gives a good idea of how relatively low elevations dominate on the continents, and the increasing gradient of the curve with increasing altitude shows that the higher the elevation, the smaller the area of land involved. If we were to look for a geological explanation, it would be easy enough to find, not only as a result of erosion, which tends to wear down the relief on the landscape, but also because of the plastic nature of the substratum, into which the landmass tends to sink. This is called *isostatic compensation.*

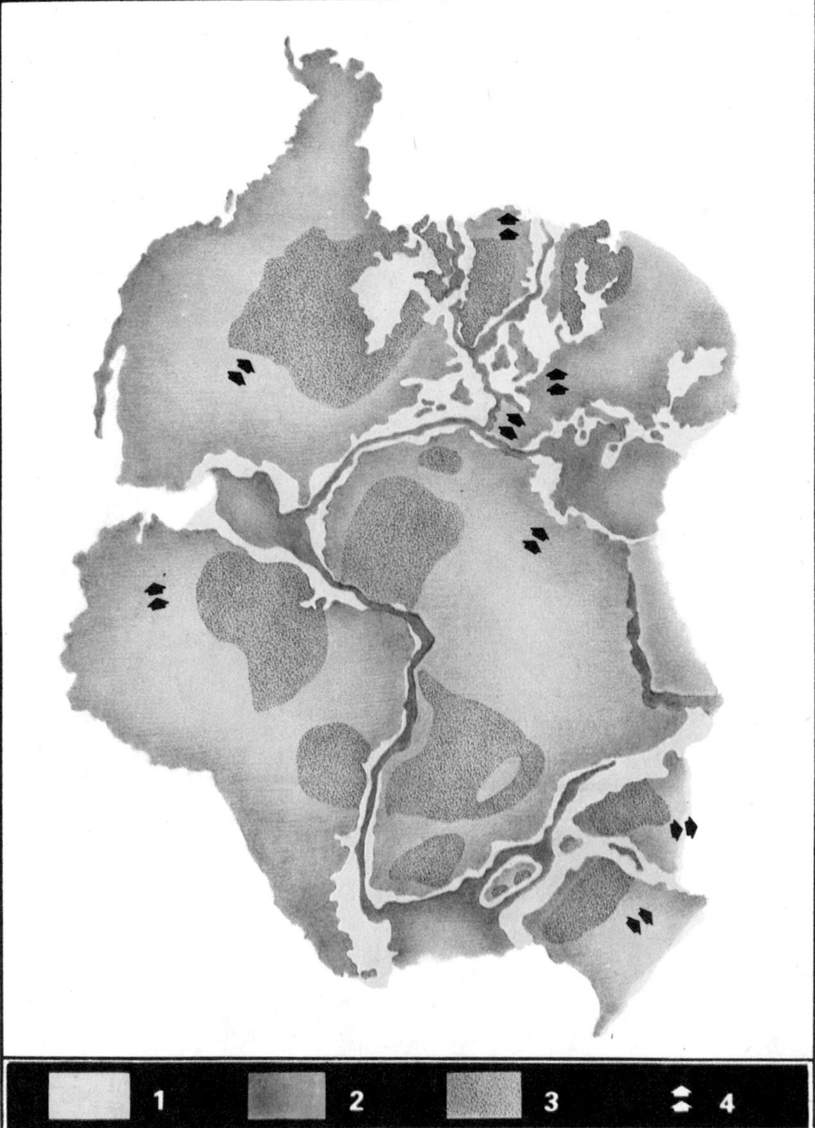

265

266

265 Anyone who takes the trouble to cut out the shapes of today's continents can have a go at fitting them together. The continental shelf **(1)** belongs to the continents. When it has all been put together, only the small areas of the ocean deeps **(2)** remain. The figure also shows the oldest part of the continents – the shields **(3).** The arrows show where today's north is on each block **(4).**

266 The planet Earth can be divided up in ways other than those we are used to. Here we are looking on the left at the 'dry' hemisphere, on the right at the 'wet' one.

Nor are the extremes of ocean depth evenly distributed, which indicates the 'youth' and exceptional nature of these formations. Modern hypotheses of the movement of plates, and of collisions and overlap, assume that the most active geological activity is concentrated in narrow, elongated zones along the mid-oceanic ridges, where the crust of oceanic plates and active continental shelves are formed, as for instance on the coast of the Pacific Ocean. The origin of mountain chains like the Alps or the Himalayas is explained by this theory as the result of a collision of two plates. Morphologically positive formations are created in this way. Overlapping of plates, on the other hand, leads to the destruction of the oceanic crust and gives rise to negative morphological features such as the ocean trenches.

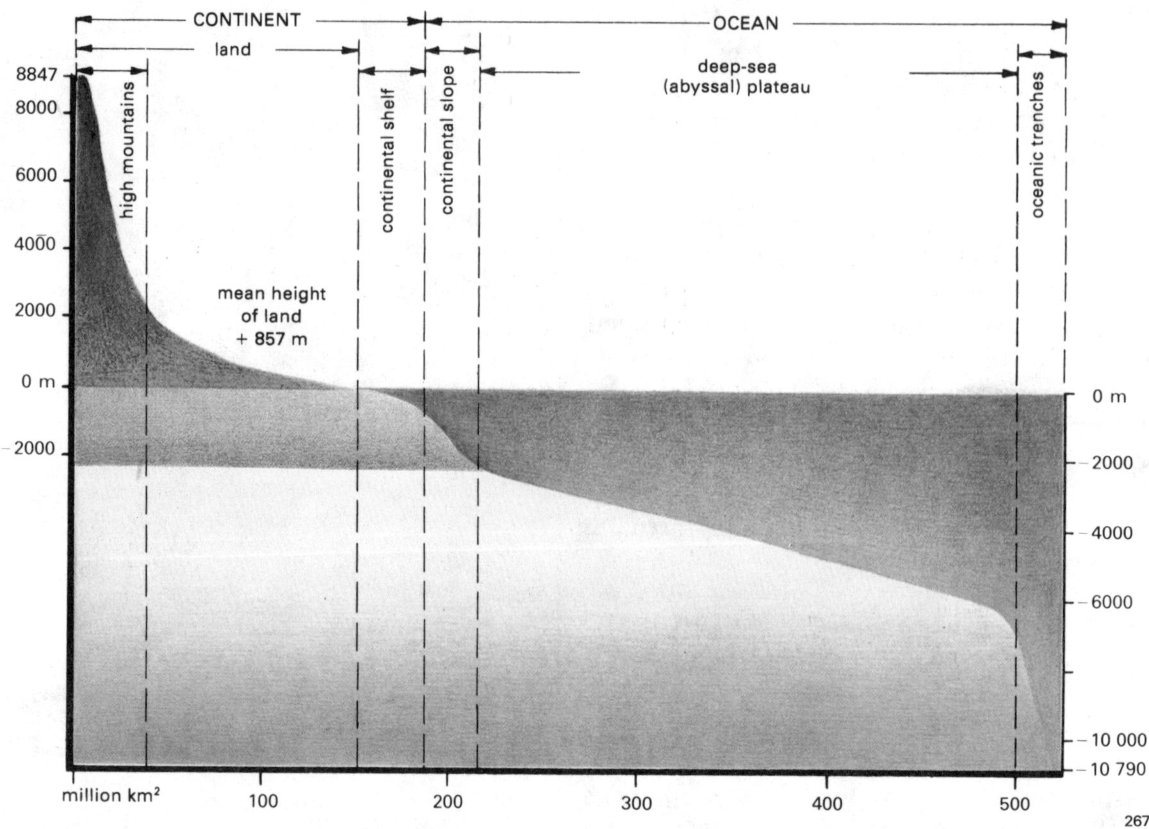

267 The hypsographic curve shows the proportions of ocean and land on the surface of the Earth. It can also be seen from it that most of the continental part consists of land up to an elevation of 2000 m (6560 ft), while the area of shallow seas is small, as is that of high mountains or ocean deeps. The most common formation is seabed from 2500–6000 m (8200–19,680 ft) deep.

Shape, dimensions and mass

When Christopher Columbus reached America, he imagined he had achieved his goal: India. For he was seeking a new, easier route to the other side of the world. His journey, like those of many of his contemporaries among the ocean voyagers, was based on the assumption that the world was round. In the midst of all the strange ideas about the form of the Earth and its place in the Universe, even weird notions of what was inside the planet, the conception of a round Earth was relatively quickly accepted and proven. Only some of the civilizations of the Arab world and of India continued in their belief that the Earth was flat, and stood on some sort of giant pedestal.

In ancient times the first of the 'round-earthers' were the members of the Pythagoran school, and the first measurements by the Greek geometricians were very accurate. The principle of measurement used by the Greeks – simple trigonometry – was until recently still used for measuring the shape of the Earth. But classical geometry has now been replaced by satellite technology, not only for measuring the Earth, but for weighing it, too.

If you take a look at the globe, the Earth seems to be an ideal sphere, which is also how it would look from some cosmic body. But even the Renaissance scholars, and later Isaac Newton himself, deduced that our world is flattened. The reasons for their suspicions were various, and some supposed the polar regions to be flattened, others the equatorial area. The dispute was resolved in a sporting manner. The Académie Française sent out two scientific expeditions, one to the equator, the other to the polar region, where they carried out measurements and came to the conclusion that the Earth is flattened at the poles, and that therefore the equatorial diameter is somewhat greater than the polar. The difference is not a big one: a mere 44 km (27 miles), but from the geophysical point of view it is important. This difference influences the gravitational acceleration (i.e. the amount of gravity) – it is different at the equator and the poles. If we were to express the difference as a percentage, it would be 0.3 per cent of the total diameter. The Earth's shape can be characterized as a rotational ellipsoid, not as a sphere. It would be more exact to call it a *geoid.* That refers to a surface which is perpendicular to the gravitational axis at any given point. The geoid is a very complex shape, and every large piece of matter, such as a tall range of mountains, influences its shape. At present the shape of the Earth is genuinely derived from the measurement of gravitational acceleration, which is carried out not from the ground, but from space. The orbit of a satellite is affected by changes in the gravitational field, so that analysis of the shape of the orbit allows one to determine the gravity and therefore also the shape of the Earth. It is, of course, roughly spherical, but it has many bumps of different sizes; if we were to exaggerate these bumps, our planet would look like a potato.

The size of the gravitational pull is associated with the mass of the Earth, which has been known since Isaac Newton defined the laws we all learnt at school. The English scientist Henry Cavendish managed to measure the magnitude of the gravitation constant in the laboratory over 200 years ago. The results of the first measurements, made under very simple conditions, do not differ from those obtained much later and with far more complex and more advanced instruments. Today the Earth's mass has been accepted as 5.976×10^{27} g. This mass exerts a gravitational acceleration at the equator of 978.038 cm/s^2. The Earth is 81 times heavier than the Moon, but 100,000 times lighter than the Sun. If we consider the dimensions of the Earth and its mass, we find that it has a high density (5.5 g/cm^3), which is much more than that of the minerals of the Earth's surface. Their density is on average 2.75 g/cm^3.

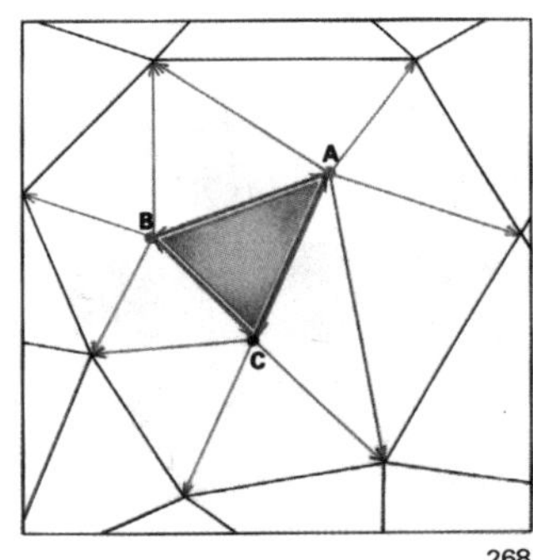

268

268 The exact measurement of the Earth's surface is the basis of a knowledge of our planet and its internal structure. Such measurement is made possible mainly by trigonometry. Though aerial and satellite photography is now available, the importance of trigonometry has not declined, since it is entirely accurate – in the sum of the angles of a triangle even the smallest error comes to light.

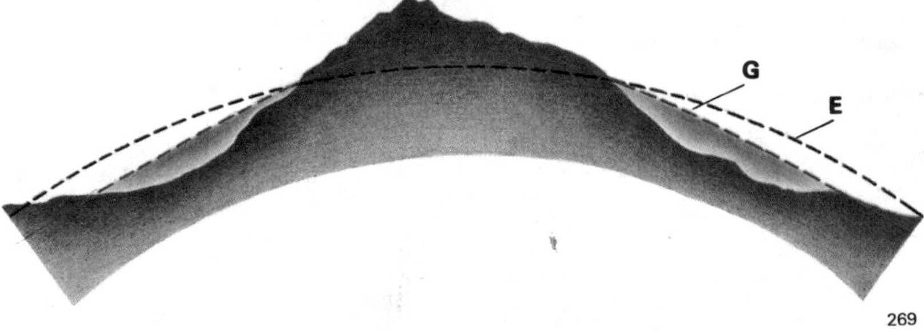

269

269 The triaxial rotational ellipsoid **(E)** is the best depiction of the Earth as a body, but the actual shape of the Earth is expressed as a geoid **(G).** At any point perpendiculars lie in the direction of a plumb-line, which is where the geoid (red) differs from an ellipsoid.

The heat of the Earth

Since early in their history, men have known that the temperature increases as you go further down into the Earth. That is where they imagined Hell to be, and where they knew hot volcanoes, geysers and springs issued from. In the temperate and northern climatic zones the inhabitants knew that although the ground froze every winter, the frost only went down a short distance, and that a few metres under the surface the temperature in winter was higher than it was at ground level. They therefore built food stores under the ground. With the development of mining, men discovered that this increase in temperature with depth is regular; it is known as the *geothermal gradient.*

There are two possible explanations for this phenomenon: either the interior of the Earth produces its own heat, or this is residual heat from the time of the formation of the hot planet. In fact, the dimensions, thermal conductivity and in particular the age of the Earth rule out the latter possibility; without its own source of energy, the Earth would long since have cooled down.

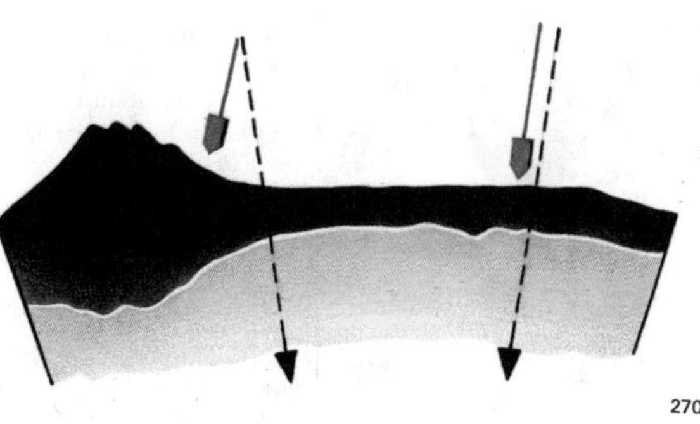

270 The great mass of mountains deflects a plumb-line from the perpendicular. The difference is exaggerated in the picture. The measurement of gravity also determines the shape of the Earth. But it can be used in the search for useful minerals, too.

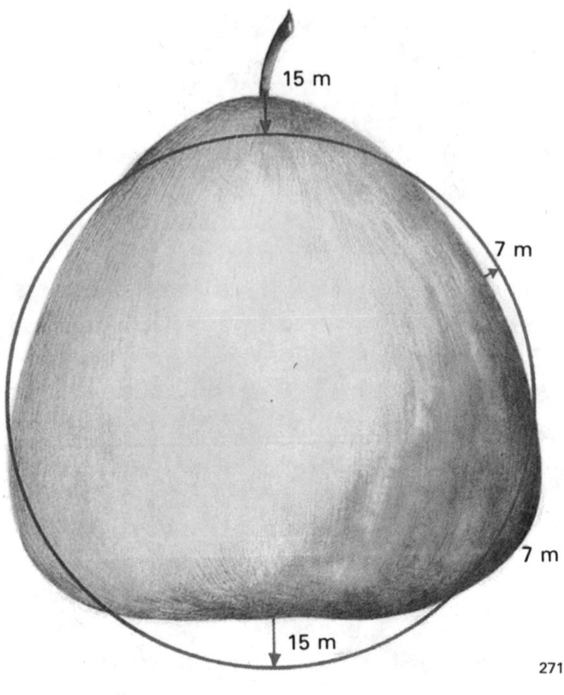

271 The shape of the Earth has in the past been compared to a sphere, an ellipsoid, a potato and a pear. In this picture the pear shape of the Earth has been greatly exaggerated. In fact the deviation is never more than 20 m (66 ft).

Making its own heat is not something peculiar to the Earth. Even small planets have their own heat supplies – in fact the 'cold' Moon is also equipped with 'central heating', in the form of decaying radioactive material. So the process of burning hydrogen which takes place on the Sun and supplies energy to much of the Solar System, together with the internal heat of the planets, form the basic energy sources of geological processes. If we compare the amount of energy falling on the Earth from the Sun (before its depletion by the protective shield of the atmosphere) with that produced inside the Earth itself, we find that the former is 6000 times greater.

The geothermal heat has always attracted the attention of those who like to bathe in the warm natural waters which occur in certain spas, but today it has become one of the 'new' energy sources on which man has pinned his hopes of solving the energy crisis. There are many places in the world where this energy is already being exploited. In New Zealand, Italy and the United States there are power stations which use it to produce electricity. Many greenhouses are heated with water warmed by geothermal energy, even in, for example, Iceland and Kamchatka.

There are plenty of places where we can see that it is hot inside the Earth, not only at volcanoes, geysers and hot springs. Geologists measure temperatures even in places which at first sight seem cool. To see that temperature increases with depth one only has to sink a bore or go down a shaft. But the increase is not the same everywhere; in geologically 'young' areas, i.e. where there are still changes taking place geologically, or were until recently, the increase in temperature with increasing depth is greater than in those places where there has been geological inactivity for a long time – i.e. hundreds of millions of years. In the case of the latter, the temperature increase is only in the order of 1°C (1.8°F) per 30–40 m (98–131 ft) depth increase.

A look at the world map with points of increased temperature marked on it reveals an outstanding region in the vicinity of the Pacific Ocean, called the 'ring of fire' in professional geological circles. It is well known from school atlases and geography lessons. But geologists were surprised to find out a few years ago that the bottom of the Indian, Pacific and Atlantic Oceans have places with increased heat production, especially in their middle regions. The heat flow there is twice the world average.

273

Somewhat later it was found that the beds of the world's oceans have volcanic rocks on them, and these are only a few hundred thousand or a couple of million years old – which makes them young, geologically speaking. This was one of the greatest discoveries of the 1960s and 1970s: there are ridges of volcanic rocks stretching right across the oceans. And since these are very 'lively' places, there are frequent earthquakes in the middle of the oceans. The heat output there is tremendous. One should bear in mind that the flow of materials in the Earth's mantle is very evident there, and that the mid-oceanic ridges are the points where the flow emerges. Other similar places are the 'hot spots' where the heat output exceeds even that of the mid-oceanic ridges. Volcanic activity is thus very great there. The Hawaiian Islands, Tahiti and the Azores are among these hot spots.

The magnetic field

Some of the planets in the Solar System have their own magnetic fields, while others either do not, or have one which is almost negligible. The Earth is among those with a very strong magnetic field, which acts as an excellent shield against cosmic radiation. Thus fewer cosmic particles fall on the Earth than on planets without a magnetic field. As Figure 277 shows, the interaction of the magnetic field and the solar wind is complicated. One might say that the Earth has several protections against cosmic radiation in the form of natural shields. As a result of the existence of the magnetic field, magnetic

272 Hot lava of basalt composition flows down the slopes of the volcano Mauna Loa in Hawaii.

273 One attempt to solve the problem of the world energy shortage has been the use of geothermal energy. Apart from pioneer methods in Italy, Iceland and North America, the geothermal power station at Wairakei in New Zealand has been operational for more than two decades.

'insulating bands' of fast-moving particles are 'wrapped' round the Earth. They are called the *Van Allen belts*, after their discoverer, and there are two of them, an inner and an outer.

The origin of the magnetic field has always been somewhat puzzling for us. We know that a magnetic field arises where an electric current passes, and since we have evidence to show that the Earth has a metal core, we may attribute the existence of the magnetic field to the currents which occur in that. The relation between the magnetic field and geological processes, and the effect it has on living organisms are not yet reliably explained. But a study of the magnetic properties of rocks shows that the Earth's magnetic field has suddenly changed completely many times in the last 200 million years (which is only a fraction of the Earth's history). The north and south magnetic poles have moved to totally different places. The possible consequences of this are clear. If the Earth is left without its magnetic shield even for a short period, cosmic radiation begins to fall on it. This radiation is similar to the radiation from radioactive materials, and some scientists believe it can cause great changes to life on the planet. But magnetism has been a feature of the Earth throughout most of its existence, and one can therefore obtain very important and useful information on the history of the Earth – such as the position of the continents – from the magnetic record in rocks.

The basic magnetic information is concealed in those rocks which contain ferromagnetic minerals; the main magnetic materials are thus magnetite, ilmenite and pyrrhotine. For if a rock crystallizes from magma, its particles are non-magnetic at temperatures over 425°C (797°F). Below that temperature, called the *Curie point*, rocks acquire a magnetic orientation. This orientation corresponds to that of the magnetic field which is in effect at the time. The orientation of the rocks remains, even if the magnetic field changes. In the sedimentary rocks the sedimentation of magnetic particles is also governed by the orientation

274

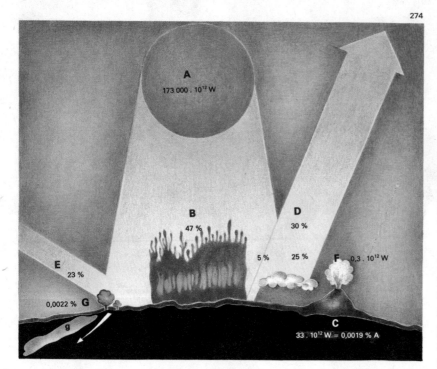

274 Much of the solar radiation **(A)** which comes to Earth is reflected directly into space **(D),** and almost half is transformed into heat **(B),** which sooner or later escapes into space. A certain amount of solar energy is consumed in evaporating the water, in wind, rain etc. **(E).** Only a very small proportion is used up by photosynthesis **(G)** and is stored in plants as an energy source **(g).** The actual energy of the Earth makes up only a tiny fraction of the energy balance **(F),** C – crust.

275

275 A cross-section of the ocean floor from the mid-oceanic ridge to the active continental border has characteristic morphological features: ridge, trench and volcanic arc. There are also characteristic heat flow values for different parts of the seabed.

of the magnetic field at the time; a component floating in water observes all the laws of magnetism. And so if we take a sample which is properly oriented, we can easily read from old rocks the orientation of the magnetic field in the past. If we look at rock complexes in detail, for instance at the order of lava outpours over a period of 1 million years, we find that in the course of a relatively short period – a few tens of thousands of years – a sudden change takes place in the polarity of the magnetic field. The north and south poles change places. If we study geological history over a long period, we find not only these 'switchings' of polarity, but also another phenomenon – the magnetic poles are gradually moving. But the continents are moving too, and since the magnetic poles are probably always close to the axis of rotation of the Earth, the magnetic record in rocks can be used as a further proof of the theory of continental drift. The magnetic properties of rocks have already offered much evidence on the dynamic, ever vital nature of the Earth.

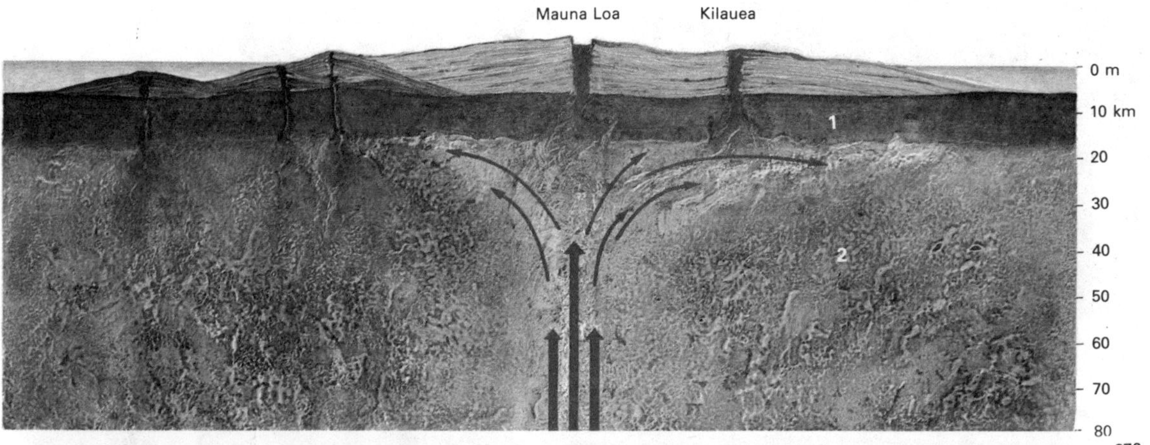

276

276 An idealized representation of a 'hot spot' beneath the Hawaiian Islands. A plate with its own volcano 'passes' the hot spot, which is why the volcanoes on the left of the picture are extinct. The oceanic crust is relatively 'thin' **(1)**, while a hot plume of the asthenosphere rises through the mantle **(2)**.

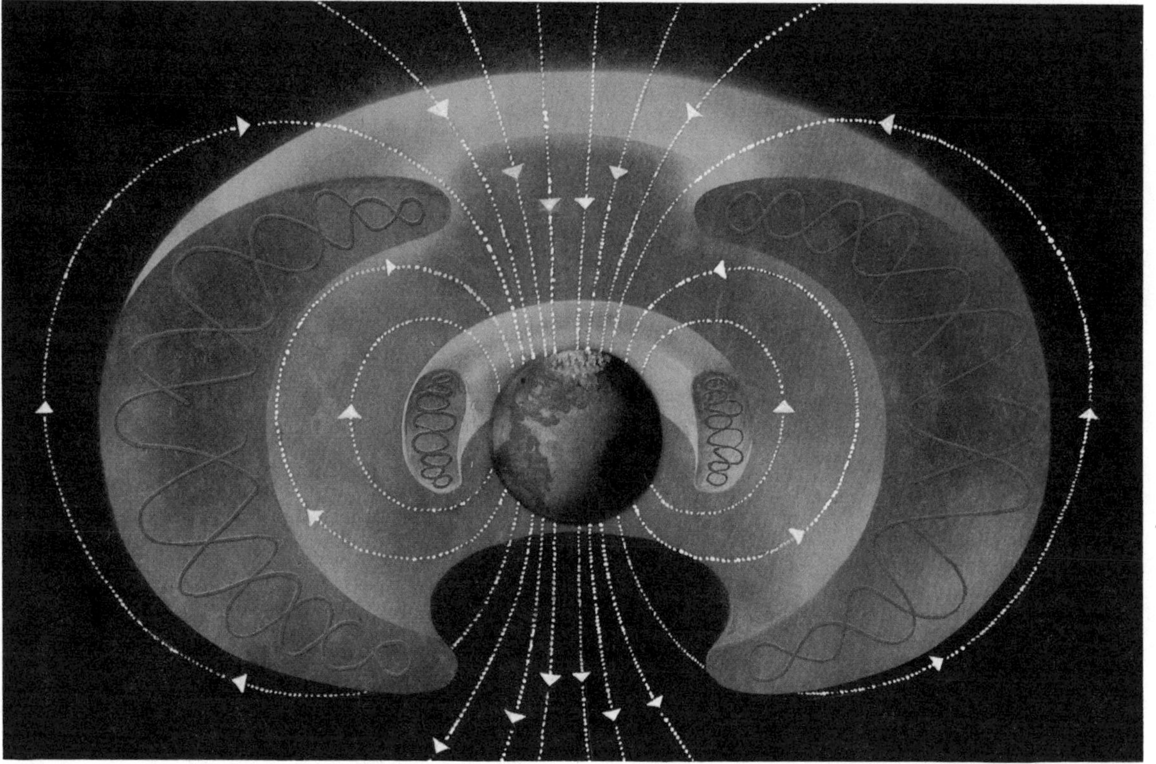

277

277 One of the first very important discoveries of the space age was the existence of magnetic belts around the Earth, called Van Allen radiation belts after their discoverer. Thanks to these the Earth is protected from dangerous cosmic radiation.

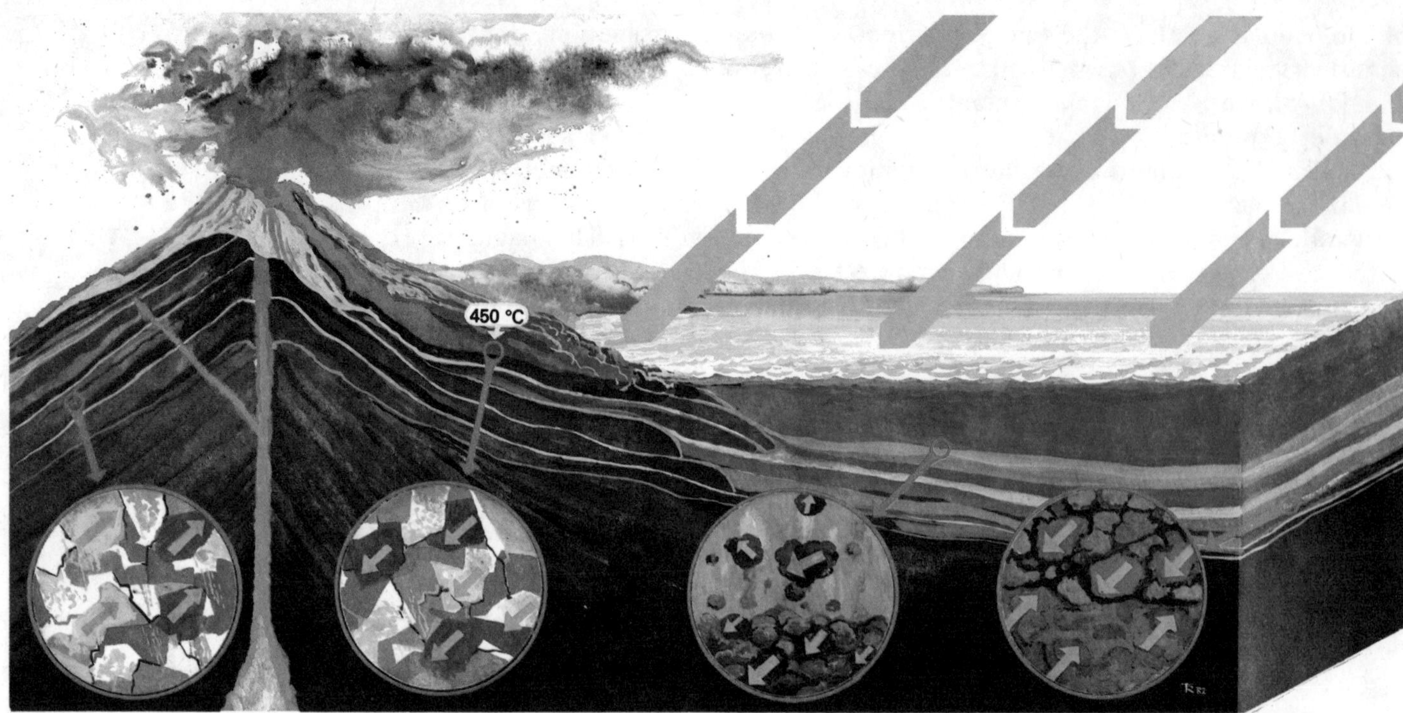

278

278 All geological processes are constantly under the influence of the Earth's magnetic field, so that in the formation of rocks the magnetic components receive the same orientation. This takes place in sediments when they fall to the seabed and in lava the moment its temperature falls to 450°C (842°F). Since the magnetic field alters in the course of geological eras, the orientation of magnetic components is different in each stratum, and may even be in the opposite direction, as is shown by the green arrows.

There exist on the ocean bed almost parallel strips of rocks with opposite magnetic polarity, following the course of the mid-oceanic ridges. This discovery was also important in the formation of the idea of a moving ocean bed.

Though we do not as yet understand the origin and changes of the magnetic field properly, we suppose that they are due to a 'dynamo' phenomenon. The matter inside the Earth moves in a different way from that on its surface, which gives rise to an electric current similar to that produced by the movement of the rotor of a dynamo inside the stator. A magnetic field is also created in this way. But we are making a careful study of the Earth's magnetosphere, and the interaction of the solar wind and magnetic storms on the Sun with the Earth's magnetic field. And we exploit the magnetism of the Earth for various purposes, from navigation to the search for mineral deposits.

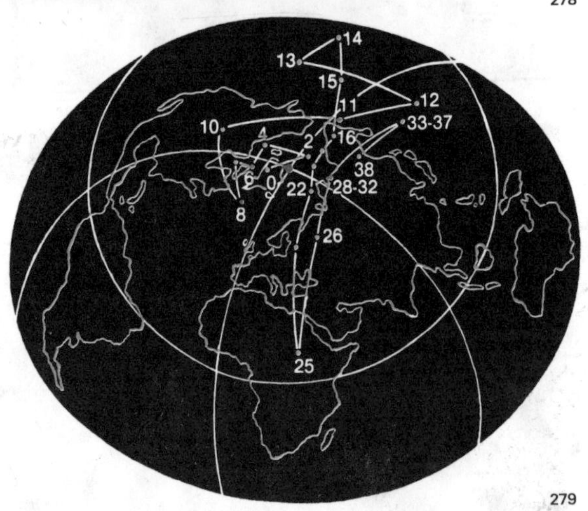

279

279 The position of the magnetic pole changes in the course of geological eras. The figure shows the movement of the Earth's magnetic pole over the last 38,000 years. But the evidence is still the subject of discussion, and the information from rocks from one location does not entirely agree with that offered by rocks from another.

The Earth's onion-like structure

A comparison between the density of the Earth and that of other planets shows that the density of the Earth is not in any way exceptional. But one can see at once that there are inconsistencies. The surface rocks cannot make up the whole of the planet, or the Earth would be much lighter. Somewhere beneath the surface there must be a material which is much denser. Other measurements also show that this must be so. The movement of the Earth around its own axis and the moment of inertia show how the matter in a rotating spherical body is distributed. Such

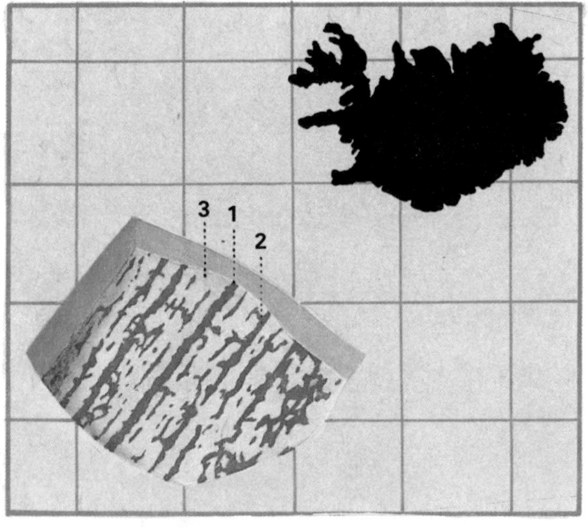

280

280 The rocks of the ocean floor in the region of Iceland, as in many other places in the ocean, are basalts. The orientation of the magnetic fields in these rocks changes; sometimes it accords with today's **(2),** sometimes it is in the opposite direction **(3).** But if the results are mapped, a striped structure of the ocean floor appears, parallel to the mid-Atlantic ridge **(1).** This is further evidence of the expansion of the ocean floor.

281

281 Natural catastrophes which afflicted men took the form of earthquakes, volcanic eruptions, etc. But among mythical catastrophes is a magnetic mountain, which caused the wreck of many a ship from which, according to the legend, it withdrew the nails.

measurements show, for instance, that Mercury must have a large nucleus made of relatively massive – heavy – material. The same applies to the Earth, Venus, Mars, and to a certain extent also the Moon.

How is one to resolve this question? One might try to explain the change in density as being the result of high pressure and temperature. You will remember the chemical analogy, that of graphite and diamond, where diamond is the high-pressure modification, with considerable density (3.5 g/cm^3), and graphite the low-pressure modification, with a low density (2.1 g/cm^3). But if we try to explain an increase in density by means of an increase in pressure we also find ourselves with a contradiction. The material we find on the Earth's surface does not exhibit the necessary physical properties even at ultra-high pressures. We therefore have to consider the chemical composition of the material of which rocks are made and find an answer which is in accordance with geophysical measurements. If we wish to study the interior of our planet, we have to accept the improbability of anyone ever looking at the centre of the Earth, or even at a depth of 200–300 km (120–180 miles). Every idea concerning the condition of the Earth's interior must be based on our geological, physical and cosmochemical knowledge. A study of meteorites suggests what the overall composition of the Earth might be, and indicates a metal core. The Earth's mantle is similar to stony meteorites (chondrites). The similarity between individual groups of meteorites and the strata of the Earth led scientists, even in the first decades of this century, to the view that the Earth's core is metallic. The proof of the existence of a core as a physical formation had been produced at the turn of the century. But even today there is no way of showing for sure that it is made of iron, although it is more than likely.

Analogy with meteorites would not be very useful for considering the structure and composition of the Earth without the information we have on the propagation of seismic waves through it. This offers more basic information on the physical condition and the behaviour of individual parts of the Earth. It is clear that earthquakes, though they bring men many worries, are also very useful tools for the scientists.

The crust

Most geological processes take place in the Earth's crust, which forms the uppermost layer of the planet. The crust is and will be for a long time the centre of interest for geologists, and also the place from which mankind gets its raw materials. The border between the crust and the further layers of the terrestrial body – the upper mantle – is called the *Mohorovičić discontinuity*

282 The layered structure of the Earth is indicated by the rates of propagation of seismic waves, changes in the mineralogical composition of rocks, changes in specific gravity (density) and changes in temperature and pressure.

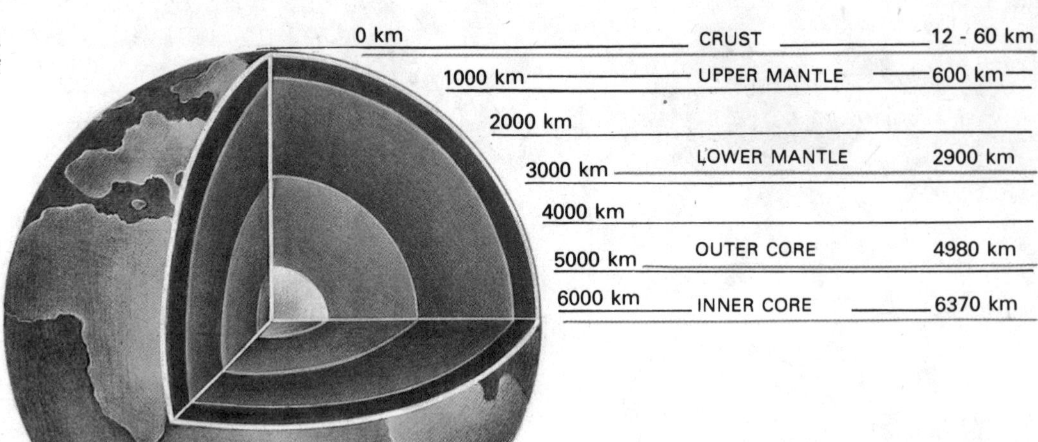

282

(*Moho*). Geophysicists speak of a discontinuity where there is a sudden change in the physical properties, such as the rate of propagation of seismic waves (Figure 283 illustrates such changes at the border between the crust and the upper mantle, and in the upper mantle itself).

A couple of decades ago scientists – geologists, geophysicists and geochemists – held a lengthy and thorough debate over the question of what was responsible for the change in the propagation rate of seismic waves between the crust and the mantle. While these waves move at a rate of less than 6.5 km/s in the crust, the speed in the mantle increases to over 8.0 km/s. After research and a confrontation of opposing views, they were presented with two possibilities:

- The change in the propagation rate of seismic waves is caused by different chemical, and thus mineralogical, composition. This means that above and below the discontinuity the chemical composition and thus also the mineralogy of the rocks is different.
- The second theory is that the change in physical properties is caused only by a change in the mineralogical composition of the rocks above and below the Mohorovičić discontinuity, which means that the chemical composition is the same and the different mineralogical composition is caused by increased pressure and temperature.

Today the main reason for the change in the rate of passage of seismic waves at the crust-mantle interface is considered to be the chemical composition. Scientists thus tend to favour the first explanation.

The difference between the chemical and the mineralogical composition of a rock can be illustrated using the example of basalt. It is a solid black rock with crystals of the green mineral olivine, or brown crystals of pyroxene in a fine-grained matrix surrounding these fragments. Beneath the microscope the petrologist or mineralogist would also find the feldspar plagioclase, further, smaller crystals of olivine and pyroxene, pieces of magnetite or ilmenite, and also pieces of a non-crystalline, solidified material – glass. This list comprises the mineralogical description of the rock. But if we were to send the same basalt to a chemical laboratory, there they would grind the sample up, dissolve it, and find that most of the rock is made up of nine basic oxides. A rock whose main components are 50 per cent SiO_2 (silicon dioxide), 1 per cent TiO_2 (titanium dioxide), 17 per cent Al_2O_3 (aluminium oxide), 10 per cent FeO (ferrous oxide), 0.2 per cent MnO (manganese oxide), 8 per cent MgO (magnesium oxide), 7 per cent CaO (calcium oxide), 4 per cent Na_2O (sodium oxide), and 1.5 per cent K_2O (potassium oxide), can have the following names: gabbro, basalt or eclogite. If the main components are feldspar, pyroxene and amphibole, it is gabbro; if they are olivine, pyroxene and feldspar, most of the rock is very fine-grained and glass is present, it will be called basalt, though the composition is the same. But if garnet and clinopyroxene are present, a rock of the same chemical composition is called eclogite.

These are just the differences in mineralogical composition which can tell the

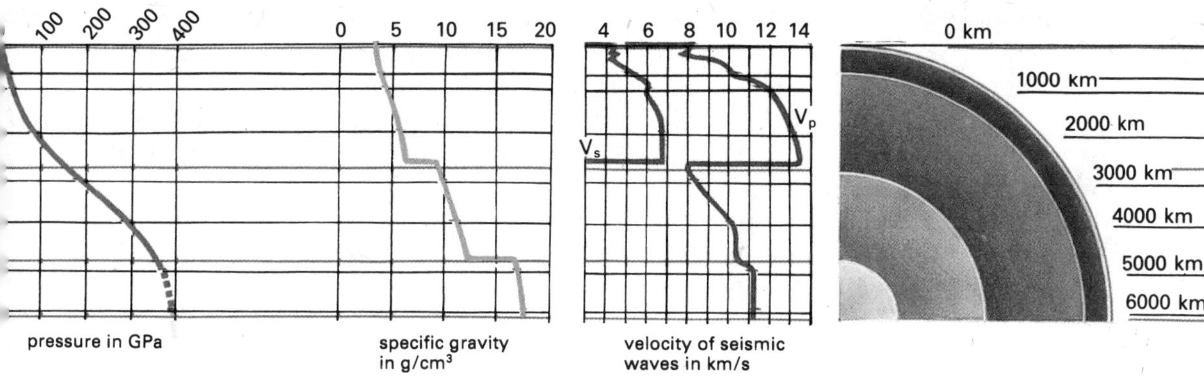

geologist in what environment the rock crystallized (formed). Thus gabbro is a rock which crystallized inside the Earth's crust, say at a temperature of 800°C (1470°F) and a depth of about 10 km (6 miles); basalt crystallized on or near the Earth's surface at a temperature of more than 1200°C (2190°F). And eclogite? Temperature does not play an important role in its crystallization; it can crystallize at anything from 400–1200°C (750–2190°F) but the pressure must be high and correspond to a depth of at least 20 km (12 miles).

These are the main reasons why geologists need to know the mineralogical composition of rocks as well as their chemical composition.

What, then, is the crust made of, and what lies beneath it? Geologists depict the Earth by investigating the rocky substratum and registering its characteristic features on a geological map. From this map one can easily read which rocks occur on the Earth's surface in a particular locality. By putting together all the geological maps for different parts of the world we can find the sum of all the mapped areas of a given rock, such as sandstone, slate, granite or basalt, and arrive at the average mineralogical and petrographical composition of the surface layer of the Earth's crust. We say surface layer advisedly, since it is probable that the topmost stratum is slightly different from the one beneath it: it naturally contains more sedimentary rocks.

If we sampled every rock occurring on the Earth's surface, taking a quantity proportional to the amount of that rock present in the Earth's crust, then mixed the samples thoroughly, ground them up, and gave them

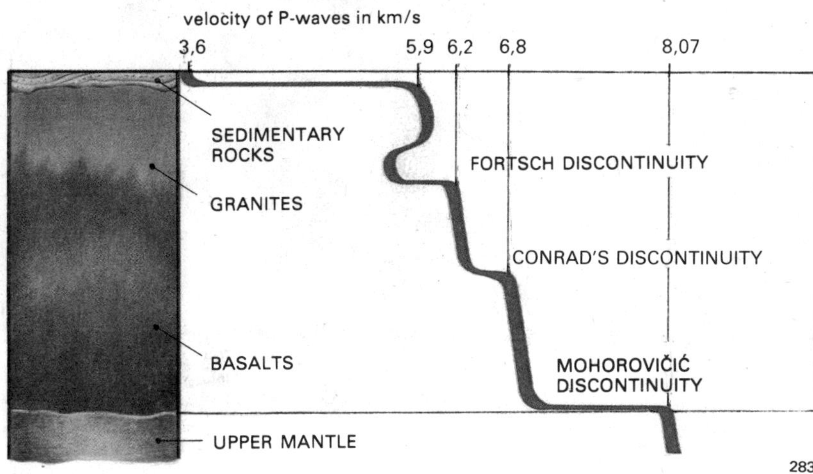

283 The rate of propagation of seismic waves changes according to the chemical composition and the physical state of the material of the Earth's interior. A zone of increased rate of propagation may, for instance, indicate that the rocks in that area are partly molten. Sudden changes in the rate are called discontinuities. The most important discontinuity in the Earth's crust is Conrad's discontinuity; the Mohorovičić discontinuity divides the crust from the mantle.

to a chemist to analyse, we should find that the average composition of the Earth's surface can be expressed using nine common oxides: SiO_2, TiO_2, Al_2O_3, FeO, MnO, MgO, CaO, Na_2O and K_2O. These oxides make up more than 99 per cent of the Earth's surface, irrespective of which continent we take our samples from. We can take them from Asia, South America or Europe, and in all cases the results will be virtually the same. If we were to seek a rock which arose from a conglomeration of all the rocks occurring on the Earth's surface, we should find it in granodiorite, and we therefore say that the top layer of the crust has, on average, a 'granodiorite composition'.

So far we have been speaking of the rocks which make up the surface of the continents.

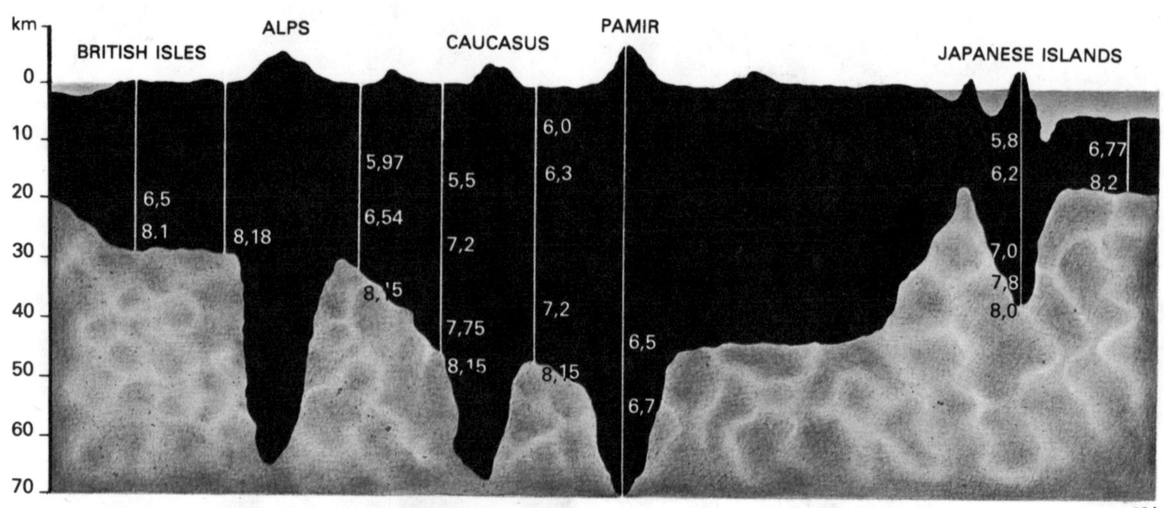

284 This vertically magnified cross-section of the Eurasian continent shows the crust in black, the mantle in brown. The cross-section illustrates the fact that high mountains have 'roots' and that the crust 'floats' on the mantle. The vertical lines show places where seismic sounding has been taken. The numbers show the rate of propagation of seismic waves.

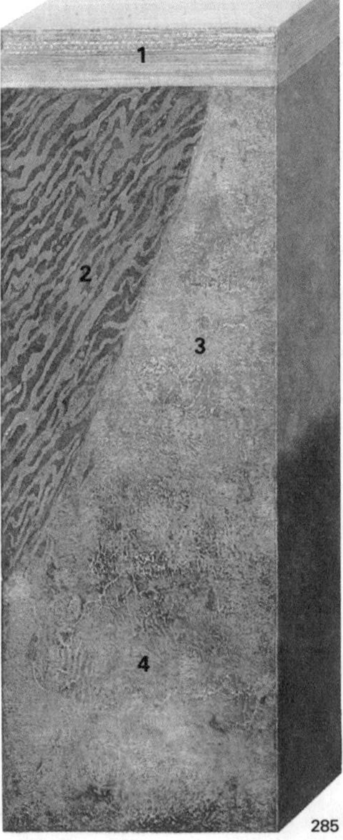

285 The Earth's crust, i.e. the top 30 km (19 miles) of the continents, is made up of sedimentary rocks **(1)** – 7.9 per cent; metamorphic rocks **(2)** – 27.4 per cent; and igneous rocks, of which those rich in silica **(3)** form 22 per cent, and those with little silica **(4)** 42.7 per cent of the total volume of the crust.

The situation as regards the average composition of the oceanic crust is quite different. Though the oceans make up two-thirds of the total area, the diversity of their rock composition is not as great. There would be no problem at all in taking the average of the composition of this part of the world, since with a very few exceptions the rocks of the ocean bed are the same everywhere. Beneath a thin layer of sedimentary rocks there is a monotonous layer of basalt. The basalts from beneath the sedimentary layer at the bottom of the Indian Ocean are just the same as those from the Atlantic or the Pacific. The results of drilling into the ocean floor and the measurements of oceanographic ships indicate that beneath the basalt there is another stratum of rock whose character we are not yet able to determine exactly. A comparison of the composition of the oceanic crust with the continental crust indicates that it contains less silicon dioxide, sodium oxide and potassium oxide, but more ferrous oxide, magnesium oxide and calcium oxide. But let us return to the continents. The whole history of our planet is recorded there, and there are still mysteries which remain to be solved.

For instance, geophysicists have found that the physical properties of the lower layers of the continental crust, such as the rate of propagation of seismic waves, electrical conductivity, etc., differ from those at the surface. Even if we subject the surface rocks to the temperatures and pressures prevailing at that depth, their properties do not correspond. And to complicate the situation of the oceanic crust. If we were to make a more detailed study of the Earth's crust we should arrive at the conclusion that there are differences in its composition in areas which are geologically younger (i.e. which came into being in the last 600 million years), or older (such as the localities of shields and platforms). Where the crust is very old, such as at the shields, it is not usually as thick, on

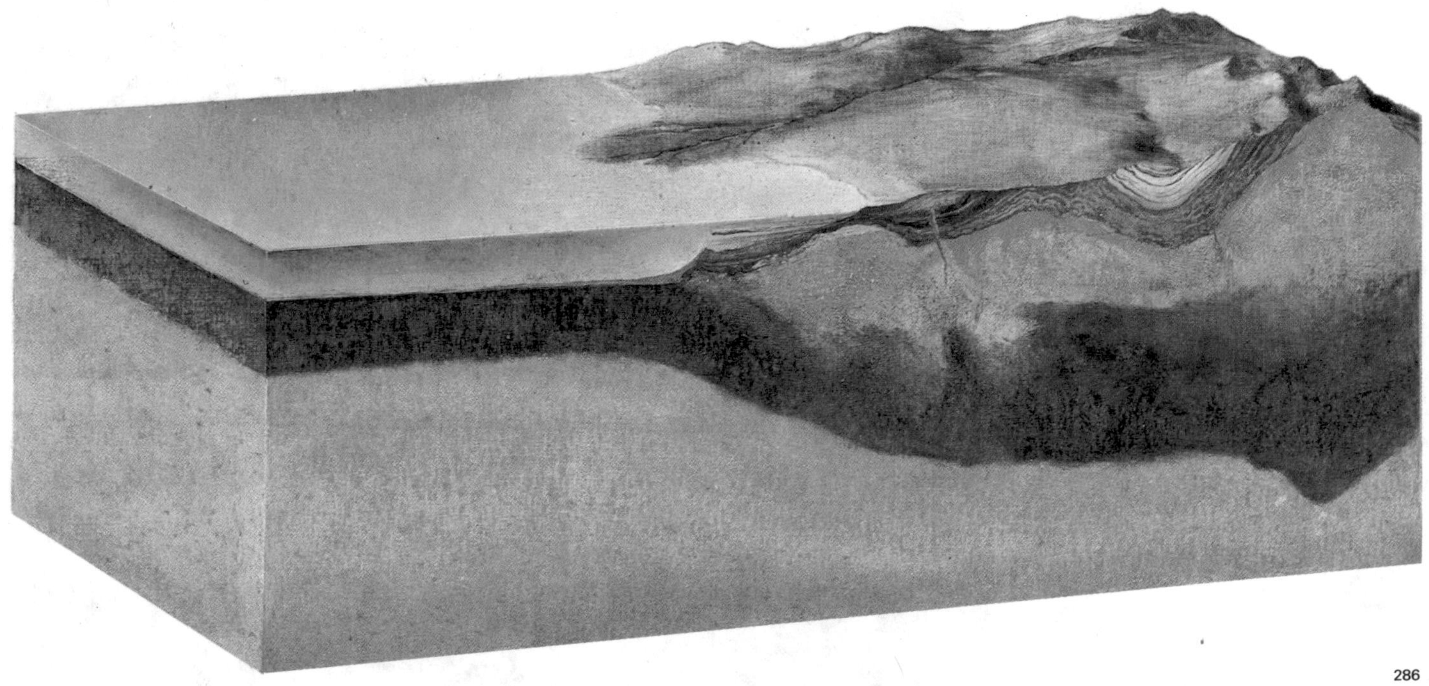

286

continental crust still further, we must add that at a depth of 10–15 km (6–9 miles) the rate of propagation of seismic waves changes. This area is known as *Conrad's discontinuity*. Present theories suppose the composition of the rocks above and below Conrad's discontinuity to be different. But no one has yet managed to bore down to that level, so that our ideas must be based on indirect evidence. And this does exist. The igneous rocks which break through the Earth's crust contain rocks which come either from the mantle or from the lower crust; they include many different sorts of rock – granulites and basalt rocks – with amphibolites most common. Since rocks with a basalt composition predominate, the lower part of the Earth's crust is said to be the basalt layer, though under the pressure and temperature conditions at that depth it cannot have the mineralogical composition of basalt. The lower part of the continental crust is often compared with the average 35 km/22 miles (distance from the surface to the Moho), while in places with recent volcanic activity, the localities of today's fold mountains, the crust is very thick, often over 50 km (30 miles), and in some parts of the Andes even 75 km (47 miles).

But a comparison of the physical and chemical properties of the continental crust with that at the ocean floor shows that these are different geological units. The two types of crust differ both in structure and in their chemical and mineralogical composition. We have already mentioned this fact, and since the differences are geologically fundamental, let us consider them once again. The thickness of the oceanic crust is much less than that of the continental one, a mere 10 km (6 miles) on average. For this reason one of the first major geological projects – drilling to ascertain the nature of the Moho – was undertaken there. The attempt was unsuc-

286 The continental shelf, the delta of a river, and a relatively thick crust are typical of the passive continental margin. The transitional zone between continental and oceanic crust is one of the areas most intensively studied at present.

cessful, but it showed that proper theoretical preparation and laboratory study can offer results which are much cheaper but just as valuable as expensive drilling. However, the oceanic crust has a much simpler structure than the continental crust. Beneath a thin layer of sediment the ocean floor contains a layer of rock of basalt composition; even that is not very thick (geophysicists say it measures a few hundred metres in places, elsewhere 1 km/0.6 miles). Below this there are rocks of ultrabasic composition (i.e. with a very low content of silica and a high content of magnesium and iron), making up the lower layer of the ocean floor, beneath which is the Moho.

The creation of the oceanic crust is in modern theories attributed to the expansion of the ocean floor. Continental drift, as we have described, actually starts with the formation of a new ocean floor. In the region of the mid-oceanic ridges all the material of the oceanic crust (apart from the thin layer of sediments) emerges from the upper mantle.

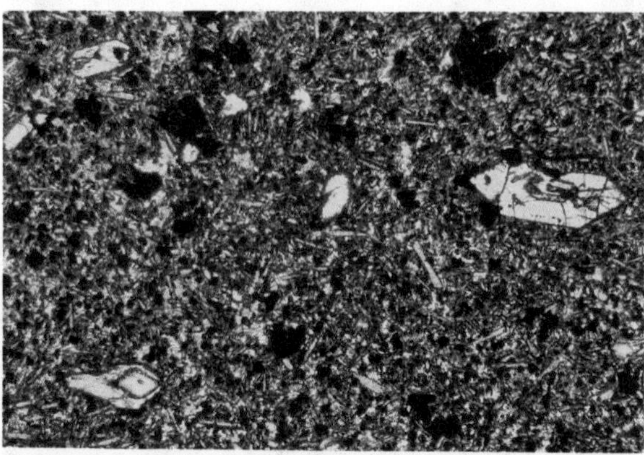

287

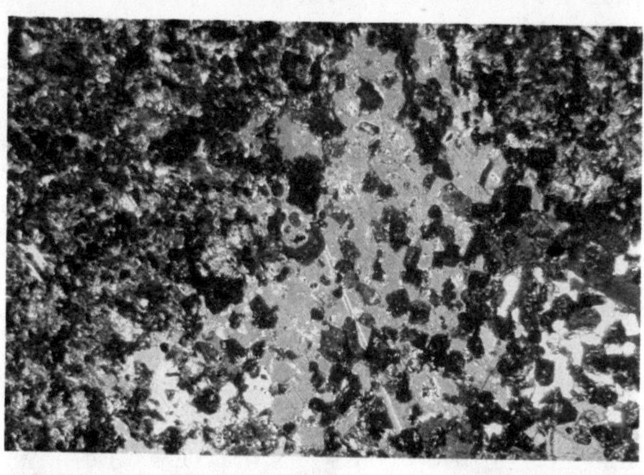

288

289

287, 288, 289 Rocks with the same chemical composition need not have the same petrographical name. They get their names from the minerals they contain. The appearance of a rock to the naked eye is quite different from that of a thin section under the microscope, where all the structural components are visible.

Basalt (top pictures) is an igneous rock which solidified on the Earth's surface, whereas amphibolite is a metamorphic rock – basalt transformed by temperature and pressure. Eclogite (bottom pictures) arose from the metamorphosis of amphibolite at extremely high temperature and pressure.

It has been found that the oceanic crust is formed at a relatively high rate, i.e. around 2–35 cm (0.8–13.8 in) a year. Figure 292 shows that the formation of the oceanic crust is a relatively simple affair.

The continental crust is formed more slowly, through a long-term geological process. For, though the oceanic crust is easily formed, it is just as easily lost. This is natural

290 One of the most important formations of the seabed is the mid-oceanic ridge, a divergent interface between oceanic plates. A hot plume of the asthenosphere passes through the rigid lithosphere and forms the mid-oceanic ridge with a central rift valley, which is filled with new igneous rocks of basalt composition which rise up from the depths.

– if the size of the Earth is to stay the same, material which comes to the surface must disappear again. The evidence for the disappearance of the crust is quite convincing. On the bottom of today's oceans the geologist cannot find rocks older than 200 million years anywhere in the world. Compared with the total age of the Earth that is not very long at all. So he supposes that the older oceanic crust has been 'lost' into the interior of the Earth.

The history of the Earth is thus much longer than that which we see on the ocean floor, and is recorded in code in the rocks of the continental crust.

The mantle

In describing both the rocks of the land surface and the ocean floor we have recalled many times that the basalt and other rocks come from the upper mantle. There a partial melting of rocks occurs, and magma rises to the surface of the Earth. The processes in the upper mantle and the movement of its matter affect what goes on in the crust. It can be said that even the formation of giant mountain chains, continental drift and the movements of the plates of the lithosphere all have their ultimate origin in the upper mantle. Our knowledge of the Earth's crust and the processes going on in it is based on direct observation. We can make a daily study of

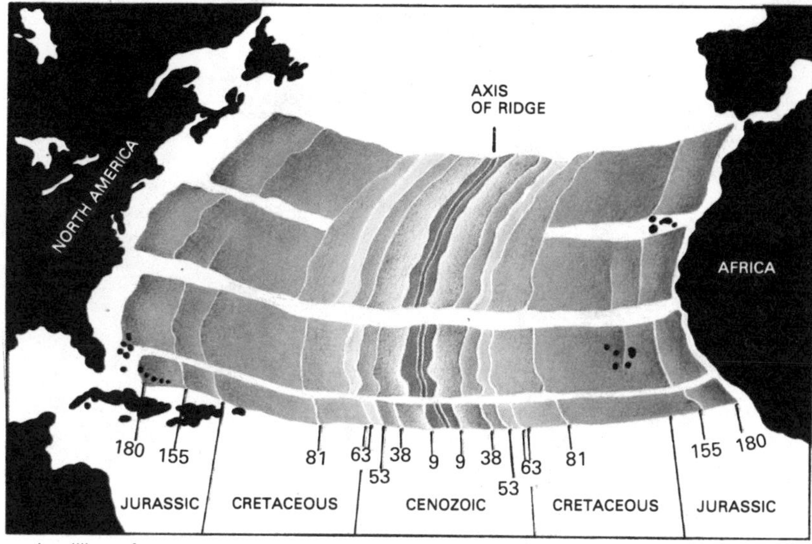

291 When the age of the rocks of the ocean floor was determined, there was new evidence of the expansion of the floor and the movement of oceanic plates from the mid-oceanic ridge to the continent. The figure shows the Atlantic Ocean. The age of the rocks here increases from the mid-Atlantic ridge (red shades) through the brown, orange and green shades, to the blue rocks close to the continent. These rocks are the oldest.

the activity of wind and water, and we also have opportunities to study gas explosions and volcanic activity – but these are only the consequences of some process. It is as if we were to watch the packing at the end of a chocolate factory's production line and remain ignorant of what actually goes into the making of the chocolate. But our present-day technology is limited, and man has not yet been able to get down to the upper mantle, so that a knowledge of its structure relies on indirect instrumentation and human ingenuity and reasoning.

Geophysicists studying the stratified structure of the Earth evaluate and measure the properties of the upper mantle. It consists of a material which transmits seismic waves at more than 8.3 km (5.2 miles) per second, and has a density of more than 3.3 g/cm^3. But we do not know exactly what it is, or its mineralogical and chemical composition. It must be a material which, when heated to the temperatures and pressures occurring at depths of 50–400 km (30–250 miles), produces rocks of basalt composition. Otherwise we should not have volcanic islands such as Hawaii, Tahiti or Iceland. One of the ways of finding the chemical composition of the upper mantle is by analysing similar basalt rocks.

292

293

292 In comparison with the complex structure of the continental crust the structure of the oceanic crust is simple. The uppermost layer is made up of clay-like silicates, sometimes also calcareous thin sediment **(1,2).** Beneath this there is usually a basalt lava layer **(3).** The next layer is made up of rocks of the gabbro **(4)** or peridotite **(5)** type.

293 Volcanic rocks give information about the areas of the Earth from which they come. Inclusions of 'olivine nodules' in alkaline basalts are specimens of the impoverished upper mantle.

A further argument in the discussion of the composition of the Earth's mantle is introduced by the cosmochemists, and it is an interesting one. They claim that the Earth as a whole does not differ a great deal from the primordial matter of the Solar System – i.e. from meteorites or, more exactly, chondrites. So we can calculate the composition of the mantle. Then there is a geological finding which must be considered. Diamonds, which are created under high pressure, certainly in the upper mantle, come from a rock known as kimberlite. It is found in South Africa, Siberia and Brazil. And in addition to diamonds, kimberlites contain fragments of other rocks they have picked up on the way to the surface. We therefore find small pieces of the rocks from the upper mantle in kimberlite. They have a high iron content, a lot of magnesium, and a small amount of silicon – they are ultrabasic rocks, such as dunite, wherlite, lherzolite and the garnet-bearing rocks such as peridotite and eclogite. All the minerals of these rocks of the upper mantle crystallized at high pressures and temperatures.

The upper mantle zone to which the above findings apply stretches to a depth of only 400 km (250 miles); but a glance at Figure 282 shows that the mantle itself extends to a depth of 2900 km (1800 miles). What there is down at that depth remains a mystery. Only indirect data exist for the composition of the mantle at great depths. We have seen that the rocks of the crust, though they have the same chemical composition, may contain different minerals. Much depends on the depth, pressure and temperature at which they occur. And the same goes for the upper mantle. Similar rules apply. With increasing depth, pressure and temperature also increase, and the materials are adapted to these conditions; the structure of the rocks is more compressed, so that the density of the constituent minerals is greater. We have proof of this from the laboratory. At extremely high temperatures and pressures conditions in the laboratory approach those inside the Earth, though only milligram quantities of samples can be used. The results of such experiments are compared with those obtained from measurement of the propagation of seismic waves. It has been found that the depths where the rate of propagation increases are likely to produce a change in the internal structure of minerals are comparable with those which laboratory experiments also show. Figures 294 and 295 thus demonstrate both the changes in the rate of propagation of seismic waves and changes in the structure of minerals in the depths of the Earth.

The core

The statement that the Earth has a core seems obvious to almost everyone today. But if we try to prove it, we shall probably get into difficulties. It is relatively hard to prove that the Earth's core exists and to show its structure. In order to prove that there is a metal core down there, and what is more a liquid one, we have to turn to that tried and tested science, seismology. Let us then put together the evidence which we do have to date.

The basic physical data of which we have spoken, i.e. the Earth's mass, density and angular momentum, indicated that with increasing depth there is also an increase in the amount of material, whose mass is greatly different from that of matter on the surface.

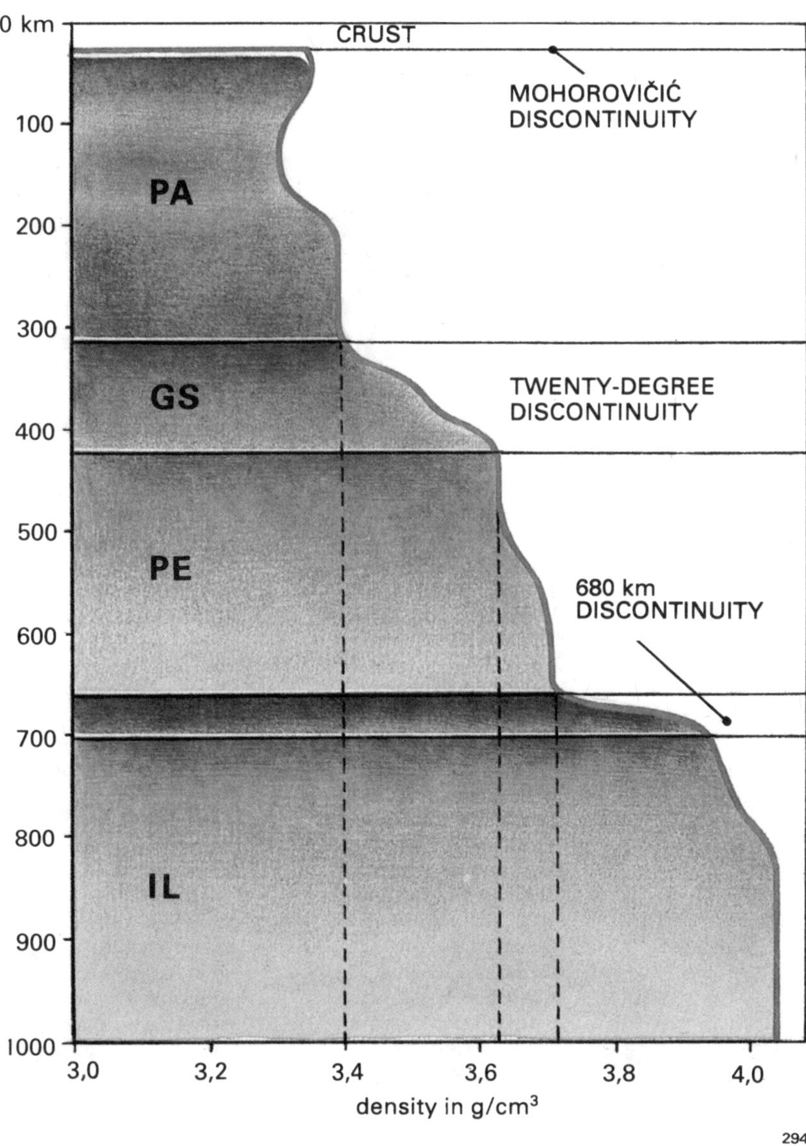

294 The structure of the upper mantle is now well known. Both experimental studies of rocks at high temperatures and pressures and an analysis of the rate of propagation of seismic waves contributed to an understanding of this inaccessible part of the Earth. The propagation rate of seismic waves changes both gradually and suddenly. Most of the sudden interfaces (discontinuities) are explained by a change in the mineralogical composition of the material of the mantle. **PA** – an area of partly molten rocks, **GS** – minerals with the structure of garnet, **PE** – the structure of perovskite, **IL** – silicates with the structure of ilmenite.

It must therefore be a much more massive sort of material than that which makes up the surface rocks. Even the rocks of the upper mantle lack the density required to account for the physical properties of the Earth as indicated by its total density. The density of the rocks of the upper mantle is only $3.3-3.8$ g/cm^3. So from the physical point of view the presumption of the existence of a heavy nucleus is the only possible answer. From the viewpoint of cosmochemistry, too, by comparing the amounts of elements in meteorites and in stars, we find that the Earth should contain many more heavy elements than the composition of its surface rocks would suggest – more iron, for instance, than either the surface or the upper mantle contains. It must be somewhere in the Earth,

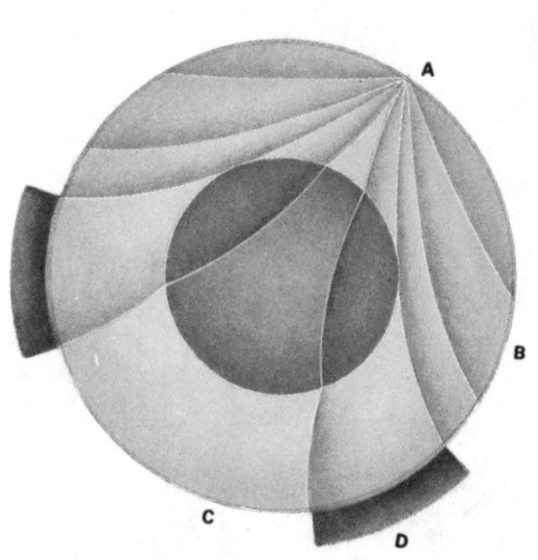

295

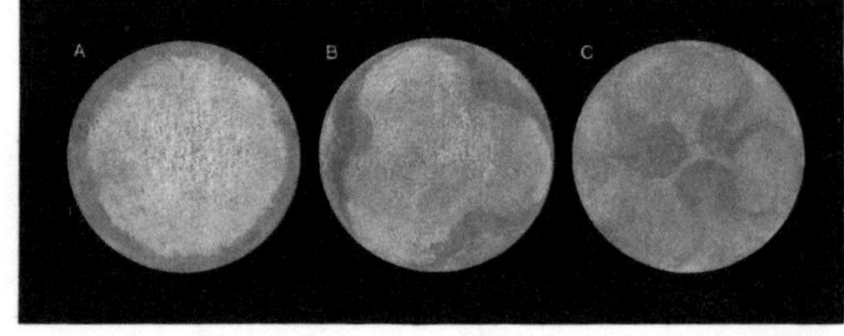

296

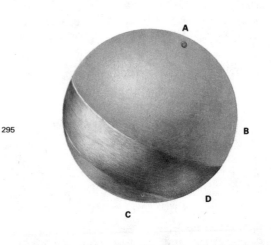

295 Evidence of the existence of a liquid core is offered by the propagation of seismic waves, which are diffracted on crossing from one medium into another; they are also reflected, and some of them do not propagate in liquids. It is these observations, and the existence of a shadow, which indicate that at least part of the interior of the Earth is in a liquid state. **A** – point of hypothetical earthquake, **B** – area where both P and S waves are seismographically recorded, **D** – the shadow of the seismic waves, **C** – the area where only P-waves are recorded.

296 Geologists consider the formation of the Earth's core to have been the most significant geological event in the history of the Earth. The formation of the core was probably accompanied by the heating of part of the planet and the separation of the sulphate component from the silicate **(A)**. The formation of pockets of metal **(B)** and their falling into the gravitational centre of the planet **(C)** are depicted here according to the notion of the New Zealand scientist Walter Elsasser.

and the same goes for other metals similar to iron, which geochemists call *siderophile,* and which behave similarly to iron. They include the metals of the platinum group, which also occur on the surface in smaller quantities than the composition of meteorites would indicate. Thus iron is the likeliest candidate: it has a high density, and there is plenty of it about both in space and on Earth.

The proof of the existence of a core comes from seismology, from the study of the course of earthquake shocks through the Earth. It was put forward at the beginning of the century. The interface between the mantle and the next layer of the Earth is at a depth of 2900 km (1800 miles). It is called the *Weichert-Guttenberg discontinuity,* and it is more marked than the border between the crust and the mantle (the Moho). There is a strong inflection and deflection of seismic

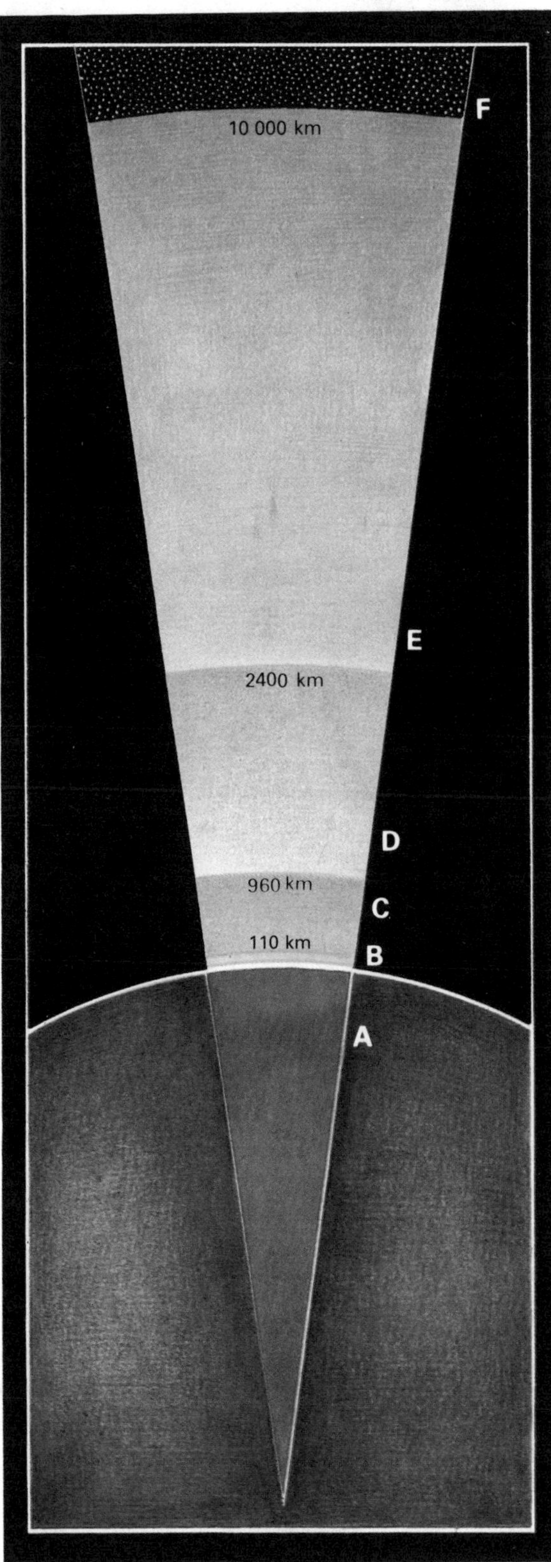

297

297 The extent of the atmosphere is more than twice that of the body of the Earth. But for man only the lowest layer of the atmosphere is important, the part we breathe. **A** – Earth, **B** – oxygen and nitrogen zone, **C** – oxygen zone, **D** – helium zone, **E** – hydrogen zone, **F** – interplanetary gas zone. The chemical composition of the bottom layer is different from the uppermost strata of the atmosphere, the places where the Earth **(A)** meets outer space.

waves. One type of seismic wave, called S-waves, does not pass across this boundary at all; this is considered a proof that at least part of the core is liquid, since S-waves are not propagated in liquids.

But that is not the geologists', geophysicists' and geochemists' last word on the state and composition of the core. Only a few successful laboratory experiments have been

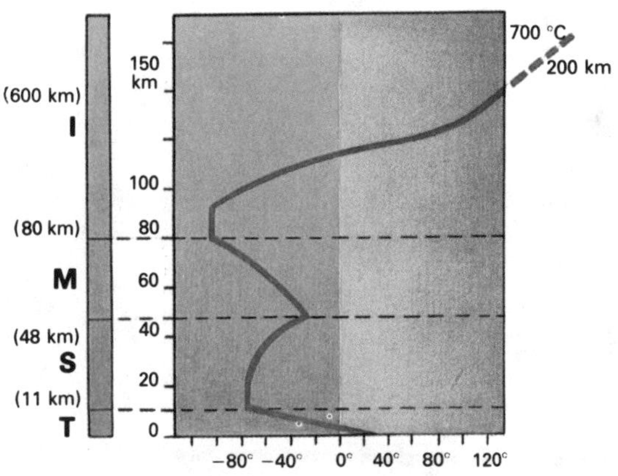

298

298 Temperature differences between the various parts are important points of reference for investigating the atmosphere. **T** – troposphere, **S** – stratosphere, **M** – mesosphere, **I** – ionosphere.

performed on the conditions which exist at the interface between the core and the mantle, i.e. at a depth of 2900 km (1800 miles) below the surface of the Earth, and such conditions cannot be maintained for long, so geologists rely on the study of what happens in big explosions. The core is thus liquid, but it is thought that it probably has its own nucleus, the inner core, which is solid. Iron on its own would not have the appropriate physical properties, so the presence of another metal, nickel, is assumed, and some scientists think that there is also a considerable content (10–20 per cent) of metallic silicon. The similarity to iron meteorites is called to mind in this context. They, too, contain a considerable amount of nickel in addition to iron. And since it is probable that iron meteorites are the remains of some small, disintegrated or smashed planets (the result of a cosmic collision), we suppose that the Earth, too, has an iron-nickel core. But these are all questions which scientists must solve some time in the future, starting experimentally in the laboratory. One day perhaps it will be possible to make equipment capable of penetrating to huge depths in the Earth's mantle, maybe even down to the core itself.

299 The variability of our planet is illustrated by the map showing the amount of precipitation in different parts of the world.
1 – deserts and semi-deserts
2 – less than 250 mm per year
3 – 250–500 mm per year
4 – 500–1000 mm per year
5 – 1000–2000 mm per year
6 – more than 2000 mm per year

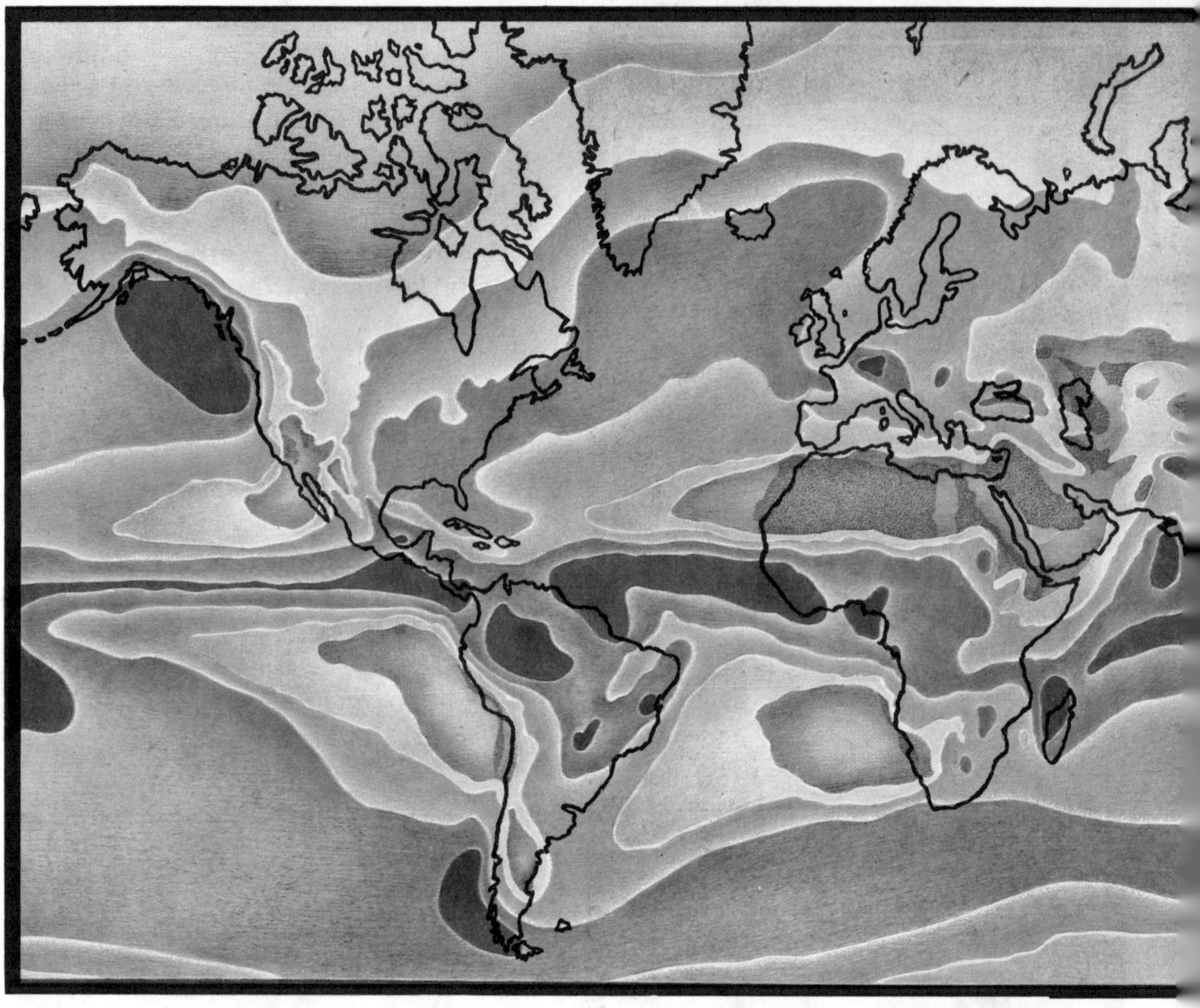

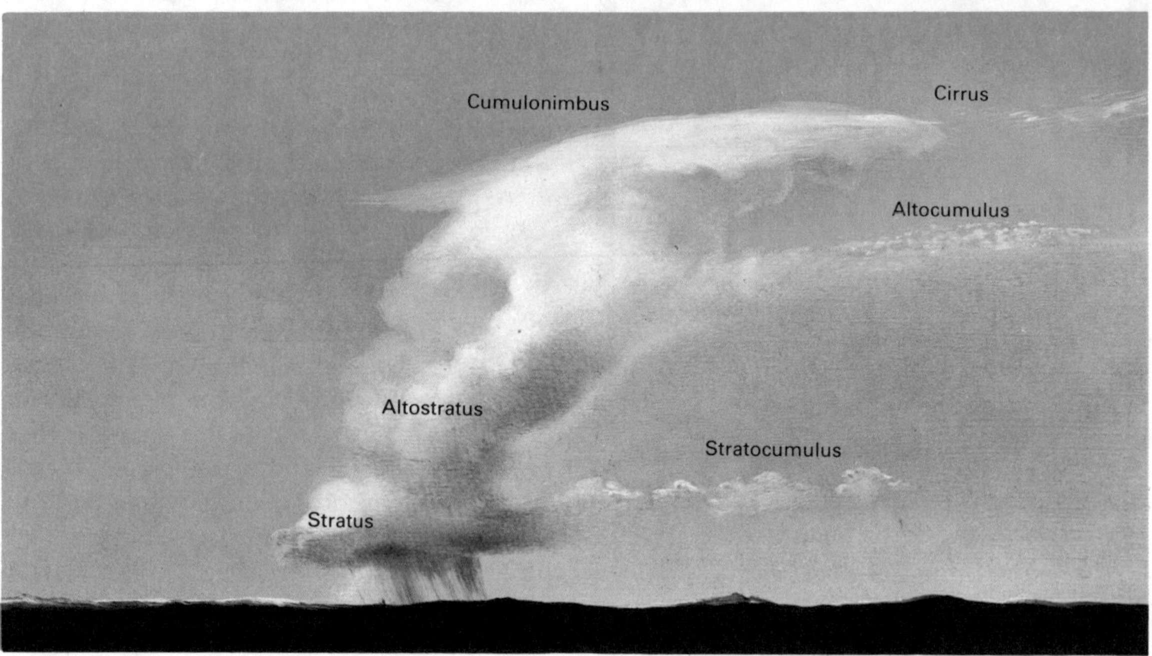

300

300 Along the warm front a varied mixture of cloud formations occurs, in a certain order and at certain altitudes. Weather forecasting according to clouds is an old, tried and tested method.

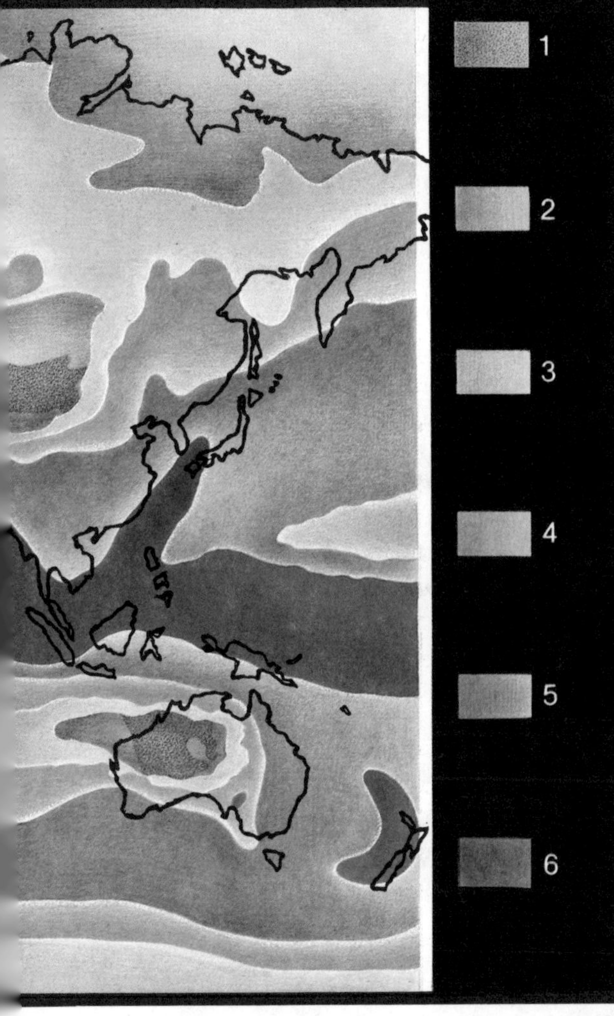

299

The atmosphere

The crust, the mantle and the core, which we have been considering, are all part of the solid strata of our planet. The following paragraphs are devoted to the atmosphere, which is invisible, but whose influence on life is decisive.

The atmosphere is, like the crust of the Earth, a product of distillation – the evolution of the Earth's mantle. It probably formed together with the crust, and we might call it a side-product. Some scientists even consider the atmosphere to be a part of the crust. In the course of geological time the atmosphere, too, has evolved. In the early stages of the Earth's evolution its composition was different from that today. It probably did not contain free oxygen. Geologists consider the coming of atmospheric oxygen to have been a milestone in the history of the Earth. The 'oxygen explosion' which took place more than 2000 million years ago brought about a huge flourishing of life, and that really is a notable event.

In studying and describing the crust, the mantle and the core, we considered their composition and structure. But can we do the same for the atmosphere? We should have difficulty in applying geological standards, it

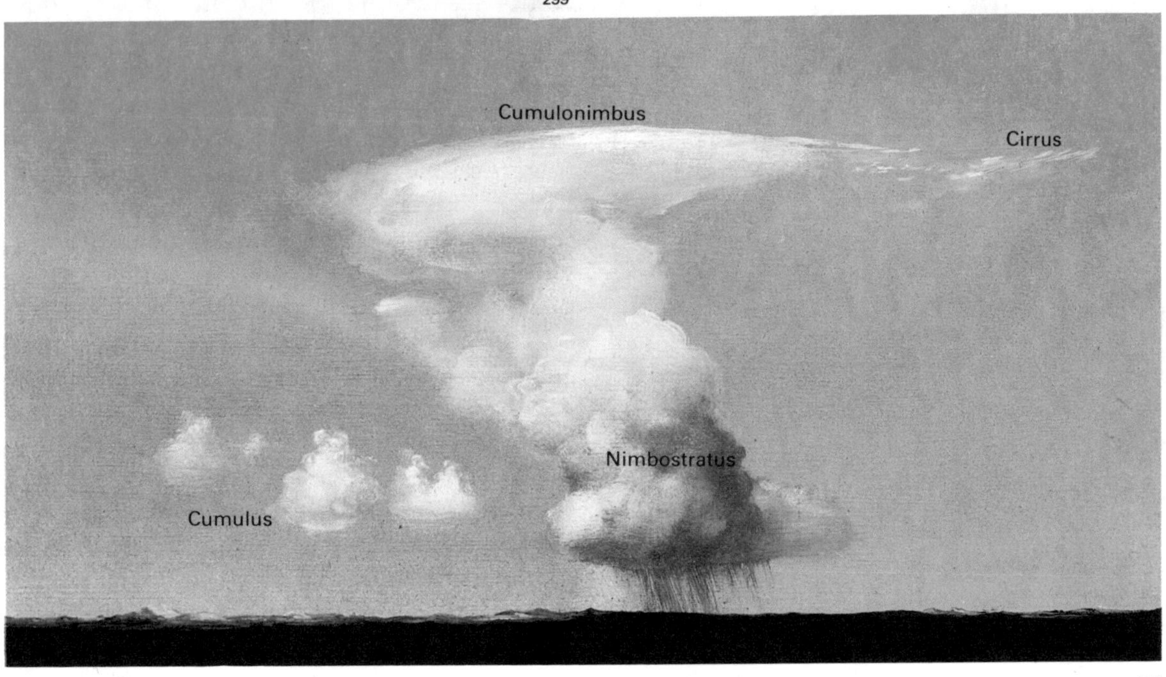

301

301 Changes in the weather usually occur when cold and warm air meet. Water vapour condenses and falls as rain. The cold front is accompanied by characteristic clouds.

302

303

302 'Cumulus' type clouds are typical of summer weather.

303 Low cloud of the 'stratus' type is usually accompanied by rain. It occurs directly at the front.

is true. The geological processes in the crust and the mantle are dynamic, and the crust is changeable, though over a long period of time. But the atmosphere changes from day to day; it is much less stable than the crust or the mantle. So most of the cosmic influences (such as solar radiation, the solar wind and the stream of energy particles from space) are reflected in the behaviour of the atmosphere. Radiation is slowed down by it and absorbed, so that the atmosphere forms a protective shield. Man, too, influences the atmosphere, and unfortunately in a negative way. He pollutes it with carbon dioxide, sulphur compounds and nitrogen, and with particles of dust. There are those in the scientific community who think that the fall in temperature of almost half a degree which the northern hemisphere has undergone in the last 25 years was caused by man himself. It may not seem very much, but it is enough to have a detrimental effect on agriculture, since it shortens the growing season.

We have already spoken about oxygen as a basic and important part of the atmosphere. It makes up 20.9 per cent of the lower layers. Nitrogen is more abundant in the atmosphere than oxygen, making up more than three-quarters of it (78.08 per cent). The remaining components are argon, which comprises less than 1 per cent of the volume (0.93 per cent), and carbon dioxide (0.033 per cent). The other gases are present in such small quantities that for present purposes we can ignore them. All the components mentioned above (oxygen, nitrogen, argon and carbon dioxide) together comprise 98.8 per cent of the atmosphere, and form the mixture we call *air*. This composition of the atmosphere has been found to extend to a height of 88 km (55 miles). There is very little hydrogen in this lower layer, because it is so light it easily escapes into space.

While we have noted that the temperature increases the further one goes down into the Earth, we find that the temperature of air falls with increasing height. Even at the altitude of 10–12 km (6–7 miles) at which jet airliners fly, it is very cold, around −50° C (−58° F). Higher up, at an altitude of 30 km (19 miles), to which radio-sonde balloon fly, the temperature is actually higher than at 10 km (6 miles). The lowest temperatures in the atmosphere have been measured at heights of 80–90 km (50–56 miles).

According to its temperature, composition and physical properties, the atmosphere can be divided into layers. The *troposphere* is the part between the Earth's surface and an average height of 11 km (7 miles). It is a relatively extensive and dense stratum, which contains most of the atmosphere's water vapour, and in which most of the atmospheric phenomena in which we, the Earth's inhabitants, are interested occur. The troposphere contains the clouds, rain, etc. The decrease in temperature per 1 km (0.6 mile) of altitude is around 6.5° C (11.7° F) Naturally, air pressure also falls with increasing height; indeed, this is the principle on which some altimeters are designed.

The boundary which divides the troposphere from the next major stratum of the atmosphere, the *stratosphere,* is called the *tropopause.* The tropopause is very cold and is not at the same height throughout the world. It is closest to the Earth's surface around the poles, while at the equator it is highest. The stratosphere is a low-temperature layer. Its composition is like that of the troposphere, but it is where *ozone,* a form of oxygen, is formed and accumulates. The ozone layer is very important for life on Earth, and has recently been the subject of much discussion, since it absorbs most of the ultraviolet radiation which falls on the Earth. If it were not for this layer, we should get tanned quicker, but the ultraviolet rays would also destroy most living things on our planet. When it was found that the ozone layer is damaged by fluoromethanes (which are contained, among other things, in aerosol sprays), scientists began to sound alarm bells. The stratosphere is a relatively calm region, unlike the lower layers of the atmosphere. Air currents like those in the troposphere are also rare in the stratosphere, and its individual strata are not mixed a great deal. Clouds are seldom seen there.

The *ionosphere* is the ionized layer of the atmosphere. It reflects high-frequency (short-wave) radio waves. The ionization of air is caused by solar radiation, especially gamma, X-ray and ultraviolet radiation, and for this reason the boundaries of the ionosphere change in the course of the day. In the daytime ionized air forms in the lower layers of the atmosphere. Short-wave radio transmissions bounce off this low 'ceiling', and cannot therefore be sent over long distances.

304

305

304 High cloud of the 'cirrus' type heralds a change in the weather. The front is still several hundred kilometres away.

305 With increasing industrialization the influence man has on the processes of the Earth increases. Dangerous pollution with gaseous exhalations brings about the creation of a 'microclimate'.

306 Our impression that the sea supplies most of the water for rain inland is not quite accurate. Only some 27 cm (10 in) of rainfall over land per year (101 cm/40 in) is of 'marine' origin, and this flows back into the sea.

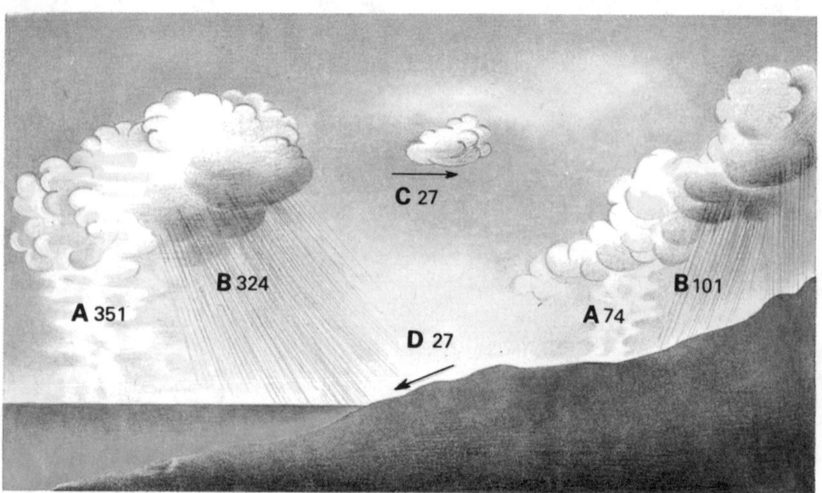

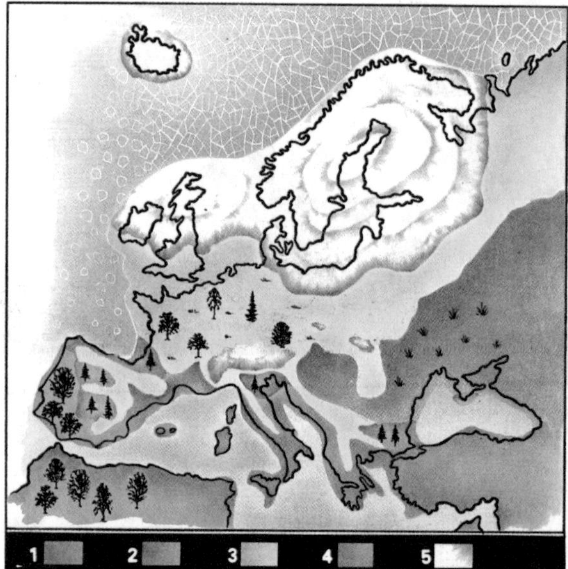

307 The map of Europe in the last ice age (25,000 years ago). While northern Europe is covered in ice **(5)**, most of the continent is covered with 'northern tundra' **(3)**. Only in the south of Europe were extensive broadleaved forests **(1)** to be found at that time. Coniferous forests **(2)** and steppes **(4)** complete the picture of Europe as it was then.

That is why we can pick up fewer stations on short-wave during the day. At night, when the ionosphere rises very high, short radio waves can be propagated for great distances, and you can tune in to a broad range of stations.

The hydrosphere

If we speak of the hydrosphere as being the liquid layer of the Earth, we are not being quite accurate. For the hydrosphere includes all the water which is held in the pores of rocks for long periods of time, and also the water which is present in the solid state in the Arctic and the Antarctic ice-caps. But if we were to confine the term hydrosphere to rivers, lakes and marshes, and the salt water of the seas and oceans, it would still be inaccurate, but we should not be far from the truth. For this water makes up more than 97 per cent of the hydrosphere's volume. Everyone is likely to take a different view of the importance of the hydrosphere to mankind – the inhabitants of coastal regions would emphasize the significance of fishing, while those who live in desert regions would speak of the necessity of fresh water supplies for life in their villages, and so on.

The hydrosphere's formation was dependent on the formation of the Earth's crust, and was in fact a by-product of that process. It seems incredible, but all the water of the oceans, like all the air of the atmosphere, comes from inside the Earth. It formed through the differentation of the Earth's mantle. This was convincingly proved in the 1950s. In the course of volcanic activity new batches of *juvenile* water (water which has not previously been involved in the hydrological cycle and since the Earth came into existence has been bound in the upper mantle) are introduced into the hydrosphere. The water of the atmosphere – water vapour – is also a part of the hydrosphere. Every molecule of water present in the atmospheric moisture remains in the atmosphere for an average of three weeks, after which it again becomes a part of the hydrosphere proper. Water vapour condenses, falls to the ground as rain, permeates the surface, gets into watercourses – such as streams and rivers – runs into the seas and oceans, and the whole process we call the *hydrological cycle* begins all over again.

The journey of the water molecules through the atmosphere, the ground and the rivers is a long one. The interplay of the climatic conditions of the Earth, evaporation and condensation of water, takes place at the intersection of the Earth's crust, the hydrosphere and the atmosphere. The binding of water, evaporation and also condensation in ice-fields are in fact the best regulators of the

308 A cross-section between North America and Europe shows the distribution of ocean temperatures. The blue colour indicates the coldest parts, the red and orange warmer areas. The Gulf Stream can clearly be seen (top left). The vertical sizes are considerably exaggerated, so that the mid-Atlantic ridge appears as a huge mountain chain.

thermal balance of the Earth. For the hydrosphere holds excess heat and cold. The balance is a delicate one, and sometimes men seem to rely on it a little too strongly, without taking much care to maintain it. For instance, they build cities in the vicinity of the sea, in spite of the fact that a fairly minor increase in temperature is enough to melt the ice-caps (which contain around 3 per cent of the volume of the hydrosphere) to such an extent as to cause flooding (if the whole of the ice-caps were to melt, the level of the seas would rise 60 m/197 ft worldwide). An increase in the Earth's temperature can also be caused by a high degree of air pollution, causing a *greenhouse effect* similar to that which exists on Venus.

In describing the individual layers of the Earth – the core, the crust and the mantle – we were interested mainly in their composition and structure. In the case of the hydrosphere these features seem simple – we are only talking about water, after all. But the hydrosphere's structure is in fact very complex and variable. Take the chemical

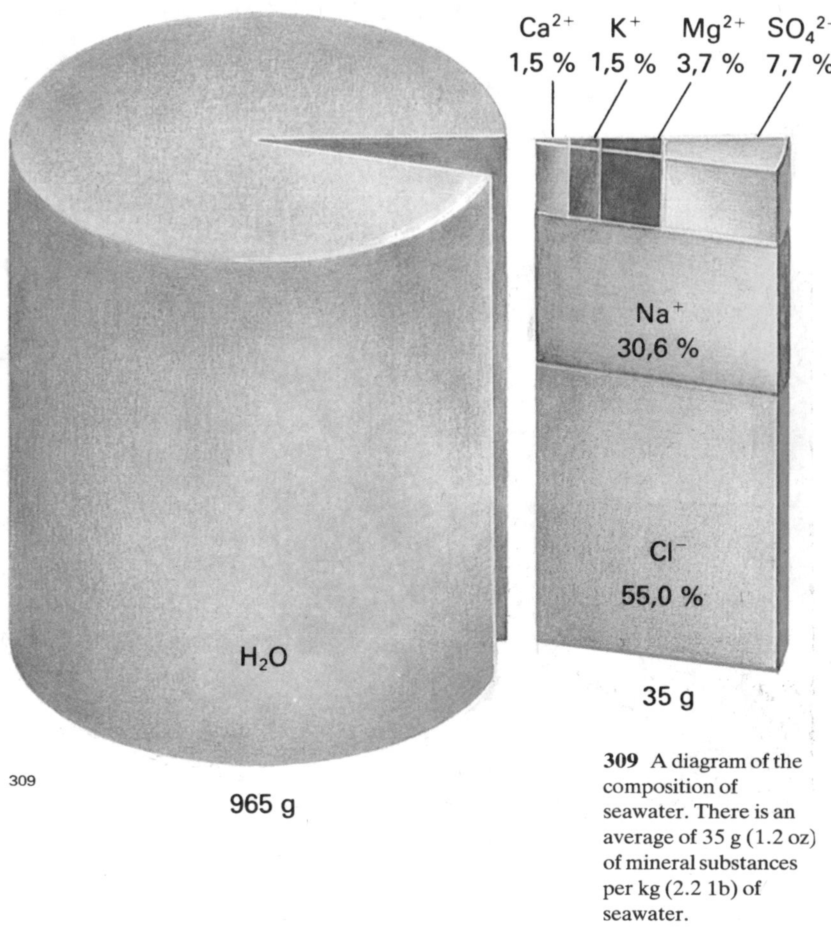

309 A diagram of the composition of seawater. There is an average of 35 g (1.2 oz) of mineral substances per kg (2.2 1b) of seawater.

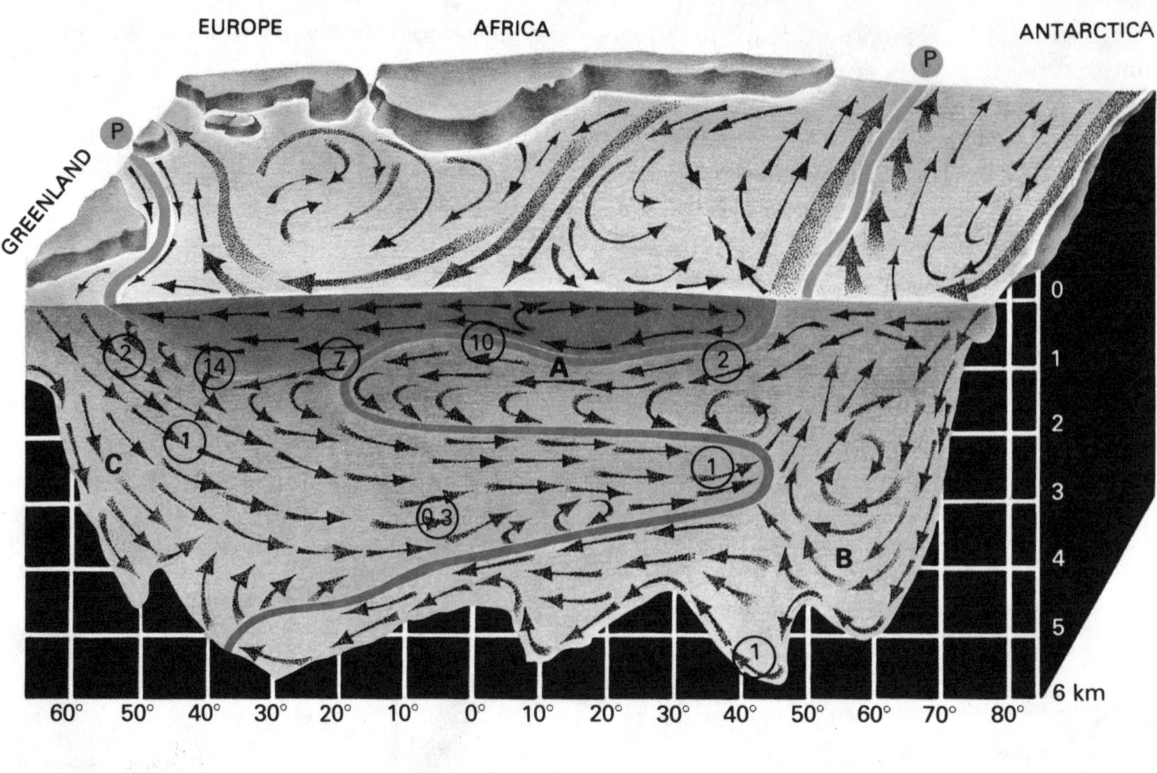

surface currents 40–150 cm/s

surface currents 5–40 cm/s

figures indicate rate of flow in cm/s

310

310 If we stood on the American continent and looked towards Africa, at the same time looking down at the ocean bottom, we should get this sort of impression of the temperatures of the water in the Atlantic Ocean. On the right is Antarctica, on the left Greenland, **A** – the Antarctic central current, **B** – the Antarctic lower current, **C** – the northern polar current, **P** – limits of the polar ice. The blue line indicates the area where the northern polar and Antarctic currents meet.

311 Underground water fills either the cracks and caverns **(A, B)** in mountains, or their pores **(C, D, E).** An example of crack permeability is presented by the slates on the left-hand side of the picture. The springs which rise there are called crevice springs **(1).** Beneath the slates is limestone, where the water flows through caves and caverns. Such springs are called karst springs **(2).** Examples of crevice permeability are montane loess and alluvial gravel; loess springs flow from the former **(3),** while drinking water is obtained from alluvial plains by sinking wells **(6).** Many domestic wells obtain shallow underground water from weathered, disintegrated cliffs **(7).**

A special case occurs where a permeable rock stratum is enclosed in impermeable rocks (right-hand side of the picture). There the water is under pressure and if it finds a way to the surface it spouts. Such springs are called artesian wells **(4).** If the permeable stratum is cut through by a valley, the underground water flows freely from it **(5).**

A

B

C

1

2

3

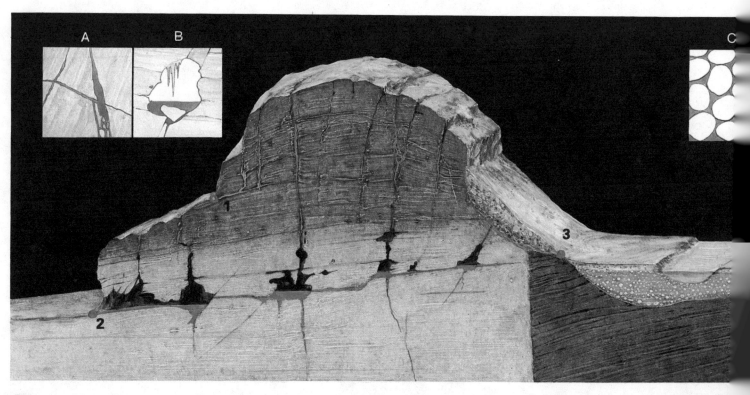

composition of seawater. Everyone knows that seawater is salty; and that this is due to the presence of common salt – sodium chloride. Seawater contains, out of 100 per cent of dissolved compounds, 77.7 per cent sodium chloride. It is also bitter, due to magnesium salts such as magnesium chloride. This makes up 10.8 per cent of the dissolved salts. Magnesium sulphate also gives seawater an unpleasant taste. It makes up 4.7 per cent of the sea's salts. Then there is calcium sulphate (after precipitation it forms gypsum anhydrite), potassium sulphate, calcium carbonate, and also some bromide salts. The other compounds which are found in seawater occur in tiny quantities only, but seawater contains nearly all known elements occurring in nature. The amount of salts present in seawater is called its *salinity*. It varies, of course, from place to place, but on average there are 35 parts of salt in 1000 parts of seawater. But there are seas where the inflow from land is high and evaporation low; such seas (the Baltic, for instance) have a low salinity and are therefore comparatively 'fresh'. Others (like the Red Sea) evaporate much more and have little input from land or other oceans, so the salinity is high.

It must be clear from these observations how the salinity of seawater comes about. Rainwater, which we might suppose to be the 'freshest', actually contains dissolved carbon dioxide, sulphur dioxide and other compounds. It is in fact a very weak acid. When it comes into contact with rocks on the Earth's surface, it reacts with them, and further substances are acquired: sodium, potassium and calcium, but also aluminium and iron. So even the freshest water flowing into the sea contains a certain small proportion of dissolved substances. In the sea, the water evaporates and effectively 'thickens'. Seawater thus has a greater density than fresh water; it is a better 'carrier'. A second important phenomenon is that when seawater meets fresh water it sinks, so that at the mouths of certain rivers the surface water is less salty than the deeper water. The salt content also affects the water's freezing point: salt water freezes at a lower temperature than does fresh water.

The temperature and salinity of seawater are in fact responsible for the structure of the oceans – of the hydrosphere. At low temperatures (in the case of fresh water 4° C/39.2° F) water is at its greatest density. For this reason the water at the bottom of lakes in winter is at that temperature. The water at the bottom of the oceans will be only approximately 4°C because the temperature depends on its salinity. The water of the sea and ocean beds is subject to quite different currents and movements from those we know from maps. The surface layer of water mixes easily up to a depth of about 100 m (330 ft) due to wind and waves, and behaves quite differently from the water at great depths, which is not in contact with the atmosphere.

311

312 Obduction is one of the relatively rare results of a collision between two plates. Unlike subduction, where an oceanic plate glides under a continental one, in the case of obduction an oceanic plate rides over a continental one (black). (**1** – basalt, **2** – gabbro, **3** – peridotites). This case is well documented on the east coast of Papua New Guinea. **P** – Pacific plate, **NB** – New Britain deep-sea trench, **IA** – Indo-Australian plate

ground is amazing – 0.31 per cent, which represents 4.2×10^{15} m^3 of water. That is the quantity contained at depths down to around 700 m (2300 ft). But there is the same amount of water at greater depths, probably right down to the border between the crust and the mantle. Bores sunk to almost 10 km (6 miles) came upon rocks which contained water in their pores even at that depth.

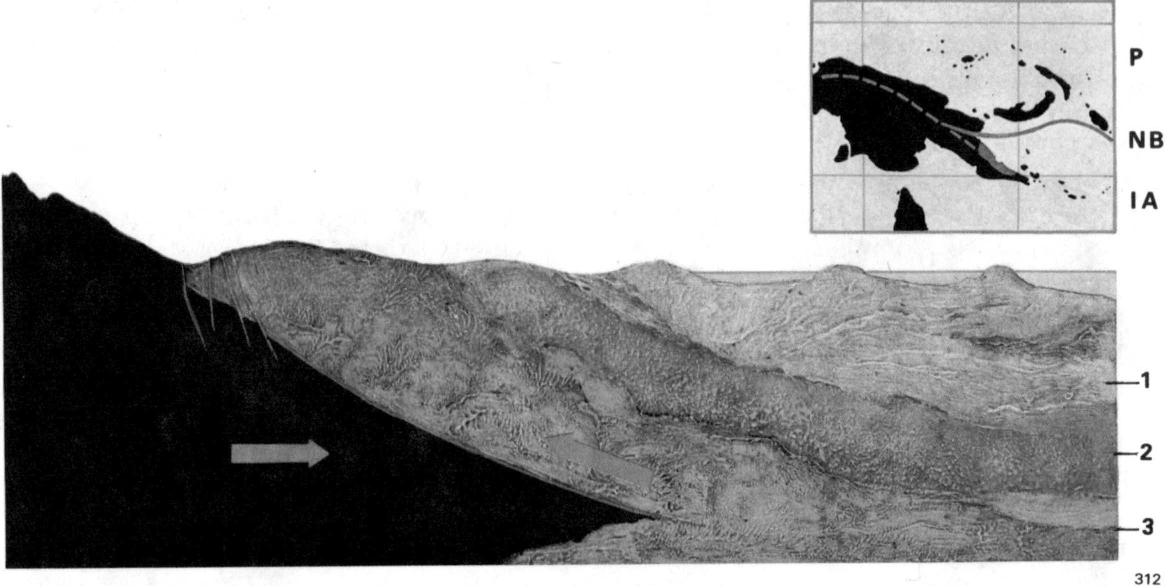

312

The waters of the equatorial belt are the warmest at the surface, with an average of 24–25° C (75–77° F), while in polar regions water freezes. The water near the ocean floor, however, has a temperature which is similar at the equator and at the poles.

There are also certain amounts of gases dissolved in water: oxygen, nitrogen, carbon dioxide. These amounts vary according to many factors: depth, temperature and the presence of living organisms whose existence depends on the consumption of oxygen, nitrogen (denitrifying bacteria) and sulphur (sulphur-reducing bacteria). If one compares the amount of water which is available for consumption by people with the total volume and mass of the hydrosphere, the result is alarming. Only a tiny proportion is usable by man. Thus all fresh water lakes together contain a mere 0.006 per cent of all water; and rivers and streams contain only 0.0001 per cent. Compared with these relative figures, the amount of water present under the

Fresh (drinking) water should not contain more than 1 g of soluble material per litre (0.04 oz per 0.2 gallon). If it contains more than that, it may be called mineral water, or industrially polluted water. The average content of salt in seawater is 35‰, which is 35 g per l water, but some seas are saltier, and others 'fresher'. Thus the Baltic has only 3–20‰ solubles, the Black Sea 13–25‰, whereas the Mediterranean has as much as 36–39.5‰. The Red Sea is the saltiest of the open seas with 38–41.5‰.

The amount of water which evaporates from the sea is surprisingly large. If the hydrological cycle were to be disturbed, 26 per cent of all water would evaporate in 1000 years. That means that you would be able to walk from Europe to Iceland or from Australia to Asia. But provided no major changes occur in the atmosphere, we can expect evaporated water to continue to condense and to get back into the rivers, which will again feed the oceans.

Rainwater, which contains carbon dioxide, and is therefore carbonic acid, is an excellent solvent. It not only dissolves salt easily, but also substances which seem at first sight to be insoluble. Limestone, for instance, is dissolved slowly but relatively easily. It is also important for the formation of seawater that rainwater dissolves certain elements from weathered minerals. An example is ordinary feldspars, which are a normal component of igneous rocks, and contain in addition to silicon, aluminium and calcium, a certain amount of sodium. This small quantity of sodium is carried into the sea by fresh water. Many other compounds get into the sea this way. Calcium, together with carbonate ions, is the most common element, and is present in quantities much greater than the amount of sodium which is brought by rivers. Here we arrive at an apparently contradictory finding. Seawater should be calciferous, not salty – but that is not the case. The calcium does not stay in the water as long as the sodium.

Geochemists use the term *residence time* to describe the time between the 'arrival' of the average atom-ion of a given element (calcium, sodium) in the seawater, and its sedimentation as a solid. And since calcium is rapidly consumed by organisms such as molluscs for building shells, it remains in the water for a relatively short time. The same goes for other soluble substances such as aluminium, titanium, chromium and iron, which are precipitated to the seabed and form new minerals. It has been calculated that these elements spend only a few hundred years in seawater. Chlorine and sodium, on the other hand, the only two elements which go into the making of seawater, have a residence time of about 100 million years. That is why the sea is salty and not ferrous or calciferous, even though these elements get into the sea from fresh water.

The greatest reserves of fresh water are contained in the ice-caps; they are of little use to mankind, however, since to recover them would be too expensive for most countries. But fresh water can be obtained by other means. In some places it flows from the ground of its own accord in the form of springs; elsewhere it has to be pumped out. It is present almost everywhere below the Earth's surface, in cracks in rocks, and especially in their pores. Thus sedimentary porous rocks such as sandstone and conglomerates are of prime importance for water supplies. Figure 311 shows how and where water is present below ground, and where it should be looked for.

The quality of water is important, especially in industrial and intensively cultivated areas. If water quality is good, we do not usually pay much attention to it; only when it is heavily polluted or smells of disinfectants do we consider the value of truly fresh water. Not all fresh water is drinkable. Rainwater, for instance, is not very suitable. Though it is the atmosphere's purest distillate, it is too soft, lacking the necessary mineral content. But as soon as it comes into contact with soil and rocks it changes its character. Mineral substances are dissolved in it. Such water is more suitable for drinking. The most common element in it is calcium, which is easily dissolved. The more minerals the water contains, the harder it is. Over-hard water also has its disadvantages, and has to be modified for industrial use. Water may contain ions other than those of calcium. Some of them, such as nitrate ions which get into water from industrial fertilizers and from the waste of farms, are dangerous. And if water contains compounds of lead or mercury it becomes a dangerous poison.

Volcanoes form the oceanic and continental crusts

If it were not for research on the ocean floor in the last two decades we should still be under the impression that the floor of the world's seas was made up of rocks similar to those of the continental crust – igneous rocks and sedimentary rocks where there used to be dry land. In the early part of this century Alfred Wegener and many other geophysicists tried to prove from gravimetric data that the composition of the ocean floor must differ from that of dry land. There are, of course, sedimentary rocks on the ocean floor, but not nearly as many as we would imagine. The main reason is that the oceanic basin is not as old as the continents. In fact volcanic activity is still occurring regularly on the ocean floor, and will probably continue to do so in the future.

313 The classification of igneous rocks is based on both their chemical and their mineralogical composition. Minerals bring the former with them from wherever they come, while the latter is the result of the place and thus also the manner of crystallization of the rock in the Earth's crust. The same rock has several names according to whether it crystallized at depth or just below the surface, or poured onto the surface of the Earth, where it quickly solidified.

There is a great deal of evidence available for volcanic activity on the seabed. The most important is the igneous rocks which form the main material of the ocean floor. Drilling rigs operating from research vessels have produced evidence not only of present-day igneous rocks, but also of the fact that volcanic activity is not exceptional there. Shallow earthquakes accompanying volcanic activity also indicate that intensive geological processes are under way along the mid-oceanic ridges, of whose very existence geologists were unaware until 40 or 50 years ago.

The rocks of the ocean floor – called deep-sea basalts – are, however, very dull and ordinary. They are usually poorly crystallized as a result of the rapid cooling they underwent after flowing out on to the ocean floor (they often form *pillow lavas*). In addition they contain a large amount of glass. This glass is very susceptible to the action of seawater, and these rare and scientifically precious rocks are very unattractive.

A chemical analysis of the rocks of the seabed shows that they are difficult to distin-

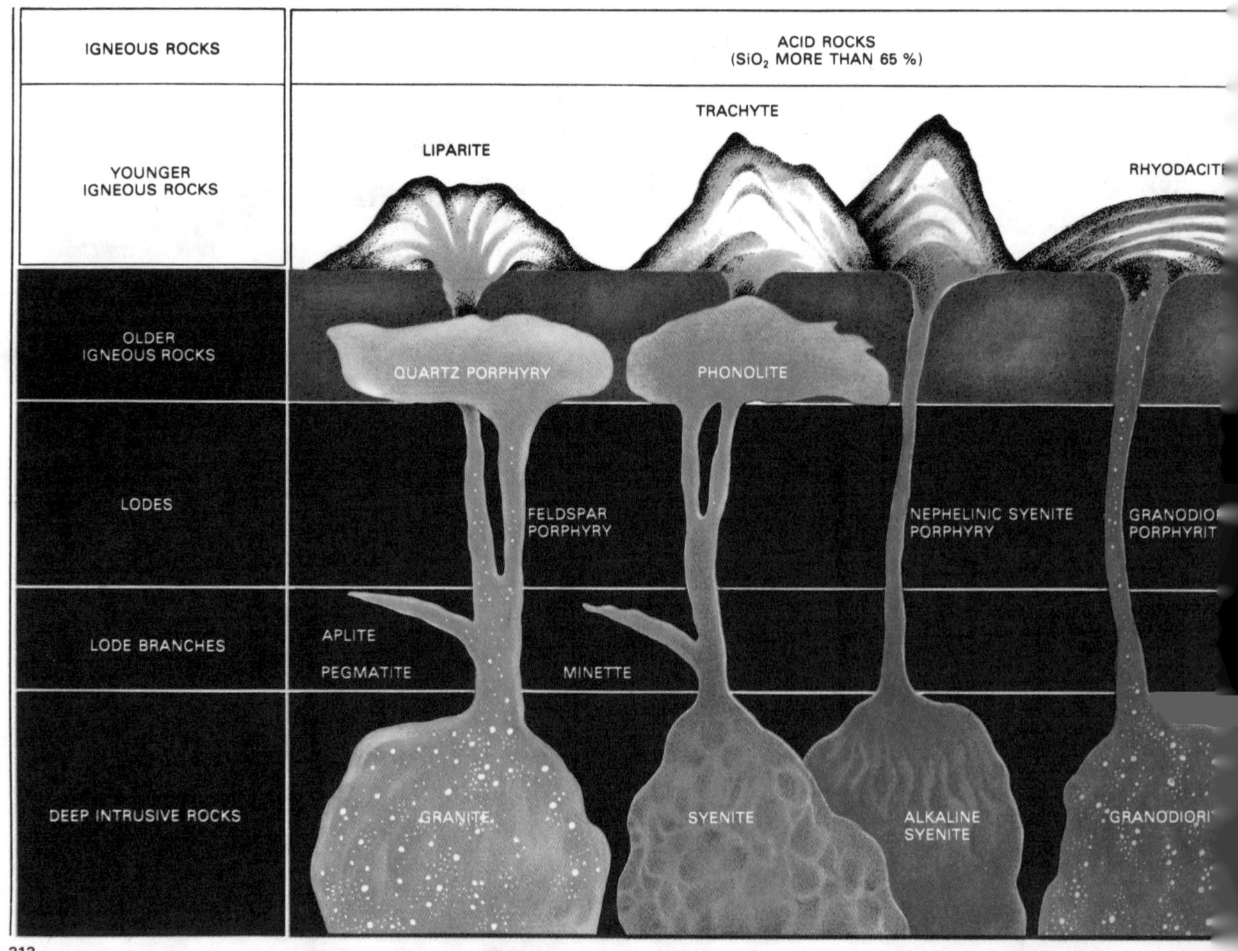

313

guish from one another – only analysis of the isotopes of individual elements shows minor differences. The rocks from different oceanic regions are very similar to each other, and those from the Indian Ocean are almost identical to those from the Pacific or the Atlantic. It has been shown in the laboratory that they arise from a melting of the rocks of the upper mantle. When the magma rises to the surface of the Earth, it is differentiated in shallow magma chambers, after which the rock arrives on the ocean floor. But we cannot say that only basalts are formed there. Some of them crystallize not far below the surface, and sometimes other rocks from the upper mantle reach the surface with them, so that the structure and composition of the seabed rocks is fairly complex. Ultrabasic rocks such as peridotites, dunites, wherlites, etc. occur there, and all manner of combinations of olivine, pyroxene and spinels. These rocks accompany the basalts. The study of the rocks of the seabed is, however, very complicated, and geologists trying to explain the structure, composition and processes of the ocean floor rely on geophysical data.

In some areas of the world 'fragments' of oceanic crust have been found on what are now continents. How these pieces of seabed got on land is best illustrated by Figure 312: in short, they slid ashore. Such pieces of seabed have been found on Cyprus, the island of New Guinea, the Alps and the Urals. Geologists are studying these rocks very carefully, not only because they bear witness to the way in which the ocean floor develops, but also because the geologists are interested in the possibility of finding deposits of nickel, chromium or copper. Scientists call the fragments ophiolite complexes, and since they also contain sedimentary rocks which formed on the ocean floor, they are considered a proof that the oceanic crust may, thanks to long geological processes of shifting and overlapping, become a part of a continent.

The rocks of the ocean's margins

The Pacific coast is lined with volcanoes. Starting with New Zealand, they continue

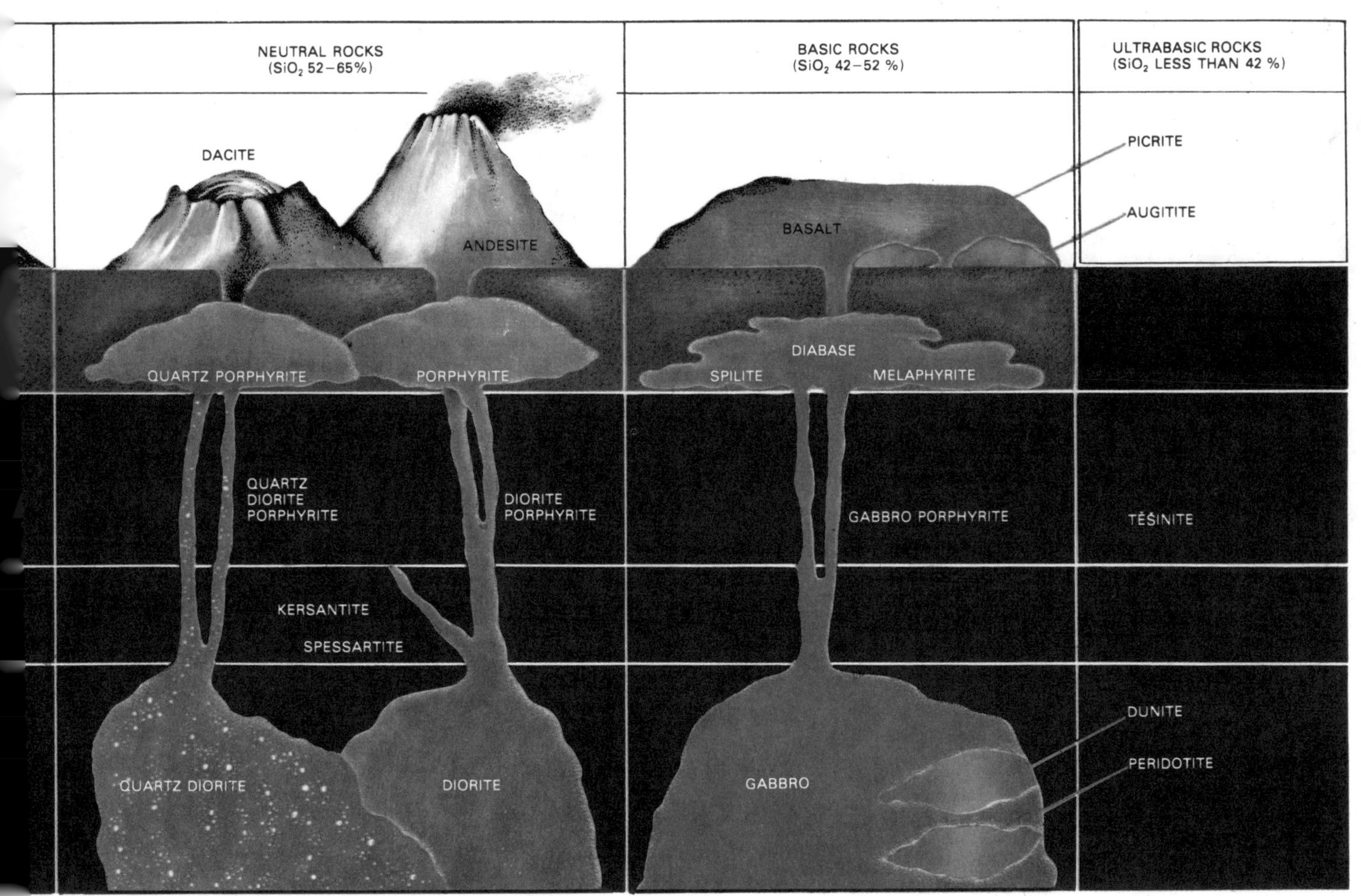

through Tonga and the Kermadec Islands, Fiji, the New Hebrides (Vanuatu), the Solomon Islands, New Guinea, the Philippines, then towards Japan through the Kurile Islands, Kamchatka, the Aleutian Islands, then the west coast of the USA, Central America to the Andes, and through the South Sandwich Islands to Antarctica. In older geological works you will find the expression Pacific 'ring of fire' used to describe this line of volcanoes, in more recent ones the term *andesite line,* since it was then supposed that all volcanoes are formed of igneous rocks – andesites.

Along the mid-oceanic ridges, as we saw in the last chapter, volcanic activity is accompanied by earthquakes. This is so, even in the area around the Pacific, whether on islands (scientifically speaking *island arcs*) or on the mainland's oceanic shore. A thorough study along with measurements shows that the epicentre is the large tectonic plateau, called the *Wadatti Benioff zone* after its discoverers, which is oriented diagonally down from the ocean under the land. Not only do huge shifts in the large continental and oceanic plates occur along it, but magma and igneous rocks are formed there. While the activity of the volcanoes of the mid-oceanic ridges remains hidden beneath the surface of the ocean (but for minor exceptions around Iceland), the volcanoes of the Pacific coastlines form a dangerous 'ornament' to it.

The rocks which occur in the area of the island arcs are considerably more varied than those of the ocean bed. The simple mineralogical composition of the seabed minerals gives way to a wide variety of types, differing also in chemical composition, in the area of the island arcs and the active continental regions. There are both simple basalts (similar to the rocks of the mid-oceanic ridges) and andesites, which are igneous rocks with a higher silicon content and a lower iron and magnesium content than basalt.

The volcanic activity of the island arcs and the continental shores is accompanied by crystallization of igneous rocks below the surface. In this way plutonic rocks arise, sometimes accompanied by veins of ore. We shall see later that most of the world's copper production comes from widely dispersed deposits (called porphyry copper) occurring only at the sites of present or recent volcanic activity in the island arcs or the continental margins.

Just as sections of the ocean floor, with their associated rock complexes, have been found on the continents, so also have the rocks of the island arcs or the continental margins been found there. Canadian geologists studying one of the oldest geological units in the world – the *Canadian Shield* – even

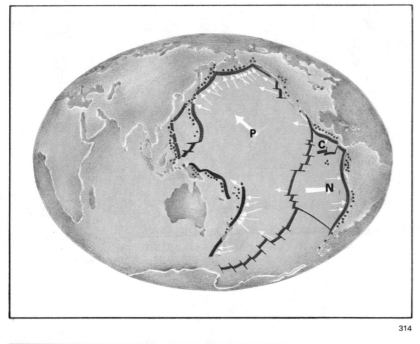

314

315

314 The 'ring of fire' in the Pacific. The Pacific plate **(P)**, the Nazca plate **(N)** and the Coco plate **(C)** move from the mid-oceanic ridge to the continent, where they slide under and melt. This is the explanation for the origin of the volcanoes on the edge of the Pacific.

315 The contact of two oceanic plates forms an island arc, an extended chain of volcanic islands. Characteristic features are a deep-sea trench in the place where plates overlap, and the occurrence of magma rocks at the point of contact of the two plates.

316 The illustration shows the most basic shapes and geological forms of igneous rocks. **1** – batholith or pluton, **2** – pluton sends out into its surroundings injections of igneous rocks, **3** – chimney, **4** – radial dyke associated with the volcanic feeder, evident on the surface as a ridge **(5)**, **6** – sill, **7** – laccolite, **8** – dyke, cutting across strata, **9** – group of dykes, **10** – lava sheet, **11** – volcanic cone, **12** – parasite cone.

317

318

317 A caldera is a mature volcanic formation and characterizes volcanoes which have been active for a long time. Part of the material falls into a magma chamber, where it is remelted. A characteristic feature of the caldera is the presence of younger volcanic cones and parasite craters. Some calderas are filled with water.

318 One of the most dangerous of volcanic formations is the volcanic dome. It consists of highly viscous lava which is pushed up through the opening of the volcano. On the surface the lava solidifies, and in places there are sharp rock protrusions called Pelean needles, and hot gas emissions. Frequently the sudden emergence of magma and the accumulation of a large amount of gas cause a violent explosion (for example Mount St. Helens on the west coast of the USA).

claim that at the time it formed, around 2000–3000 million years ago, conditions similar to those now found in the island arcs prevailed at the site of Canada today, a stable continental shield. The remains of rocks of the island arcs are found almost everywhere in the world, in the fold mountains and in the stabilized parts of the crust. Since this type of volcanic activity is generally accompanied by a raising of mountains – *orogenesis* – the whole of the community of rocks from basalts to andesites is known as orogenic igneous rock.

The cause of the formation of these rocks was investigated in the past by many geologists and geophysicists, who studied their mineralogical and chemical properties. They even tried to simulate, in the laboratory, the conditions under which they formed and reached the conclusion that they arise relatively deep in the upper mantle (100–200 km/62–124 miles down) by the melting of the rocks of the upper mantle, but also by the melting of those of the oceanic plate (the lithosphere) which gets down to that depth by *subduction* – sliding under. The temperature in that region is up to 1000° C (1832° F). Since the rocks which are melted contain water, the melting is made easier. At high pressures the addition of water to magma lowers the melting point. It is ordinary water which causes volcanoes to erupt so violently, the same water that under favourable circumstances plays the chief role in creating veins of ore.

Igneous rocks where they ought not to be

The igneous rocks we have described correspond clearly to the places bordered by individual plates of the lithosphere. The volcanic activity of the oceanic regions is confined to divergent edges of plates, i.e. to the places where two plates are moving away from each other. The volcanic activity of the ocean shores and island arcs, on the other hand, is associated with convergent edges of plates, i.e. places where plates meet and collide or overlap.

But there are also areas with volcanoes and igneous rocks which are unconnected with either divergent or convergent plates. Among outstanding examples are the Hawaiian Islands, Tahiti and the Canaries. And there are also inland volcanoes, which are not at the edges of plates, but in the middle of them. If we look back into the past we find huge volumes of igneous rocks, extensive basalt caps, for instance, in India, the western part of the USA, and in Antarctica and Tasmania. These show that there was intense volcanic activity on the continents in the past.

It has been proved that most of these continental volcanoes, and the intra-oceanic ones, have basalt lava. Experiments with rocks at high temperatures and pressures in the laboratory indicate that the temperatures required to melt basalts are so high that these rocks must come from the upper mantle. There are no temperatures that high in the crust. In an attempt to explain the presence of basalts inside the plates, places have been found in the upper mantle where the temperature is considerably higher than the surrounding area. There is a heat flow from the interior of the Earth towards the exterior. Because of their cylindrical shape these places have been named *thermal plumes*, or *hot spots*. In them the material of the upper mantle slowly, in the course of geological eras, moves up towards the surface. In the course of this movement a partial melting of the rocks occurs. This leads to the formation of alkaline basalts with high potassium and sodium contents. Some of the components of the upper mantle are easier to melt than others, the latter forming a residue. The former, easily melted components, are alkaline elements such as potassium and sodium.

The thermal plume theory has offered evidence of the fact that the ocean floor – in effect a thick plate – is moving. For if we study the age of individual islands in the Hawaiian archipelago, we find that the centre of volcanic activity has moved over the last 10 million years. But we can reverse the situation, and assume that the volcanic activity centre does not move, is stable, and that it is the whole oceanic plate with the older and newly forming islands which is moving. You can judge for yourself – Figure 319 shows both possibilities.

319 **319** The concept of lithospheric plates explains the long-known phenomenon of the 'migration' of the volcanic activity of the ocean islands, such as Hawaii or the Galapagos Islands. The hot spots of the upper mantle remain in the same place, but the plate carrying the islands moves.

320 **320** Cones of volcanic rocks in the middle of the ocean are typical of hot spots. The creative activity of the Earth's interior is destroyed at birth here by water erosion.

Volcanic rocks without volcanoes

Igneous rocks are very important for the metabolism of the Earth: they give rise to the crust. So the rocks forming the ocean crust leave the interior of the Earth, the mantle, on the mid-oceanic ridge, and on the continental margin and the island arcs. The igneous rocks could be found everywhere in the continental

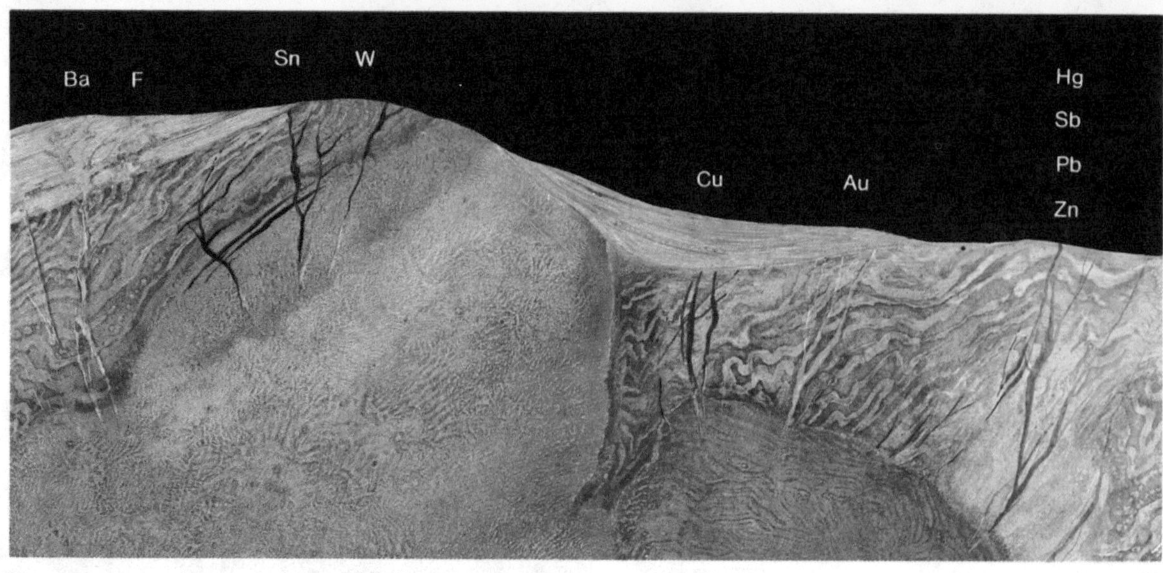

321 Magma was often thought to be a carrier of ore components. The figure illustrates one of the most successful hypotheses of the prospectors – the idea of the American geologist Ebenezer Emmons about the relation of ores and bodies of igneous rocks.

nuclei and the orogenic zones. It is clear that igneous rocks are the main source of both sedimentary and metamorphic rocks.

So far we have spoken only of the igneous rocks which come mainly from the upper mantle. We have talked about basalts, but also about the alkaline basalts or andesites of the island arcs. The Earth's crust, which has a very different composition from the mantle, also produces igneous rocks, even some with a much lower melting point than basalt. At first sight it might seem that this contradicts our statements in the previous text. How is it possible? Relatively speaking, there is quite a lot of water in the crust, and we have said that water lowers the melting point at high pressures. This means that the different material of the Earth's crust and the higher water content might cause large amounts of molten rocks (magmas) to occur in the crust. Along some continental coasts, such as the west coast of the USA and in South America, it has been proved that the rocks which arrive at the surface in the form of volcanic lava originated in the crust. Some rhyolites and dacites, for instance, which are very light-coloured igneous rocks, are the product of melting of the material of the crust. If we consider their chemical composition, with a predominance of silica (there is also a large amount of alumina, sodium oxide and potassium oxide), we find that these igneous rocks are chemically indistinguishable from granites, granodiorites and other rocks we call plutonic.

But the difference lies in the fact that rhyolites or dacites were carried to the surface, where they very rapidly cooled (the speed of cooling is shown, for instance by the presence of glass, and by the mineral structure of these rocks), while plutonic rocks cooled slowly, relatively deep down.

If we observe how quickly weathering and erosion wear volcanoes down, we understand why there are more plutonic rocks than volcanic ones in the older formations such as the Palaeozoic or even the Pre-Cambrian. It takes only a few million years for a large volcano, rising proudly above the surrounding countryside, to be worn right down to the ground by the action of wind and water. Volcanic plutonic rocks which cover a large area are very probably the eroded remains of what was previously volcanic terrain. In many places the igneous rocks did not reach the surface, however, and so plutonic rocks or batholiths represent not only the roots of volcanoes, but also independent formations with no relation to volcanism. Plutonic rocks are found all over the world, in central Europe and France, in the western USA, and also in the Himalayas. Plutonic igneous rocks are distinguishable from other igneous rocks by the very fact that they are well crystallized, that the dimension of the minerals are usually very similar, and that the various components can relatively easily be recognized. Among them are granite and other types of rock of the same family, differing either in the smaller amount of silicon they possess, or

322 The first block shows the evolution of the central European brown soil. The second block shows podzol, which occurs where rain robs the soil of mineral substances. In the third block is the opposite process, taking place in tropical regions, where a long period of drought follows the rainy season, and capillary transport brings minerals to the surface. (The symbols **A, B** and **C** are accepted designations of soil horizons.)

in the larger amounts of what are called the dark components – such as mica and amphibole – or sometimes in the presence of pyroxene. Apart from potassium feldspar there is also a sodio-calcic variety of feldspar. Gabbroes and diorites, which are considerably darker than the granites, are also plutonic rocks.

Geologists seek out plutonic rocks because they are often accompanied by mineral deposits. Granite, for instance, is often found together with tin deposits (and stannous granite is recognizable by its composition), whereas in other places copper is found with granite. The plutonic rocks, granite, diorite and others, are also a favoured building material used in external walls and valued for their durability and appearance.

The soil – most precious of rocks

An individual may consider that the most precious thing he has is his family, his friends, or perhaps great wealth. However the most precious thing mankind holds in common is the soil. It nourishes man, and the animals and plants on which he depends. If one were to give a somewhat rude description of man, one might say he was a parasite on the soil – but then so are all terrestrial creatures. Without soil there would indeed be no life at all.

It might seem an exaggeration, but the destruction of every acre of arable land is a permanent loss for the generation responsible. The character and composition of the soil in any locality is connected with the geological substratum, the climate, and many local factors. Under European climatic conditions, for instance, the soil horizon – the soil which can be cultivated – has been evolving for hundreds and thousands of years. The situation is better only in humid regions. The soil develops more quickly there. On the famous volcanic island of Krakatoa, in Indonesia, for instance, a soil layer 35 cm (14 in) thick formed in only 45 years, and on the slopes of some volcanoes the farmers sow crops a mere five or 10 years after an eruption. The soil develops from the parent rocks of the substrate by weathering, in which not only the climate and precipitation are involved, but also micro-organisms. And since it is a very slow process, each little piece of soil should be treasured. (In dry, desert climates the soil horizon forms even more slowly than in our temperature zone, and is more rapidly eroded – usually by the wind.)

The preceding paragraph has already outlined the requirements for the creation of soil. It is then subject mainly to the effects of the climate, since the rate of weathering depends on the amount of precipitation and the temperature differences. The main soil types are shown in Figure322. Laterite soils, for instance, occur in places with a tropical climate, while chernozems are the remains of past interglacials.

There are many apparently less significant factors which affect the character of the soil. The most important role is played by micro-organisms, which are able to break down 'insoluble' components of rocks such as quartz and feldspar. Given time, it is no problem for nature to break down apparently resilient and indestructible rocks using the agents of water, wind, frost and micro-organisms. All the other factors can be seen in Figure 322 which shows a typical soil profile. Pedologists (Soils Scientists), whose work it is to study the soil, divide it into several *horizons*. These, too, are clearly shown in the picture, and many of us have seen them. Have you never noticed? You only have to look at the side of a deep trench or the wall of a quarry: over the hard, unweathered rock there is weathered rock, the more so, the nearer one gets to the topsoil. According to the proportions of parent rocks in the soil, various types are distinguished. For a soil horizon to form, the lie of the land is important. You have probably noticed that on some deforested slopes water rapidly carries away the soil, and after heavy rain you can see where the soil was taken from and where it has been deposited. Water seldom

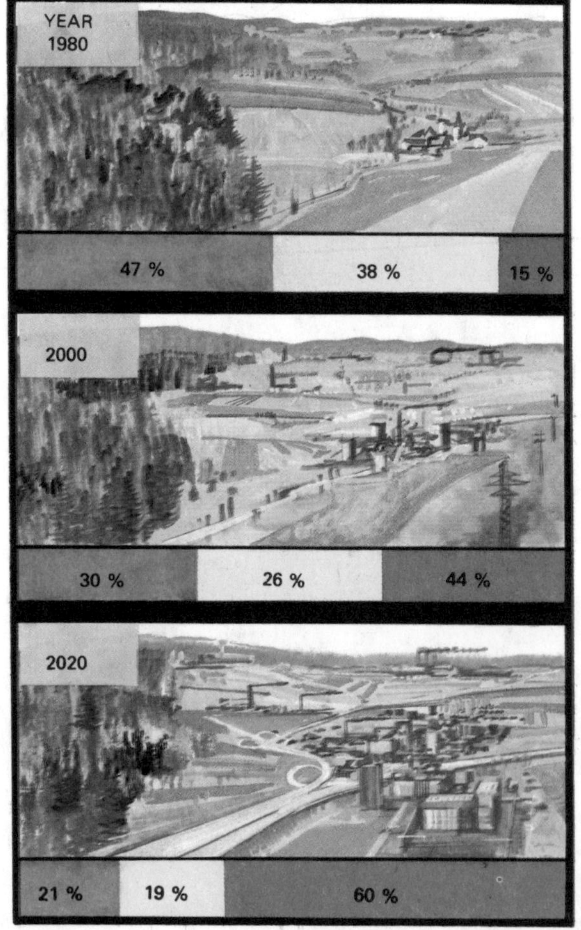

323

323 One can characterize the effect of civilization on the countryside in the course of human history as deforestation and the obtaining of arable land. But the most recent years of industrialization have caused not only deforestation, but also a loss of agricultural land.

erodes soil in the fully treed forests, but in places where trees have been felled there is a price to pay. Soil loss in meadows is minimal, but where maize or grain crops have been planted, for instance, there is a major soil loss before the crop comes up. If such a loss is repeated several times over, the soil begins to lose its function as a provider of nourishment.

Every soil type requires its own sort of treatment. Some do not tolerate irrigation (a layer of salt is deposited), while others suffer because of drainage (nutrients are easily carried away). In some places more fertilizer must be used, in others less. In those parts of the world where fertilizers are not yet used, the soil must be left to recover for some time before it is fertile again (in tropical regions this takes up to 15 years).

The rocks we have described in the preceding chapters were characterized by their chemical and mineralogical composition. Soils, too, can be typified in this way, according to the minerals and the chemicals contained in them. But their most fundamental characteristic is the fact that they are the most precious of all rocks, and quite indispensable.

Sedimentary rocks

The igneous rocks of which we have spoken could be described as primary rocks. They may come from the upper mantle, chiefly in the form of basalts, or may be a product of the partial melting of older, existing rocks, as in the case of granite. Melting erased many of the features of the original material from which they formed. Sophisticated methods are required in order to determine the history of such a material prior to melting. In the next few paragraphs we shall be considering a group of rocks which we might call derived, or secondary rocks. These are the sedimentary rocks and the metamorphic rocks. In discussing what is primary and what is secondary we might easily find ourselves in the 'chicken and egg' situation – which came first? For many sedimentary rocks formed from igneous rocks, many metamorphic rocks from sedimentary and igneous rocks, and many igneous rocks are the result of the melting of existing sedimentary and metamorphic rocks.

Let us start with a simple case, which should serve to illustrate the way in which sedimentary *clastic* rocks were formed. Consider a basalt or andesite volcano, somewhere on the edge of the Pacific Ocean. Lava runs out from it, volcanic ash flies into the air, and the slopes are covered with a clearly 'primary' material. Rain and snow fall on the lava and ash, the wind blows, and the rocks are weathered, usually mechanically, to form smaller pieces. The weathered fragments from the slopes are carried away by water into streams and thence into rivers, and to the bottom of the nearby sea. There the fragments of rocks and minerals pile up. In this way sedimentary rocks are formed. The rock and mineral pieces, together with gravel, sand and clay undergo a process known as *diagenesis,* which leads to the formation of solid sedimentary rocks from rocks which

324 If one studies the flow of water from the high-mountain glaciers to the deep seabed, one sees a great diversity of sedimentary rocks.
1 – glacial sediments,
2 – gravitational (slope) sediments,
3 – lake sediments,
4 – river sediments,
5 – wind sediments,
6 – delta sediments,
7 – shelf sediments,
8 – deep-sea sediments.

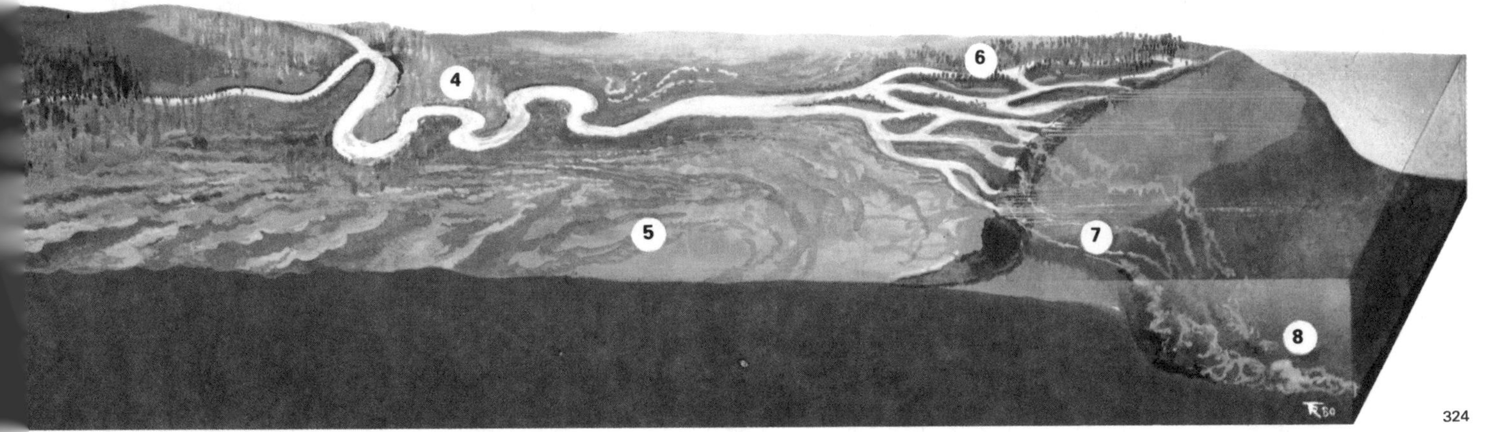

324

were originally not solid at all, but perhaps granular. In the case of sand diagenesis leads to sandstone, in the case of clays to claystone or slate, and in the case of gravel to conglomerate.

The process of diagenesis is difficult to imagine. If you were to take a sack of wet clay and weight it down with a heavy object, the excess water would run out, and without our even being aware of it, the components would arrange themselves in parallel. The rock is solidified through the action of the liquid which remains in it. (In the process of diagenesis neither temperature nor high pressures are involved. If these influences do act on a rock, the process is called *metamorphosis* (or metamorphism). This chapter is concerned with those sedimentary rocks in whose formation metamorphosis is not involved.)

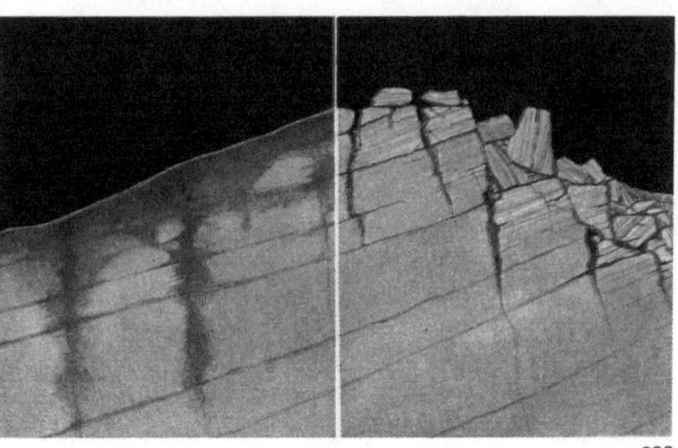

326 The difference between chemical weathering (left) and mechanical weathering (right) in the same rock. In the case of mechanical weathering, such as frost cracks, the rock merely disintegrates into smaller pieces, while in the case of chemical weathering its chemical composition changes.

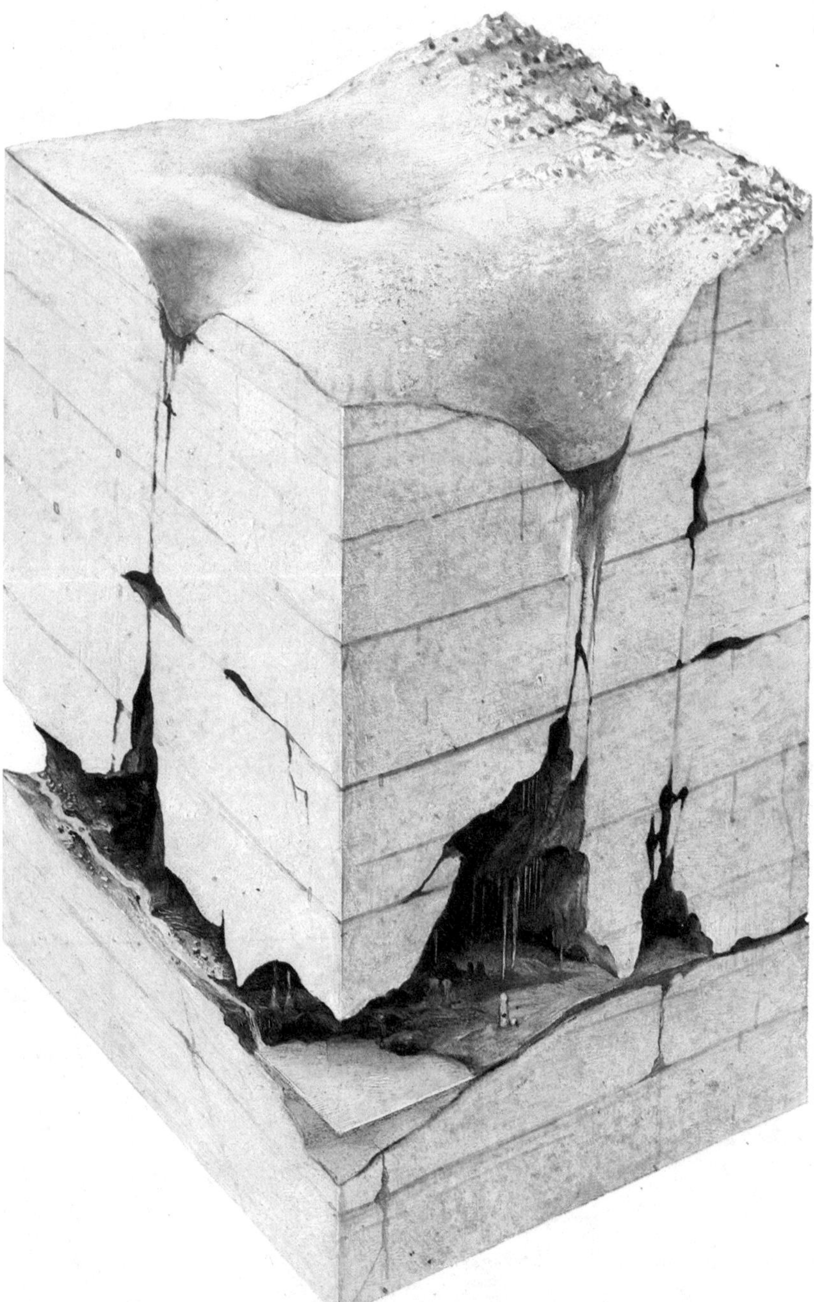

325 A special case of weathering is chemical weathering, quite simply the dissolving of rocks. In nature one finds the dissolution of limestone, which gives rise to karst formations. Rainwater on the surface forms sinks or conical, funnel-shaped formations (sink-holes), the bottoms of which are filled with the remains of dissolved limestone. Underground systems of caverns are formed, usually along cracks; where the water circulates for a long time, a cavern storey is hollowed out.

Let us return to the rock and mineral fragments deposited somewhere in the vicinity of our volcano. Weathering is the main factor involved in the creation of material suitable for the formation of sedimentary rocks, and everything depends on its manner and the course it takes. The source may, of course, be any rock which is already in existence. A comparison of the chemical and mineralogical composition of the newly created sedimentary rocks with that of the volcano would not reveal any major differences, and this is because in the case we described there was merely the rapid mechanical weathering and immediate removal of the fragments into the sedimentation environment.

If there is chemical weathering – where,

for instance, water enriched with carbon dioxide or humic acid arising from the decay of organic particles acts on the primary rock – then the chemical composition of the components changes. It is clear that some rock components, such as feldspar or mica, are more susceptible to chemical weathering while others, such as silica, are resistent to ordinary chemical weathering.

Chemical weathering in itself alters the composition of the original rocks, and these are therefore quite different from the secondary sedimentary rocks. Thus from granite, made of quartz, feldspar and mica, one may quite easily obtain, due to the effects of chemical and mechanical weathering and of sedimentation, sandstone. Feldspar is easily broken down, weathered into clay-like material. But in the process a water solution enriched with feldspar components, especially sodium (which gets into the sea in this way), is produced. In fact most of the sodium in seawater arises from the chemical weathering of sodium feldspar.

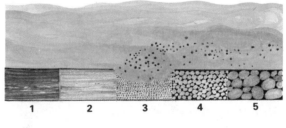

But let us follow the fate of the weathering products of granite still further. The clay components have quite different properties in a water suspension to the particles of quartz. The water carries the clay further, and while the quartz particles sediment out, the clay remains in suspension. Thus there is a sorting of components according to their mechanical properties. These play a key role in the formation of sedimentary rocks. And we should not, of course, forget another aspect – the rate of flow and the amount of water in the sedimentation environment.

327 The particles which are carried by wind or water are not carried in a straight line, but parabolically. They often fall to the bottom, to be carried away again by the current. This type of movement is called saltation.

327

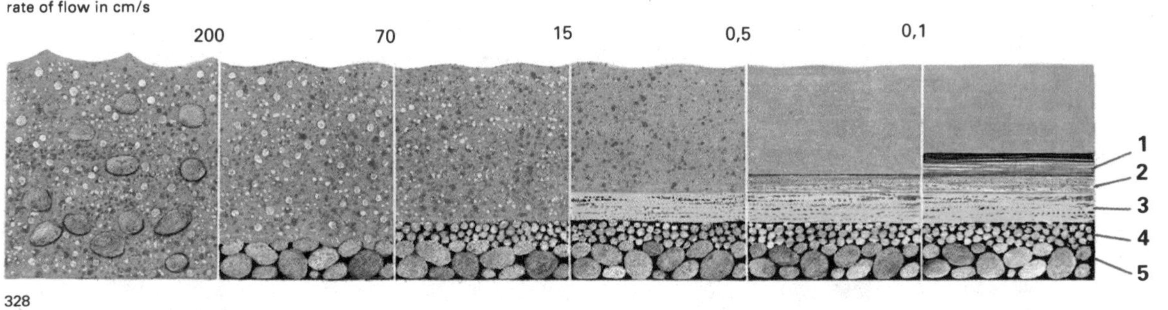

328

rate of flow
20 cm/s

1 2 3 4 5

100 cm/s

329

328, 329 These diagrams help to illustrate sedimentary geology. The first shows the dependence between the rate of flow and the deposition of grains of various sizes. As the rate of flow falls, first gravel **(5)** is deposited, then fine gravel **(4)**, sand **(3)**, dust grains – silt **(2)**, and finally, in stationary water, grains of clay **(1)**. In the second diagram water flows over the various fractions. At a speed of 20 cm (8 in) per second sand is lifted, at a speed of 1 m (3.3 ft) per second dust granules and fine gravel are lifted. Clay and gravel are the last to be carried away.

A mountain stream in full flood can carry off boulders weighing tens or even hundreds of kilograms, while the lower reaches of a great river cannot even shift a small particle, say a millimetre across; it can carry only a fine suspension of clay. In order to form a picture of the rivers of the distant past, the currents of sea water and the sediments of deltas, geologists have tried to simulate the conditions which exist in rivers and at their mouths, and the size of particles which can be carried. This gave rise to a very well-known diagram, which is shown in Figures 328 and 329.

The climatic conditions are further important factors in the weathering of rocks, which takes place quite differently in the mountains, where frost take effect, or in the tropics, where there is constant rainfall and rocks are more easily dissolved and leached. The topography is also important, affecting not only the speed of streams, but also the processes on slopes. Then the medium in which the rock sediments is naturally a factor. Sedimentation takes place differently in mountain torrents and broad river valleys, on the seabed, etc.

Rocks are classified according to various properties: particle size, the degree of wear during transport, the degree of sorting, and the chemical composition of the components.

None of this is undertaken without a purpose. Sedimentary rocks must be classified in such a way as to help geologists to tackle

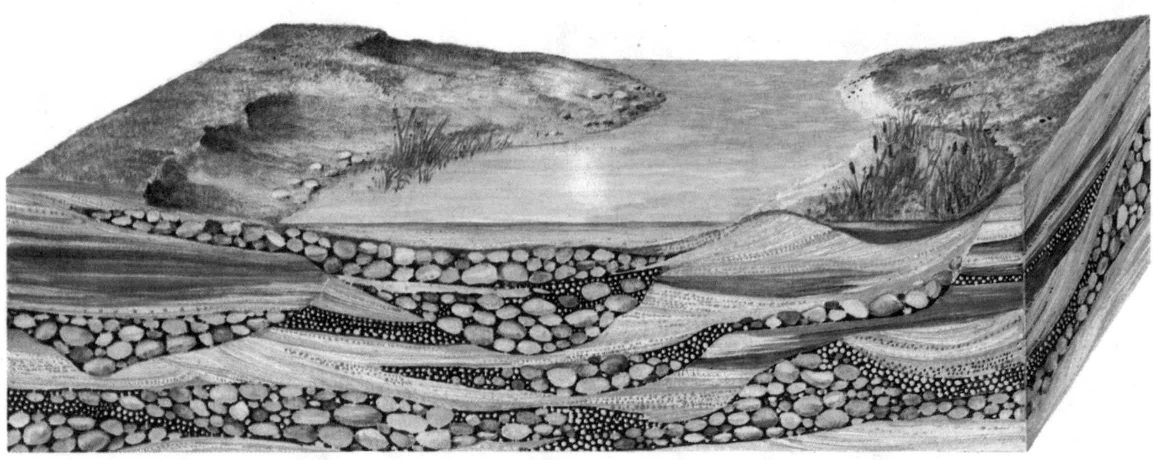

330 A cross-section of the bed of a river illustrates how the force of the river's current changed, and how its bed moved. The main beds are filled with coarse-grained gravel, which shifts only when the river is in flood. The calmer places are covered in sand, and in backwaters clay-like mud is laid down.

330

331

331 The amount of material carried in a year by rivers in the form of mud and particles is incredibly large. The picture shows the amounts carried off by some of the world's largest rivers:
1 – the Mekong, **2** – the Amazon, **3** – the Euphrates and the Tigris, **4** – the Ganges, **5** – the Zambezi, **6** – the Nile, **7** – the Rhône, **8** – the Danube, **9** – the Rhine, **10** – the Indus, **11** – the Volga; the Pyramid of Cheops **(Ch)** is shown for comparison.

practical tasks. We should hardly expect to find the clay sediments used in the manufacture of pottery to occur in the old beds of mountain streams. It is also probable that a coal seam in a small, fresh-water basin will be smaller than one in a larger basin beside the sea. All these are reasons for geologists to classify sedimentary rocks according to the criteria we have mentioned.

The basic division of sedimentary rocks is shown in a very simplified table in Figure 333. You will surely find in your neighbourhood rocks which you are able to classify according to the table and you will then be able to say how they were formed. But pay attention to the texture, which is also included in the table. Criss-cross stratification is typical of the movement of bed or delta slope sediments, while fine alternation of strips indicates seasonal changes in the weather. An unordered mixture of boulders and clay may, occasionally, be the sediment of what was once a glacier.

The way in which a rock formed is only one of the aspects of classification for those who study sedimentary rocks. But if we want to categorize them from other points of view, we must go into the question of where the rock formed, and make a study of the surroundings or the region in which the rock took shape. The sediments of the sea, for instance, may form from material carried from the mainland or from the continental shelf. This is usually sedimented on the continental slope. But similar rocks cannot form in the middle of the ocean, because mainland *(terrigenous)* material is not carried that far. The sediments of the deep ocean have a quite different character. They are called *pelagic,* and usually contain less than 30 per cent terrigenous material (usually the finest clay). In scientific terminology they are called green, red or brown muds. Pelagic sediments often form from the animals living in the deep-sea chasms, far away from land. The rate of rock formation at the mouths of rivers or in areas washed by surf is much faster than in the deep sea, where only 1–10 cm (0.4–4 in) of sediment is formed in a thousand years. By contrast, at the mouth of a river there may be several metres of new sediment every year.

The sites of lake sedimentation have quite different features. Here, for instance, characteristic features may be rapid and huge floods

332 High-mountain glaciers are a very effective means of transport for fragments of rock torn from the bed **(2)** or falling on to the glacier **(3).** At the point where the glacier melts, called its head **(1),** end-moraines are formed **(4, 5).**

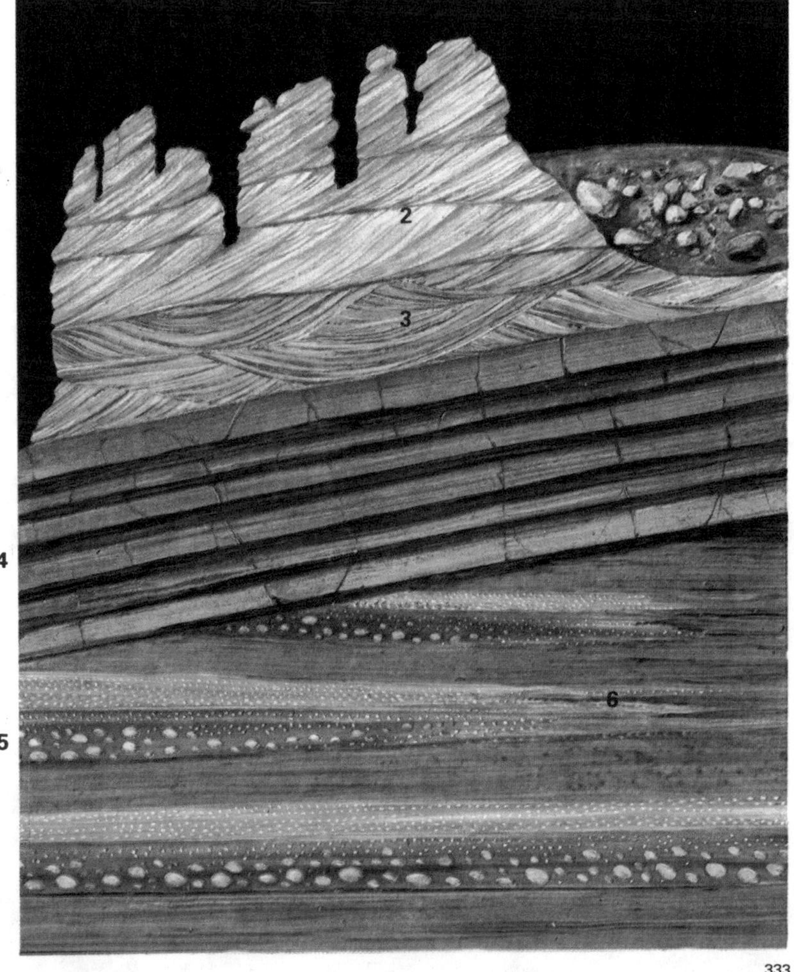

333 Sedimentary rocks are geological 'books' which one could browse through almost indefinitely. So, for instance, the cross-stratification of sediments in the upper part of the cross-section indicates sedimentation in a delta or on some slope. The coarse-grained, unsorted sediments along with fine-grained ones in the small valley at the top indicate that the rock sedimented after the recession of a glacier **(1).** The regular, plate-like deposition suggests calm sedimentation on the sea bottom, while the alternation of coarse-grained and fine-grained sediments is a sign of rythmic **(4)** and cyclic **(5)** sedimentation, caused, for instance, by movements of the crust. But if the strata are inclined towards each other, one speaks of cross or diagonal stratification **(2, 3);** if there is a pause in sedimentation (a hiatus) it is angular discordance. Gradual or facial transition **(6)** indicates a change in sedimentation conditions.

334 A very effective means of transportation of weathered material is the wind. This is illustrated by a picture of the desert, where the material is in constant motion. From stony desert fine particles are carried off to sandy desert, where they are laid down in the form of dunes.

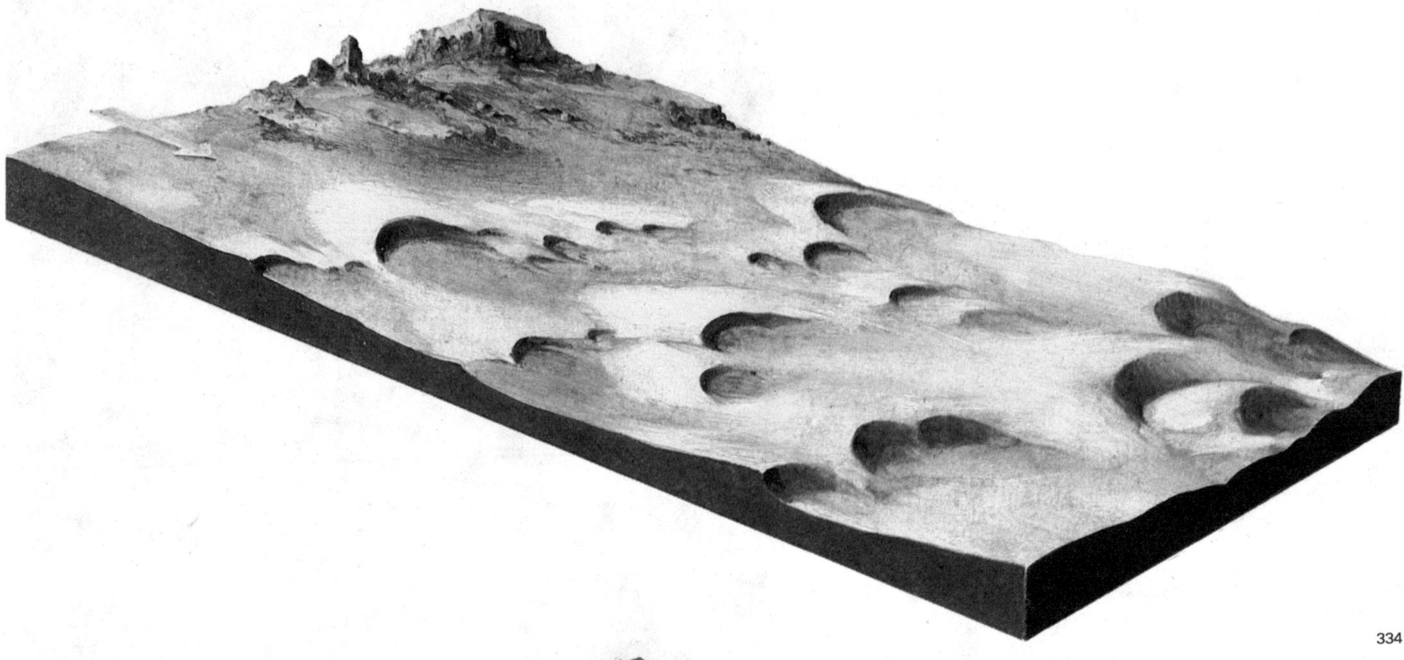

334

335

335 The formation of limestone rocks is one of the natural regulators of carbon dioxide in the air and of calcium in seawater. Coral reefs **(1)** or the shells of small animals are natural moderators of the creation of calcium carbonate. When animals fall **(2)**, however, they dissolve at a certain depth **(3)**. In such areas calcium-rich clay – deep-sea mud – occurs **(4)**.

of material, and the alternation of coarse, fragmentary sediments with fine-grained material. If there are glaciers in the vicinity of such lakes, the sediments have quite unmistakable features. They are disordered, and apart from coarse fragments there are clay-like and loamy components. This is a phenomenon which can scarcely occur in an aquatic environment with water flow, since the movement of the water readily sorts the components: coarse from fine, clay-like from sandy, etc.

Wind transport also forms sediments. Their characteristic feature is precise sorting by size, and even by composition, of individual components. Some sands – coastal dunes and desert sands – are typical of wind sediments. In Europe and North America *loess* soils are the result of wind transport at the time of the ice cover.

In speaking of sedimentary rocks, we have so far confined ourselves to fragments of igneous rocks, and have said something of the types of weathering – mechanical and chemical. All these processes give the sedimentary rocks their own peculiar character. We should bear in mind that biogenic processes play a very important role in the formation of sedimentary rocks. Even the components – fragments – of the future fragmentary clastic sedimentary rock may be of organic origin, such as the shells and bones of animals and the remains of plants. These types of rock sediments are called *organogenic.* In the deep-sea regions, for instance, far from land, where material cannot be transported by rivers, there are only organogenic rocks, the pelagic rocks of which we have spoken. In red deep-sea mud we find almost exclusively organogenic material – the shells of diatoms.

In describing clastic rock sediments we mentioned the binding material which sticks rocks together. It is a material precipitated from the solutions which either circulated through the rocks or were left in them. The very existence of chemical weathering (i.e. the dissolution of mineral components in water or other solutions) shows that an opposite process – their precipitation – exists in nature. If there was no 'degrading' mechanism of this kind, seawater would just get more and more salty. This process, which is not as unusual as one might suppose from its complexity, also gives rise to sedimentary rocks, known as *chemogenic*. This is a large group, and clear examples of it are salt or gypsum deposits. A glance at Figure 335 shows that the solubility of the calcic shells of marine animals depends on the water depth (and thence pressure) and on temperature. We can deduce from such a picture when a calcium carbonate, in this case chemogenic limestone, can and cannot form. If we consider the importance of sedimentary rocks from the point of view of the needs of man, we naturally conclude that they are just as important as the volcanic group. Suffice it to say that all deposits of conventional fuels lie in sedimentary rocks, and that the greatest sources of nuclear power, i.e. radioactive fuels, are found there, too. The same goes for precious minerals such as gold. Building materials like limestone for manufacturing lime and cement, gypsum, clays and sands are all sedimentary rocks. And because they are so important, their study is a daily task of geologists and sedimentologists.

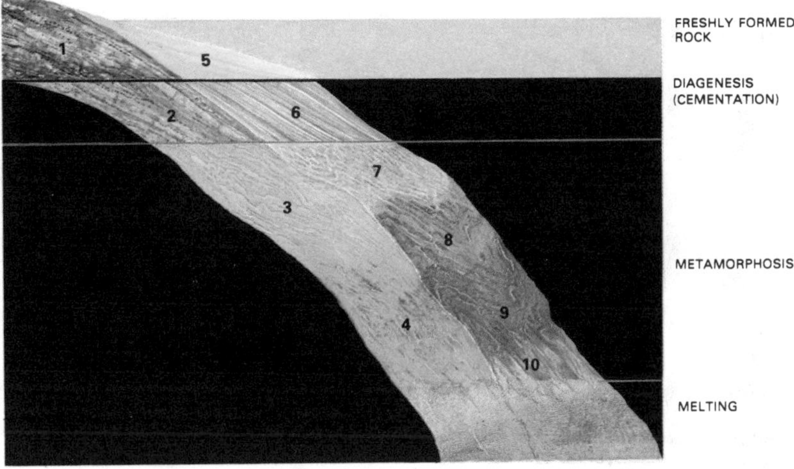

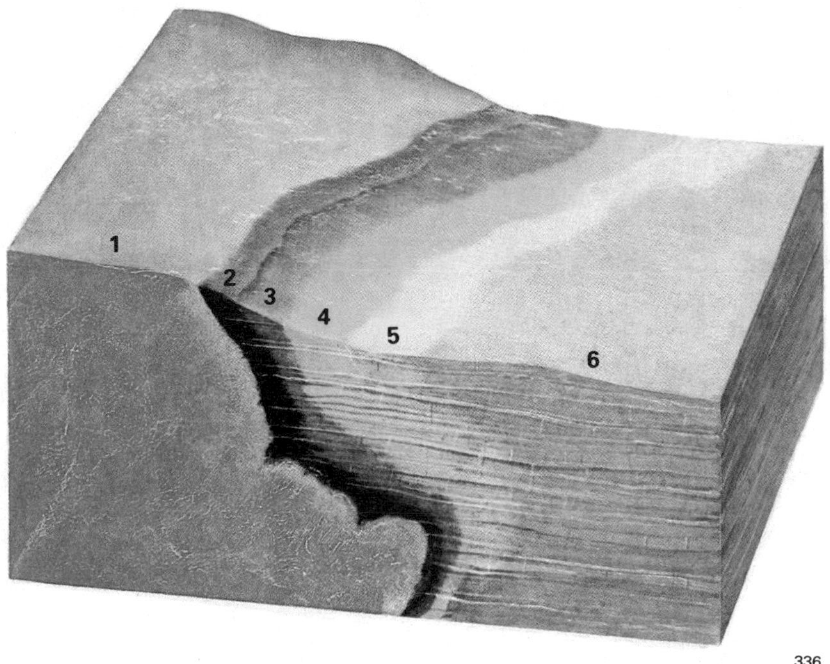

336 The heat of the huge mass of igneous rocks, pluton or batholith, affects their surroundings – in this case a stratification of sedimentary rocks. This is called contact metamorphosis. The picture shows the characteristic rocks and minerals which are created by this process in sedimentary rocks. **1** – granite pluton, **2** – hornfels with pyroxene, **3** – hornfels with sillimanite, **4** – with biotite and andalusite, **5** – with chlorite and muscovite, **6** – unchanged rocks.

337 The transformation of rocks as a result of changes in temperature and pressure is called metamorphosis. The rock adapts its mineral composition to the changed conditions. By the consolidation (diagenesis) of volcanic ash **(1)** tuff is formed **(2)**; slate **(6)** is consolidated clay **(5)**. With increasing temperature and pressure these rocks change into green schist **(3)** and amphibolite **(4)**, or in the case of clay to schist **(7)**, mica schist **(8)** or gneiss **(9)**. At greater depths there is partial or complete melting of the original rocks, and migmatites **(10)** are formed.

Metamorphic rocks

We have already seen how granite may form by the melting of some existing rock, perhaps a sedimentary one. The process calls for high temperatures. Anyone who has ever tried to melt a piece of rock will know, success is unlikely. The melting must take place at a relatively great depth, and the presence of water lowers the melting point at high pressure. We have also described the conditions under which sedimentary rocks form through diagenesis. This occurs in those parts of the Earth's crust which have neither high temperatures nor high pressures. If we take a look at the diagram of the temperatures and pressures in the crust (Figure 338), however, we see that the conditions for the creation of both igneous and sedimentary rocks are exceptional. They are the extremes of the diagram, the very highest and the very lowest values. All the other values of temperature and pressure represent the area of metamorphosis (transformation). Let us add that the majority of the rocks in the Earth's crust are of this type.

Rocks (whether they were originally sedimentary or igneous) which arrived in the Earth's crust and stayed there for any length of geological time are metamorphic rocks.

Every metamorphic rock has its ancestors, which were either igneous or sedimentary rocks. We could classify metamorphic rocks according to this ancestry. It would not be a difficult classification, but an impractical one. The final form and properties of rocks depend on the temperature and pressure which acted on them. We have already mentioned that diamond and graphite, for instance, are chemically one and the same material – quite ordinary carbon. But the arrangement of the building blocks of the two forms is different. In the case of graphite the carbon atoms are arranged in leaves with a hexagonal symmetry, while in diamond the atoms are in more complex, spatially centred cubic matrices. This 'tighter and firmer' arrangement is due to high temperature and pressure. Diamond and graphite are two different structural modifications of carbon, diamond being the high-temperature and high-pressure version. So if we were to sink a lorryload of graphite to a depth of 300 km (190 miles) and leave it for the conditions there to take effect, we should end up with a little pile of diamonds. For the time being we can't do that, but scientists have been trying for some time to imitate these conditions in the laboratory, where the effects on rock of high temperature and pressure are studied. Nature offers geologists finished products. The information on temperature and pressure is contained in these, waiting to be deciphered. And we should need a laboratory to imitate the conditions of metamorphosis; we could then compare natural minerals with those which crystallized at given temperatures and pressures.

In the chapter on terrestrial heat we discussed how temperature increases with increasing depth; for the crust and the upper mantle this increase is considerable. This fact is illustrated by the diagram in Figure 338. We still do not know exactly the temperature of the centre of the Earth, but we suppose it is no higher than 5000–6000° C (9000–10,800° F). The increase in temperature with depth is called the geothermal gradient. From deep bores, even from deep mine-shafts, we know that the temperature

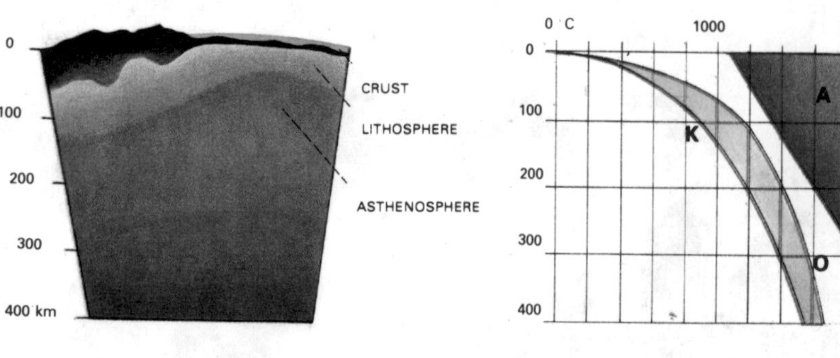

338 With increasing depth the temperature of the Earth also increases. This rise in temperature is slower beneath the continents than in oceanic areas, where the temperature at the same depth is rather higher.
K – thermal gradient below continents,
O – below oceans,
A – zone of completely molten basalt rocks.
The figure also indicates that the conditions for the formation of basalt magma do not exist in the crust. Thus scientists take the view that most basalts originate in the deeper regions of the Earth, in the mantle.

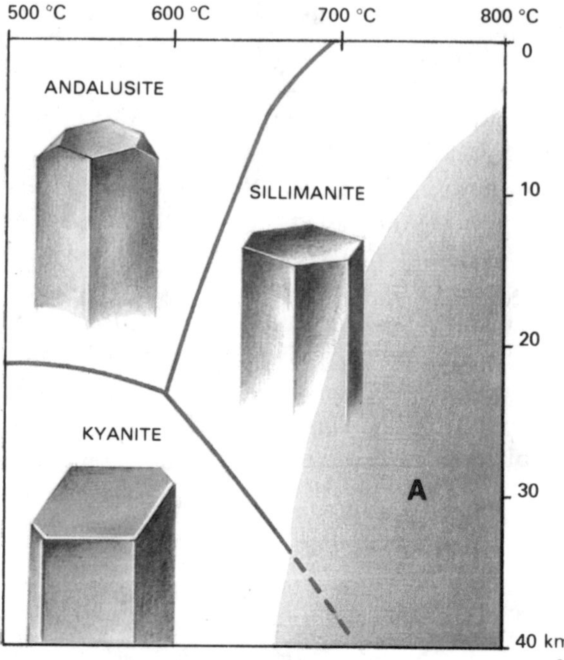

339 Petrologists studying metamorphic rocks are very familiar with the trio of minerals andalusite, sillimanite and kyanite. This is a chemical compound which has three different modifications, depending on temperature and pressure in the place where it crystallized. Andalusite occurs in places with low pressure, sillimanite at high pressure and high temperature, while kyanite is characteristic of rocks which crystallized at high pressure. Region **A** is the area where melting occurs.

rises about 1° C for every 30 m (98 ft) of depth. But the Earth does not work according to a strict pattern. It has been found that the rise in temperature with increasing depth is not uniform throughout the world. Where igneous rocks occur, the increase is steeper than in areas where there are none, or where sedimentary rocks accumulate very fast. One of the most important features studied in metamorphic rocks is the distribution of temperatures and pressures at the point where the rock occurred. This is in fact one of the main goals in studying metamorphic rocks. It is only from a knowledge of temperatures and pressures involved in the crystallization of rocks that the geological history of a region and its evolution can be described. It is then possible to predict the type of minerals which may occur in that area.

Many factors are important for the formation of metamorphic rocks. Perhaps the most important of all is the composition of the original material, after which temperature, pressure, speed, and of course time have their own roles to play. The metamorphic processes take place slowly in the solid state. If the temperature and pressure conditions are changed there is a slow recrystallization, the transformation of one mineral into another which is stable under the given conditions and corresponds to the temperature and pressure which act on it. The speed of the reaction is also affected by the presence of water and other fluids. Some minerals act as geological thermometers and manometers (pressure gauges), since they occur under precisely defined conditions. They are known as index minerals. Among them are garnet, biotite and disthen-kyanite. If we find them, we can be almost sure of the conditions to which the rock was subjected, because they are just the conditions required for these minerals to occur. Figure 339 shows one of the most important groups of index minerals, modifications of Al_2SiO_5. This compound becomes, according to pressure and temperature, either andalusite (high temperature, low pressure), kyanite (low temperature, high pressure) or sillimanite (high pressure and temperature).

8. THE METABOLISM OF THE EARTH

A summary of anatomy

In the previous chapters we have described the basic units from which the Earth is made. We know that atoms are built into molecules, and that molecules or ions form the basis of the crystal lattice of minerals. Minerals then make up rocks, rocks constitute still higher geological units, and these form whole chains of mountains and other landscapes. We know of the existence of the crust, the mantle and the core, and we know that the atmosphere and the hydrosphere are part and parcel of the uppermost layer of the Earth, and that there is a lively exchange of material and energy between the individual parts of the Earth.

We have already mentioned several times that in the distant geological past the continents changed their positions. Indeed, the Earth is a very vital organism, and what we sometimes call the non-living part of nature is full of movement and change, although these take place over long periods of time.

Our explanation so far has been a little bit like an account of the structure of the human body. A doctor must first know all the organs thoroughly, understand their functions, and only then can he describe the working of the body as a whole. The same applies to our mother Earth. We first learned about individual parts of it: geological units, layers, rocks and their properties; we also spoke of the interactions between the various 'organs', but it is only now that we are going to try to put our knowledge together to make a larger whole.

At the beginning of the 20th century a new hypothesis of geological events on Earth was formed, and its details worked out by scientists, especially in central Europe and North America. It was a hypothesis widely held until the 1960s, and it was called the *geosynclinal hypothesis*. It interrelated the processes of sedimentation with those of *magmatism* – the formation of igneous rocks – and the processes of folding and mountain formation, together with weathering. But in the 1960s intensive research was begun on the ocean floor; its magnetic properties were measured, the gravitational effect on land and at sea was investigated, and samples were taken from the seabed. These observations brought so many new findings that the existing conception, until then considered unassailable by geologists, began to fall apart. It was found that the ocean bed is very young, that the rocks there are no more than 200 million years old, and that there are even rocks which can be proved to be in the process of formation to this day. Take a look at the map of South America and Africa. You can see how the two of them 'fit' into each other. All these phenomena can be put together to form a cogent whole, which is what we aim to do in the following chapters.

In addition we shall see that even the formation of such rarities as oil-bearing sediments, limestone rocks with karst caves, the igneous rocks of continental drift and plutonic rocks with ore-bearing solutions, can be found a place in the heating system of the Earth.

We shall see that even the composition of the atmosphere is related to the rocks of the surface, and we shall demonstrate how the smoke and exhalations of volcanoes, though they may seem deadly and unbreathable, were, in the course of geological history, a source of the atmosphere we breathe today.

The organism we are about to study in operation is in dynamic equilibrium. There is an exchange of substances taking place, and though we have no wish to examine a sick Earth, like a doctor we must first consider the age of the patient. Questions of the age and origin of the Earth and of its evolution must be considered as being of foremost importance.

340 The planet Earth is a very lively organism. What we call non-living nature is, in fact, a process full of movement and change, though it is spread over a long time. Before we can understand the Earth we must know its composition, structure and age.

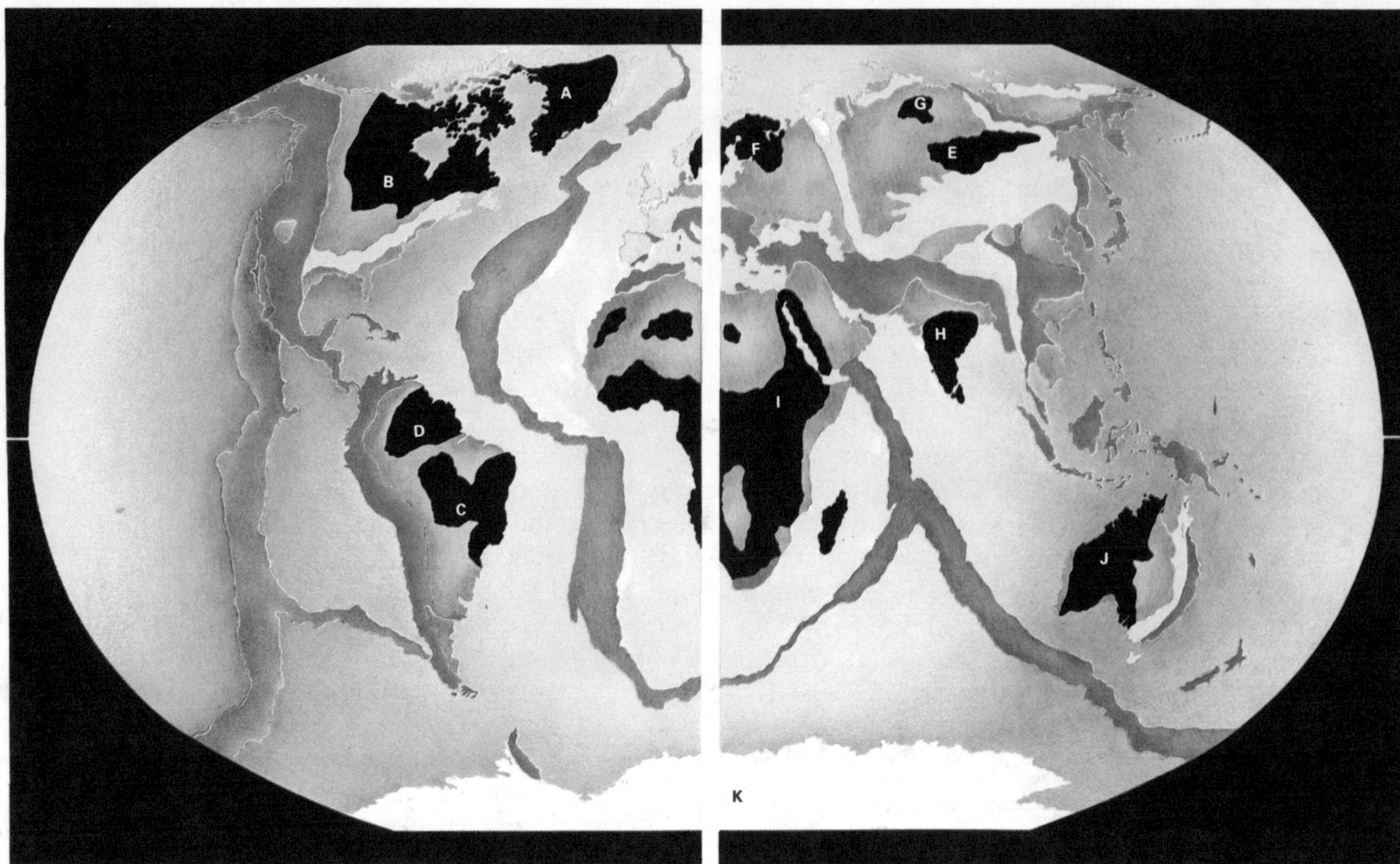

341 Map of the world seen through the eyes of a geologist. The oldest parts of the Earth, called shields, are: **A** – the Greenland, **B** – the Canadian, **C** – the Brazilian, **D** – the Guyanan, **E** – the Siberian, **F** – the Baltic, **G** – the Anabarian, **H** – the Indian, **I** – the African and **J** – the Australian. The areas shown in green are only slightly younger; the areas of Palaeozoic folding are in yellow, and the regions of the last major folding activity, the Alpine, are shown in red, together with current geological activity. The youngest areas of the Earth are those of the mid-oceanic ridges (brown).

The 'creation' of the Earth

To understand the function of an organism, we must know from where it came and how. And that is the next question in our investigation of the metabolic system of the Earth: what the Earth as a whole is made of, and by what mechanism such an enormous and complex body came into being.

It is only in the last few decades that we have been able to answer such questions at all. In fact people only rarely asked them, since they knew that the answer lay beyond the frontiers of the Earth, to which they were securely bound by the chains of gravity.

A study of cosmic material and its composition, for instance of Moon rock together with meteorites, helped to clarify some of the unknowns. Meteorites are definitely the basic building material of the planets of the Solar System, as we know from the little information we do have about the surfaces of Venus and Mars. Even the composition of stars corresponds to what we find in meteorites and in Moon and terrestrial rock. All these bodies have a common denominator, a certain shared source, from which they all originated. It is only natural that we should consider it likely that they all come from a single event in the Universe, which formed the matter of the Sun and its planets. There are clear differences between individual planets. The giant planets Jupiter and Saturn, and the Sun (a star), contain large amounts of hydrogen, helium and other gases. The Earth, the Moon and the other planets and some meteorites have a smaller content of these gases. We even speak of the depletion of the *volatile component* in these bodies: rare gases, water, and some volatile metals such as bismuth or caesium. Nonetheless, the planets of the Solar System and the Sun have a common basis, carbonic chondrites. They contain all the materials present on planets, from water and carbon, through silicates, to metals of the platinum group.

At present we can consider the mechanism of the origin of the planet Earth in much more precise terms than we might have 20 years ago. Here, too, space research has brought answers. Though we do not at

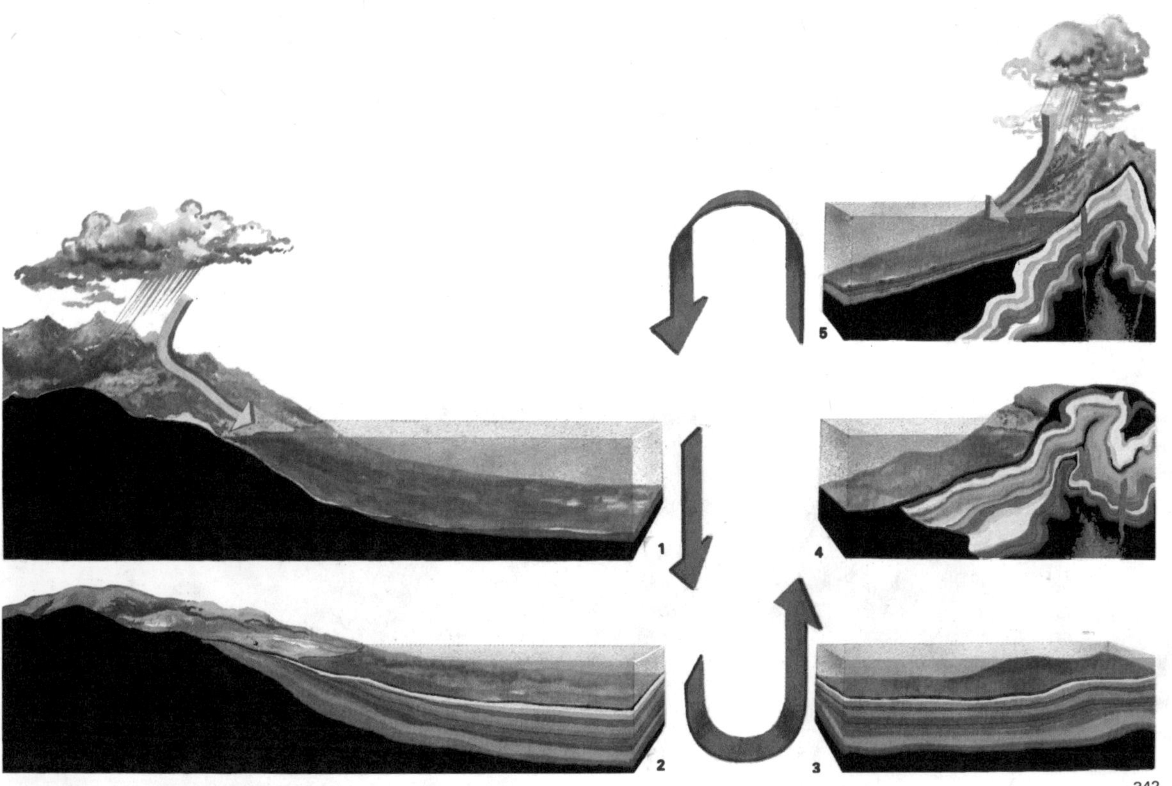

342 The eternal conflict between the external and the internal forces of the Earth:
1 – atmospheric influences break down rocks and water carries them to the sea.
2 – on the seabed a thick layer of sediments accumulates.
3 – under orogenic pressure these unstable strata bend and fold.
4 – mountains start to rise from the sea surface, and magma penetrates into weakened zones.
5 – the raised sediments and igneous rocks are disintegrated by atmospheric influences and taken away to the sea, and the cycle starts all over again.

present see planets being created in our vicinity, a study of the surfaces of minor planets, moons and meteorites has made it possible to formulate views on how a body like the Earth might be formed in space. We have already mentioned that mankind has known the answer to this question since the time of the German philosopher Immanuel Kant and the French mathematician and philosopher Pierre Laplace (second half of the 18th century and first half of the 19th, respectively), who were of the opinion that planets come into being through an agglomeration of the components of a solar nebula. With minor modifications the theory holds to this day. Evidence which Laplace and Kant could not produce comes from a study of meteorites and the surface of planets. We see an agglomeration of the tiniest components and their condensation from the original cosmic gas into solid matter in the meteorites. This is indicated by their structure and chemical composition. The meteoritic bombardment of the planetary surfaces then gives an indication of what the actual *accretion* – sticking together – of the larger bodies and existing small fragments of cosmic matter was like.

In contrast to the older hypotheses, our current ideas can account for some of the processes involved in the formation of a planet. We know the temperatures of the condensation of solids from gases of solar composition, and we know that condensation took place over a broad range of temperatures. We also know that the actual accretion of larger components was relatively rapid. More and more matter of meteoritic composition built up around a small gravitation centre. The bodies which struck this changed their kinetic energy into heat. The planet grew warmer. The actual settlement and compression of the planet also released a large amount of heat, and so it is probable that in the early stages of the evolution of the Earth (and of the other planets) these bodies grew so hot that the original, primitive matter was able to separate out in the form of the molten state of pure metals and sulphides. Since these have a high density, they fell to the centre of the body and formed a metal nucleus to the planet. This released further heat. From this body, still in fact being formed, an 'easily melted' silicate component, probably basalt, was formed, and it was this which gave rise to the first, most primitive upper layer of the Earth. (Such molten material has a low density, less than that of its crystals, and so it would rise to the surface.) But at this stage of the formation of the Earth its surface was very intensively bombarded

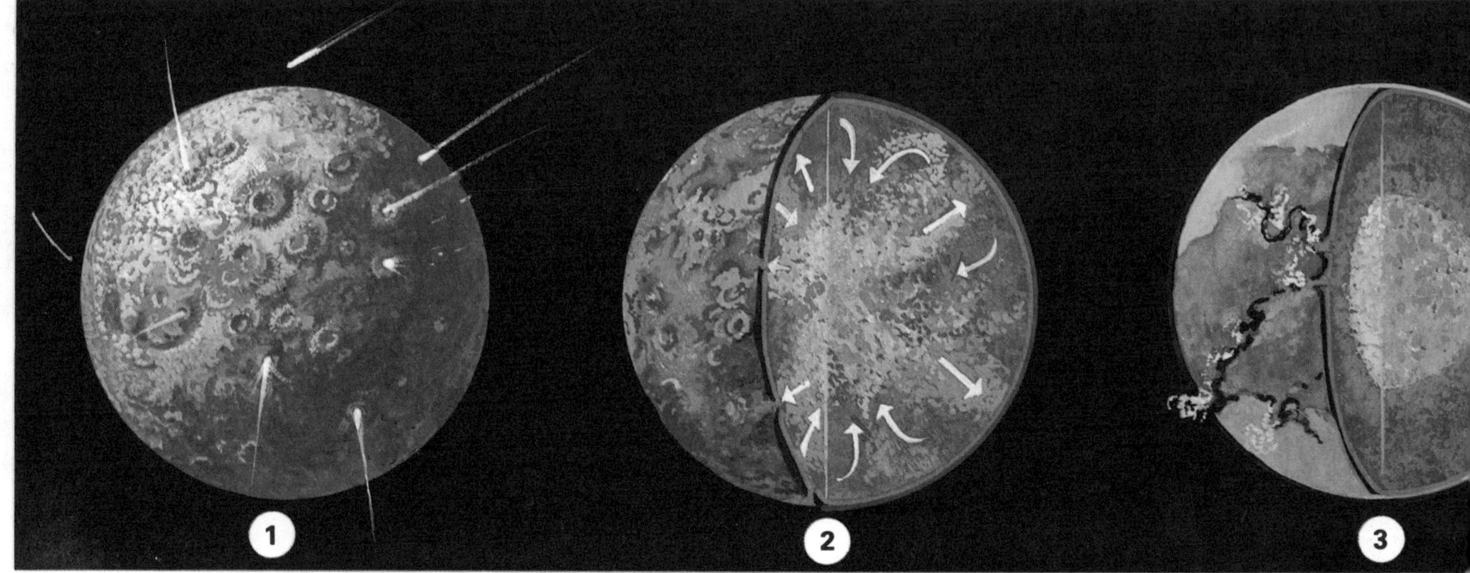

by meteorites. Naturally, with the separation of the easily melted component from the main part of the Earth, i.e. with the separation of iron and molten silicates, a large amount of gas was released. Thus the Earth's first atmosphere was formed. It had a composition very different from that of today, probably similar to that of the gases emitted by a present-day volcano on erupting. This primitive, original Earth was still a most inhospitable place, with a hot surface and an unbreathable atmosphere. But it already had all the features of a planet – a skin-like structure, a core, mantle, crust and atmosphere. There was as yet no hydrosphere, since at that early stage the temperature was still too high. But from the moment the layered structure and the core came into being we can already speak of the true existence of the planet Earth.

A working hypothesis

In any branch of human activity, such as physics, if we wished to find the idea which contributed most to its development, we should be unlikely to alight on one alone. But there are certain milestones: the ideas of Isaac Newton and Albert Einstein, or Henri Becquerel's discovery of radioactivity. In geology the development of the discipline has been much more leisurely, and we can say that there have been no revolutionary discoveries in the field, perhaps because most of the ideas and hypotheses had to be tested over long periods. In physics it has been possible to verify many new observations or hypotheses experimentally soon after their appearance. But even geology has its milestones, especially in methodology. Following the medieval scholasticism the return to direct observation in nature was a great step forwards. From the disputes between the Neptunists, who regarded all rocks as being products of water, and the Plutonists, or Vulcanists, who thought they all originated from fire, the ideas on the processes of the formation of rocks on the surface and within the Earth materialized.

Nevertheless, geology does have one single hypothesis which led to an exceptional flourishing of the science. All the fairly modern ideas in geology which have caused the so-called revolution over the last quarter-century are in fact based on the *actualistic (uniformitarian) principle.* It has been known since the time of Charles Lyell, James Hutton and Mikhail V. Lomonosov, and states that the processes taking place on the surface of the Earth and on the Earth in general at present also took place in the geological past. In other words, it is not necessary to seek an extraordinary explanation for the geological past, nor any peculiar processes; everything can be explained in terms of the present.

The principle applies to volcanic activity, orogenic processes, and also to mechanical, chemical and biological weathering, the transport of material on the Earth's surface through gravitation, watercourses, wind or glaciers, the formation of sediments of rocks on riverbeds, lake bottoms and the ocean floor, the formation of ore deposits, and the effect of lunar or solar gravitation on the tides. The actualistic principle seems to us an almost naive observation, something to be taken for granted. But let us recall that years ago people were unwilling to believe that fossils are really the petrified remains of extinct life, and that the idea of the evolution of organic life from primitive forms to complex ones took a long time to achieve general acceptance.

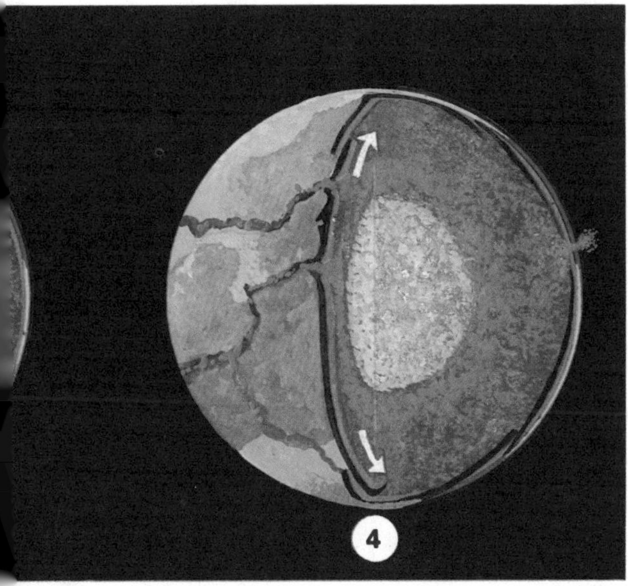

343

343 The early history of the Earth is one of the most important chapters in its life. Following intensive bombardment of its surface with meteorites **(1),** the first great transformation of this planetary body took place **(2).** The metallic core, silicate mantle and later the surface crust formed **(3).** The Earth's metabolism, the shifts of the lithospheric plates and the configuration of the continents as we know them today are of still later date **(4).**

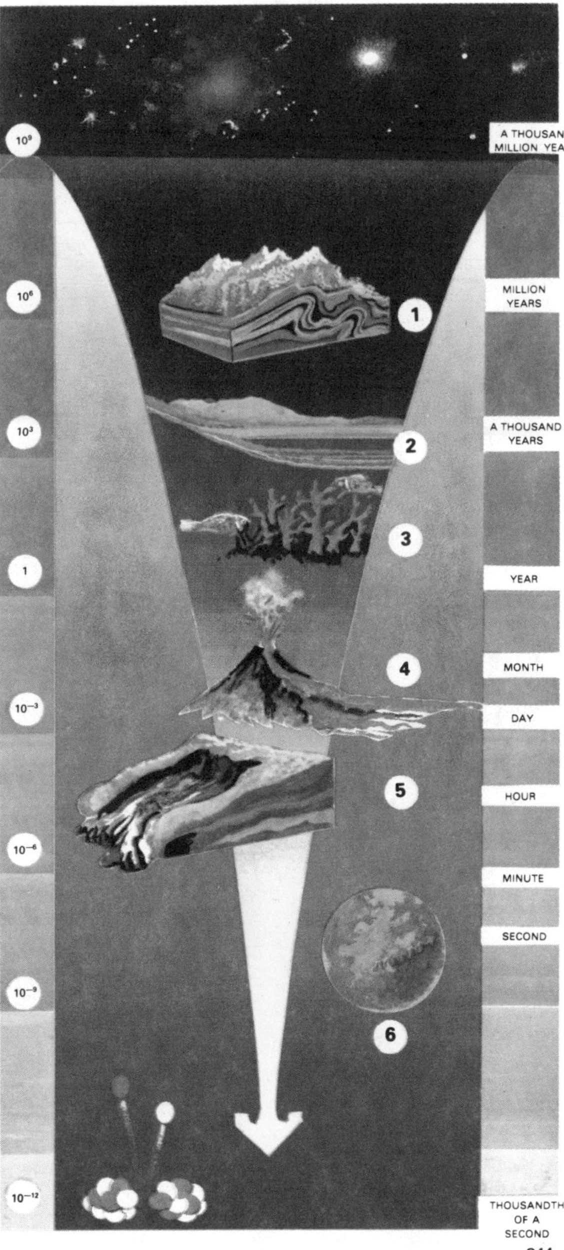

344

344 One of the most important quantities in geology is time. Many geological processes take millions of years. But geologists also study events which take place much more rapidly than, say, the folding of mountains **(1)** or sedimentation **(2).** So, for instance, they can observe the growth of coral reefs, which covers a period of hundreds to thousands of years **(3),** while the eruptions of volcanoes take a matter of days **(4),** landslides last minutes or hours **(5),** and the rate of propagation of seismic waves is measured in a matter of seconds **(6).** (The unit of time on the left is the year.)

But we should not apply the actualistic principle too literally or dogmatically in the study of the history of the Earth. We have seen, for instance, that the early stages of the evolution of the Earth were marked by high temperatures on the surface, that there was no hydrosphere, and that therefore the geological processes which took place must have been different from those we observe today. The same applies to the chemical evolution of the planet. For instance atmospheric oxygen only became a breathable product in the course of a geological era – the Proterozoic – through the breakdown of more complex compounds.

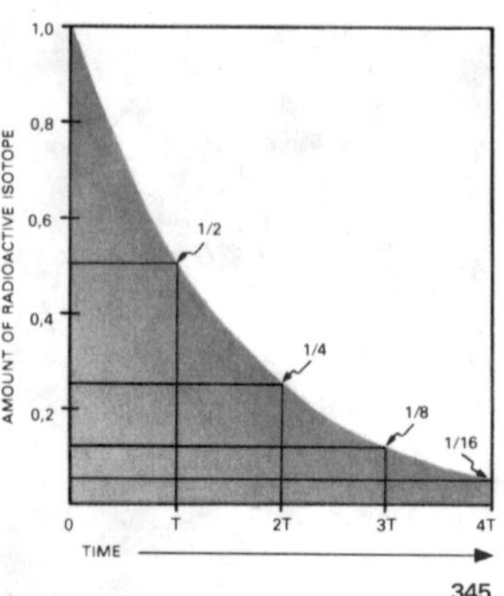

345 The half-life of a radioactive isotope is the time it takes for half of it to decay, or disintegrate. It is the most important unit in the radiometric determination of the age of rocks.

346 The Palaeozoic and the Mesozoic are the last two evolutionary stages of the Earth. In fact oxygen only appeared in the atmosphere in a breathable form some 2000 million years after the planet began its development, and animals with a rigid outer shell only 600 million years ago. Since we know only the more recent geological era reasonably well, the diagrams from this are more detailed.
1 – origin of the Earth,
2 – the oldest known rocks on the Earth,
3 – the Pre-Cambrian, the presumption of the start of life, **4** – oxygen appears in the atmosphere,
5 – the Palaeozoic,
6 – the Mesozoic,
7 – the Caenozoic (the Tertiary),
8 – the Quaternary,
9 – the Cambrian,
10 – the Ordovician,
11 – the Silurian,
12 – the Devonian,
13 – the Carboniferous,
14 – the Permian,
15 – the Triassic,
16 – the Jurassic,
17 – the Cretaceous,
18 – the Caenozoic (the Palaeocene, the Eocene, the Oligocene, the Miocene, the Pliocene),
19 – the Pleistocene,
20 – the Holocene.
(Dates in millions of years.)

The age of the Earth

We can understand the evolution of the Earth only when we know its composition, structure and age.

It is no simple matter to find out precisely how old terrestrial rocks are. If we are to determine their age reliably and exactly, we have to find out the quantities of an original element and of an element (isotope) which arose through radioactive decay. Men only learned to tell the age of rocks after the discovery of radioactivity. All methods of determining *absolute or isotopic age* are based on that principle.

It may be that someone will ask why people should need to know the age of rocks, or of individual geological formations, and the age of the Earth, the Moon, or a meteorite. What is the point of it all? There is a point, of course. Look at Figure 346. Most petroleum originated in the Tertiary era, and is 15 million years old, and most black coal is from the Carboniferous or Permian eras, with an average age of 300 million years. Similarly in the case of iron ores – the most extensive deposits are the Pre-Cambrian ones. So there is clearly a good practical reason for geologists and geochemists to know the age of rocks.

Establishing the age of the Earth is one of the most dramatic chapters in the history of the Earth sciences. At the time when men became aware of the need to know it, most scholars were of the opinion that the planet has been created by a supernatural force. There were even attempts to calculate its age from biblical history. But to those who were the least bit observant, it was clear that all geological features could not have been formed or 'created' over such a short period. Scientists gradually inclined towards the view that the geological forces which are forming the Earth today were also at work in the past, and that there is nothing supernatural about geological events. They also grasped the fact that the complex and slow geological processes could not have lasted a mere couple of millenia. With increasing knowledge, the age of the Earth 'grew'. While in the 17th century it was thought to be 4000 years old, half-way through the 18th century its age was estimated at tens of thousands of years. And today? It is thought (and there is much evidence to support the view) that our planet is 4600 million years old!

If we speak of the Earth's age, we mean its existence as a separate body, divided from the remaining material in the Solar System. Since then it has developed independently. So all rocks on the Earth must be 'younger' than 4600 million years. And they really are. Geologists are constantly looking for the oldest rocks, but there is a long gap in the geological calendar between the beginning of the Earth and the oldest rocks found on its surface, which are only 3600–3700 million years old, leaving 1000 million years without any geological record. The old igneous rocks have much to say about volcanic activity and its character in the past, while sedimentary rocks indicate the conditions which previously existed on the seabed and the beds of rivers and lakes. Rocks as old as 3600 million years are, however, quite exceptional on the Earth's surface. They have been found in Greenland and in South Africa. But most of the rocks on the Earth are less than 3000 million years old.

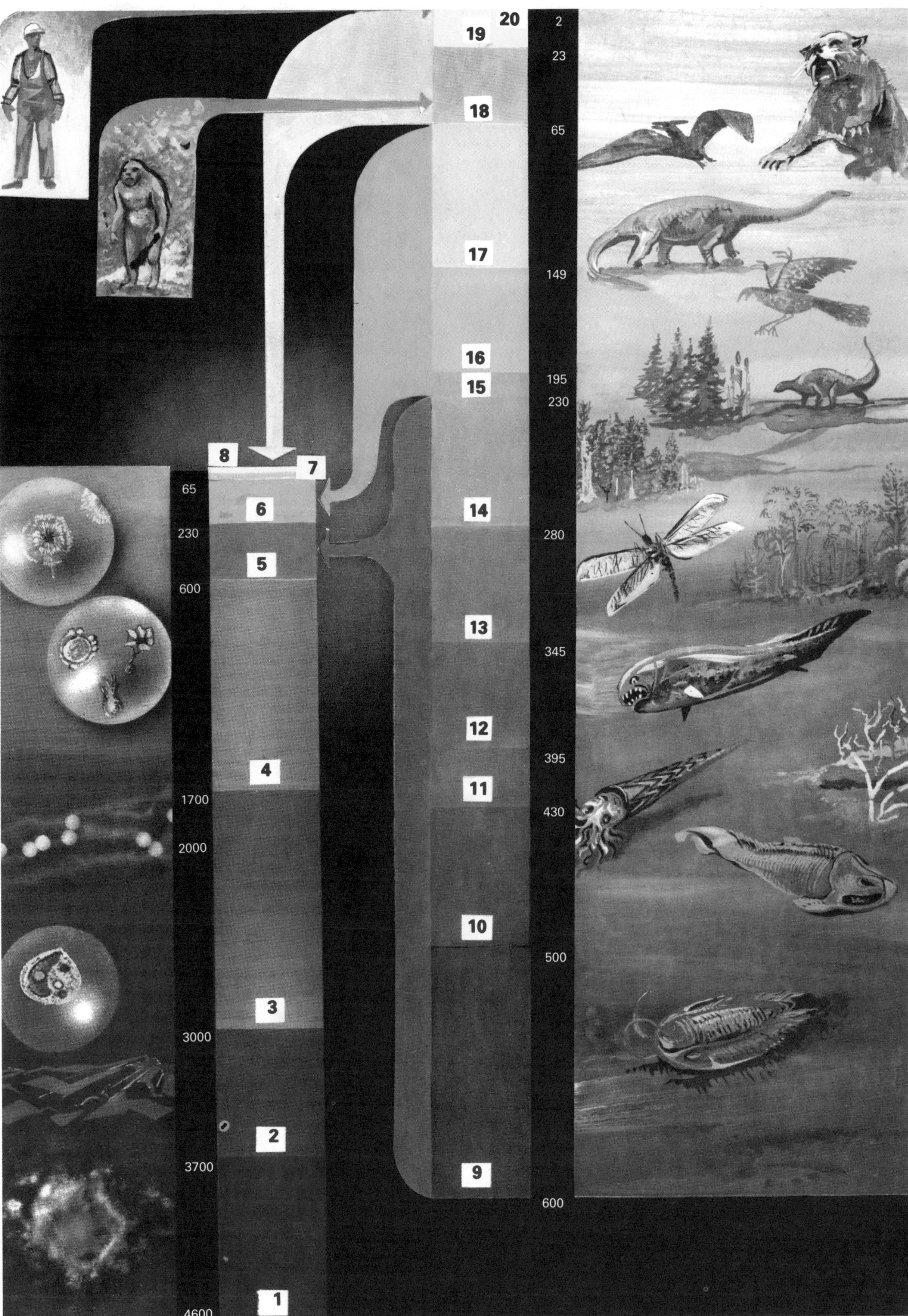

20
19
2
23
18
65
17
149
16
15
195
230
8
7
65
6
14
230
280
5
600
13
345
12
395
4
11
1700
430
2000
10
500
3
3000
2
3700
9
600
1
4600

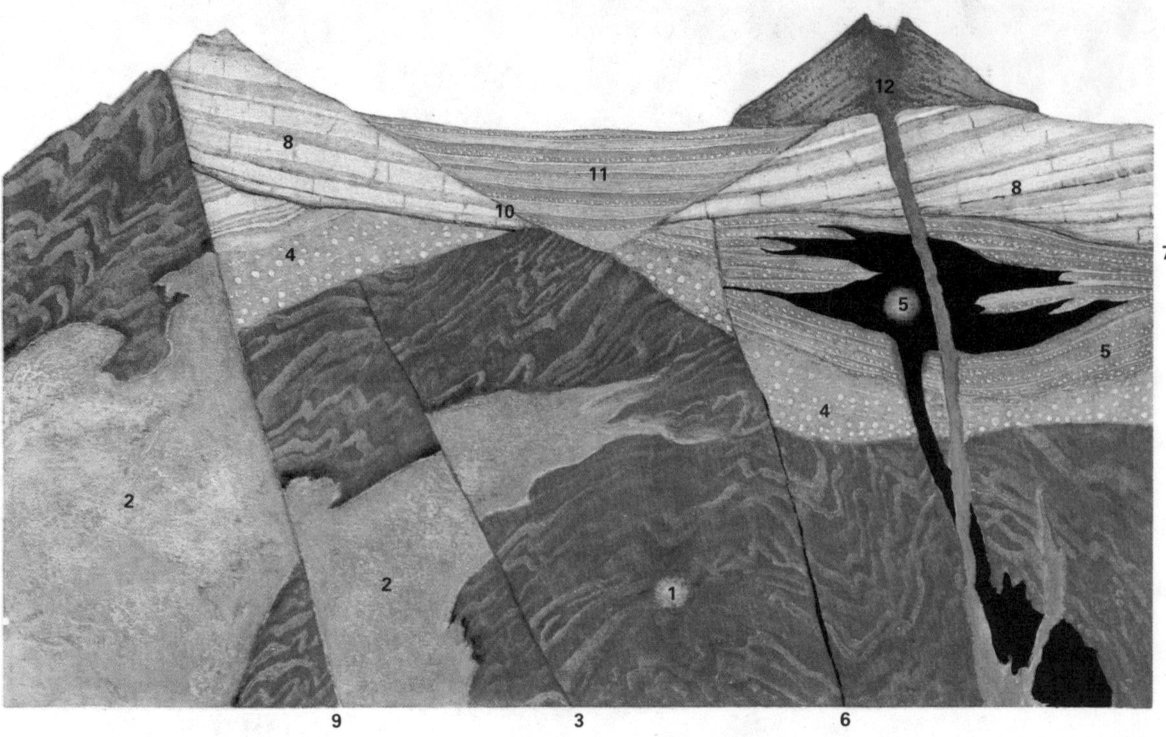

347

347 The order of various geological events is shown in this geological section by numbers. The oldest rocks – the metamorphic **(1)** – are cut into by granite intrusions **(2),** which are dislocated by tectonic movements **(3),** followed by sedimentation **(4),** accompanied by the extrusion of igneous rocks and intrusions of the laccolith type **(5).** There is further tectonic disturbance **(6),** long-term erosion **(7)** and sedimentation **(8).** At the end of the whole sequence of geological events **(9, 10, 11),** there is volcanic activity **(12).**

348 The rate of expansion of the ocean floor is estimated from the distance of rocks whose age is precisely determined from the mid-oceanic ridge. The black lines show the rate of expansion in the west Pacific **(A),** the south Atlantic **(B)** and the north Atlantic **(C);** the yellow bands show the periods of reverse magnetization, which were called after their discoverers (Matuyama, Gilbert).

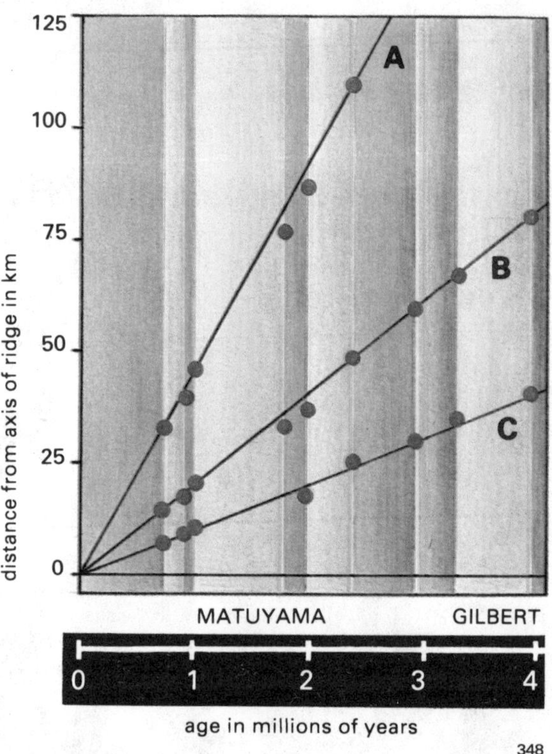

348

Geologists are intrigued to find that the oldest rocks, say around 3000 million years old, are similar to the youngest, those forming today. If it were not for the methods used to determine their age, we should scarcely be able to tell one from the other. This suggests that the same geological events were involved in their formation. Take igneous rocks, for instance. Their remains can be found in

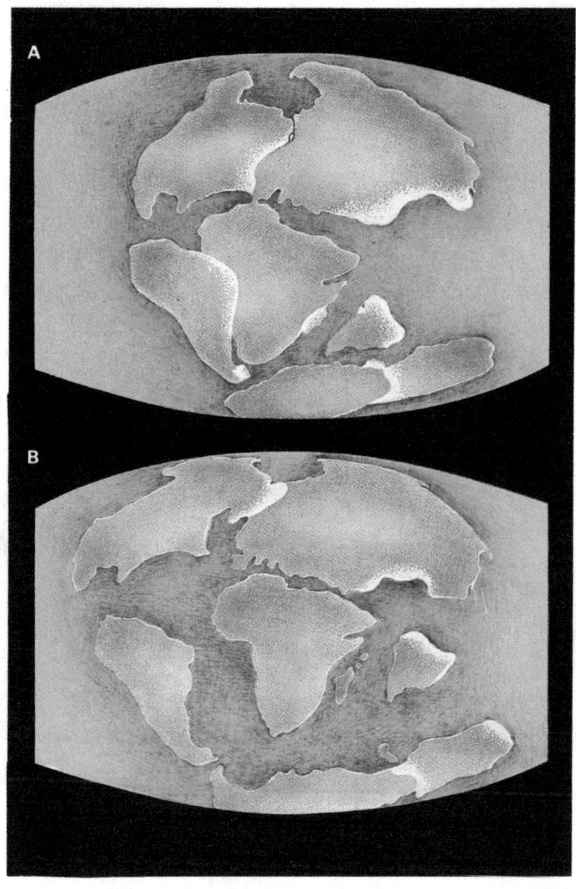

349

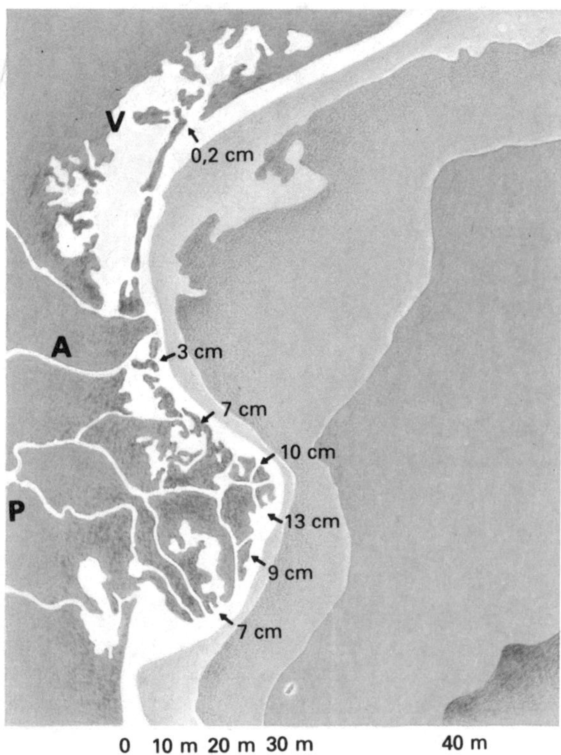

350

349 The configuration of the continents has changed in the course of geological eras, and will continue to do so in the future. Today's configuration is the result of the last million years of activity.

350 The Po delta is an example of formative but also destructive geological activity. A large amount of material brought by the rivers (**P** – Po, **A** – Adige) forms numerous small islands, and the landmass here seems to 'grow'. But the huge mass of these sediments causes a large part of the delta, including the region of the famous city of Venice (**V**), to sink gradually into the sea.

Canada, Siberia and elsewhere. Their composition is just like the igneous rocks of today. So it is probable that volcanoes such as those we now know in Kamchatka, Japan, the Philippines, New Zealand and Iceland, which have been there for 5 million years at the most, existed many hundreds of millions of years ago where Canada, Siberia and Australia are today. Even the deltas of rivers such as the Nile or the Amazon existed in the geological past, long before Africa or South America were independent continents.

Movement throughout the world

A map of the world showing the sites of earthquakes over, say, the last 20 years, would show the Earth divided differently from the geographical or political maps we are used to.

There are huge areas of land and ocean without so much as a trace of an earthquake, while there are other areas, or rather long, narrow strips, covering a tiny proportion of the globe, where the majority of earthquakes are concentrated. The vertical distribution of earthquakes shows that our horizontal projection of them is not truly accurate, but that earthquakes really occur only in limited zones, along some sort of planes (Figure 351). Here we must ignore the fact that the effects of earthquakes can be felt over long distances, since we are interested in where they originate. The most important feature, and for the layman the most surprising, is that most earthquakes are centred on the ocean floor. Every ocean, the Atlantic, the Pacific or the Indian, has an earthquake zone running down its centre. This zone corresponds to the morphological features called mid-oceanic ridges. These have enhanced thermal activity, and it has been found that in the very places where earthquakes originate there are submarine volcanoes. The interior of the planet is pouring hot basalt lava on to the ocean floor.

Where the Pacific Ocean meets the land there are significant earthquake zones, accompanied by volcanic activity, and in some places there is an increased heat flow.

Geophysicists who study the magnetic properties of the rocks of the ocean floor have shown that igneous rocks form extensive linear strips of evenly magnetized material, symmetrical to the mid-oceanic ridge. This means that these strips are of the

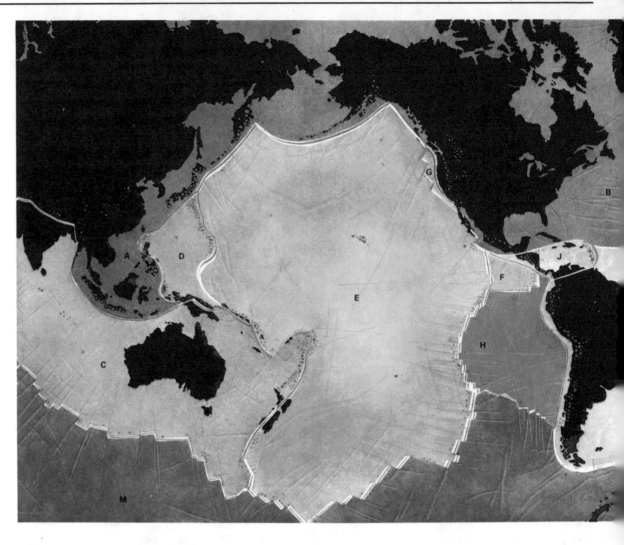

351 Modern geologists divide the world into a system of lithospheric plates, whose interfaces are the major regions of geological activity. The map illustrates one variant of the 'plate division' of the planet. **1** – epicentres of earthquakes, **2** – active volcanoes, **3** – Alpine-Himalayan collision zone, **4** – subduction zones, **5** – mid-oceanic ridges. **A** – Eurasian plate, **B** – North American plate, **C** – Indo-Australian plate, **D** – Philippine plate, **E** – Pacific plate, **F** – Coco plate, **G** – Gorda plate, **H** – Nazca plate, **J** – Caribbean plate, **K** – South American plate, **L** – African plate, **M** – Antarctic plate.

same age and were poured out from the mid-oceanic ridge. The mid-oceanic ridges are thus important features in the structure of the Earth's crust. The identical position of areas of increased volcanic activity, thermal activity and earthquake activity led geologists to divide the world into huge units called *plates.* The earthquake zones are in fact the borders between these plates. According to the ideas formulated in the 1960s and 1970s, these plates move in relation to each other – one might say they float.

The hypothesis of continental drift is very old, and was voiced by several scientists at the end of the 19th and the start of the 20th centuries. The best-known version of the theory was that of the German scientist Alfred Wegener, dealt with above, which stated that pieces of continental crust, the continents *(sial),* floated on the plastic matter of the *sima.*

Today's theory is rather different. Plates are not defined in the chemical sense of the word, as was the case with Wegener's lighter sial (made of light silicate) and heavier sima (made of heavier silicate). They are defined mainly according to their mechanical, elastic properties. The plates are rigid and firm, and are called lithospheric; and they move on a plastic layer which is called the *asthenosphere.* The lithospheric plates may carry continental crust or oceanic crust. While the thickness of the crust is small, the Moho is at a depth of 35 km (22 miles), so that the plates are made up of both crust and mantle.

Thus the points of contact of the plates are the most important geological places on the Earth's surface. Plate margins can be of two sorts: divergent and convergent. Along the mid-oceanic ridges, as can be seen in Figure 351, there is expansion, an opening up and separation of plates, which is known as a divergent margin. In these places a new ocean floor is being formed, and new pieces

351

of the lithospheric plate are coming into existence, which shift symmetrically outwards from the centre. If the Earth is not to increase in size, this newly formed matter must go somewhere. The places it goes to are called convergent margins. At this type of plate margin two lithospheric plates meet in such a way that either one overlaps the other, or they collide. Centres of earthquake activity can also be seen clearly on the map at points where plates collide, and there is an increased thermal flux, with the occurrence of volcanoes. In somė places, where the two plates collide and there is no overlap, there are orogenic regions, such as the Alpine-Himalayan line. Apart from these zones marking the edge of the plates, an important role is played in the plate tectonic theory by *transform faults.* There are also important shifts along these, and their significance in evening out the movement of lithospheric plates can be seen in Figure 351.

352

352 After the end of the ice age, the Earth's crust in the Scandinavian region was relieved of the weight of the ice sheet which had lain on it. And since the crust and upper mantle behave like a highly viscous fluid, the crust is domed there to this day. The bold figures show the number of centimetres per 100 years by which parts of Scandinavia are rising. The figures on the contour lines show the total increase in the elevation of the ground since the ice-cap melted (in metres).

This rather dry account of the movement of lithospheric plates in fact conceals almost the whole mystery of modern geology – *plate tectonics.* Most of the other geological phenomena can be 'related' to the plate theory. We shall therefore take another look at it from a slightly different, let us say more practical, point of view, that of the recycling of matter on Earth.

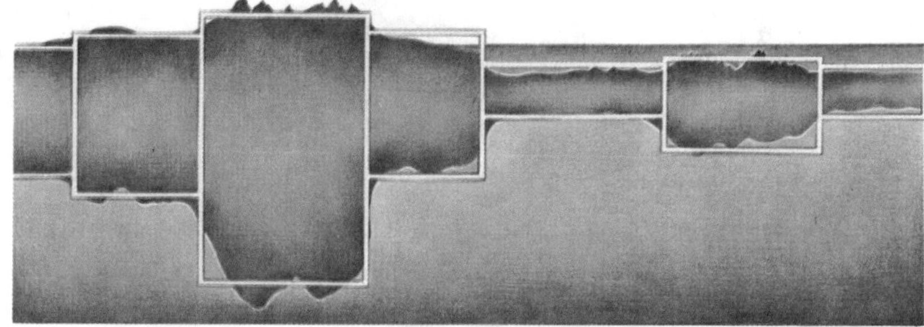

353

353 High mountains tend to have deep roots. The Earth's crust is thus expecially thick under mountain chains. The illustration shows the isostatic model of the British mathematician and astronomer Sir George Biddėll Airy.

The mid-oceanic ridges are the plates where new areas of lithosphere are being formed; a new oceanic crust. Matter is rising from the depths of the upper mantle and differentiating; molten basalt is formed, which flows out on to the seabed and forms the residual phase of ultrabasic rock which makes up the lower part of the structure of the ocean floor. The rocks of the ocean floor are subject to metamorphosis, and react with seawater, and in some cases there is a leaching of metallic components by the solutions which circulate on the bottom of the ocean, and their precipitation elsewhere. This may give rise to deposits of base metals: copper, zinc and lead.

In the places where two lithospheric plates meet there is a collision between them, or they overlap. Important geological processes occur in the course of these events. The rocks of overlapping plates, for instance, fall into an area of increased temperature and pressure, where they are subject to metamorphosis. The water content of the rocks decreases, escaping into the superstratum, where it may cause transformations which may even lead to local melting. Together with the molten matter, water reaches the surface, as well as such components as hydrogen sulphide, carbon dioxide, etc. Different rocks form by melting here compared with the mid-oceanic ridges. Volcanoes appear on the surface. At the points of contact of the plates structures called island arcs are formed. A present-day example is the 'ring of fire'

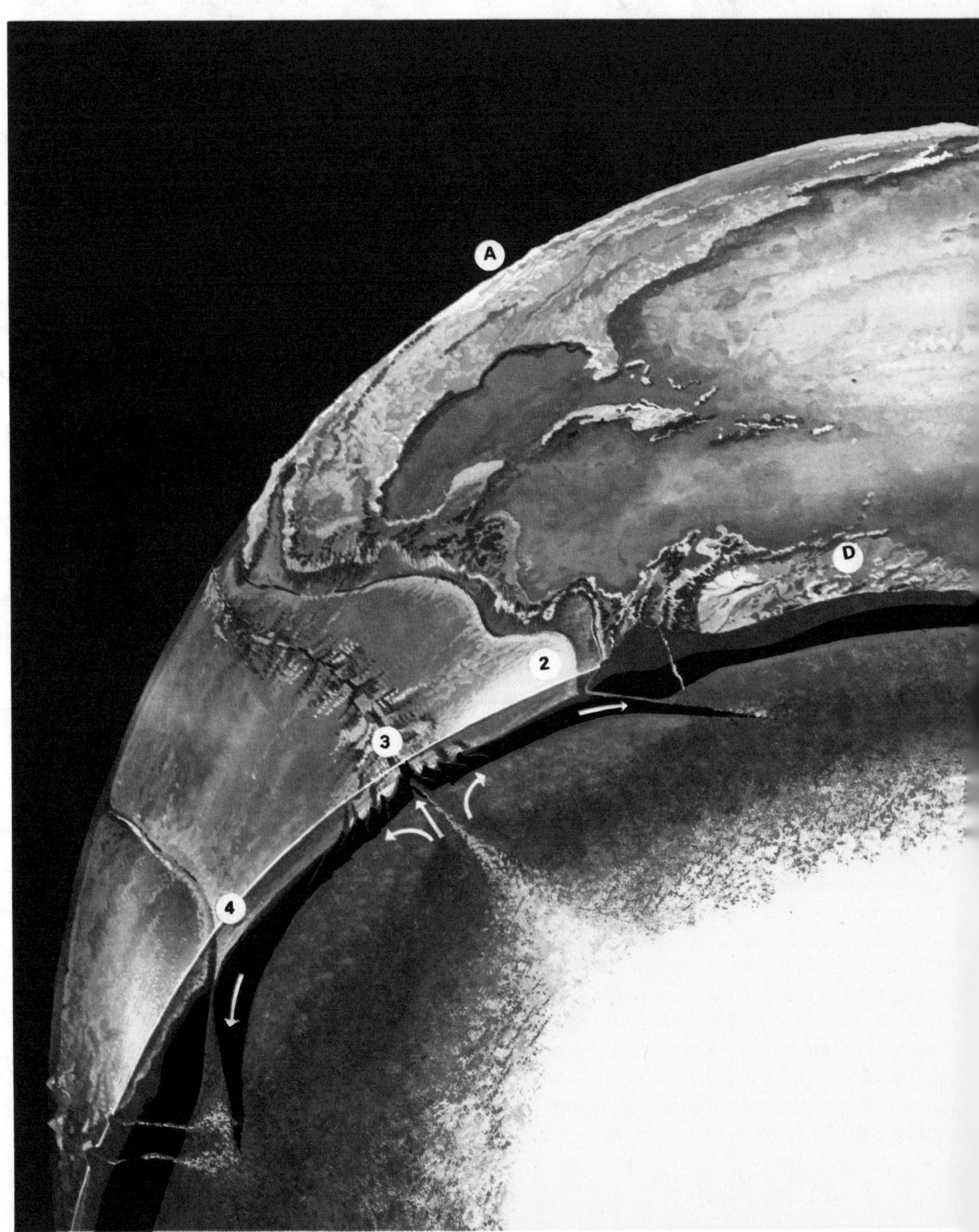

around the Pacific Ocean. Another example is where a continental plate meets a simple oceanic plate, such as the border between the oceanic plate and the continental plate of South America; here volcaic phenomena occur.

Such events are naturally accompanied in the depths by rocks which never reached the surface in the form of lava, but solidified in the form of plutonic rocks beneath it. In the course of further geological processes, such as folding caused by the collision of two plates, and subsequent erosion, these rocks may come to the surface.

The collision and separation of plates offer in the peripheral seas a mechanism for the appearance of oceanic basins or smaller peripheral sedimentation basins. Where continent borders on ocean, but there is no border between plates and thus no overlap or collision (such as the coast of the Atlantic Ocean), a different, calm type of sedimentary

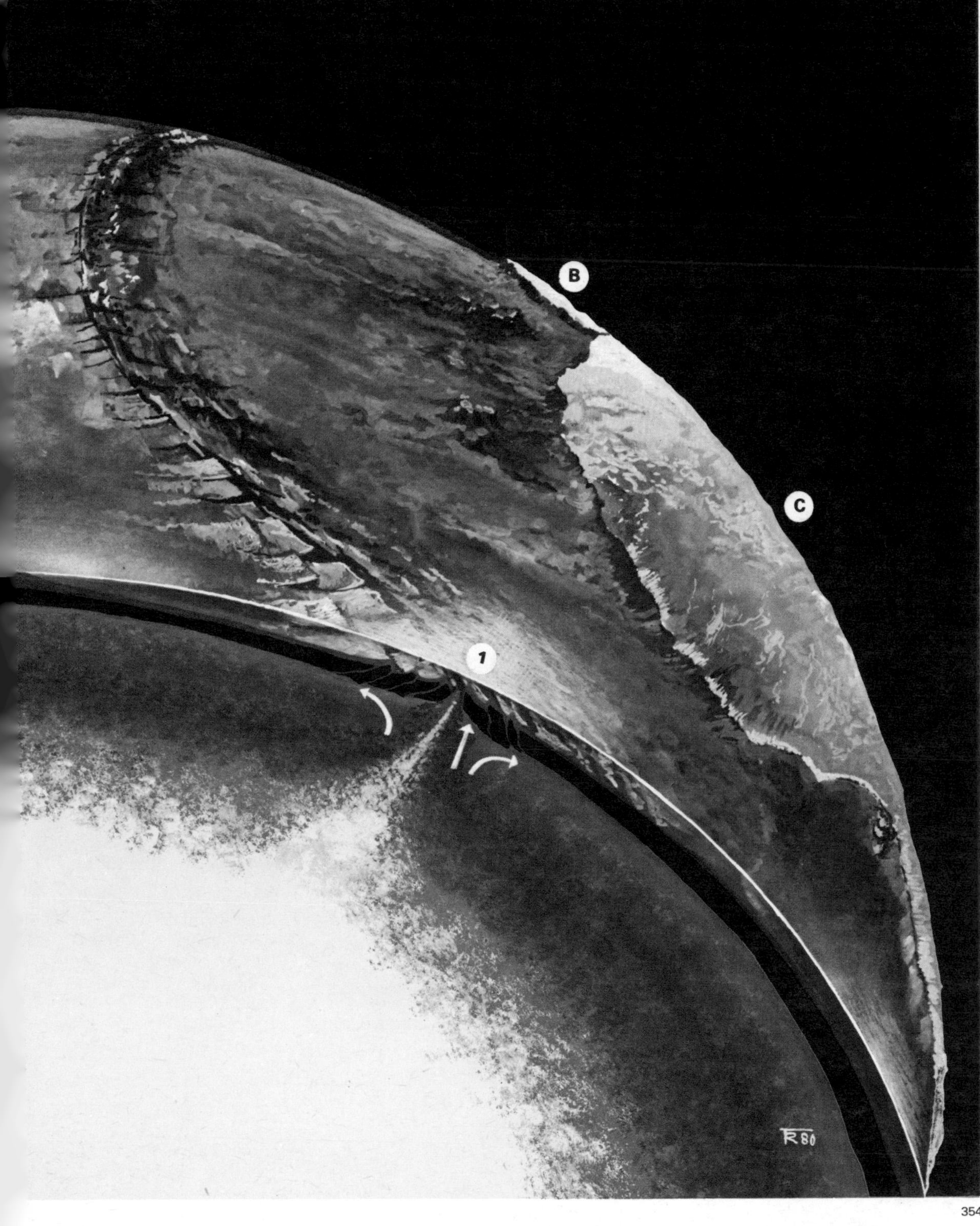

354

354 The metabolic system of the Earth is very complex. The mid-oceanic ridges are points where the material of the upper mantle comes to the surface. **1** – the mid-Atlantic ridge, **3** – the eastern Pacific ridge. At the continental margins or volcanic island arcs (Peruvian-Chilean trench – **2,** Tonga trench – **4**) there is a descent of material into the Earth's mantle. **A** – North America, **B** – Europe, **C** – Africa, **D** – South America.

basin occurs, and conditions suitable for the accumulation of petroleum deposits are created.

The plate theory offers geologists many solutions. Because of its simplicity, and because many of the predicted phenomena have since been discovered, it has offered many advantages over previous theories. In fact, for many geologists it has become a working hypothesis for the search for mineral resources. For those who study the atmosphere and the hydrosphere, it explains the mechanism of the formation of the gaseous and liquid layers of the Earth, known as its degasification.

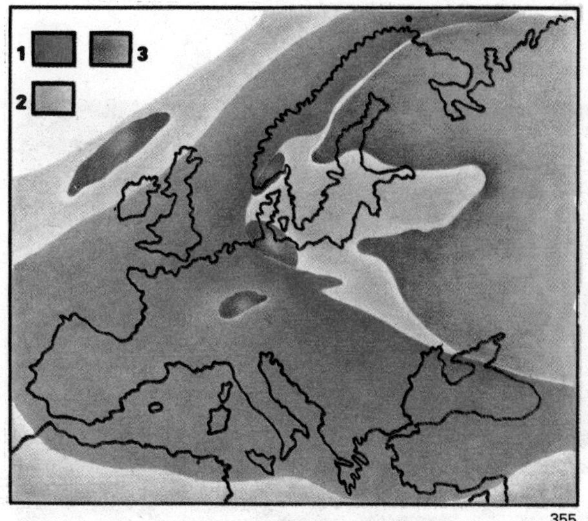

355 The continent of Europe at the time of the Palaeozoic era was a quite different shape from today. **1** – shallow marginal sea, **2** – deep sea, **3** – landmass.

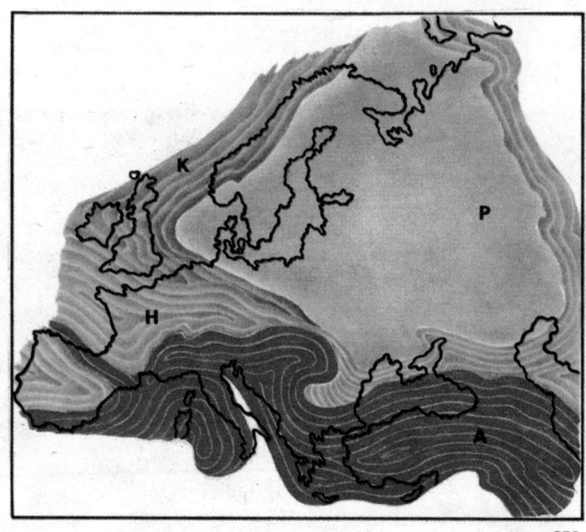

356 The geological structure of Europe illustrates its long history. The oldest part is the Baltic and Ukrainian shield **(P).** The Caledonian fold at the start of the Palaeozoic era formed the first skin of the landmass **(K).** The Hercynian fold in the middle of the Palaeozoic era formed a further skin **(H).** The youngest part of Europe originated at the time of the Alpine fold in the Tertiary era **(A).**

The mechanism of folding and the formation of huge mountain chains, the process which is considered one of the most important in geology, was also difficult to explain in the past. In such massifs as the Alps, the Dinaric Mountains, the Caucasus and the Himalayas, it was not easy to explain the phenomena by the classical theories. Here there is the folding of huge complexes of rocks, accompanied by the displacement of whole geological units by many kilometres. Using the theory of plate tectonics, in the case where two plates carrying continents collide, one can find a certain logic in most of the processes of this type, and today most geologists are agreed that the Himalayas formed as a result of the collision of the plate of the Indian peninsula with that of Eurasia.

The Earth and its neighbours

Our look at the metabolism of the Earth, where we have examined the relationship between its individual parts and the movement and exchange of materials and energy, can now continue. The Earth is part of a larger whole, the planetary system of our star, the Sun. The exchange of energy between the Earth and the Sun is very significant; were it not for the energy of the Sun, many geological processes would cease, and others would exist in another form. The amount of energy the Earth obtains from the Sun is more than 6000 times the energy it produces itself. But the Earth also communicates with the Solar System in other ways. Its atmosphere, its top layer, allows hydrogen molecules and ions to escape, since the Earth's gravity is insufficient to keep them within its gravitational field. On the other hand, meteorites, cosmic dust and energy-charged particles of cosmic radiation fall on the Earth from space. It is therefore clear that we have here a further exchange of materials, a sharing of heat and matter.

The mechanical effect of the bodies of the Solar System also has an impact on the Earth. Here we are not referring to some indefinable and indefinite influence of heavenly bodies, but the measurable movements of the Earth, the ebb and flow of the tides, and the movements of the top layers of the crust caused by the other members of the Solar System.

The Moon, the Earth's companion, and the Sun, our star, have a very definite impact on the movement of all matter on Earth. This movement is governed by the generally applicable laws of gravitation. It is something which is familiar to the inhabitants of coastal regions, and all those who visit the seaside; it is the ebb and flow of the tides. Even those ancient seafarers the Phoenicians made use of the tides to get to and from their protected harbours. In recent times men have been seeking ways of harnessing the considerable force which moves the oceans, of making electricity out of the energy of the ebb and flow. For fishermen, a knowledge of the tides is very important. The inhabitants of the Pacific shores know how to exploit the incoming and outgoing tides for the gathering of food. The height of the high and low tides depends on the phase of the Moon and the position of the Sun. In some parts of the globe the difference can be many metres. The record is held by the Bay of Fundy, Nova Scotia, where the difference between high and low tides has reached 16 m (52 ft).

It follows from the familiar laws of gravitation established by Isaac Newton 300 years ago, that the effect of the Moon on the size of tides is twice that of the Sun. Though the latter is much more massive, it is also much further away. But apart from the movement of the sea, the gravitational forces of the Moon and the Sun also act on the solid crust. This effect is not, of course, as obvious, but using sensitive instruments it can be recorded. Due to the Sun and Moon the whole Earth expands and contracts, and the difference between the 'ebb and flow' of the crust is about 20 cm (8 in) in Europe. It is clear from the position of the Sun and Moon that the extent of this effect varies from day to day. The extremes are shown in the two diagrams in Figure 357.

Many phenomena occur on the Earth because of the attraction of the Sun and Moon. Apart from the ones we have described, an effect has been demonstrated on day length, and on the slowing down of the Earth's rotation about its axis. This is called *tidal friction.* It acts, for instance, on the ocean floor between the solid crust and the waters which move as a result of the tides. This causes the speed of rotation of the Earth to change in the course of geological periods. It is not something which is apparent from day to day. Since the speed of rotation decreases, days are getting longer, but only at the rate of 1 second every 100,000 years. Palaeontologists studying the remains of animals from ages long past have even found evidence of the fact that in previous geological periods, such as the Palaeozoic, in the Devonian, days were only about 22 hours long, and the year lasted around 400 days. This has been shown on the basis of a study of the evolution of some fossil animals which not only have growth increment zones for every year (like the rings of trees), but even daily 'growth rings'.

THE MOON IN ITS LAST QUARTER

NEW MOON

357

357 The gravitational pull of the Moon and the Sun causes the ebb and flow of the tides in the oceans and seas. In spite of the huge mass of the Sun, its effect on the surface of the waters is smaller than that of the Moon, since the latter is much nearer. But when the two act together the effect is greatest.

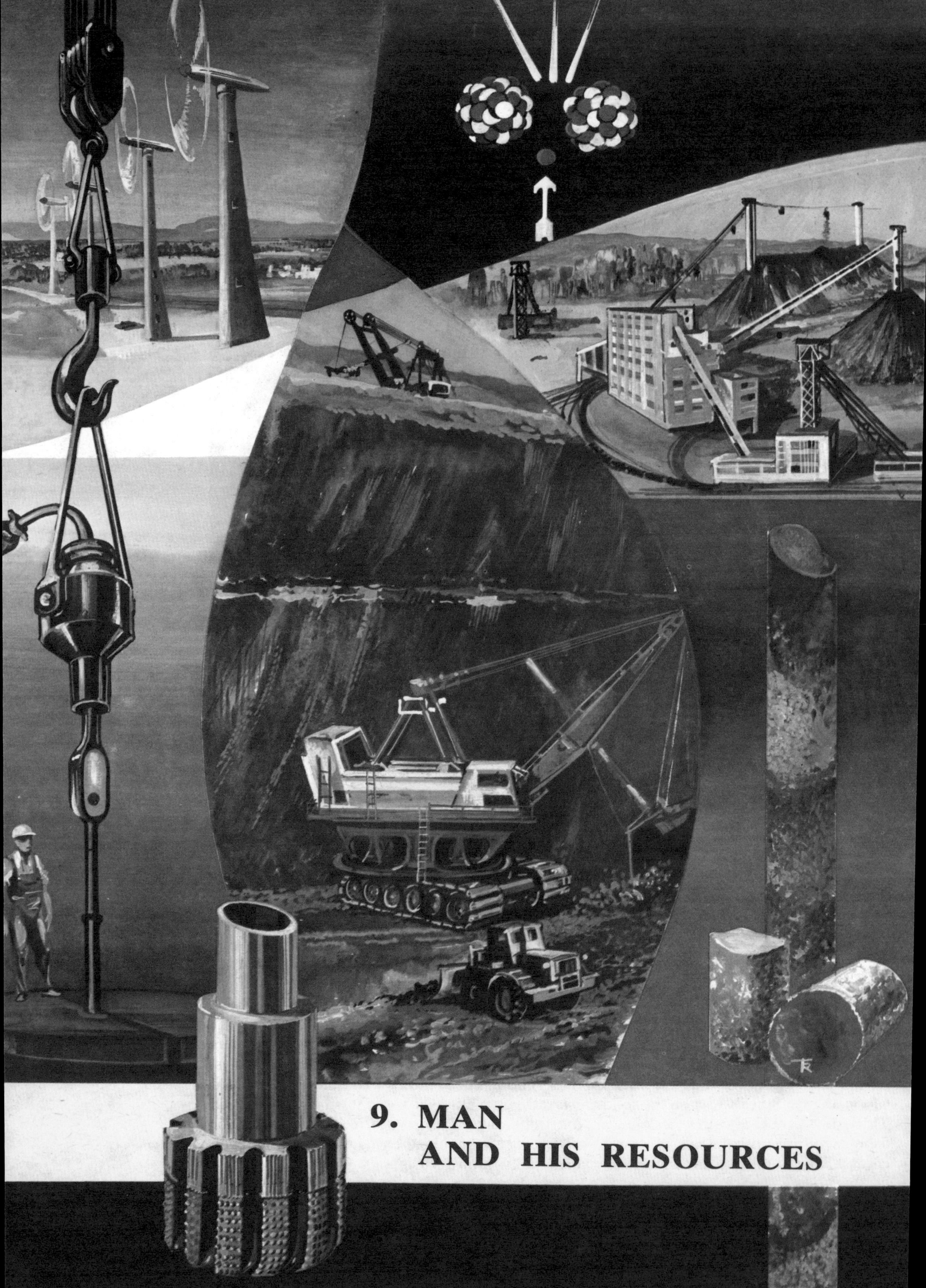

9. MAN AND HIS RESOURCES

Mineral resources

Not a day goes by without our reading or hearing from the mass media about the problems that face mankind. Whether it be diminishing natural resources or pollution, we must solve the problems if our planet is to survive. And how? In a sensible way. First and foremost, we must get to know our own planet thoroughly, and if possible even our neighbouring planets and stars. We must understand the role of all the mechanisms of the lithosphere, the atmosphere, the hydrosphere and the biosphere; we must learn about the workings of the whole Earth, and events such as the formation of mountains and the ocean basins, in order to understand how deposits of minerals come about. We must understand the problems of the formation of soil and air, learn how to estimate the potential food production of sea and land, and study the extent to which the energy resources and mineral wealth of our world can be exploited.

There is virtually no straightforward answer to questions of shortage and plenty. They have to be tackled in a coordinated way by economists, politicians, geologists and technologists. Some would say that the mineral resources will soon run out. Optimists point to the fact that every rock has a certain number of metal elements, so that in the future we shall exploit 'ordinary' rocks – for instance basalt – to obtain metals.

In this section of the book, where we are going to consider the very practical needs of man – supplies and resources – we shall very soon run up against a very complex problem. What are resources and what are supplies of minerals? The geologist would immediately reply that supplies are those minerals which, at a given level of technology and cost, can be recovered and used to produce some sort of material, perhaps to make iron, or to produce energy. Any such definition must, however, take into account many variable factors, and these have altered a great deal in the course of man's history. Was a lump of uranium ore a resource or an energy source for primitive man? Of course not. But a piece of meteoric iron or limonite (iron oxide), from which he could make a couple of kilograms of forgeable metal, that really was a resource for Neolithic man.

It is generally accepted that air, water and soil are the most important natural resources. They are called *renewable*. Other renewable resources include forests.

But mineral resources are *non-renewable*. Once copper ore has been mined, all that is left is an unsightly hole in the landscape and a heap of spoil. But just because a resource is non-renewable, it does not mean that it is irreplaceable.

Among the elements that are indeed irreplaceable are some that are essential to life, to basic biological reactions. They are hydrogen, oxygen, carbon, nitrogen, calcium, phosphorus, chlorine, potassium, sulphur, sodium, and magnesium. They are truly indispensable, and it is fortunate for man that supplies are inexhaustible; apart from calcium, that is, which modern agriculture uses in vast quantities, and phosphorus, which may become a problem in the future.

Apart from these clearly irreplaceable vital elements, mankind also makes use of those which seem essential, given the present state of technology. Examples would be copper (as a good conductor, without which electronics would be difficult to imagine) or nickel, chromium and tungsten (as additives to steel). But we should note that even these apparently indispensable metals are only essential for the time being. The example of mercury shows this. It is the only metal which is liquid at normal temperatures, and in addition it is a good conductor. It might seem that these qualities were really indispensable in their field. There is no other liquid metal. But mercury has one major disadvantage – it is poisonous. What is more, very little of it exists in nature. Most of the good deposits of mercury are exhausted. Is that a catastrophe? Not at all. A potential crisis was averted with the development of semiconductors which made mercury arc rectifiers redundant. They replaced mercury, which had seemed irreplaceable. Perhaps sooner or later the same will apply to all the apparently indispensable elements – ways may be found of getting by without them. There are still many raw

358 The aim of the practical work of geologists is to find the mineral raw material resources without which human society would not be able to function. But since these resources are finite and non-renewable, modern geology is faced with the urgent task of resolving the question of new energy resources.

materials for which we have as yet no apparent replacement, such as iron. But supplies are still so extensive that for the time being we are not obliged to look for one.

Ores

If we go back several centuries, to the time of mediaeval miners, we find that their requirements for a good ore differed a great deal from those of today. For it to pay them to mine an ore, it had to have a high metal content and not be too far below the surface. Ore was mined by means of a tunnel dug in such a way that a man could crawl along it; it was manually sorted underground, and only the best, metal-rich pieces were brought to the surface. The amount of metal obtained in that way was, however, small. The ideal ore of the late 20th century is poorer, but must be mined in large quantities. The best way is for it to be collected by means of a mechanical shovel and transported by huge dumper trucks through galleries the size of railway tunnels. But even this method of mining brings problems.

If we consider what has been meant by 'ore', even during the 20th century, we find that this has not remained constant. Take copper, for instance. In 1925 its ore was expected to contain an average of 2.5 per cent of copper – that was an ore which was worth mining, to make metal which was saleable at world prices. A little later ore with a lower content of metal started to be mined, because with the expansion of industry and the growing use of electricity, the demand had risen. Today ore is mined which, on the world average, contains about 0.7 per cent copper. This is because the demand has risen

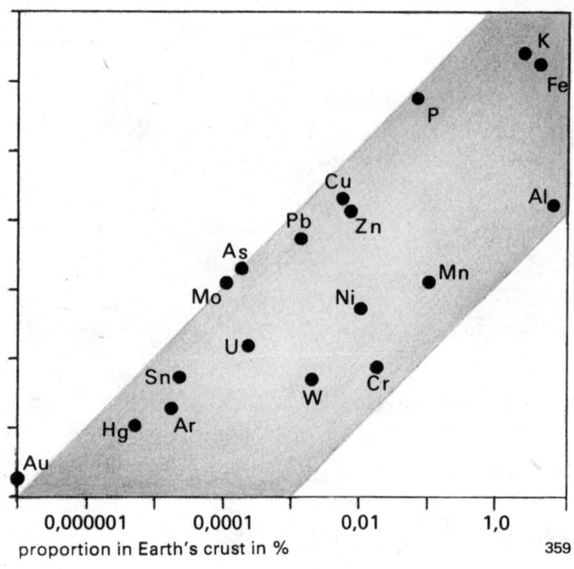

359 Resources of mineral raw materials are limited, and their quantity corresponds to the amount of the given element in the Earth's crust. The vertical axis shows the resources of different metals in the USA, the horizontal one their amounts in the Earth's crust.

360

360 The mining of minerals has been a part of man's life since early in his history. Mediaeval mining did much to stimulate the development of artisans' trades. In spite of primitive methods, surprisingly high quantities of silver and base metals were mined in central Europe.

many times over, and the price with it. But what is implied by a decrease in the content of copper in the ore is shown in Figure 361. That is the other side of the 'coin': in order to extract 1 tonne of metal from a rich ore (2.5 per cent), one requires a certain amount of energy. To obtain the same quantity of metal from a lean ore (0.7 per cent), more than three times that energy is needed. And that is to ignore the fact that to make use of poor-quality ores one must transport an enormous amount of material, build new roads, use huge vehicles, and burn a large amount of fuel. A copper mine also creates an ecological scar on the landscape. If we were to look at other types of ore, such as that of tin, we should arrive at even more dismaying figures. Today we are mining ores which are almost 50 times less rich than those we used at the start of the 20th century.

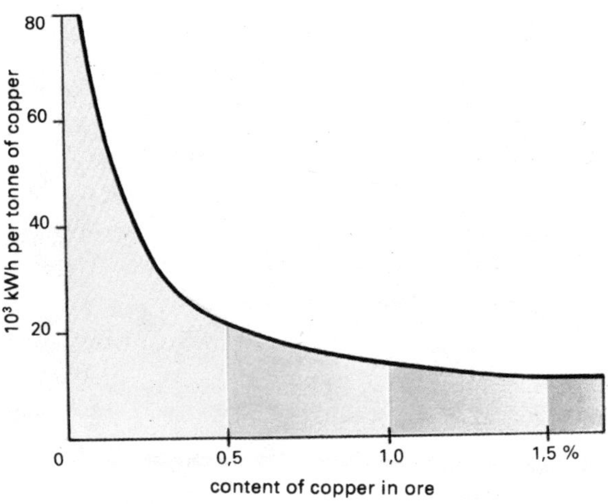

361 The processing of ores to obtain metals requires a certain amount of energy. The ore must be mined, separated, milled, prepared and turned into metal. The amount of energy required increases with lower grade ore. The figure shows the amount of energy needed to produce one tonne of copper from ore containing different amounts of copper.

This is a field in which geologists frequently pool their knowledge with economists. Trade in mineral raw materials and fuels is extremely active. Geologists thus examine the prospects for the supply and demand of energy. But such estimates are based on current qualities of ore. If we mine poorer ores in the future, for some metals at least, the duration of supplies will be prolonged considerably. But this also means greater investment and a higher consumption of energy for processing.

Let us start with the most ordinary of metals, iron. Men have known it for many millenia, and it seems that we shall not be able to do without it for some time yet. Geologists know even today of the existence of supplies of iron that will last for the next 250 years, and more will certainly be discovered. It is a quite ordinary metal, but one should add that the production of iron, like that of other metals, is considerably cheaper if secondary resources (scrap) are also used.

362 The modern mining of copper calls for explosives, huge machines, transport and the expenditure of large amounts of energy.

While the prospects for the supply of iron ore are good, steel additives such as vanadium, molybdenum, tungsten or manganese are rarer, and known reserves are only sufficient for a few decades. An exception in this respect is chromium. There are considerable reserves of chromium ores at present, enough for half a millenium. Nonetheless, the situation with this metal is a delicate one. Large deposits exist in only a few parts of the world. Other nations have to import chromium, so that it can become an instrument of political pressure.

Another very abundant and widely used metal is aluminium. The reserves of aluminium ore are almost inexhaustible, but its manufacture is very energy-intensive, so that production is dependent on the availability of cheap energy.

The other metals, among them the base metals lead, zinc, copper and tin, are very rare, and most known reserves (those with a metal content which makes them worth mining) will be exhausted by the end of this century.

It is therefore likely that the level at which the mining of ores pays will fall, the energy requirements rise, and the prices of these metals increase. But another question arises here: what is the limit of 'does it pay to mine'? Will we obtain our metals from ever poorer ores, and in the end from ordinary rocks? A tonne of basalt, for instance, contains 70 kg (154 lb) of iron, 170 kg (375 lb) of alumina, 20 kg (44 lb) of sodium, 1/2 kg (about 1 lb) of copper, and so on. Can we extract these metals?

There are no technological limits to recovering metals. Technically it is possible to recover from every kilogram of rock all the metals which are in it, but at an enormous cost. So most of the problems concerned with raw materials for the building industry, ceramics or even artificial fertilizers are centred around energy. Mineral resources are inseparably linked with energy resources which, as we shall see in the following sections, are linked with mineral ores.

363

363 For oil to occur the geological conditions in the Earth's crust must be right. One often speaks of oil traps. These are anticlinal depressions with an impermeable superstratum, which contain not only oil **(1)** and water **(2),** but also natural gas under pressure. Oil traps are usually formed by porous sedimentary rocks (which have been omitted from the picture to improve its clarity).

364 Coal formed in periods with a suitable climate. On the left is an example of a coalfield from the end of the Palaeozoic, into which seawater penetrated periodically, and on the right a Tertiary coalfield, which formed in a freshwater lake. A condition for the creation of coal from plants is the rapid deposition of clay and sand in the basin. These sediments (depicted in black) form a matrix which must be exposed and removed in order to recover the coal. What it took nature millions of years to form is removed by mining machinery within a few dozen years.

Fossil fuels

If you want to know how the Sun shone many millions of years ago, you only have to burn some coal or oil. These mineral fuels are 'canned' solar energy. The process of fossilization was very slow, even in geological terms. The debris of plants piled in thick layers was subjected to pressure, temperature and chemical changes. Large amounts of water and volatile substances escaped from the plants. First of all the pile of debris was changed into lignite, then brown coal, and finally black coal – anthracite. In the process an ever-increasing amount of energy was compressed into an ever-decreasing amount of matter. But the creation of coal demanded special conditions. The plant remains must not decay, they must very quickly be covered with a layer of sediment, have minimal contact with the atmosphere, and so on.

The conditions required to produce that other type of fossilized sunshine, petroleum, were even more special. The organic remains in this case had to be protected from oxidation; chemical reactions then converted the plant debris into gas and oil. But because these are volatile, they had to be trapped in the right sort of structures – in the pores of

364

sedimentary rocks. Nor was it possible for this stratified structure to move a great deal in the course of orogenic processes. If that had happened the oil and gas, and also the water which often holds the oil in the rock, would have escaped. The temperature, too, would have had to remain below a certain level in the oil-bearing strata, otherwise the oil would have been converted into less useful carbonic material. But such conditions do exist on Earth. This is proved by the hundreds of oil deposits which have been discovered in the crust of the continents, and especially of the continental shelves.

It is clear that the use of coal (practised in ancient China, as Marco Polo relates) led European civilization to a period of enormous growth in the 19th century. It was coal that formed the basis of the industrialization of countries like Great Britain, Germany and the United States. A few decades ago it seemed that oil would entirely replace coal. Anyone who says that oil is cleaner, burns better and with fewer by-products of combustion, and even flows out of the ground of its own accord, is right, of course, but a glance at Figure 365, showing how much energy is available from known oilfields and how much from known coalfields, clearly shows that coal wins hands down. Coal and lignite make up 88.8 per cent of the energy available from fossil fuels. The remainder comprises oil, natural gas and every other sort of 'petrified' energy.

Where to look for mineral deposits

When we defined deposits of minerals, it was not clear just how special such a deposit in the Earth's crust really is. The average content of metal in the crust, such as mercury, uranium, silver or lead, for instance, is low. In the case of lead it is 12 g (0.4 oz) per

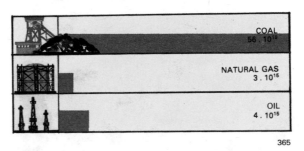

365 The energy resources which are stored in the form of 'canned' sunshine in fossil fuels are finite. The relative amounts of coal, natural gas and oil which remain in the world are shown in the picture (values in kilowatt-hours).

tonne of rock, in the case of chromium 110 g (3.9 oz), in the case of mercury 0.09 g (0.03 oz), in the case of tungsten 1.1 g (0.04 oz). But if we are to mine lead ore, for example, there must be a concentration of about 40 kg (88 lb) per tonne, 1 kg (2.2 lb) of mercury, or 4.5 kg (10 lb) of tungsten, and so on, apart from the fact that the ore must be present in a particular place in sufficient quantities, to make mining a worthwhile proposition. These concentrations thus represent a considerable deviation from the average.

The search for mineral deposits has its own logical rules. Not even a layman would go looking for salt in granite; it is geological nonsense. He would go to the sea shore or to rocks which formed in the sea. No one would search for coal in a place where only igneous rocks occur. The activity of the mediaeval prospectors and probably of others even

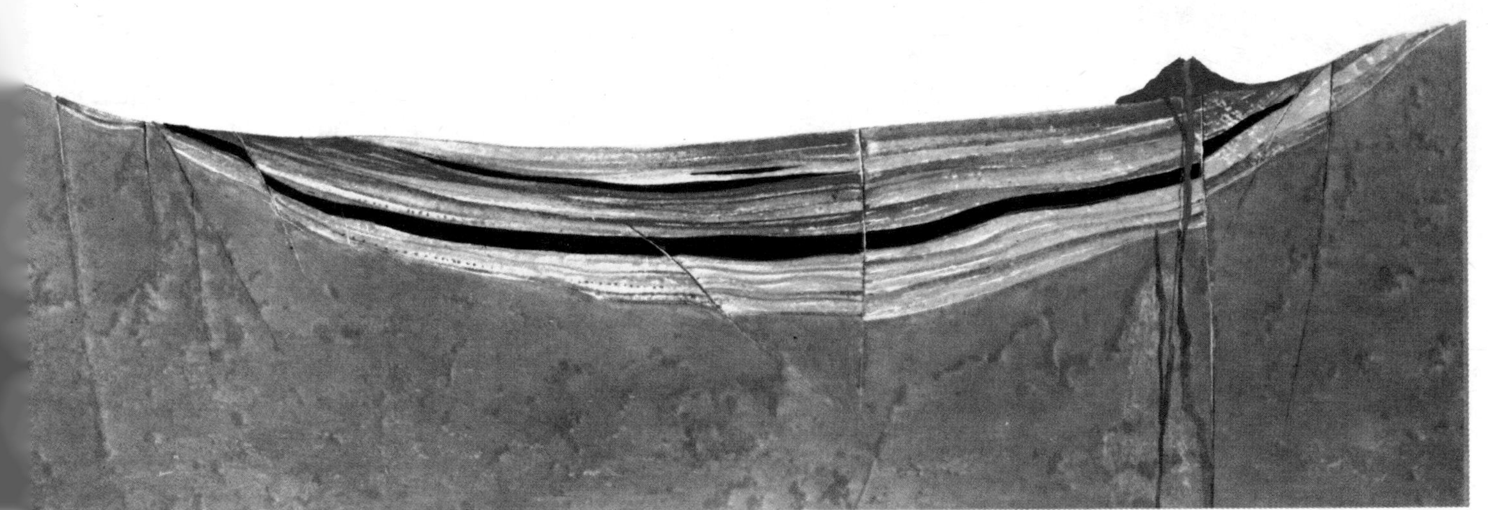

366 The concept of lithospheric plates describes the metabolism of the planet Earth. It even predicts the places where mineral raw materials form and are to be found. The most important of these are illustrated in the picture, which shows the mid-oceanic ridge, the active continental margin, and the passive margin.

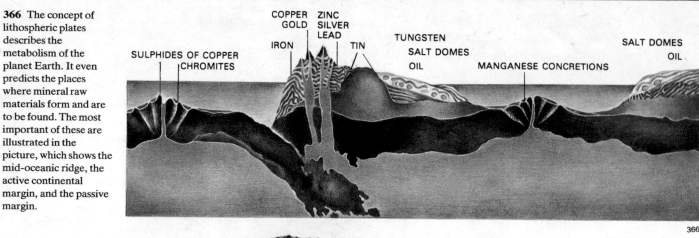

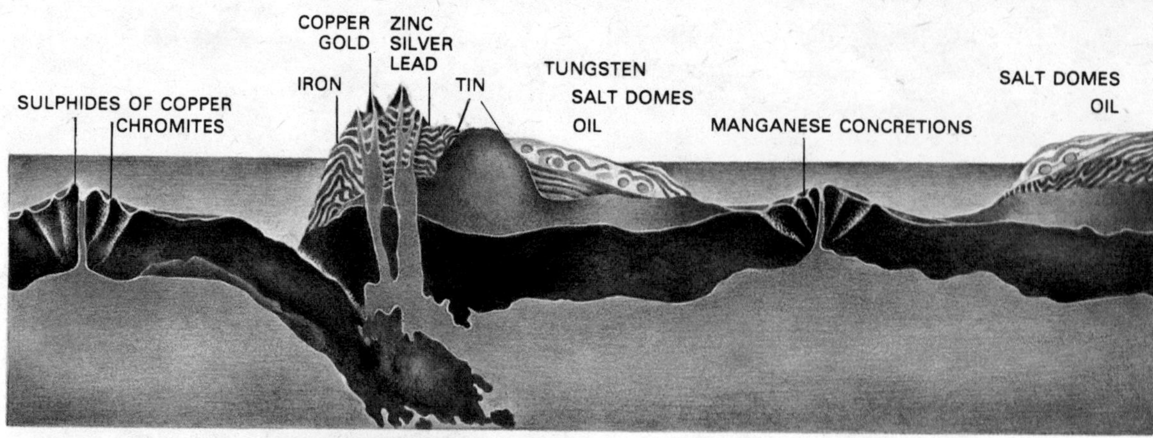

367 Hot springs, which are to be found in the vicinity of both extinct and active volcanoes, came into being through the penetration of rainwater – meteoric water – to magma hearths, or into deep parts of the Earth through a system of cracks.

before them, taught men to recognize the patterns in the occurrence of minerals in rock complexes. So, for instance, tin ores are never situated in basalts, but they always have some relation to granite rocks. Chromium ores, on the other hand, are not found in the vicinity of granites, but occur only together with deep-layer igneous rocks of 'basalt' composition.

The theoretical explanation of the occurrence and origin of mineral deposits is much less simple. It includes arguments both physical and chemical, the behaviour of elements during melting, during the formation of solutions in the Earth's crust and the mantle. In the case of some deposits an important role is played by the weathering processes and metamorphosis; in other cases by the environment and its mechanical qualities.

In the Middle Ages, and even later, the most important deposits were lodes and bodies of ore. Ore lodes form by means of hydrothermal activity. It is a very complex process, but is easy to imagine if we look at how hot springs or geysers arise (Figure 367). The surface water – possibly rainwater – penetrates to depths of from several hundred metres to several kilometres, where it is heated up, comes into contact with hot magmatic rocks, or even with solutions released from the crystallizing magma. The hot water then dissolves and leaches ore (metallic elements) from the rocks. In other places, usually in shallow parts of the crust, where the hot solution rises, the content of material dissolved in the water precipitates on cooling in cracks or in porous rocks.

In this way hydrothermal deposits are formed. Recently the seabed in the area of the mid-Atlantic and mid-Pacific ridges and

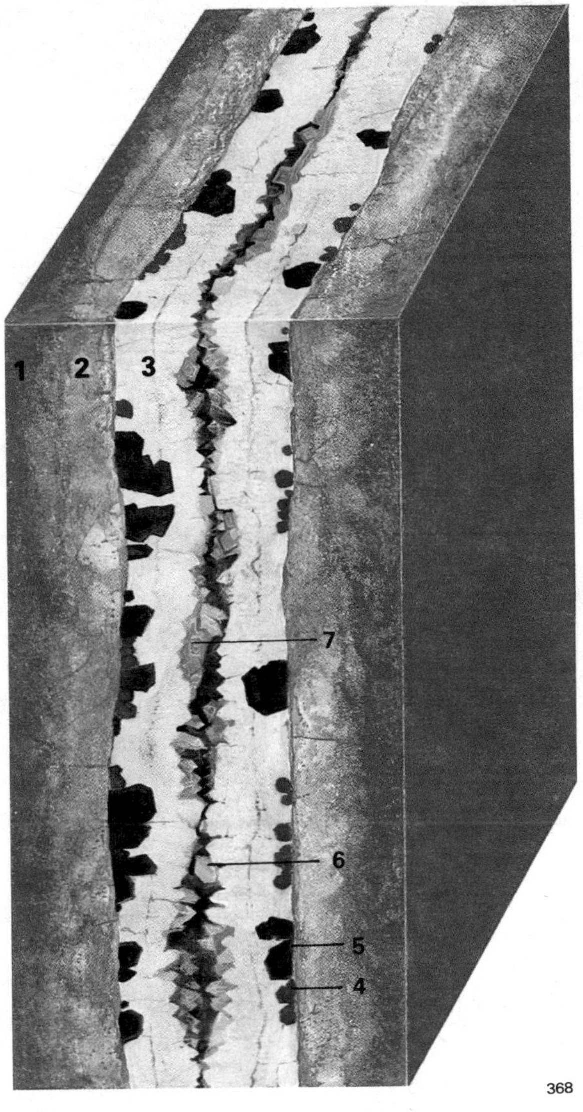

368

369

368 Lodes of ore arise through the crystallization of hot solutions in weakened zones or cracks, and you can tell from the illustration how such crystallization takes place. First of all the tin ore cassiterite was formed **(4)**, and the tungsten ore wolframite **(5)**, then the fissure was sealed by quartz **(3)**, and only at the end, in the remaining cavities, did crystals of fluorite and scheelite grow **(6, 7)**. The hot solutions also acted on the surrounding granite **(1)**, changing it to greisen **(2)**.

369 Some rocks are especially subject to disintegration or replacement (metasomatism) by ore components arriving from deeper parts of the Earth's crust. An example of such good collectors is carbonate rocks – limestone. Ores of zinc are shown in black.

in the vicinity of the Galapagos Islands has been found to be producing ore concentrations. Hot springs with a high concentration of sulphur spurt from the ocean floor, leaving 'precipitates' of ferrous, cuprous and other sulphides on the surrounding bed. Many scientists now think that seawater, penetrating the ocean floor, leached from its rocks ore components, which it deposited elsewhere. In other hydrothermal deposits it is clear, however, that water, together with ore components, is released direct from the magma, forming larger or smaller lodes in its vicinity.

Hydrothermal lodes and deposits are usually very rich sources of ore, but they are small, and therefore suitable for drift mining. But they often go down great distances. The lodes may be from a few centimetres in diameter, up to a metre. In hydrothermal lodes ores of lead, zinc, copper, silver and gold usually occur, but also arsenic, bismuth, uranium, tungsten, tin and mercury. The variability of hydrothermal lodes and bodies is considerable, so that the results of mining are sometimes a commercial disappointment.

A type of deposit similar to the hydrothermal is the disseminated deposit. These were formed with the participation of volatile components such as water. The best-known and most sought-after deposits of this kind are of copper. They are bound up with volcanic rocks which crystallized not far below the surface, containing finely dispersed particles of copper sulphides. These deposits are usually very extensive, but have relatively low metal contents. Thus a large amount of material has to be mined, which is then milled and processed in a very complicated manner. But the means of mining – using mechanical shovels in large, open pits – is considered to be more productive than many other methods.

The best deposits for mining are those which formed by means of sedimentation. The simplest example is limestone, which is used for the manufacture of cement and lime and as an agricultural fertilizer. Washed clays, arising from the weathering of granite rocks, are used in the ceramics industry, in

370 The cross-section of a volcano is also a cross-section of a deposit of copper 'porphyry ores'; it shows the features accompanying volcanic activity, and post-volcanic activity also. The zone of extreme hydrothermal alteration **(5, 6)** can be seen where copper-rich strata occur **(4, 5).** Only with denudation, the long-term erosion of the volcano **(2),** shown by the white line of the new surface **(3),** are the deposits themselves revealed. The volcano is associated with subvolcanic (plutonic) rocks **(1).**

the case of kaolin for the manufacture of bone china. Pure sand is used to make glass. But one can also find deposits of copper, and especially iron, in sedimentary rocks. In fact most of the iron ore mined today comes from sedimentary deposits, and these are all of about the same age – they are found in Pre-Cambrian strata. Much has been published on the question of these deposits in geological literature, and it is clear from this that the processes in the Pre-Cambrian, when the content of oxygen in the atmosphere was lower than it is today, allowed the transport of ferrous oxides into the then shallow sea, where the iron precipitated from the solution. So most of the Soviet ores from the Ukraine, the Indian ores and those from Brazil are from this period of the Earth's development.

How to look for mineral deposits

The simple examples we have used to illustrate the relationship of some mineral deposits to their parent rocks (such as the connection between iron ores and Pre-Cambrian shallow-sea rocks, or that of dispersed copper ores with andesite-type igneous rocks) are the basis of the search for mineral deposits. It is not, then, just a question of knowing the relationship between the rocks which occur in each other's vicinity and having a perfect picture of the geological conditions in an area, but it is also necessary to have a deep understanding of its geological development. So, for instance, in the case of the iron ores, we must know the age of the strata complex in which we are looking for the ore, and we must be able to determine on the basis of geological and petrographic observations whether the rocks formed in a deep-sea or a shallow-sea environment. In the case of igneous rocks we must know both the composition and the approximate depth of crystallization.

As well as these general geological aspects there are methods of locating mineral deposits in any such promising areas. In this book we have already shown how rocks display basic physical properties. For instance, minerals containing a large amount of iron and ferrous oxides such as magnetite or haematite, or even sulphides, differ greatly from other rocks in their magnetic properties. If we cross such a territory with a magnetometer or other instruments which register the intensity and direction of the magnetic field, a strip of strongly magnetic rocks will be evident even if the land is covered with a thick layer of sand and clay.

Elsewhere the Earth's pull – gravitational acceleration – is measured. If there are rocks of high density below the surface, such as

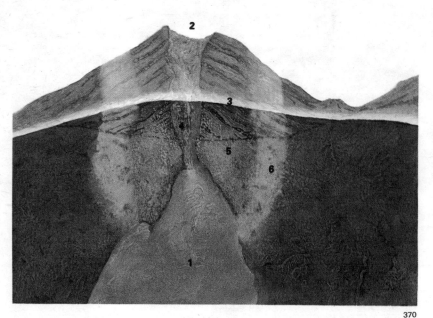

371 The mining of mineral raw material, such as sand for the manufacture of glass, leaves behind in the countryside scars which are not easily healed.

those with a high content of the heavy mineral chromite, detailed measurements of the gravitational acceleration will indicate that the locality has a greater 'attraction'. Localities with light rocks have a smaller gravitational acceleration. In such cases deposits can even be located at relatively great depths, since the irregularities of the gravitational field will be measurable there, too; geophysicists are even able to calculate the dimensions of such deposits. Places with an exceptionally low density are also sought in this way, especially in sedimentary complexes, since oil or gas deposits may occur there.

Measuring the electric currents in the Earth's crust is a complex procedure, but often produces excellent results in the search for ore. For example, rocks containing sulphides are more conductive than those without them; even places with a higher water content are better conductors than those with less water. And these are all properties which can be exploited in the search for ores.

Chemical properties can also be exploited. We have already mentioned how the content of metal in an ore deposit is many times the average for ordinary rocks. But since ores are weathered in the same way as other rocks, elements from this concentration get into the surroundings. Water dissolves even ore components, and springs or streams in the area tend to have higher concentrations of metals or their associated compounds. Geochemists therefore sample water and use very sensitive instruments to analyse the metal content, detecting locations with greater concentrations.

The elements of ore bodies, lodes or dispersed deposits also get into the soil, so that analysis of this can be a good location technique. From the soil they may get into the vegetation, thus the bodies of some plants may contain significant amounts of metals. Analysis of the ash left after burning such plants can indicate a high content of some element or another. Clearly the geochemist engaged in prospecting requires the services of many different experts. A single car battery discarded upstream can mean that he will search in vain for a lead deposit if he goes only by the analysis of water. The same goes for other metals. He must therefore know the geological, geochemical and geophysical data in order to say whether a locality is worthy of investigation. Only after thorough laboratory tests, the interpretation of geophysical measurements and thorough geological study can the drilling rigs move in to verify the geologist's assumptions regarding the occurrence of mineral deposits. And if the truth be told, out of a thousand promising sites, maybe a hundred prove to be worth more detailed investigation, and hardly a score of these will actually be exploitable.

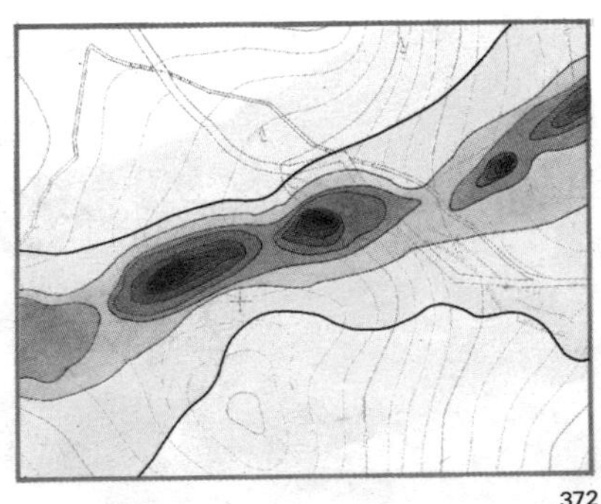

372

372 The modern search for ores relies on geochemical methods. Deposits concealed beneath soil and other accumulations are revealed by an increased content of the element in the soil. Only on the basis of such data can a more detailed survey be undertaken.

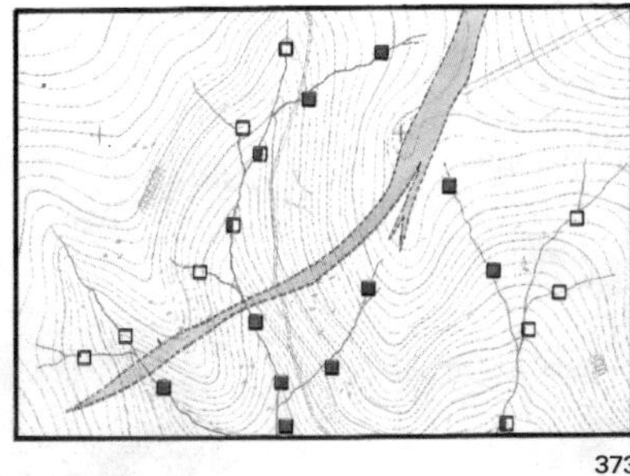

373

373 An effective way of seeking deposits of useful raw materials is the old prospector's method of panning. The insoluble mineral – for instance gold – is concentrated in the sediments of streams, where it is sought using a special 'pan'. The figure shows (in red) sites of successful prospecting, which can be joined together to show the course of the vein.

Marine resources and reserves

As well as utilizing the land, man has, for most of his history, made use of the sea. Until recently the main bounty from the sea was fish, but in the last couple of decades the recovery of oil from the seabed – or rather the continental shelf – has played an important role in the economies of some countries whose shores include these oil-rich areas. The salts dissolved in the sea are also exploited, and now projects designed to use the raw materials of the ocean floor are taking shape. Dozens of research vessels ride the waves. The resources of the oceans are often spoken of as the saving of mankind. The seas and oceans which cover more than two-thirds of the globe will, it is hoped, resolve the energy, mineral and food shortages of an expanding world. It is only natural to ask whether such hopes are well-founded.

It would seem reasonable to assume that the salt people use every day comes from the

374 The 'mining' of manganese concretions – 'nodules', from the seabed is still at the stage of testing and technological experimentation. But it is clear that the content of metals in these raw materials is so high that we shall have to take them into account in the future. On the left you can see the present method of collecting these concretions, on the right a project for future large-scale exploitation.

375 The continental shelves of Europe, North America and Australia are rich in oil supplies. The platforms which perform research and drilling in the waters of the Arctic Ocean belong to the world's leading companies.

sea, but that is not in fact the case. Less than one-third of all salt (sodium chloride) is obtained by the evaporation of sea water, while the remainder is mined inland or obtained from the evaporation of the *salinas* – mineralized waters which accompany oil deposits. Seawater is a chemical raw material. But the most precious chemicals obtained from it are not salt at all, but bromine (used mainly in the photographic industry) and magnesium. In the case of both of these the sea provides more than two-thirds of the world supply. Seawater contains many other dissolved compounds, and from time to time we can read in the papers how much uranium or gold is dispersed in the sea. The figures relating to how much of which metals we might obtain from the sea are indeed amaz-

375

374

ing, but we are limited by the fact that for the time being we do not have the energy available to exploit these resources. Many of the processes of obtaining materials from the sea are governed by nature herself.

Some metals, such as copper, manganese, cobalt or nickel need not be recovered from seawater at all, since they crystallize on the bed of the ocean basin in the form of manganese concretions. These bodies vary from the size of a walnut to that of a football, and are abundantly distributed over the ocean floor. They were found in the Pacific and the Atlantic as long as a hundred years ago by the famous British ship *Challenger.* These chemical deposits of seawater are formed of layers of ferrous oxides and manganese, whose crystal structure very readily binds heavy metals such as nickel, cobalt and copper. The total content of these metals in manganese concretions is up to 2.5 per cent. So research vessels map the ocean floor, submarines photograph, and scientists analyse the metal content of these nodules. As yet their content is low, and the cost of obtaining raw materials from the seabed high. There is hope for the future, though the legal problems are as yet unresolved.

A rather more successful venture has been the recovery of heavy minerals from coastal regions. The mediaeval miners and modern prospectors both obtained gold by means of 'panning' in streams. The water washes away the lighter silicates, leaving behind the heavier material which, given a little luck, may include gold nuggets. In many places the same method has been used in sea surf and strong marine currents. Heavier minerals such as cassiterite (tin ore), zircon (zirconium ore), rutile (titanium oxide), monazite (a complex phosphate containing elements from the rare earths group) and even diamonds are released from rocks in weathering and, since they are more resistant than many other minerals, are washed out to sea. There they are sorted like the contents of the prospector's pan: the lighter, usually silicate and quartz, material is carried away, leaving on the shallow seabed the heavier, useful

376

376 Today, marine oil drilling is widespread, though the costs of finding and recovering oil at sea are several times higher than on land. Nonetheless, marine rigs overcome the oil shortage in many countries.

377

377 A relatively small inflow and heavy evaporation bring about a concentration of seawater and the formation of salt deposits. Apart from common salt (sodium chloride) **(1),** salts of magnesium, potassium and gypsum **(2)** occur.

fraction. Minerals are mined at the meeting of sea and land in numerous parts of the world.

Non-ore raw materials

The shortage of mineral raw materials such as tin, chromium, molybdenum or vanadium ores, like the lack of fuels – oil and coal – causes confusion in the world economy. The prices of mineral ores, as we have seen, are mainly governed by availability, distribution throughout the world, the energy required to process them, the amount of metal they contain, and so on.

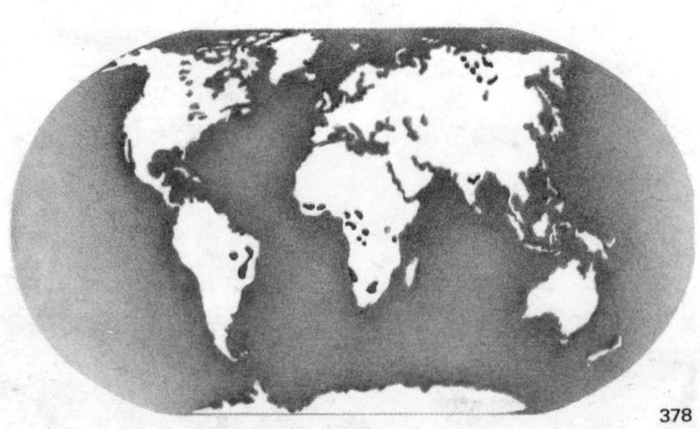

378 Only the oldest parts of the Earth, the Brazilian, African, Indian and Siberian shields contain diamond-bearing rocks (in red).

One does not hear about a shortage of non-ore raw materials very often, but they are no less important. Fortunately, some of these substances are so common that the governing factor is transport costs. Among the rarer materials, occurring in only a few parts of the world, are diamonds – though industrial diamonds have been produced in factories in recent years. Among the common materials are limestones for the manufacture of lime and cement; they occur in many geological formations where sedimentary rocks are found. But even with limestones geologists are faced with problems – not every limestone is suitable for quarrying, and not every limestone region can be subjected to the terrible scars caused by it; indeed, the geomorphological formations which occur in such areas – karst regions – are often interesting and attractive, and frequently protected by law.

While it is usually the case that ores must be processed (they are milled, and a concentrate is obtained from the milled material), with non-ore materials the opposite is normally the case. For the most part they are used in the state in which they are mined or quarried. Though there are plenty of such materials on the world scale, there are very strict quality requirements and regulations for their use. For the manufacture of glass, silica sand is necessary; but one must use particular types of sand. The presence of oxides of iron, manganese or titanium produces undesirable colouring. Even the size of individual grains of sand must be uniform, or problems may occur during smelting. And that is another criterion which excludes the majority of silica sands. There are similar technological requirements in the case of clays. These include kaolin for the manufacture of bone china and clays for other ceramics. Here, standards of manufacture affecting the quality of the finished product apply.

The quarrying of building materials also poses some problems; not even acquisition of ordinary gravel for the construction of roads is the simple affair it might seem. Modern structures require strong concretes, and the choice of the rocks on which they are based is no easy matter. A rock may seem strong enough, but if it contains even a small amount of sulphides, it is not suitable. For these disintegrate through the action of atmospheric conditions, forming sulphates and sulphuric acid, which not only cause the quality of the concrete to deteriorate, but acidify water, which can then corrode the steel reinforcement.

Non-ore materials are of great importance in the chemical industry. Can you think of any examples? Let us start with sulphur. There is a world shortage of pure sulphur, but the amount of sulphides such as pyrites in rocks is so great that it will be no problem in future to supply the sulphur required for the manufacture of fertilizers, sulphuric acid, and other requirements of chemical technology. Nor is there any danger of a salt shortage, though the consumption is enormous. This applies both to table salt (sodium chloride) and to the potassium salts which are important in modern agriculture as artificial fertilizers. The supplies of nitrogen fertilizers or saltpetre, on the other hand, are finite, and have largely been exhausted, though there are synthetic substitutes available.

There are, however, some non-ore deposits which are at present very rare. For example, the basis of the manufacture of phosphate fertilizers, the mineral apatite, is in very short supply, and will soon be exhausted. There are only a few regions, like the Kola Peninsula, where there is still enough. Because phosphorus has no substitute in nature, we must expect to obtain it in the future from raw materials which are at substitute for a material like plaster, which is made from mineral gypsum, or anhydrite – calcium sulphate. In spite of a rising demand, the present supplies of gypsum seem to be virtually inexhaustible.

Even the assertion that there is a sufficiency of non-ore raw materials should not mean that we no longer look for them, or especially that we should not mine and quarry them

379 A salt dome (diapir) is one of the strangest of geological formations. Its shape is due to the 'plasticity' of salt, the ease with which it is deformed. The characteristic features of such salt diapirs (salt deposits) include associated minerals – gypsum, forming the cap **(6)**, anhydrite **(4)**, or the potassium salt carnalite **(3)**. The substratum is made up of slate and gypsum **(1)**, and the main part of the dome of rock salt **(2)**. The dome is usually bordered by a diverse series of sedimentary rocks, of which the most abundant are salt clays **(5)** or sandstone **(7)**.

present considered uneconomical sources because of the low concentrations of the element they contain. Some çountries intend to use secondary sources such as animal bones. Phosphorus is an example of where the problems of ore and non-ore raw materials have been met, and at the same time is a case in point of the need for recycling.

A list of the non-ore materials we have not mentioned here would surprise even many experts. One example would be asbestos, which occurs as a secondary mineral in ultrabasic rocks, and is difficult to replace in many areas of human activity. It is an excellent thermal insulator, and in the past asbestos-based products have provided good heat conservation at low expense. We should also find it hard to come up with a simpler with discretion. Even though the volume of these materials is smaller than that of ores, they must be sought and recovered with consideration for the environment we live in, and localities we know today should be protected and conserved for generations to come.

Raw materials for the 21st century

The second millenium AD is drawing to a close. Some people expect revolutionary changes to take place as the figure 2 slips into place at the left-hand side of the date; but there is no fear of that. Geological time goes on running its steady course just as it did many centuries ago, and it is most unlikely that some great natural catastrophe will occur around the turn of the century – provided, of course... The proviso is a very important one. Man must ensure he does not cause a disaster by his irresponsible behaviour and indulgent use of precious resources.

If one compares the graphs of the energy consumption of industrialized states over the last 50 years against the population increase (chiefly in the developing countries) over a similar period, it is clear why the alarm bells are ringing. Add to these statistics the figures relating to loss of arable land and air and water pollution, and the picture of the future starts to look grim. Are we moving towards a catastrophe or not? It is possible to take either an optimistic or a pessimistic view of the coming millenium. In some countries environmental pollution really has been brought under control. Let us consider the question through the eyes of the geologist. We have spoken of how the chief and final aim of his research is to get to know the metabolism of the Earth and to use its natural resources in the interests of favourable living conditions for its inhabitants.

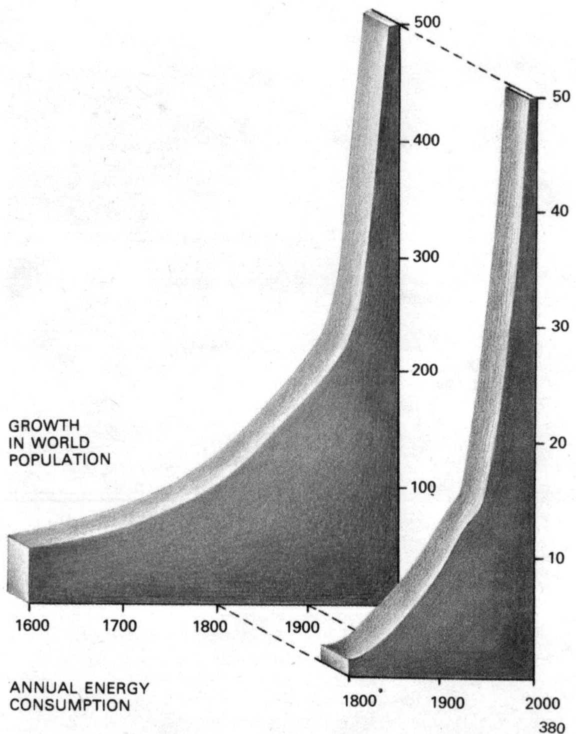

380 The consumption of energy is increasing mainly in the industrialized countries, while the population growth is concentrated in the developing countries.

The first demand a geologist has to meet is that for new energy supplies. The resources of mineral raw materials, i.e. coal, oil, natural gas, and even radioactive materials are finite, exhaustible and non-renewable. For this reason people are turning to renewable sources of energy such as solar energy, tidal and wind energy and geothermal energy. It may be that our problems will be solved by nuclear energy. But what has the geologist to say about this? Will he be out of work in the coming centuries? Will he look for radioactive materials only? It is clear that such materials as coal and oil will be more vital to the chemical industry, since, for example, the production of plastics is entirely dependent on oil, and there are some chemicals which must be made either of coal or of oil. It will therefore be sensible to put our money into other energy resources, as has been suggested in previous paragraphs. But this will require new materials for the exploitation of these alternative resources. So, for instance, the exploitation of solar energy and its conversion into electrical energy call for *photovoltaic cells,* which have to be manufactured. The best material from which to make them is compounds of the rare element gallium, although less efficient solar cells can be made from silica compounds. But gallium does not occur in any great concentration in the Earth's crust, and because of this lack of suitable mining localities, it has to be obtained laboriously in the course of making aluminium. And there is precious little gallium in aluminium ores. The world production of gallium is only a few dozen tonnes a year.

In the case of geothermal energy there are increased demands on the piping used to distribute it, since hot water from inside the Earth is often corrosive or strongly mineralized. Nor is the production of energy from nuclear fuels without its problems. Here special steels are required, which in turn call for additives such as molybdenum, vanadium, tungsten, tantalium, niobium and other elements which produce the necessary qualities.

Some economists rely on other nuclear solutions, but these, too, bring problems for the geologists. There would be a huge rise in the consumption of lithium and many other metals. The tube of a colour television, for instance, contains europium and other rare elements, without which it would not function.

The increasing prices on the world market,

and not only of oil and energy in general, are accompanied by increased costs of other mineral raw materials, from lithium, through scandium, to rare earth elements such as lanthanum, cerium, neodymium, samarium and europium. Man has uses for all of them; there is scarcely an element in the periodic table for which the demand will not increase in the 21st century.

Conclusion

This chapter can have left no one in any doubt that the Earth sciences offer much work for the future. Every technological advance, every modern society, brings new demands on mineral resources and energy – even on drinking water.

Though we assert that most of the elements we use today are dispensible, future generations will hardly be able to manage without the basic chemical elements to which we are accustomed. Though it is possible to replace copper or iron, for instance, with other elements, the latter are for the time being more expensive and more difficult to obtain. What, then, are we to do? We shall continue to require large amounts of material whose supplies we know will be exhausted in the future. Someone will have to find the answer. One view is that we will use poorer-quality ores, which means mining them in greater amounts. This requires more energy, not only for mining, but also for processing.

There are those among both economists and geologists, however, who think we shall find new sources of rich ores at as yet unexplored depths. But the journey into the Earth is a difficult one, and it, too, requires a large amount of energy. The resolution of this complex situation cannot be left to those whose task it is to find new sources of primary materials. In future the role of secondary materials – recycled waste – will be greater, and waste energy will also be exploited. This is something which depends on the understanding and responsibility of everyone; we must learn to use what we have. It will take a long time, but nature herself offers a guide to how we should behave on our mother planet. In order not only to conserve, but to help form our environment, we need a detailed understanding of how the metabolism of the Earth works, how the individual elements in nature behave; we need to find the natural place which man and his needs should occupy in that great metabolic system. It is only in such harmony that man can survive on his own planet.

381 The proportion of different types of energy used by man has changed over the years. It is difficult to predict how this development will continue in the future. But it is clear that the share of fossil fuels will fall. **1** – muscular energy, **2** – the energy of water and wind, **3** – burning of wood, **4** – energy from coal, **5** – from oil, **6** – natural gas, **7** – nuclear energy, **8** – solar energy.

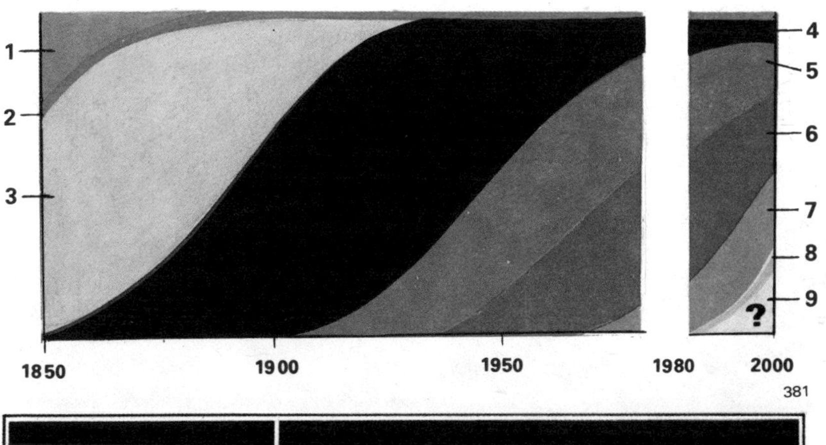

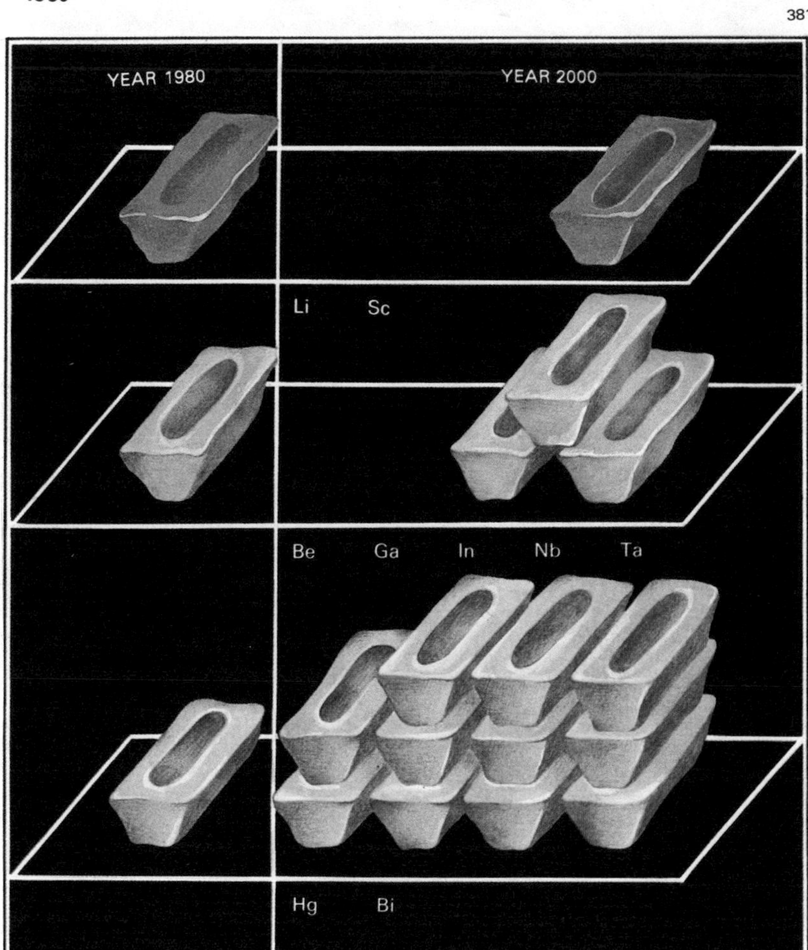

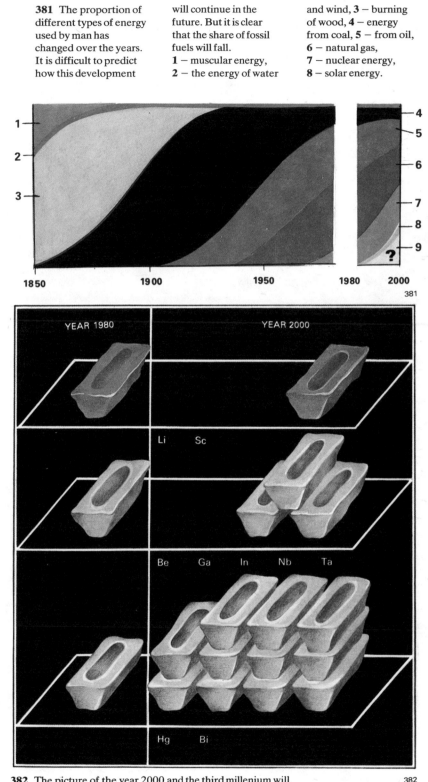

382 The picture of the year 2000 and the third millenium will partly depend on the mineral raw materials which are available. The annual consumption of some precious metals is rising rapidly, while that of others remains unchanged. On the left you see the consumption of the metals mentioned, in 1980, on the right the forecast consumption for the year 2000. 382

383 The Sun gives us energy in the form of radiation. Green plants capture and store it by means of photosynthesis, making sugar, starch, cellulose, etc. It is then consumed by the animal and human organisms in the form of food, to be converted into the work of muscles and brain. Solar energy absorbed by green plants in the past has been preserved underground in the form of coal and oil.

10. THE SUN, THE EARTH AND MAN

For us terrestrials, the planet Earth is our cradle, and will always remain our home. Our bodies grew out of its atoms. Together, the Earth and the Sun create a suitable environment not only for man himself, but also for millions of other animal and plant species. The Earth supplies all living organisms with the atoms they need in order to grow, while the Sun provides the energy which is essential to life.

Of the total energy the Sun emits into space every second (3.8×10^{26} watts), only one two-thousand-millionth falls on our planet (1.8×10^{17} watts). It is like a drop of water from Niagara Falls; but that 'drop' of energy is vitally important to the Earth. What would happen if the Sun were suddenly to disappear?

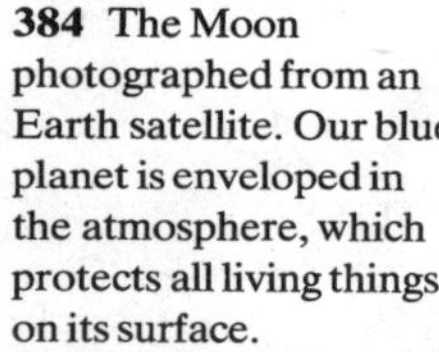

384 The Moon photographed from an Earth satellite. Our blue planet is enveloped in the atmosphere, which protects all living things on its surface.

385 The Earth photographed from a lunar satellite. This is the most beautiful photograph man has taken, and shows the true nature of our home.

384

385

The Earth without the Sun

Let us try to imagine the grim and hopeless situation the Earth and its inhabitants would find themselves in if they were deprived of the Sun's rays.

There would never be another dawn, and we should never again see a blue sky or the countryside around us. There would be no moonlight, nor would the planets appear, or any of the comets, for there would be nothing to provide their light. No clouds would form in the sky, with no Sun to evaporate water from the sea. No winds would blow, for the Sun causes them, too. Plants would be unable to grow, and men and beasts would go hungry, since all food is in fact sunshine hidden in grain, fruit, vegetables, meat, eggs, etc. All living things would be condemned to die of starvation. The Earth would irradiate its accumulated heat without gaining any new heat to replace it. Frosts would come, and all water would freeze; air of the atmosphere would freeze and become solid ice, covering the oceans and continents. There would be no air: all living things would long since have disappeared, leaving the once-beautiful planet devoid of all life...

Scenarios such as these can be found in works of science fiction, but scientists can assure us that such extinction is something of which we need not be afraid, even though mankind has invested so much of its energy and invention into creating the means of its own destruction. Man should constantly remind himself of his own dependence on nature, of his being subject to the laws of the Universe, part of an order he cannot afford to disturb.

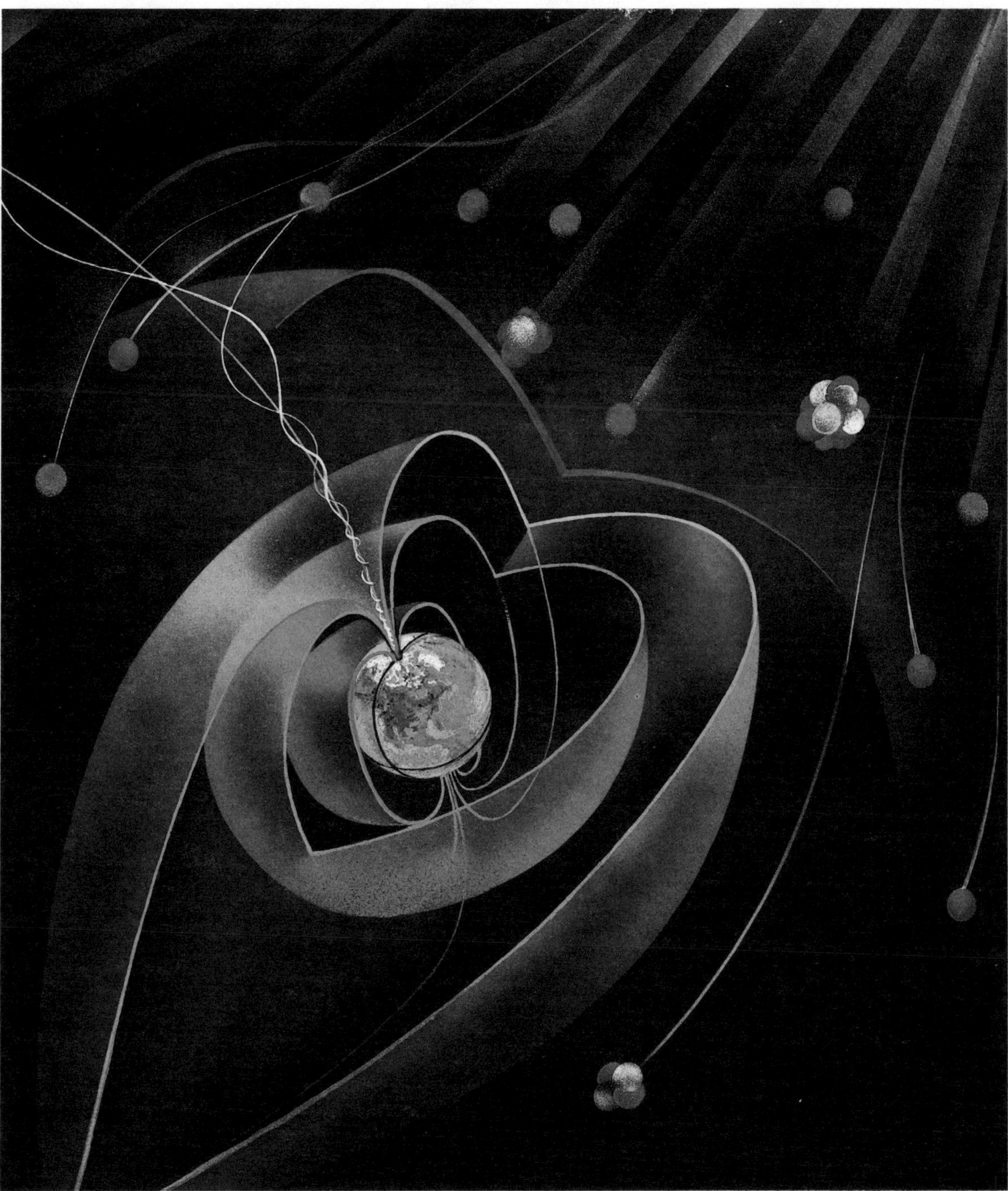
386

386 The magnetosphere of the Earth is a huge magnet whose force lines reach many thousands of kilometres out into space, protecting the Earth from the solar wind. The magnetosphere gives elastically under the strain of the solar wind. This creates changes in the magnetic field on the surface of the Earth (e.g. magnetic storms). Without the magnetosphere there could be no life on the surface of our planet.

Solar energy on Earth

An energy of 1.8×10^{17} watts falls on the Earth from the Sun every second. Its three different components are very different from one other. You have learnt from the preceding text that neutrinos, which carry around 4 per cent of the solar energy, are quite unsuited to the task of transferring energy from the Sun to the Earth. The electrically charged particles of the solar wind (protons, electrons and atomic nuclei – Figure 386) carry with them much less energy than the neutrinos – only one two-millionth of the energy falling on the Earth (10^8 kW). These

387 Solar radiation on the Earth. Neutrinos pass through, cosmic radiation is partly absorbed in the atmosphere, and high-energy particles penetrate to the surface and below it, where they are absorbed. X-ray, ultraviolet and to some extent also infra-red and radio photons are absorbed in the upper atmosphere **(1)**. One-fifth of the light is absorbed in the troposphere **(2)**, one-third is reflected from clouds and the Earth's surface **(3, 4)**, and almost a half is absorbed by the surface **(5)**.

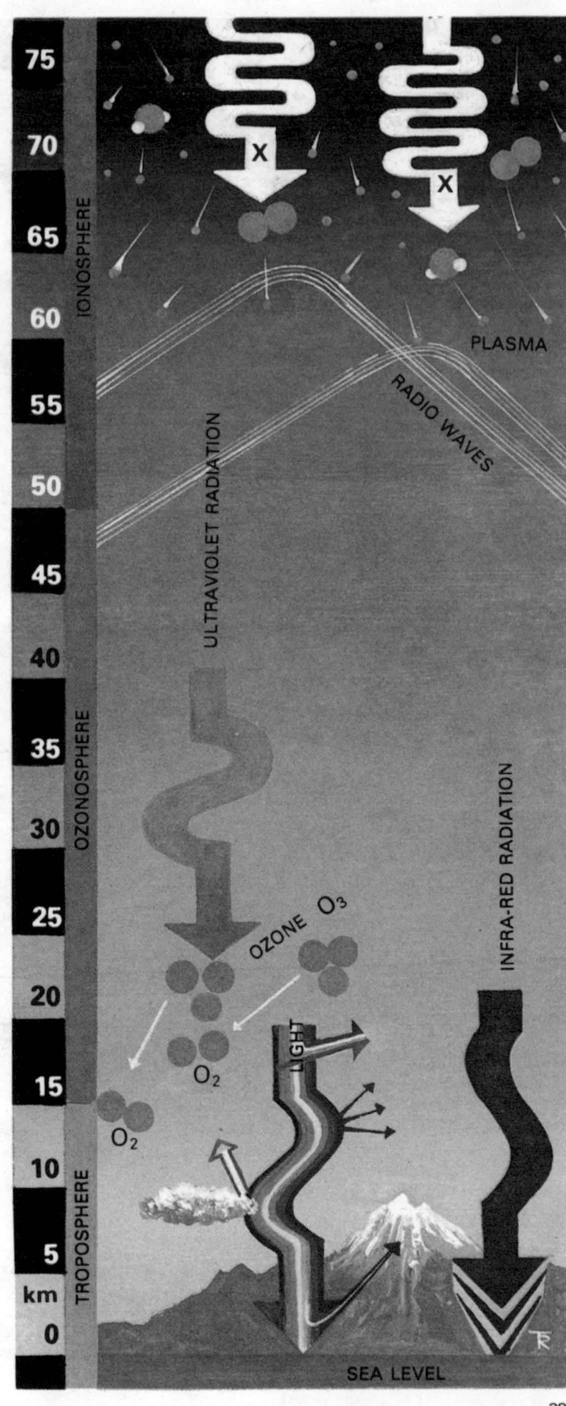

388 The electromagnetic radiation from the Sun in the Earth's atmosphere. X-radiation is absorbed at heights above 50 km (30 miles). There it ionizes the air and thus forms the conductive ionosphere. Ultraviolet radiation is absorbed by ozone (O_3) between 15 and 50 km (9 and 30 miles). Light and infra-red radiation penetrate the troposphere to the surface of the Earth.

389 The planet Earth photographed from the Meteosat satellite. The weather over equatorial Africa is cloudy, over the Sahara clear. On the left the coast of South America can be seen. Europe is at the top.

particles of the solar wind move at up to one thousand times the speed of a bullet. They are very dangerous to life, but the Earth is protected from their destructive effects by its magnetosphere. The solar wind can, to some extent, especially following large flares, slip along the lines of force into the polar regions, where it causes the phenomena we call *aurorae* high in the atmosphere (over 100 km/62 miles up).

The main part of the solar energy is brought to Earth by the photons. It amounts to 180,000 TW (terawatts, or billions of watts). These have no electric charge, and pass unobstructed through the magnetosphere into the atmosphere.

The atmosphere allows through only some sorts of photons (for example light photons), while absorbing others (Figure 388). At altitudes of over 60 km (37 miles) the air absorbs X-rays and ultraviolet radiation, which are dangerous to life. There the absorbed radiation ionizes and heats up the oxygen and nitrogen. The ionized air is a good conductor and reflects radio waves. The layer over 60 km (37 miles) up is called the ionosphere.

Below the ionosphere, at heights of between 15–50 km (9–30 miles), photons of near ultraviolet radiation are absorbed (they have only a little more energy than violet light photons). The absorbed ultraviolet photons form ozone at such altitudes; its molecule consists of three atoms of oxygen. The layer is accordingly known as the ozone layer or ozonosphere.

The photons in the visible and infra-red parts of the spectrum reach the Earth's

390 Europe and North Africa photographed from Meteosat. Meteorologists use such satellite pictures to help forecast the weather.

surface. The layer of air immediately above the latter is the troposphere. It reaches an altitude of from 8 km (5 miles) at the poles to 16 km (10 miles) at the equator. In the troposphere light meets molecules of air, drops and crystals of water, and the dust raised from the surface of the Earth. This meeting causes some of the light to be

390

reflected, some to be diffused, some to be refracted and some to be absorbed. This produces various phenomena in the troposphere such as the blue sky, the red Sun at sunset, solar pillars, light crosses, mock suns, rainbows, and so on (Chapter 11, Figure 411).

The photons which pass right through the atmosphere fall on the Earth's surface. The land and seas reflect a small number of them back to space – which allows spacemen to photograph our planet (Figure 385). But for the most part photons are absorbed by the surface and converted to heat.

We can generally sum up the fate of the photons which fall on the Earth as follows: of the total radiation, carrying 180,000 TW, about one-third are reflected from the atmosphere or the surface, about one-fifth are absorbed in the atmosphere, and almost a half are absorbed by the surface and converted into heat (Figure 387).

The Sun, the climate and the weather

We live in the troposphere (Figures 236 and 388), and its properties are of vital importance to us. The temperature, humidity, cloud cover, winds and precipitation all affect our lives. The immediate condition of the troposphere is the weather, while its average state over several centuries is the climate, which may be *oceanic, continental, tropical,* etc. Both weather and climate are mainly determined by solar radiation.

The extent to which the surface of the Earth is warmed up is not the same everywhere. Much more radiation falls on each m^2 of the equatorial regions than at the poles. The reason is clear from Figure 391. In the polar regions the Sun always falls at a very oblique angle, thus having to pass through a thick layer of air. This causes considerable diffusion and absorption before they even reach the surface, and when they do, the snow-covered polar regions reflect much of the radiation back into space.

The equatorial zone of the Earth receives most warmth. The warm air rises, and its place is taken by cooler air which moves in along the surface. In the upper troposphere the rising equatorial current flows outwards towards the north and south (Figure 392). Similar air currents occur in the troposphere in the temperate zones and at the poles. The flow of the whole troposphere is 'driven' by solar radiation, and is called the *general circulation of the troposphere.* The general circulation cools the tropical regions, and brings warm air to the cooler parts of the Earth's surface.

The amount of solar radiation absorbed by the Earth's surface also varies from place to place. The oceans and continents are distributed unevenly over the globe. The general circulation is not, therefore, as regular as would seem from Figure 392. The landmasses heat up more quickly than the seas, but they also cool more quickly at night. That is why the wind blows inland from the sea during the daytime, and seawards from the land at night (Figure 393). So even those winds which break the overall pattern are

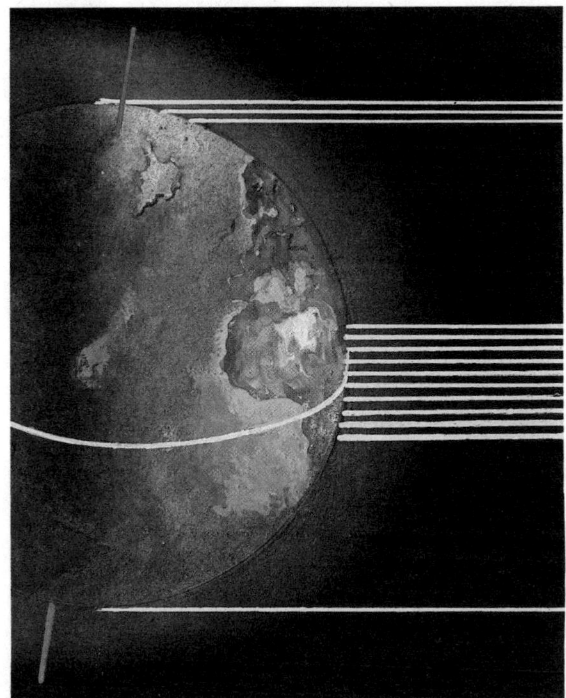

391 Much less radiation falls at the poles than at the equator, per given area.

392 The uneven heating of the Earth causes the circulation of the whole atmosphere. Hot air rises at the equator and then flows northwards and eastwards at high altitude. At 30° latitude it descends and returns to the equator over the surface (returning to the equator, it is deviated westwards); such winds are called tradewinds.

393 In the daytime the land is warmer, but at night the sea is warmer. This determines the movement of air, which flows from sea to land in the daytime and from land to sea at night.

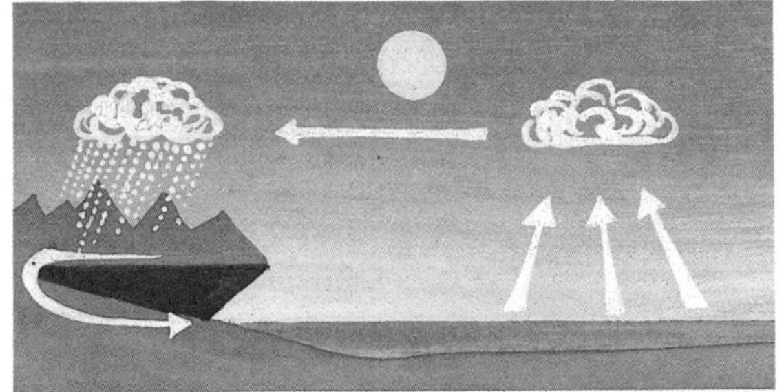

394 The Sun evaporates water from the oceans and drives the winds which carry it landwards.

powered by solar energy converted into heat. All the movements in the troposphere are in fact part of a huge thermal engine, which is changing the heat of the Sun into the kinetic energy of air. The Sun continuously supplies around 1000 terawatts of energy to this atmospheric engine, which is 120 times more than the present total energy consumption of humanity. It would be feasible to transform 10 terawatts of this wind energy into electricity.

The water in the troposphere has a strong influence on climate and the weather. The solar radiation heats up the surface of the oceans, the seas, lakes, rivers and aquatic parts of the continents. In the process, the energy of the incoming photons is converted to that of the movement of water molecules. Those that move fastest escape from the heated water and fly off into the air, where they form a gas – water vapour. The number of water molecules in the air is usually smaller than 1 per cent of all air molecules. In the whole of the troposphere there is about one hundred-thousandth of the water which is in the oceans and seas. (If it were all to fall in the form of rain, it would form a layer 2.5 cm/1 in thick over the whole of the Earth.)

Warm air is light and therefore rises from the surface of the water (Figure 394), taking with it a large number of water molecules. But up in the troposphere the pressure and temperature are both lower. The rising air expands and cools. The water molecules in it begin to condense into drops or into crystals of snow. In the rising stream a cloud is thus formed. The condensation of 1 kg of water from vapour releases 600 kcal of heat. This condensation heat warms the higher layers of the troposphere.

The rising currents and winds carry water vapour from the seas and oceans inland (Figure 394). There it falls from clouds as rain or snow. Part of it runs into lakes, rivers and streams, part is soaked into the soil *(ground water),* and a certain amount is consumed by plants and animals. But most rain water returns to the sea. There it evaporates again and is lifted up by the air and taken over the land, and the cycle is repeated. This repeated movement is called *water circulation.* The content of water in the troposphere has a turnover rate of about 10 days.

The circulation of water is for us very useful and important not only from the point of view of the supply of drinking and industrial water, but also as an energy source. A large cloud may contain up to 300,000 tonnes of water (3×10^8 kg). The average height of a cloud over the Earth's surface is around 4000 m (13,000 ft). For the Sun to

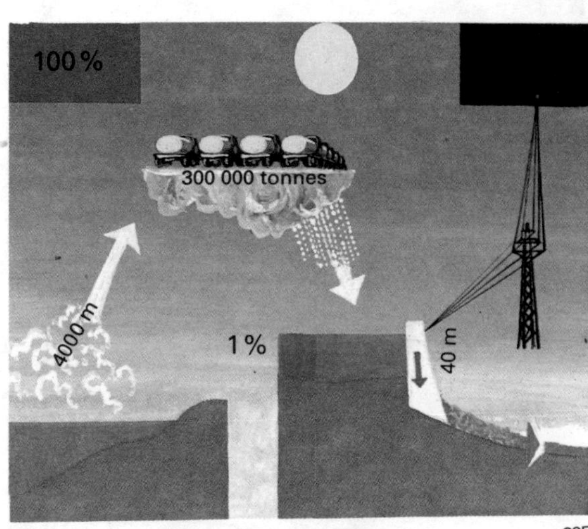

395 The energy required to raise a large cloud from the sea is 1.2×10^{13} joules (3.3 million kilowatt-hours). Only a small part of this energy is captured by the dam shown (about 1 per cent), where it is converted into electrical energy.

raise such a cloud it must perform work equivalent to 3×10^8 kg $\times$ 10 m/s^2 $\times$ 4 $\times$ 10^3, which is 1.2×10^{13} joules, or a little over 3.3 million kW hours. The quantity 10 m/s^2 is gravitational acceleration. Such is the enormous gravitational energy of a large cloud. We can capture only a small fraction of it in a reservoir. If the dam is 40 m tall, we capture only 1 per cent of it (40 being 1 per cent of 4000). Then the gravitational energy is converted to kinetic energy in the inflow to a hydro-electric power station. The turbine utilizes this kinetic energy to drive a generator, which converts it into electricity. Thus the electrical energy generated by a hydro-electric power station is in fact converted solar energy. As a matter of interest let us add that only about 0.2 TW is obtained this way at present, while it would be possible to produce 3 TW.

The total circulation of the troposphere and the circulation of water are driven by the constant (invariable) component of solar radiation (i.e. light, infra-red and ultraviolet radiation).

The variable component of solar radiation (X-ray, shorter ultraviolet and radio radiation) brings much less energy, and fluctuates greatly. But this component also, closely dependent on solar activity, affects climate and weather. Research shows that solar activity influences air pressure and temperature, cloud cover, the circulation of the troposphere, the amount of precipitation, the formation of ice, the occurrence and movement of cyclones and anticyclones, and so on. But it must be emphasized that the effects of solar activity on the weather are very complex and are felt in different ways in different parts of the world.

The Sun and life on Earth

Every living creature on the planet Earth is a part of the biosphere. Each organism requires energy to live. Plants derive theirs directly from the Sun. By means of chlorophyll they convert solar energy into chemical energy. They use it to change carbon dioxide (CO_2) and water (H_2O) into sugars, starches, fats, proteins and other energy-rich substances. This process is called *photosynthesis.* One might say that photosynthesis is the 'gateway' through which solar energy enters the biosphere.

The most important element for photosynthesis is carbon, since it has the capacity to form chains. On land photosynthesis takes place mainly through green plants. At sea, solar radiation is captured by monocellular organisms called *phytoplankton,* which also contain chlorophyll. Phytoplankton use carbon dioxide dissolved in seawater for photosynthesis.

In the course of one year photosynthesis accounts for the recovery of a total of 200,000 million tonnes (2×10^{11}t) of carbon from carbon dioxide. The atoms of this carbon are built into complex organic molecules, and become part of the *biomass.* One tonne of dry, organic matter represents a chemical energy of around 1.5×10^{10} joules. This means that green plants and phytoplankton convert 3×10^{21} joules annually into the chemical energy of the biomass worldwide. A year is just over 3.1×10^7 seconds. If we divide the amount of energy produced by the number of seconds in a year, we find that photosynthesis converts continuously about 90×10^{12} joules per second (90 TW) of solar energy into the chemical energy of the biomass.

The biomass is the material of which the biosphere is composed. It contains a total of about 8×10^{11} tonnes of carbon, so that it represents a total of 12×10^{21} joules of chemical energy. Every second 90×10^{12} joules (90 TW) of this energy is consumed by the breathing of living organisms, by their death and decay, and the burning of wood and plant and organic waste. But the same amount of solar energy (90 TW) enters the biosphere by means of photosynthesis (Figure 396). It is interesting that green plants on land capture 60 TW of solar radiation, while phytoplankton capture only 30 TW, although the oceans take up a greater proportion of the Earth's surface and receive more solar energy than land.

The energy of the biomass is being exploited by man to a greater extent in order to replace the dwindling resources of oil and coal.

Organisms release energy from the organic matter of the biomass (from food) through breathing (Figure 397). This consumes oxygen from the air and forms carbon dioxide and water vapour. Thus breathing is the opposite process to photosynthesis. Photosynthesis and breathing are mutually complementary and interdependent. Neither can exist on its own. Some 2000 million years ago the two processes arose simultaneously.

The Sun and man

Man is a natural part of the biosphere. From it he obtains the energy he requires in the form of food. The average adult has a daily intake of around 3000 kcal (12.6×10^6 joules). Children need less, but people engaged in hard physical work need most of all. In our food we take in chemical energy concealed in molecules of protein, carbohydrates and fat, from which our organisms obtain the necessary building materials to create and maintain body tissues.

A day lasts 86,400 seconds. By dividing the daily energy consumption by this number, we obtain the consumption per second, which we find is about 150 joules, or 150 watts. Thus our bodies require the same amount of energy as a 150-watt light bulb (Figure 400). The energy contained in food is chemical energy, while the bulb takes electrical energy and converts it to light energy. But both are of solar origin.

Green plants convert the energy of solar photons into the chemical energy of complex molecules. We then take in this plant energy either directly (in fruit, vegetables or grains) or indirectly through herbivorous animals (milk, meat, butter, etc.) and omnivorous animals (pork, poultry, eggs, etc.). It would take two large railway wagons to carry the

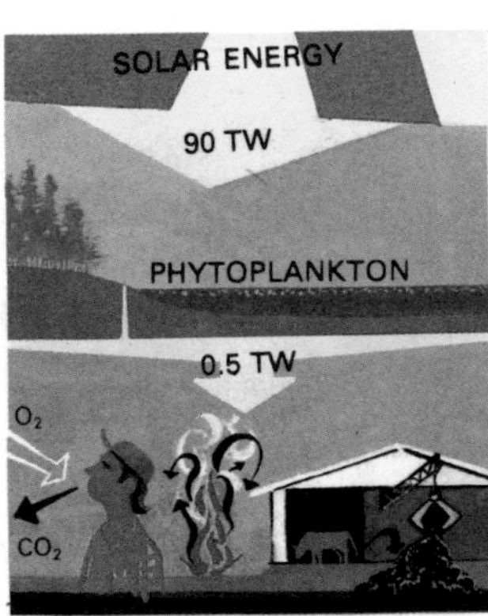

396 Photosynthesis converts solar energy into chemical energy (90 TW for the whole of the Earth). Only a half of 1 per cent of this (about 0.5 TW) is consumed by man in the form of food.

396

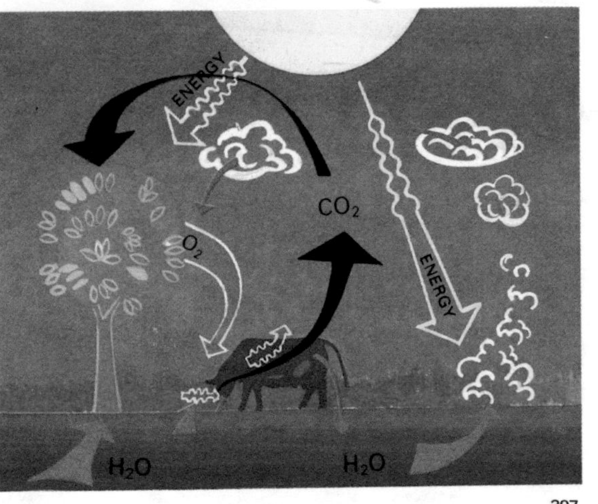

397

397 Green plants store solar energy using photosynthesis. From carbon dioxide and water they make sugar, starch and other substances. These are passed on to herbivorous animals, and from them to carnivores, and so on.

food a man consumes in his lifetime (Figure 398). For the Sun to provide that amount of energy it has to convert only about half a gram of hydrogen to helium in its interior.

One often hears of the effect of solar activity on man and other denizens of the biosphere. Here we have in mind the variable component of solar radiation. Thousands of scientific works have been written on the effect of solar activity on man and on the biosphere in general. The effect is not, however, a direct one, since the variable component of solar radiation does not reach the surface of the Earth. Following large flares, for instance, there is an enhancement of X-rays and ultraviolet radiation. Both of these are caught in the ionosphere. There they cause high ionization (the number of free electrons increases). The ionosphere then reflects long radio waves (around 10 km/6 miles) very well. Such waves occur with each lightning flash in a thunderstorm. They are called *atmospherics*. During flares the ionosphere reflects atmospherics so well that they are propagated to a distance of 10,000 km (6200 miles) or more. But at any given time there are several hundred storms in progress in the world, and every second lightning strikes about 50 times.

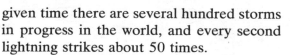

398 Throughout his life man consumes food enough to fill two large railway wagons. To provide the energy for these nutrients the Sun needs to convert only 0.5 gram (0.02 oz) of hydrogen into helium.

During a solar flare the atmospherics from lightning spread so well that those from tropical Africa, the Indian Ocean, the south Atlantic and even Indonesia can be recorded in, say, Europe and North America. This means that at the time of the flare everyone there is exposed to powerful atmospherics. The result is the same as if all the storms in the whole hemisphere were suddenly to appear in our vicinity. Numerous studies have shown that such an increase in atmospherics which may occur during solar flares affects living organisms. The atmospherics represent an alternating electric field, which interferes with the electrical pulses governing the nervous system. One of the consequences of this is a prolongation of the reaction time – the interval which elapses between the perception of danger or other situations calling for action, and the performance of the action required. The situation is one typically encountered by drivers. The extension of their reaction time produced by an increase of atmospherics is reflected in a rise in traffic accidents, particularly in cities (Figure 399).

Let us recapitulate in brief the chain of events induced by a flare on the Sun: 1) the magnetic field in a group of sunspots changes; 2) the magnetic energy thus released causes a flare; 3) the flare heats the corona to a high temperature (up to 50 million kelvin); 4) the hot corona sends out a huge burst of X-radiation; 5) the X-rays ionize the ionosphere and the electron density thus increases greatly; 6) the high electron density in the lower ionosphere makes it a good reflector of the very long waves emitted by lightning (atmospherics); 7) we are suddenly heavily irradiated by atmospherics; 8) our reaction time is prolonged.

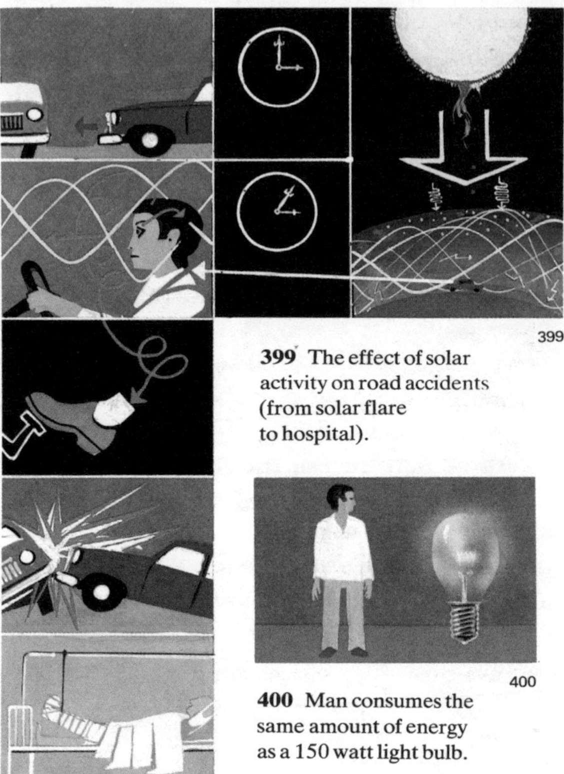

399 The effect of solar activity on road accidents (from solar flare to hospital).

400 Man consumes the same amount of energy as a 150 watt light bulb.

Professor Cizevsky from the USSR has shown that the incidence of cholera in the last century correlated with solar activity. But there are many other diseases which occur in apparent association with solar activity: jaundice, heart attacks, depression, certain mental illnesses, and others.

The Sun in the service of man

The human body requires for its survival a relatively large amount of energy, but compared with the energy required to run households, industry, transport, agriculture, and so on, it is a negligible amount. For all of this the whole population of the Earth consumes every second, 8000 million kilojoules (8 TW). With 4000 million inhabitants, that makes 2 kilowatts per person. That is the average for all countries, rich and poor, industrialized or developing. In the most industrialized countries the consumption is up to 14 kW per person, whereas in the poorest it may be less than 200 W. The statistics made by the United Nations clearly indicate that the higher the standard of living a country has, the greater is its energy consumption per person.

Everyone seeks to live better. This means that the energy consumption of the world is and will be rising, especially in the developing countries. But the Earth has only limited supplies of energy, and they are diminishing rapidly.

At present the most important energy sources are fossil fuels: coal, oil and natural gas. Through photosynthesis plants accumulated a huge amount of solar energy, many millions of years ago. This remained hidden beneath the ground in the form of chemical energy until modern times. Whenever you warm yourselves in front of a natural gas or solid fuel fire, you are taking advantage of solar energy from times long past. We cook, light our homes, streets and workplaces, and drive around thanks to all that sunshine which came to Earth many millions years ago. The vast majority of the energy used by man today is, in effect, fossilized sunshine.

There are other sources of energy on Earth, which are used to a lesser extent: a) energy stored by water behind barrages; b) wind energy; c) the heat of the oceans (exploited in the OTEC scheme, standing for Oceanic Thermal Energy Conversion); d) the chemical energy of the present biosphere (that of draught animals, wood, vegetation waste, camel dung, etc.). All these sources receive energy from the Sun. They are indirect solar energy resources – in cases a) and b) in the form of kinetic energy; in case c) heat, and in case d) chemical energy. Since this energy is constantly refurbished, its sources are known as *renewable energy sources.* Solar energy has been embodied in these renewable energy sources in recent times. When you warm yourself in front of a bonfire you are using solar energy which was captured by the chlorophyll of trees a few decades ago.

In the last few years increasing use has been made of nuclear energy. Atomic power stations use nuclear fission of uranium to release part of the rest energy of all the nucleons. Nuclear reactions are much more efficient in releasing energy from a material than chemical ones (burning). But where does the energy concealed in the uranium atom come from? It is not of solar origin. These heavy nuclei have their origin in the presolar period, since they came from the catastrophic destruction of the stars which then existed. As we have seen, such stellar destruction is known as a supernova, and it was in such events that heavy and radioactive elements were formed, at temperatures of many thousands of millions of kelvins (Figure 152 number 12). The Earth's own heat (geothermal energy) also comes from the radioactive decay of elements formed in the explosion of supernovas. The heat of volcanoes and mineral springs is a relic of the gigantic infernos of supernovas, stored in the form of the nuclear energy of radioactive elements.

In recent years scientists throughout the world have been working to try and develop a thermonuclear reactor more efficient than the fission reactor. It would employ *fusion* of light nuclei into heavier ones. Large sums of money are being invested in the project; but nature has had her own perfect thermonuclear reactor for 5000 million years – the Sun. It is perfect for the following reasons:

- The conversion of hydrogen into helium is the most effective means of releasing energy in the Solar System. No other nuclear or chemical transformation sets free such a large amount of energy as this process which goes on inside the Sun.
- The Sun is a perfectly safe reactor; it can never explode, since it has a perfect safety device. Any dangerous overheating leads to expansion, and thus to immediate cooling.
- The Sun is an almost infinite source of energy, since it will go on releasing it for at least another 10,000 million years (see page 131).
- The Sun delivers to the Earth a huge amount of energy (180,000 TW), much more than mankind is capable of consuming. It seems paradoxical to speak of an 'energy crisis' when the Sun is providing us with 20,000 times more energy than the whole population of the Earth uses.

• The energy the Sun gives us is absolutely clean. It does not damage the environment either chemically or radioactively.
• Solar energy is entirely free, it costs nothing.
• The Sun is so far away, and so immensely powerful, that no one can abuse it.
• The perfect solar reactor is used exclusively for the good of all living creatures. In the hands of men nuclear energy has also become a weapon of potential mass destruction.
• Solar energy in the form of photons is very valuable. It can easily be converted into all the forms of energy we require at home, in industry, in transport and in agriculture.

Solar energy can be converted directly into heat, electricity or chemical energy. Note that solar radiation is not converted directly into mechanical energy. The branch of energetics dealing with the use of solar energy is called *helio-energetics.* There are all sorts of helio-energetic devices in many countries throughout the world. We shall mention a few of the most outstanding ones here.

Heat from the Sun

Photons which fall on dark surfaces are absorbed, or in other words their energy is converted into the kinetic energy of the molecules of the object they fall upon. The more photons the object absorbs, the warmer it becomes. This process is going on all around us on a large scale. Nearly 90,000 TW of the infalling solar energy is being changed into heat by the Earth's surface.

Man converts solar radiation into heat for a variety of purposes: to grow vegetables and flowers (greenhouses), to heat water in the home, in industry and in agriculture (water heaters), to dry fruit, vegetables, grain and fodder (solar driers), to cook food (solar cookers), to heat homes (solar houses), to obtain electricity (thermal generators), or to smelt metals and produce high-temperature chemical reactions (solar furnaces). Every form of heat can be acquired by the absorption of solar radiation. Solar radiation is transformed either as it comes (Figure 401, right), or after concentration (Figure 401, left). It depends on the temperature we need to attain. Accordingly solar collectors are referred to as either *flat* or *focusing* (Figure 401). In flat collectors temperatures up to 100° C (212° F) are produced, in focusing collectors temperatures from 150–4000° C (about 300–7200° F). Here we shall describe only the flat type of collector, which is the most widely used device for exploiting solar energy.

A solar collector consists of the following main parts:
• An absorber, where the solar radiation is converted to heat. It is usually a black metal plate, made of some metal such as copper, iron, aluminium or steel, which is a good thermal conductor.
• An insulated case for the absorber. Otherwise the heat would be lost to the cooler surroundings.
• A top cover for the absorber, which allows solar radiation to reach it but prevents heat from escaping. The best material in this case is glass.
• A working fluid which takes heat from the absorber, leads it out of the collector, and transports it to where it is used. It may be water, oil (for instance used sump oil) or various anti-freeze mixtures. It may also be air, as in driers.

A water heater is used to heat water for the home, industry and agriculture. Its main component is a solar collector in which the water is heated. The hot water is lighter and rises of its own accord into an accumulator, while colder, heavier water flows in to take its place. The water thus circulates of its own accord and heats up. Hot water (40–90° C/104–194° F, according to the amount of water, the insulation of the absorber, the intensity of sunshine and the size of the collector) is drawn off from the accumulator. This is the simplest form of solar heater.

Air can also be used as the heated material in a solar collector. The black metal absorber in an air heater is usually corrugated. This increases the area of contact between the circulating air and the hot absorber. The transfer of heat from absorber to air is thus faster. The hot air obtained is used for various purposes in agriculture, for both drying and heating.

A simple solar drier for agricultural products consists of a few flat collectors and one drying chamber. The air is warmed in collectors, and rises of its own accord into the drying chamber. The latter consists of an enclosed space with grids on which the product to be dried is spread. The hot air absorbs the moisture, cools, and descends. Its humidity gradually rises. On additional cooling at the bottom of the drying chamber the moisture condenses and the water is removed. The cool air containing a small quantity of water vapour again flows into the heater, and the process is repeated.

Flat water and air collectors are also used for heating houses. A simple but effective type of heater is the Trombe wall. Professor Trombe was a world-famous French expert on the use of solar energy. He was also responsible for the design of the well-known solar furnace at Odeillo in the Pyrenees.

The Trombe wall is a flat air collector. The absorber is the south wall of a house, which must be black. In front of it is a glass panel set in a wooden or metal grille. The cold air from the house flows through an opening at the bottom (close to the floor) in between the wall and the glass. On coming into contact with the black wall it is warmed and rises. At the top, near the ceiling, is a further opening, sometimes two, where the heated air flows back inside. It transfers its heat to the walls and other objects in the room and cools down, thus falling to the floor again. There it is again drawn through the opening, between the heated wall and the glass, where it again rises, and so on. The heating process is spontaneous, and no fans are required.

In summer a house with a Trombe wall could get unpleasantly hot. So a valve at the top is opened to let the hot air out instead of in. At the same time a flue in the north wall is opened, which causes cold air from the shaded side of the house to flow into the room. Thus in summer, instead of heating the house, the Trombe wall actually helps to cool it. Experience dating from 1956 shows that a Trombe wall can save about half the annual fuel bill. Conventional heating need be used only on overcast, cool days.

With the help of solar radiation we can also get drinking water from seawater or impure fresh water. The amount of drinking water in the world is falling, and yet the demand is rising. Man's activities are polluting water at every turn, and in many areas there is already a shortage. There are several ways drinking water can be obtained. They include distillation from seawater in stills, chemical treatment of river water, the use of the huge reserves contained in the ice-caps, or using solar radiation in a simple device – a *solar still.*

Nature supplies us with drinking water by means of the evaporation of the oceans which results from solar radiation. The solar still is an imitation of this natural process (Figure 403). Salt or polluted water pours into a wide, shallow trough. The bottom and low walls of this trough are black. Solar radiation heats the trough, and with it the water. The water molecules rapidly evaporate, but impurities, salt or microbes are left behind. The evaporated water condenses on the glass roof, which is cooled by contact with the outside air. It slopes at an angle of at least 15°, causing the drops of water to run into

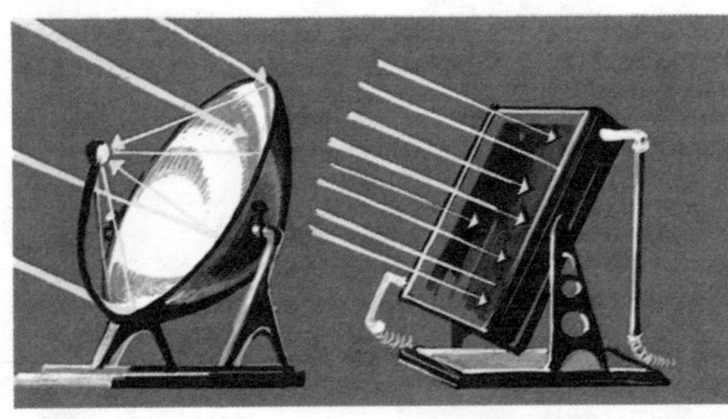

401

401 A focusing collector (left) first concentrates the Sun's rays at a focal point, where it then converts them to heat. A flat collector (right) utilizes solar rays direct.

402

402 A solar house near Aachen, FRG. The solar collectors on the roof heat water. The hot water heats the house. Excess heat is stored in the cellar for use in bad weather.

collection channels on either side of the roof. These carry the pure water into collection vessels.

This means of obtaining drinking water using solar radiation has been used for over a century in the Las Salinas mines in Chile. Some countries (such as Sri Lanka) not only use the method for obtaining drinking water from the sea, but also make use of the salt which is left behind. The residue after evaporation contains more than 20 sorts of mineral salts, which are an important raw material in the chemical industry.

The Sun can help us obtain drinking water in places where we might otherwise go thirsty. All one needs is a sheet of polythene and some sort of vessel (such as an empty can). Dig a hole in the ground and place the vessel at the bottom. Then stretch the polythene over the hole and seal the edges with earth or sand to prevent air from escaping. Put a stone in the middle of the polythene. The moisture inside the hole will condense on the polythene and drip into the vessel. To increase the yield of water you can place shredded plant matter in the bottom of the hole. The method will produce up to a litre of drinking water per day even in the desert – with the help of the Sun (Figure 404, left).

403

404

403 A solar still makes drinking water out of seawater or impure water.

404 Obtaining drinking water from the desert using plastic sheeting. More permanent devices are built using glass (right).

The Sun and electricity

Life without electricity would be difficult for us to imagine. The electric current we use in our homes, in industry, and so on, is largely of solar origin. It usually comes from fossil fuels or watercourses. Only a few per cent is produced by atomic power stations. But in the future electricity will be produced directly from solar radiation.

There exist several ways of obtaining electrical energy from solar radiation. Figures 405 and 406 show the transformations which take place for photons from the Sun to become electricity.

Photons are changed directly into electricity in solar cells. They are also called *photovoltaic cells,* which means that they convert light into electrical current. The best-known sort of solar cell consists of thin sheets of silicon. On one side phosporus atoms are added, on the other side boron atoms. When solar radiation falls on such a sheet it divides the negative electrons from the positive ions. This gives an electrical potential of 0.5 V, which tries to return the electrons to positive charges. If the illuminated side of the sheet is connected with the shaded side by means of a wire, the electrons return, which means that a current flows through the wire. About one-fifth of the radiation falling on the sheet is converted into electrical energy (1 square decimetre of a solar cell yields about 1 watt). Solar cells have proved their worth as sources of electricity in satellites and space probes, lighthouses, repeater stations in inaccessible places, as power sources for telephones in isolated localities, and so on. There are already dwelling houses with electricity provided by solar cells on their roofs. The electricity obtained in the daytime is stored in accumulators for use at night and in inclement weather.

405 Design for an orbiting power station. It is to be launched into geostationary equatorial orbit at a height of 36,000 km (22,000 miles). The large panels are covered in solar cells, in which solar radiation is converted into electric current. This is then converted into radio waves, to be sent to Earth by the antenna.

The solar energy falling every second on an area of 1 m^2 (above the atmosphere and perpendicular to the Sun's rays) is about 1.4 kJ. This quantity (1.4 kJ/s = 1.4 kW) is called the solar constant, and it was measured with great precision on board man-made satellites. In one year (31 million seconds) 1 m^2 above the atmosphere receives the energy of 12,000 kWh. The annual energy falling on 1 m^2 of horizontal surface in Europe (or in North America) is only 1200–1400 kWh. In space there is no day or night, no clouds and no atmosphere to absorb the solar radiation. The thought thus occurs to us whether it would not be possible to collect the solar energy outside the atmosphere and to convert it to electricity by means of solar cells. There are already several projects in existence for satellite power stations, and work is under way to implement them. One of the best known is the project of Peter E. Glazer.

According to Glazer's design the solar cells

406 Thermal solar power station. By means of rotating mirrors (heliostats) solar radiation is concentrated on a boiler at a height of many metres. There steam is generated, which is led (see red pipe) to an ordinary steam power station (extreme left).

solar radiation. A few of these geostationary power stations should suffice to provide all the electricity required by a medium-sized state.

Electricity can be obtained indirectly through heat or chemical energy. In a solar thermal power station (Figure 406) solar radiation is used instead of coal. The solar radiation is reflected by means of rotating mirrors (called *heliostats*) and concentrated on to a boiler placed on a tower. The boiler provides high-pressure superheated steam, which is led down into a turbine. The steam turbine drives a generator, which converts the kinetic energy into electricity.

406

would cover an area of 5 × 12 km (3 × 7.5 miles), which is 60 km^2 (22.5 sq miles), or 60 million square metres. The amount of radiation energy falling on them would thus be approximately 60 million m^2 × 1.4 kW/m^2 = 84 million kW. This considerable area of panels would be constructed at a height of about 250 km (155 miles) by astronauts in a space shuttle, and from there transported by a space tug to a height of 36,000 km (about 22,000 miles). At that altitude satellites orbit once every 24 hours. If the orbit of the satellite is over the equator, it remains in the same position relative to the Earth – it is *geostationary*. Photovoltaic power stations will therefore be geostationary satellites. The electrical energy they produce will be converted to decimetre radio waves, transmitted from the power station to a receiving aerial back on Earth (Figure 405). The energy received at the aerial would then be transformed into alternating current and put into the normal electric grid. After all conversions there should be around 10 million kilowatts left out of the original 84 million kilowatts of

To a lesser extent solar radiation can be changed into electricity through heat, in two different ways: either through a *thermo-electric cell* or through a *thermo-emission cell.* The thermo-electric cell consists of two wires of different material. One connection is at the focal point of a collector, the other in a cool place. An electric current flows through the circuit.

The thermo-emission cell consists of two metal plates (electrodes). These are very close to each other (within a fraction of 1 mm). One electrode (the cathode, or emitter) is heated at the focal point of a parabolic collector to a temperature of 1600–2500 kelvin. The other electrode (the anode, or collector) is relatively cool – around 500 kelvin. Electrons are emitted from the glowing cathode and fall on the anode. They bring not only heat, but also an electric charge. If the two electrodes are connected, the electrons return to the cathode – a current flows through the wire. The voltage of thermo-emission cells is low – about 0.5 volt – and 1 cm^2 gives an output of several watts.

Finally, let us mention the conversion of solar radiation into electricity through chemical energy. Solar radiation can be used to break water down into oxygen and hydrogen. This means that the original energy of the photons is stored in the chemical energy of the two gases. When they combine, water is again produced, and the stored energy is released, either as heat (on burning) or as an electric current (in a *fuel cell*). A fuel cell is a device in which chemical energy is changed into electrical energy.

The hydrogen-oxygen fuel cell is commonly used. On one side oxygen is admitted, on the other hydrogen. The resulting water is led away. The hydrogen gives its electrons to an electrode (cathode), from which they move through an external circuit (wire) to the oxygen anode, where they are taken over by the oxygen. The negative ion of oxygen then combines with the positive ion of hydrogen to form water. Fuel cells convert chemical energy directly into an electric current, and they will probably become an important source of electricity in the future. Hydrogen and oxygen can be obtained from water using solar energy in several different ways, and above all in unlimited quantities, since both water and sunshine are plentiful. Fuel cells are also clean and noiseless in operation. They can be used for domestic electricity supplies (with outputs of up to 12 kW). But fuel cells with an output of more than 10,000 kW are already being built.

The Sun and mechanical energy

Mechanization and automation increase our standard of living, but they call for a large amount of energy. At present the main source of energy for mechanization is fossil fuels. But there are only limited amounts of these available, and what is more they cause air pollution. This has led experts to seek ways of obtaining mechanical energy from solar radiation.

We have already mentioned that solar radiation cannot be changed directly into mechanical energy. Sunshine has first to be converted to heat (Figure 406), after which a thermal engine is used to change this into mechanical energy. Another way is to obtain electricity from solar radiation by means of solar cells, and then use an electric motor.

Devices for changing solar radiation into heat and mechanical energy are called solar engines. They are in fact a combination of a heat engine and a solar collector. The collector may be a flat one (as in the case of a solar pump), a focusing one, or the surface of a tropical ocean. Heat passes from a warmer body to a cooler one of its own accord. A fraction of the transferred heat can then be transformed into mechanical energy (work). The part of a solar engine where a material is heated by solar radiation is called an evaporator. The cold material into which the heat passes is in a condenser. The heating of the evaporator is performed by a solar collector, while the condenser is cooled by cold water. The greater the difference in temperature between the evaporator and the condenser, the more of the heat can be converted into work.

A solar engine works as follows: the heat from the collector warms the working fluid (ammonia, propane, sulphur dioxide, etc.). The warm fluid evaporates and becomes a hot vapour at high pressure. The vapour drives a turbine, performing mechanical work. By expansion in the turbine the vapour cools and is led into the condenser. There it is cooled further and liquefies, after which it is pumped back into the evaporator.

A *solar pump* is a solar engine for pumping water from a well. This water also cools the condenser. There are about one hundred solar pumps in operation in different parts of the world. They are particularly useful in areas with a large amount of solar radiation, and in deserts and semi-desert regions they are indispensable.

A natural collector of solar radiation is the surface of tropical oceans. It absorbs solar radiation, heats up, and then contains a huge amount of heat. The temperature of the surface water of tropical oceans reaches up to 28° C (82° F). But at a depth of around 400 m (1310 ft) the water temperature is only about 5° C (41° F). This temperature difference of 20–23° C (36–41° F) can be used to drive heat engines. The warm surface water heats the evaporator, while the cold water from lower down cools the condenser. This idea of a Frenchman, J. D. Arsonval, dating from 1881, was first put into practice in Cuba in 1929. In 1956 French experts built a power station at Abijan on the Ivory Coast driven by an oceanic solar engine. Its output was 3.5 MW. The OTEC (Oceanic Thermal Energy Conversion) system which is being tested by the Americans is based on the same principle; this research was inspired by Arsonval's idea.

Every motor vehicle is in fact driven by solar energy, since petrol and diesel fuel are distilled from oil, a source of fossil solar energy. Even aircraft are powered by the same source, since the fuel they run on is also obtained from oil.

The price of crude oil on the world market is going up (in the long term). The remaining reserves are sufficient only for two to three generations. In addition, burning petroleum products pollutes our air. Exhaust gases contain substances which are harmful to health. All these are good reasons for replacing fuel oil with solar energy. Several types of 'solar' vehicles have already been tried:

- In Brazil there are around one million cars which carry the inscription 'movido a alcohol' at the rear. They have engines adapted to burn ethanol (ethyl alcohol). The ethyl alcohol is obtained by the fermentation of sugar cane, cassava and other plants. Solar cars are driven by renewable energy, and their exhausts contain only water and carbon dioxide. The oxidation of ethanol (C_2H_5OH) cannot produce anything else.
- The hydrogen car. We have already mentioned the fact that solar radiation can be used to break water down into hydrogen and oxygen. Hydrogen can be used instead of petrol to power motor cars. Because it is highly explosive, it is bonded to metals for this purpose (in the form of hydrides). Gaseous hydrogen is released from these as required.
- Electrical power. Hydrogen and oxygen obtained using solar radiation can be changed in a fuel cell into electricity (see page 292). This is then used to run an electric motor. Such a car accelerates smoothly, runs quietly, and produces only pure water as a by-product. It uses energy which is available in unlimited quantities.

Electricity to drive cars and 'sporting'

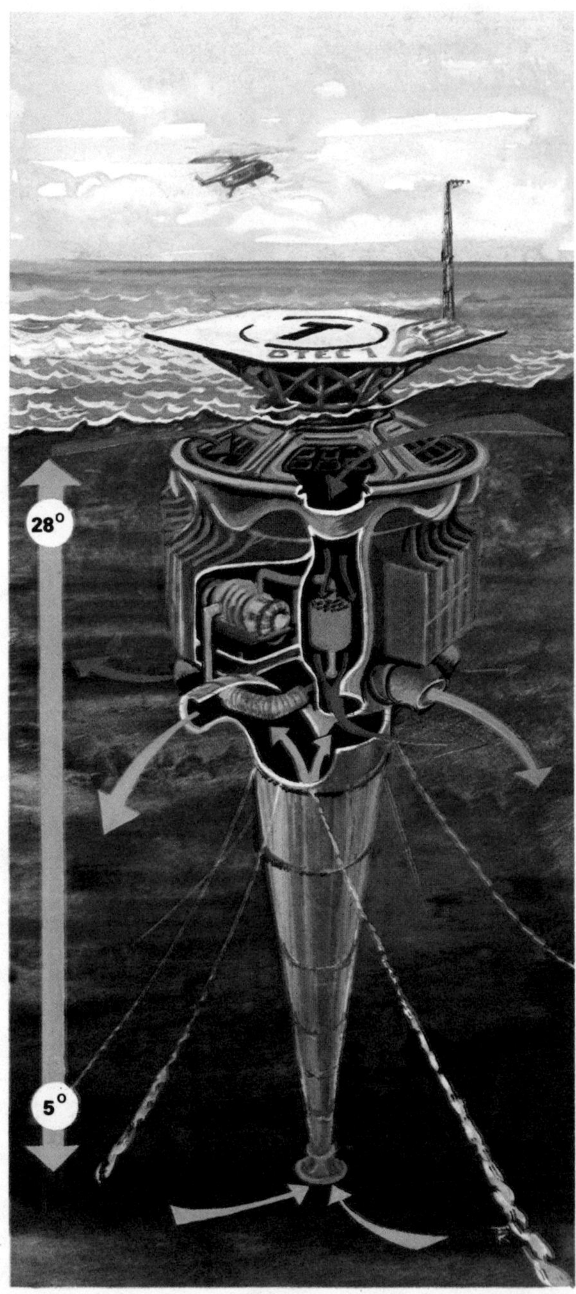

407 The heat of the oceans can be used to drive huge motors. The OTEC is one of the systems.
Warm (25–28° C/77–82° F) water is led from the surface, cold (5° C/41° F) water is pumped from greater depths. The mechanical energy gained is converted into electricity.

aircraft can also be obtained by means of solar cells. These are located on the roof of the car or on the wings of the aircraft. Several such cars and aircraft have already been constructed. But the current obtained this way is so weak that the car goes slowly, and only in sunny weather. It is, however, possible to put solar panels on the roof and store the electricity in accumulators. The car is then driven by electricity from the accumulators, which are changed as required. Cars like this are especially useful in cities, where it is not possible to drive very fast anyway. They are quiet and pollution-free.

We have not been able to go through all the changes which solar energy undergoes here on Earth. Nor has it been possible to examine in detail all that sunshine makes possible for man. But even these few examples have proved just how useful the Sun is for man and all other living things.

11. THE DAYTIME SKY AND THE NIGHT SKY

How the Universe looks from Earth

We can only see a tiny part of the Earth's surface around us. Our view ends at the line we call the true horizon. Above the horizon is the vault of heaven, our window into the Universe. The appearance of the sky depends on whether the Sun is above the horizon (daytime sky), below it (night sky), or just on it (sunrise or sunset). The sky also changes its appearance according to where the Earth is on its orbit round the Sun, and to the season of the year (Figure 105, and constellations pages 307–318).

The Earth rotates about its own axis. Every point on Earth alternates between light and darkness (Figure 409). At dawn the Sun rises above the eastern horizon; in the morning it climbs, reaching its highest point at noon, and in the afternoon it gradually sinks towards the western horizon, where it sets at dusk. In the night the Sun is below the horizon and we cannot see it. It thus divides time into a period of light and a period of darkness – day and night. Its movement divides the day into dawn, morning, noon, afternoon, dusk and evening. That is how our ancestors saw it; it is also how we see it today, and how our descendents will go on seeing it.

But spacemen have seen our planet quite differently (Figures 242, 243 and 385), half lit by the rays of the Sun and half in darkness. The inhabitants in the part which is illuminated have daylight, and those who live in the part where the sunlight is not falling have night; for them the Sun is below the horizon.

The Sun's radiation, especially the blue component, is diffused by the molecules of nitrogen and oxygen of the troposphere. It is as if each molecule in the troposphere were irradiating its own faint blue light. Together they give the impression of a continuous blue dome over the landscape (Figure 26). The light of the stars is weaker than the radiation of the sky, and we should see it in the daytime only if we were high above the troposphere. In daylight we see only the closest part of the Universe, our own planet's troposphere.

The stars and other bodies in the Universe (planets, comets, nebulae, galaxies, etc.) can be seen at night, when the troposphere does not diffuse sunlight, because the Sun is below the horizon. The sky seems dark, even velvety black (provided there is no moon).

The Earth orbits the Sun once every 365 days and six hours. In the course of this time (the *tropical year*) the seasons change here on Earth. The changing seasons (spring, summer, autumn, winter) are caused by the fact that the Earth's axis is not perpendicular to its orbit (Figure 187, upper picture).

In June and July the northern hemisphere is inclined most towards the Sun, which is high above the horizon at noon, its rays falling on us almost perpendicularly. That is why it is warm in summer, even though the Sun is at its furthest from the Earth. In December and January the northern hemisphere inclines furthest away from the Sun. At that time the Sun is low over the horizon; its rays fall upon us at a shallow angle, and are not very warm by the time they reach us. At the coldest time of the year we are actually closest to the Sun in the northern hemisphere.

It is not just the daytime sky, but also the night sky which changes in the course of the seasons (Figure 105). In winter we see different constellations from those in summer. The Milky Way occupies a different position; in winter the nights are long, in summer short, and so on.

From the surface of another planet we should see the sky in quite a different aspect (Figure 410). On Mercury, for instance, there is almost no atmosphere, so there would be no blue sky. There the sky is dark even in the daytime and the Sun seems several times larger than we see it from the Earth. On Venus, however, neither the Sun nor the stars can be seen, since our neighbour is constantly covered in very thick clouds. We should easily be able to distinguish day and

408 The Universe seen from the surface of the Earth is called the sky. In the daytime we see the Sun and a blue sky, and sometimes a solar halo (left). At sunset the Sun appears red (bottom right). The light of the Sun diffused at the interplanetary dust appears at night as the Zodiacal light (bottom left). At night we see, against a dark sky, stars grouped in constellations, planets, moons, comets and nebulae (top edge of picture). Long shadows (bottom edge of picture) trail behind the Earth and the Moon, causing solar and lunar eclipses.

night, since the Sun's light shines through the Venusian clouds brightly enough to allow photography on the planet's surface in the daytime. From the giant planets Jupiter and Saturn we should see a much smaller Sun than we do here on Earth. It is rather senseless to ask what the sky looks like from the surface of Jupiter or Saturn, since these planets have no firm surface. They have a solid core of metallic hydrogen, but not a single ray from the Sun can get down that far. From the farthest of the planets, Pluto, the Sun looks quite small, and the amount of radiation reaching the planet's surface at noon is similar to that reaching the Earth from the Moon when it is full.

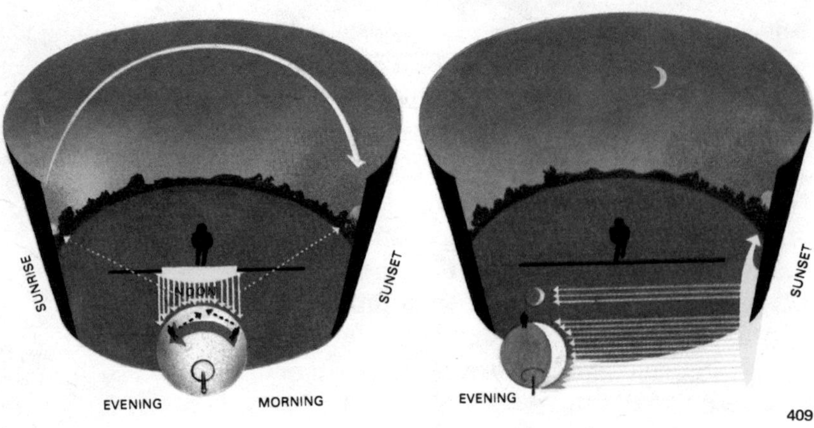

409 The alternation of day and night is due to the rotation of the Earth.

If we managed to get to the middle part of the globular stellar cluster M 13 in the constellation of Hercules, we should see a sky thickly set with shining stars (about one million). From the Earth we can see less than one-hundredth of them (about 6000). From the space between galaxies we should probably see no stars at all, only the galaxies, looking like misty little clouds.

The blue sky

The sky above us is constantly changing. On a clear, moonless night it looks like a black velvet dome set with shining stars. If the Moon is in the sky, the Sun's light is reflected on to the dark side of the Earth. We see the outlines of our surroundings, and only the brightest of the stars are visible in the sky.

In the daytime the sky is often covered with clouds of various shapes and sizes, from thin, wispy ones to dark storm clouds (Figures 301–304). Even a blue, cloudless sky may be countless different shades in different parts, changing with the time of day and year. In summer, after a long period of drought, many tiny particles of dust are carried into the air; the countryside in the distance is blurred, and the sky more whitish than blue. But after heavy rain has washed the dust from the air the sky is deep blue.

The colour of the sky cannot be due to the light irradiated by the atmosphere, or we should see a blue sky even at night. We have already said how the molecules of air diffuse the white light of the Sun particularly in the blue and ultraviolet parts of the spectrum. Red and infra-red radiation, on the other hand, is only slightly diffused (Figure 26).

Solar radiation falling on the Earth's surface is thus deprived of blue and violet rays. That is why the Sun seems a pale yellow colour instead of being white. The closer the Sun is to the horizon, the longer the path of its rays through the air, and the more blue radiation is lost through diffusion. That is the reason the sky is blue and the Sun orange or even red as it sets.

410

The cloudy sky, and what happens to light in icy clouds

Clouds come in all shapes, shades and sizes. They appear, grow and alter before our very eyes. Thin *cirrostratus* (amongst the highest clouds) diffuses the Sun's rays only slightly, while dark storm clouds *(cumulonimbus)* blot out its light altogether.

A cloud is a large cluster of tiny droplets of water or fine crystals of ice. According to its height above the ground it can belong to one of three groups: low clouds, medium-altitude clouds or high clouds (Figure 300). In low clouds the water is in the form of droplets. High clouds, where the temperature is very low, consist of fine crystals of ice. The highest of all, the cirrus (5–13 km/3–8 miles up), are fleecy or feather-like in appearance.

While the water droplets in clouds are shaped like tiny balls, ice crystals have various regular shapes, depending mainly on the temperature in the cloud. The crystals of ice in high clouds are ten to a hundred times bigger than the droplets in low clouds. The size of crystals varies from a few tenths of a millimetre to several millimetres. Out of the immense variety of beautiful shapes we are most interested in the following: the hexahedral prism, the hexahedral star, and the hexahedral pillar with two small plates. In clouds with these shapes of ice crystals various light effects called *halo phenomena* occur around the Sun. In them sunlight is refracted and dispersed, which causes splendid rings, brightly shining spots (called *parhelia*), or pillars or crosses of light to be visible around the Sun. These halo phenomena are among the most beautiful sights in nature's repertoire.

Medium-altitude clouds (about 3–8 km/2–5 miles) do not give rise to any halo effects. Light is diffracted by their fine droplets. This explains the *corona* (also called aureole) – a coloured ring around the Sun or Moon. The ring ranges in colour outwards from the Sun or Moon from blue, through green and yellow, to red.

The corona differs in size from halo rings: the radius of the corona is 10° at the most, while that of halo rings is 22° or 45°. (Let us recall that the apparent diameter of the Sun or the Moon is of about $\frac{1}{2}$°.) The sequence of

410 The Sun from Mercury, Venus, the Earth, a satellite of Saturn, and from Pluto.

colours of the halo rings is also the opposite to that in the corona. The corona is particularly easy to observe if the Sun or Moon is behind a cloud layer which covers all or part of the sky. Cloud of this type is called *altostratus,* and behind it the Sun looks as if we were seeing it through ground glass.

Very rarely all the halo phenomena shown in Figure 411 occur simultaneously. Usually only part of some circle, arc or pillar is visible. Some halo phenomena can also be seen around the Moon.

According to the shape and position of the crystals, but also according to whether the light is reflected or refracted, different halo phenomena occur in high cloud, mostly in the form of a halo of 22°. This is a coloured ring around the Sun with a radius of 22°. It is caused by hexahedral ice prisms refracting the rays of the Sun (Figure 411). The white light of the Sun is decomposed (i.e. refracted) into rainbow colours. For this reason haloes of 22° are seen on the inside (closer to the Sun) as reddish and on the outside as bluish.

On each side of the Sun, close to a halo of 22°, a parhelion appears. These are due to crystals similar to a hexahedral table (a hexahedral prism with plates at each end).

A circumhorizontal arc occurs through reflection on hexahedral prisms and completely encircles the sky at a constant height above the horizon. Reflection from plate-like crystals forms a *sun pillar* above and below the Sun. A similar effect is produced by the reflection of a light on the surface of a pond or river at night. The combination of a sun pillar and a circumhorizontal arc gives a *solar cross,* a phenomenon which in former times gave rise to fear and superstition.

The appearance of halo phenomena, especially the frequent halo of 22°, often heralds rain. This is understandable, since high clouds are often followed by medium-altitude clouds and then by the low clouds which carry rain (Figure 300).

The best-known optical effect in the lowest cloud layer (below 2 km/ $1\frac{1}{4}$ miles) is the rainbow. It is caused by raindrops dispersing sunlight into a sequence of colours (Figure 412).

411 Halo phenomena.

The rainbow — sunshine in the rain

In low clouds the tiny droplets collect into larger drops, which are heavy and fall to the ground as rain. About one million of the tiny droplets which make up clouds, combine to make a single raindrop. When a ray of sunshine enters a raindrop it is refracted and dispersed into the colours of the spectrum; inside the drop it is reflected, emerging as an array of coloured bands — a rainbow (Figure 412). The angle between the ray of sunshine falling on the drop and the coloured array which leaves it is approximately 42°. The rainbow is visible on the side opposite the Sun. If there is a wood, a mountain or a rock etc. behind the rainbow, it seems close to you, and indeed, the rain which caused it may be a matter of metres away from you. If you have ever watered the garden with a hose-pipe in the morning or evening, you may have made your own rainbow. Sometimes the rain which gives rise to the rainbow may be up to 2 km ($1\frac{1}{4}$ miles) away.

The author once saw an interesting phenomenon from the window of an aircraft. On the port side the sun was shining, and on the starboard side rain was falling from a large cloud. To the right, obliquely below in the rain, was a large coloured circular band, in the centre of which the plane's shadow was moving. The outer edge of the band was red, the inner blue, and in between were the other colours of the spectrum. It was a *circular rainbow*.

On the ground you can only see a rainbow as an arc (Figure 412). The lower the Sun is over the horizon and the higher our vantage point, the larger the rainbow appears. If we were viewing from the top of a tall tower and the rain was near to us (on the opposite side of us to the Sun), we should see a circular rainbow.

The rainbow we have been speaking about is the *primary rainbow*. Above this is often a *secondary rainbow*, which is fainter and has its colours in the opposite order, i.e. with red on the inside. The secondary rainbow is caused by double reflection in the drops of rain.

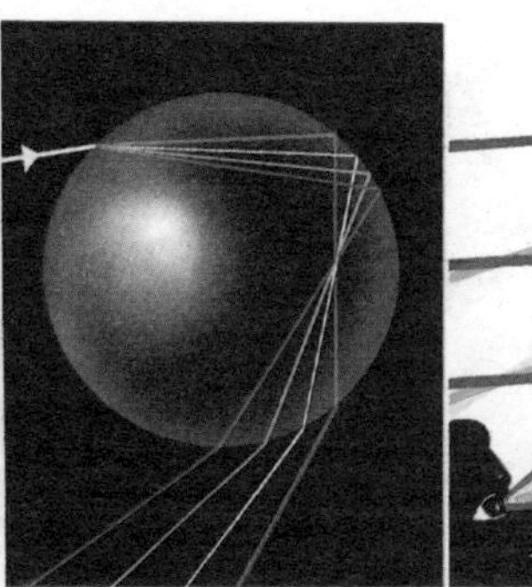

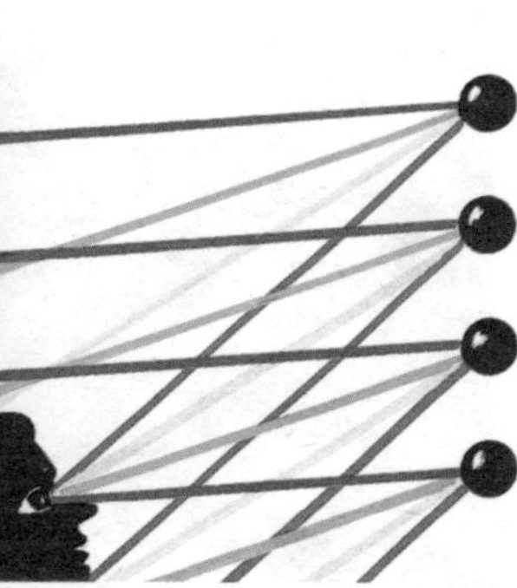

412

412 A rainbow – the decomposition of the Sun's rays by raindrops.

Evening twilight

At night we and the atmosphere above us are in the shade of the Earth. The transition from day to night is not a sudden one. Following the setting of the Sun, we on the ground are in the shade, but the atmosphere above us is still illuminated by the Sun's rays for another hour or so. Though it is night on the Earth's surface, it is as if part of the daytime sky remained up there above us. The border between the shadow and the Sun's rays rises rapidly, until it is beyond the atmosphere altogether. The transition from day to night is called the evening twilight, that from night to day the morning twilight. After sunset the Sun's rays continue to be diffused on the molecules of nitrogen and oxygen, and fall towards us. Along the western horizon where the Sun has set, an array of horizontal coloured bands appears. Their colours change with height, from crimson at the horizon, through orange, pale yellow and light blue, to dark blue. These broad strips of colour have a strangely calming effect on most of us. Gradually, as the Sun sinks below the horizon, these bands change colour and become duller.

On the opposite side, near the eastern horizon, you can see the shadow of the Earth projected into the atmosphere. It is a blue-grey strip just above the horizon, which slowly rises, and is visible to a height of about 6° above the horizon, after which it disappears.

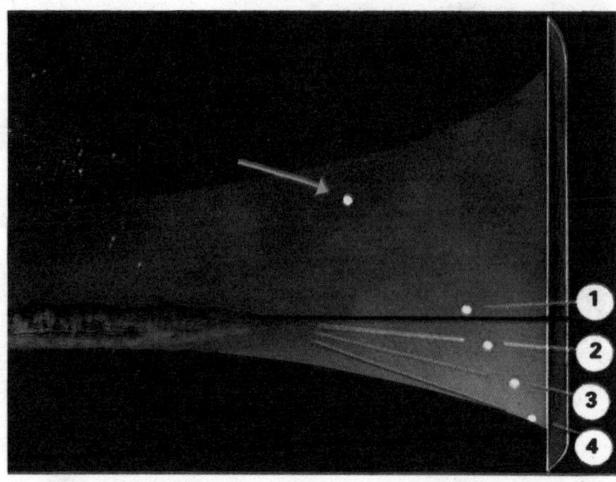

413 When the Sun is on the horizon **(1)**, the day ends and twilight begins. According to the angle of the Sun below the horizon we distinguish: civil twilight, up to 6° **(2)**, nautical twilight, up to 12° **(3)**, and astronomical twilight, up to 18° **(4)**.

When the Sun has sunk 18° below the western horizon, its rays illuminate the atmosphere so weakly that the radiation of the night sky itself is equally intense. At that moment the astronomical twilight ends, and the black velvet night suitable for astronomical observations begins. When the Sun is less than 6° below the horizon, there is civil twilight. Nautical twilight is taken as the time until the Sun is 12° below the horizon (Figure 413).

If you are out in the open somewhere, you should never fail to watch the sunset and the beautiful play of colours, light and darkness which occurs at dusk.

The night sky

When the Sun has sunk 18° below the horizon, astronomical twilight ends and true night begins. (Let us recall that 18° is one-fifth of the distance from zenith to horizon.) Against the velvety black background of the sky you can see very faint stars, for the atmosphere is no longer illuminated by the Sun. But even the night sky emits light. The light of the night sky consists of various components:

1 The light of towns and cities diffused by the molecules of air, droplets of water and fine dust. This artificially produced diffused light is a major nuisance to astronomers. They therefore build observatories far from centres of civilization, which produce light, smoke, dust, chemical impurities, heat and radio interference.

2 Various chemical reactions take place in the atmosphere; there is a recombination of ions which produces energy in the form of light. It is a very weak radiation, not perceptible to the naked eye. But it does matter to astronomers, since they have to use long exposures when photographing faint objects.

3 The diffused light of the Moon in the troposphere, provided of course the Moon is 'out'. The light of the Moon is of course much less bright than that of the Sun in the daytime. On a moonlit night only the brightest stars are visible. Moonlight is refracted by the ice crystals in high cloud, and halo effects are produced around the Moon, though they are not as colourful as those around the Sun.

4 The polar aurorae (luminescence of the atmosphere at heights of over 100 km/62

miles), in the shape of pulsating arcs, draperies, rays, a diffuse luminous surface, and so on. Various colours from pale green to dark red make these phenomena interesting and spectacular. The aurorae are excited by clouds of particles emitted by solar flares. After huge solar flares these aurorae are visible even in temperate zones, making a sight of rare beauty.

5 The Zodiacal light is a tall, faintly glowing cone stretching on either side of the ecliptic (Figure 112). On a moonless night it can be seen in the west after sunset or in the east before sunrise.

Along the plane of the ecliptic, far beyond the Earth's orbit, there stretches a large cloud of fine dust – the *meteoroid complex.* Every speck in this extensive cloud is in fact a tiny planet, each orbiting the Sun like the big planets. The meteoroid complex is a flattened cloud. It mainly consists of the original material from which the planets and their moons arose 4500 million years ago. The Sun's light is reflected and diffused by this fine dust, and we can therefore see it in the night sky. It can best be seen in February and March after sunset or in September and October before sunrise, when the Zodiacal light is nearly perpendicular to the horizon and reaches almost to the zenith.

6 Particles from the meteoroid complex very often collide with the Earth. But they do not reach the surface. The journey through the upper atmosphere heats them to such an extent that they glow white hot and burn up at an altitude of around 100 km (62 miles). They are called shooting stars by laymen. This glowing of a tiny particle of dust (around the size of a grain of sand) high in the atmosphere is called by astronomers a *meteor.* The light of a meteor comes from the kinetic energy of its particles, which is considerable, since they are moving at a very high speed, about 30–70 km (19–44 miles) per second.

All the types of radiation of the night sky mentioned come from the closest region of the Universe: from the Earth's atmosphere (1–4; 6) or from nearby interplanetary space (5) closer than a quarter of a light hour away. The planets we see in the night sky and the comets which sometimes appear are also in near cosmic space. Their light (more precisely the sunlight they reflect) takes less than an hour to reach us. Light from Saturn, the most distant of the clearly visible planets, takes about one hour and 20 minutes to reach us.

With the rotation of the Earth the stars rise and set in the sky, their relative positions remaining the same. In actual fact stars are moving very rapidly indeed, from a few to several hundred kilometres per second; but due to their enormous distances, they always seem to be in the same place.

Some of the photons from the night sky are thus only a fraction of one-thousandth of a second old when they reach our eyes – such as those from the aurorae, meteors, and so on. Others may be up to several dozen minutes old – like those from the planets, comets and the Zodiacal light. Most of them originated in the Sun and are only reflected from these bodies.

On the other hand photons from the thousands of stars in the night sky may be up to several thousands of years old when we perceive them. They are photons emitted from the photospheres of stars, and are not reflected, as in the case of the planets or comets. The photons come to us directly from stars which emitted them many millenia ago. We have no idea what their parent stars look like today. When we gaze into the depths of space we are looking into the distant past, a past more distant the further away the star we are observing (Figures 28 and 29).

The constellations – groups of stars in the sky

The names of stars and constellations came into use some seven thousand years ago. The Greeks, the Chinese and the nations along the Nile, the Euphrates and the Indus associated stars into patterns and named them after their gods, heroes and animals. Their ideas have survived to modern times. In ancient times a knowledge of constellations was much more important in practical life than today. For the farmer, the herdsman, the sailor and the traveller the starry sky was clock, calendar and compass.

It was not only practical considerations which led ancient men to learn about the stars, but also their natural inquisitiveness – the same thirst for knowledge that we have ourselves. We want to know what stars are, why they shine, how they came into being and how they cease to exist; we ask the many questions which occur to us when we look up at the night sky. Many of these age-old queries remained unanswered until recent times, when modern science was able to use huge radiotelescopes and acquired new knowledge of the fundamental particles of which the Universe is built.

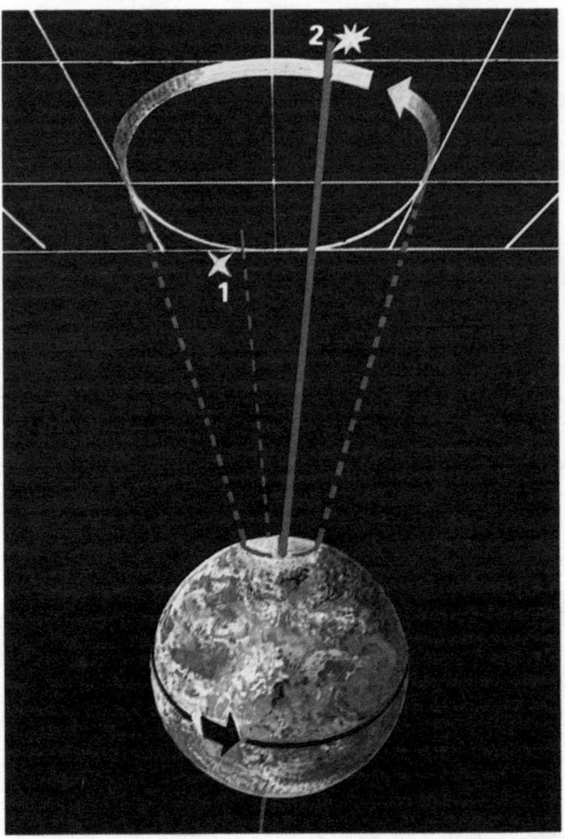

414

414 The Earth's axis gradually changes its position in space. The change is called precession. The North Pole describes a circle in the sky once every 26,000 years. This is due to the gravitational effect of the Moon and the Sun on the Earth. Today the North Pole is close to the Pole Star **(2),** but in 13,000 years it will be near the star Vega in the constellation of Lyra **(1).**

The names of 48 of the constellations in today's sky come from the catalogue of the Greek astronomer Ptolemy, 'Almagest' (dated AD 137), which was in use until half way through the 15th century. The work was translated from Greek into Arabic in the ninth century – the Arabic word *almagest* means 'the greatest'. In the 12th century it was translated into Latin. The book gives a review of astronomy in Ptolemy's day and among other things contains a list of 1022 stars. But the names of the constellations and of many of the stars are much older than Greek culture. The constellations had been known long before the time of Ptolemy.

Experts can show that all 48 constellations were defined in the sky over a relatively short period by one group of ancient observers, probably in Mesopotamia or Asia Minor. How? The 48 constellations do not cover the whole of the sky, since a small area (the *spherical cap*) in the southern hemisphere does not contain any of those in Ptolemy's book. From the size of the cap one can deduce the latitude at which the ancient astronomers who named the constellations lived. The radius of the spherical cap gives the latitude, 35°. The area of the sky which lacks ancient constellations was not visible to observers at a latitude of 35° North (Figure 415). The lower the latitude from which one observes the sky, the smaller the spherical cap centred on the South Pole we cannot see. At the equator we should see from pole to pole, or in other words all the constellations. If you look at the map you will see that the latitude of 35° North corresponds with the area of the oldest civilizations: those of the Middle East, Mesopotamia, Persia and northern India.

When were the constellations defined? Here we are assisted in our enquiries by a phenomenon we call precession. The North and South Poles slowly shift across the celestial sphere (Figure 414). They describe a large circle every 26,000 years. At the time the 48 constellations were defined, the South Pole was in the centre of the invisible spherical cap with a radius of 35°. We know the centre of the spherical cap; in other words where the pole was at that time. It is the cap on the sky where none of the original 48 constellations exists. Today the centre of the invisible cap (the South Pole) is in a different place altogether; it has moved about 96°. Since the pole moves 360° in 26,000 years (in other words describes a full circle as a result of precession), one can calculate that the movement of 96° took about 7000 years. That means, according to experts, that the 48 original constellations were defined and named around 5000 BC.

The constellations are quite random groupings of stars. They look different today compared with 100,000 years ago, and they will look different again in another 100,000 years, since the stars move slowly across the sky. The appearance of the constellations also changes, because some stars die, and new ones appear. Finally, the grouping of stars in constellations is only valid for us, the inhabitants of this part of our Galaxy. If we could ask intelligent inhabitants in the planetary system of a star, say, 10,000 light years away to describe the constellations, we should find that their views are quite different and that the stars in their sky are not ours at all. But they would see the same galaxies, such as M 31, which we see, though in quite different constellations. This is because galaxies are much farther away than stars.

The whole of our sky is today divided into 88 constellations, each of which has its own vernacular name, such as the Great Bear, the Virgin, the Eagle, etc. But astronomers throughout the world use their Latin names, such as Ursa Major, Virgo, Aquila, etc. Thus in the next chapter the English name is followed by the Latin name and genitive, for example Ursa Major, Ursae Majoris. After the Latin name you will find the abbreviation for the constellation, for example UMa, Aur, Aql. It is usual to use Latin genitives and abbreviations when denoting the stars.

Bright or important stars have a name of their own, such as Sirius or Polaris. But the most common designation of individual stars is that using the Greek alphabet. Here alphabetical order corresponds to descending order of brightness in each constellation. The capital letters of the Latin alphabet from R onwards are used for variable stars, such as R Coronae Borealis.

Nebulae, stellar clusters and the brighter galaxies either have their own names (like the Veil Nebula or the Beehive) or are given a serial number according to the Messier Catalogue (abbr. M), so that M 44 is the Beehive, M 31 is the galaxy in Andromeda, etc. In 1784 the French astronomer Charles Messier published a list of nebulous objects in the sky, in order to avoid their confusion with comets. Many of the 109 objects in his catalogue later turned out to be stellar clusters or galaxies, while others were nebulae. A much more comprehensive list was published one hundred years later (1888) by the Danish astronomer J. L. Dreyer at the Armagh observatory in Ireland. It contained nearly 8000 nebulae, stellar clusters and galaxies, and was called the New General Catalogue (NGC). This abbreviation is followed by a serial number, so that, for instance, NGC 1976 is the Great Nebula in Orion. The brighter objects have two designations: such as NGC 1976, which is M 42; and NGC 224, which is M 31, etc.

In order to compare the size of constellations every such map in the following text shows a 10 degrees scale which is approximately 20 lunar diameters.

The following pages deal with the best-known constellations, their histories, and the most interesting objects to be seen in them. The maps contain symbols as listed on page 306.

415 The movement **(1)** of the pole **(0)** as a result of precession determines which constellations are invisible. The constellations were defined and named at a time when the oldest civilizations could not see the area marked in red. Today we are unable to see the constellations in the yellow rings **(2).** The shift has taken 7000 years.

415

Very faint star, but visible to the naked eye ●
Faint star ✦
Medium bright star
Bright star
Very bright star
Variable star invisible to the naked eye ○
Variable star visible to the naked eye ⊙
Stellar cluster
Planetary nebula
Diffuse nebula
Galaxy
Radiant of a meteor shower

THE CONSTELLATIONS AROUND THE NORTH POLE

The pole is that point in the sky around which all the stars turn. This turning motion is, of course, an illusion created by the rotation of the Earth. The North Pole of the sky is close to the star Polaris (called the North Star). The constellations we see above the horizon all the time are called *circumpolar*. Those which 'turn' around the South Pole of the sky are not visible here in the North, or only partly so. They are known as the southern circumpolar constellations.

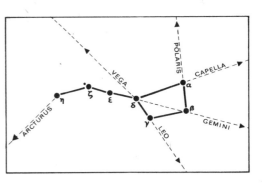

The best-known of the northern circumpolar constellations is certainly the Plough, also known as the Great Bear (Ursa Major), King Charles' Wain (wagon), or the Big Dipper (in the USA). It looks rather like a large ladle, or a wagon with a broken shaft. On a May evening you can see it high above your heads; the map will help you to find it and will make it easier for you to find the other constellations.

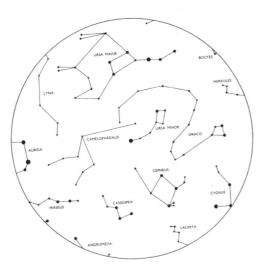

The 'shaft' of the Plough points towards the bright star Arcturus in the constellation of the Ploughman (Boötes), then towards Spica, in Virgo. If we extend the back wheels (Merak-Dubhe) we find Polaris, the North Star, and the constellation of the Little Bear (Ursa Minor) or Lesser Wain. The pole star shows us where the North Pole is, which is important for orientation at night. It is also easy to find Cassiopeia, on the opposite side from the pole star to the Great Bear. It is shaped like a broad W or M, depending on when you see it.

Between the Little Bear and Cassiopeia is Cepheus, which is not very striking, and in order to find its main stars we have to use the stars of the surrounding constellations, such as α and β Cassiopeiae. If the line joining these is extended to four times its length, one comes across α Cephei. This, along with a star map, enables one to find the other stars of the constellation.

The Great Bear
(The Plough, Big Dipper)
Ursa Major, Ursae Majoris, UMa

The Great Bear is the best known of the constellations in our northern sky. Its seven brightest stars make up the shape of a wagon or a ladle. The nations of Mesopotamia, northern Asia, Phoenicia, Persia and Greece saw it as a bear. It is fascinating that the North American Indians did, too. Is this a strange coincidence, or did the Indians bring the name with them when they moved from Asia to America across the Bering Straits? We do not know.

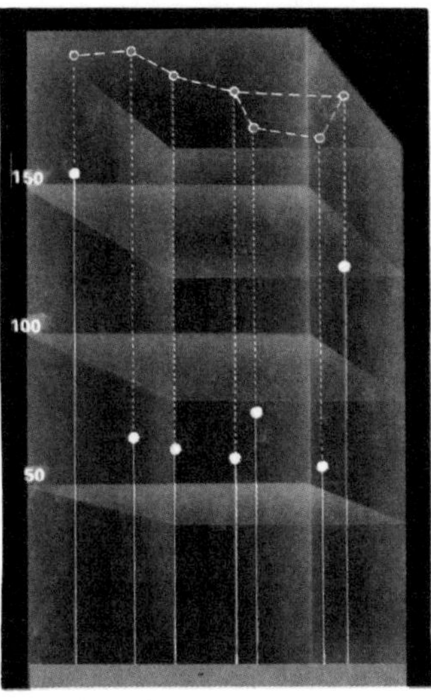

The Greeks had the following myth to explain the origin of the Great Bear. The Arcadian princess Callisto's beauty aroused the envy of Hera, spouse of the mightiest of the gods, Zeus. So the angry goddess changed her into a bear and drove her into the forest. Callisto begged in vain to be able to retain her human shape; Hera was adamant. So the erstwhile princess wandered through the woods, afraid of the wild animals she met, for she forgot she was one of them herself. But she soon learned to fear the hunters and their dogs. One day, recognizing her son Arcas among her pursuers, she approached him and tried to embrace him. Arcas aimed his spear at her in an attempt to defend himself, but at the last moment Zeus intervened to prevent Arcas slaying his mother Callisto. He changed Arcas into a bear as well – the Little Bear – and because he liked the pair of them, he placed them in the sky.

The central star in the shaft of the Plough is called Mizar (the Arab word for a belt). It is 78 light years away from us. Not far away from it is the fainter Alcor, which is often used for an eyesight test. A small telescope reveals that Mizar is in fact a binary. Actually, we see three stars through a telescope – Mizar, its companion, and Alcor. But all three of these are binaries, though their components are so close that they can be distinguished only spectroscopically – they are spectroscopic binaries. Thus Mizar is an example of a sextuple star. The six stars are held together by gravitational force and follow complex orbits around a common centre of gravity. It is worth mentioning that Mizar was the first spectral binary observed (in 1889). The spectral lines of such stars periodically split and recombine.

The stars Merak (bear's kidneys) and Dubhe (bear's back) point to the pole star. If we project the distance between them four times in an upwards direction, we find the North Star.

The most interesting property of the bright stars of the Great Bear is their movement. Five of them (β, γ, δ, ε, ζ), as well as Alcor, move across the sky at the same speed and in the

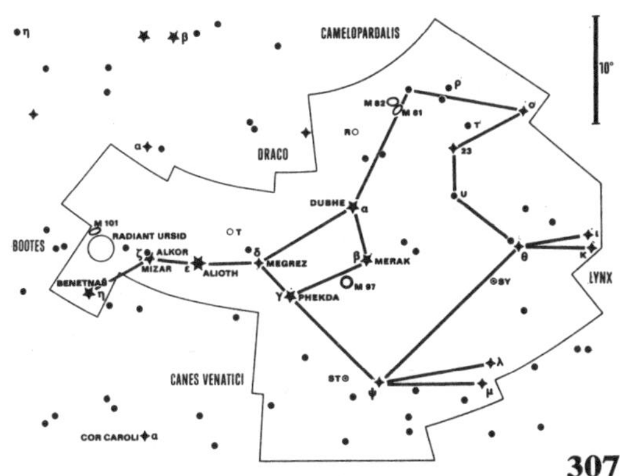

same direction (along a line joining β and δ). Dubhe and Benetnash move in the opposite direction. The shape of the Great Bear (as in the case of the other constellations) is changing (Figure 120). But the alteration is noticeable only in the course of several tens of millennia. The Babylonian and Egyptian astronomers saw the constellation in virtually the same shape as we see it today.

The stars of the Great Bear are at various distances away from us. The figure on page 307 (bottom left) shows the shape of the constellation as we see it on Earth. The distances of stars are shown in light years.

The Little Bear
Ursa Minor, Ursae Minoris, UMi

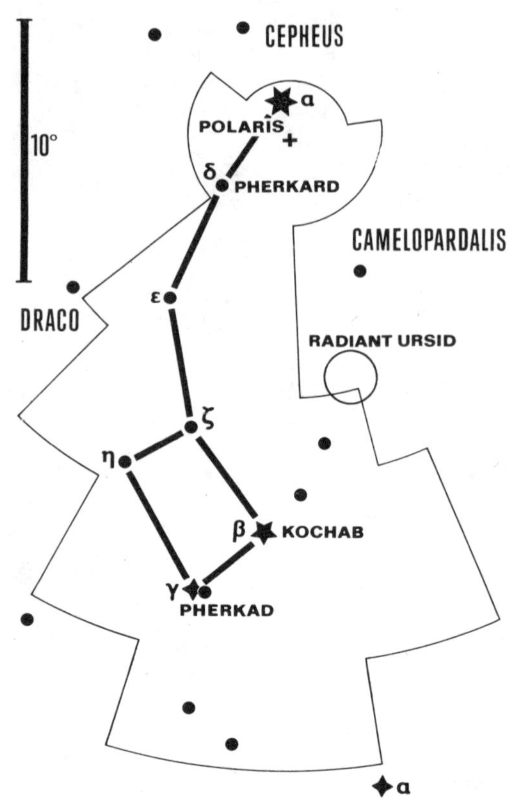

After Zeus changed Callisto and her son Arcas into stars, the goddess Hera was outraged to see how their forms had reached the heavens, so she made the god of the oceans, Poseidon, promise never to allow them to rest beneath the waves as the other constellations did. So neither of the bears ever sets; they constantly revolve around the pole as circumpolar constellations.

But there are other legends as to how the two constellations of the bears came into being. According to one of them the creatures had looked after the child Zeus when he was hiding from his father Cronus (Time) on the island of Crete. Cronus ruled Mount Olympus and, because of a curse, ate his own children. His wife Rhea, fearing for the life of her newborn son Zeus, hid him in a cave on the island, putting him in the charge of two bears. Zeus later showed his gratitude by granting them immortality in the sky.

There are few interesting features in this constellation. The most important of them is the star Polaris, because of its closeness to the North Pole. It was once known as the leading star, since it acted as a guide. At sea, in the desert and far from habitation it was the only means ancient travellers had of finding their way.

The pole star is rather less than 1° away from the pole (almost two lunar diameters), and so it describes a small circle around it. This is easy enough to demonstrate if we point a camera at it on a clear night and take a time exposure of several hours.

It is worth mentioning that not even the celestial pole itself (the point around which the whole sky seems to turn) stays in one place. Once every 26,000 years (a *Platonic year*) it describes a circle in the sky (Figure 414). In another 13,000 years it will be near the star Vega in the constellation of the Lyre. Then Vega will be the pole star. 13,000 years after that it will be back where it is today. This movement of the pole is called *precession of the Earth's axis.*

Polaris has a volume about one million times greater than the Sun. Its brightness and magnitude change regularly every four days as it pulses. It is over 600 light years away from us. From that distance we should not see the Sun even with a small telescope. This huge pulsating star is a binary, but its companion has never been observed directly. It orbits once every 30 years, and is detectable in Polaris' spectrum (a spectral binary). The two of them are orbited by three fainter companions (B, C and D), so that it is in fact a quintuple star. Like the Great Bear, the outline of the Little Bear is made up of seven stars. The two brighter stars, β and γ, which make up the back wheels of the wagon are called the 'sentries' of the pole.

Cassiopeia
Cassiopeia, Cassiopeiae, Cas

This circumpolar constellation represents the Ethiopian queen on her throne.

Cassiopeia was very beautiful. She boasted of being even lovelier than the water nymphs, the Nereids. This angered Poseidon, the sea king, since his wife was one of the Nereids. On her advice he sent a sea monster to lay waste the lands of Cassiopeia's husband, Cepheus. In order to save their kingdom, the unhappy parents were obliged to follow the instructions of a seer and have their daughter Andromeda chained to a cliff as a sacrifice to the monster. But she was rescued by Perseus who, enchanted by her beauty, made her his wife.

Cassiopeia was known as the 'Ethiopian star queen', because after her death she was raised up among the stars. But her enemies the Nereids at least made sure she was close to the Pole, so that she has to spend half the night head downwards to teach her modesty and to rid her of her conceit.

This constellation is on the opposite side of the North Star to the Great Bear. It shines above our heads in the autumn months, and is easily recognizable as a letter W. Cassiopeia lies in the Milky Way, and is rich in stars. A pair of binoculars is useful for observing it.

Among the interesting features of the constellation is the white giant Cih, which occasionally expands its gaseous envelope to between 10 and 18 solar diameters. There are other such variable stars, and they are known as eruptive variables.

On November 11, 1572, a bright new star appeared in the constellation of Cassiopeia

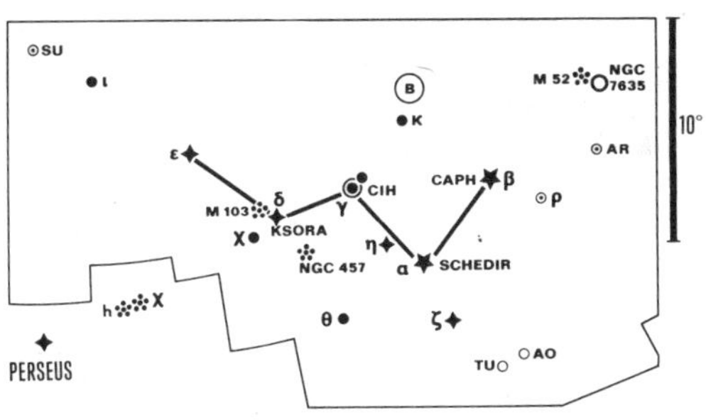

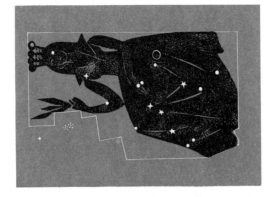

(denoted B on the map). It was observed in particular by the astronomer Tycho Brahe, so it is known as Tycho's Star. For several days it was brighter than Venus, and was visible even in the daytime. Then its brightness fell. The remains of this supernova are today still expanding at a velocity of 4000–5000 km per second and are a massive source of radio radiation. Radio-astronomers study them using radiotelescopes. They are known as the radio nebula Cassiopeia A. This rapidly expanding nebula is also a strong source of X-rays. There is so much energy in its movement that by collision with the interstellar gas it will shine for at least another 10,000 years.

Other objects in Cassiopeia worth observing through binoculars are the open stellar clusters M 52, M 103 and NGC 457, together with the nearby double stellar cluster χ and h Persei.

THE SPRING SKY

You can get your bearings in the spring sky by means of a triangle of bright stars: Regulus in Leo, the bluish Spica in Virgo, and the orange Arcturus in the Boötes.

The Boötes, Leo and Virgo have characteristic forms and are easy to remember. Beneath the shaft of the Plough you will find the Hunting Dogs, and below them, above Virgo, a hazy group of faint stars

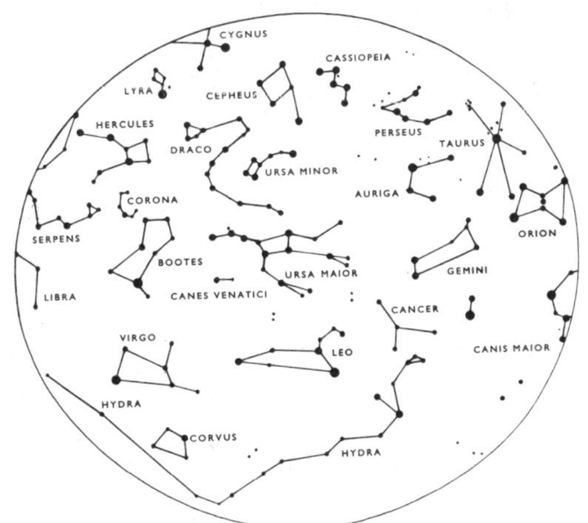

– Coma Berenices. Below Leo, Virgo and Libra the winding Hydra runs along the horizon, its head raised northwards. Between Virgo and Hydra is Corvus, with Crater leaning towards it. Below Hydra is Pyxis, Antlia and the top of Centaur. To the east of the Ploughman is Corona Borealis. To the west of Leo is another Zodiacal constellation, Cancer. Northwards of Leo is Leo Minor, and to the south, Sextans.

Cancer
(The Crab)
Cancer, Cancri, Cnc

This constellation is supposed to resemble the great crab which assisted the Hydra in its struggle against Hercules. One of Hercules' labours was to destroy the nine-headed Hydra, which lived in the swamps near the city of Lerna and was laying waste the surrounding countryside. The Hydra was helped by a huge crab with sharp claws. When the crab bit Hercules' leg, the latter's companion, the shepherd Iolaus, killed it with a well-aimed arrow. The Hydra turned to face him, and at that moment Iolaus burned off one of its heads with

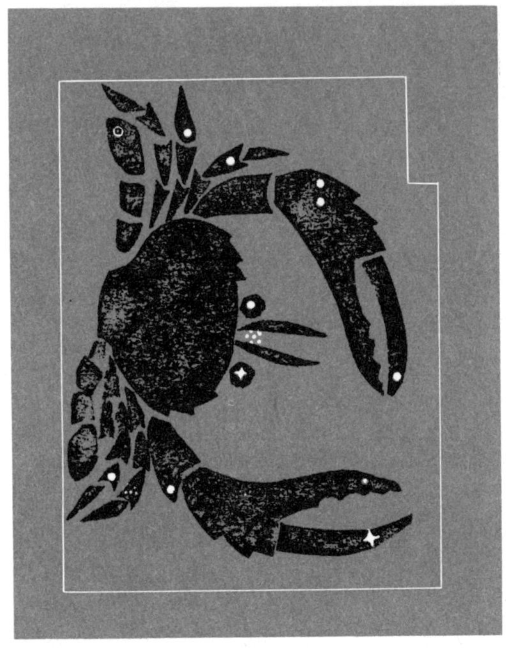

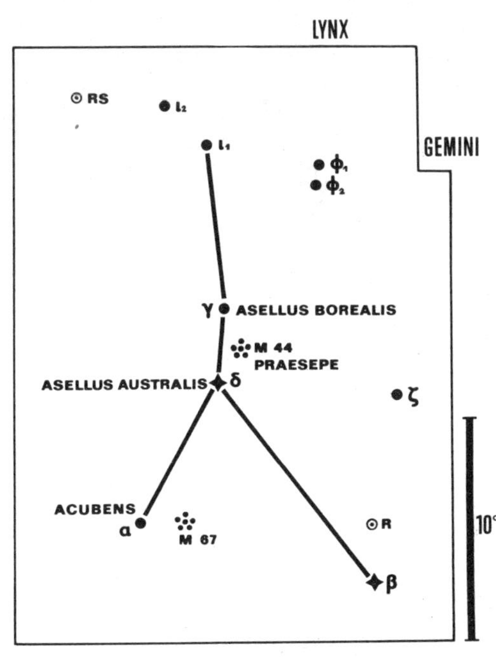

a blazing tree. While severed heads grew back on again, the burnt-off one could not. So the crab which was intended to help the Hydra actually led to its downfall. Nonetheless, the goddess Hera took it into the heavens, since she was a supporter of all Hercules' enemies.

The ancient Chaldeans called this constellation 'the gate of the people'. This was said to be where the spirits left the heavens when they came to assume human form. The Yucatan Indians believed that if the Sun got into the constellation it came down to Earth in the form of a fiery bird and took sacrifices from the altar.

Cancer is an inconspicuous constellation in the triangle of stars Pollux, Procyon and Regulus. It is the dimmest constellation of the Zodiac. Once it was its northernmost sign, so that the Sun was in it at the time of the summer solstice. The most northerly latitude on Earth where the Sun is directly overhead at noon (at the summer solstice) is still called the Tropic of Cancer, though the Sun is actually in Gemini. This shift is due to precession (Figures 105 and 415).

The most interesting object in Cancer is the open stellar cluster Praesepe (the Beehive). It is visible with the naked eye as a misty, luminous cloud looking rather like the straw in a manger. With binoculars you can make out the more prominent stars. This cluster is 500 light years away, has a diameter of 15 light years, and contains more than one hundred stars.

On either side of Praesepe are the stars δ and γ Cnc, called the Aselli (Donkeys). The southern one (Asellus australis) is exactly on the ecliptic, so that the Sun covers it once a year. Asellus borealis feeds from the manger from the northern side.

Leo

(The Lion)

Leo, Leonis, Leo

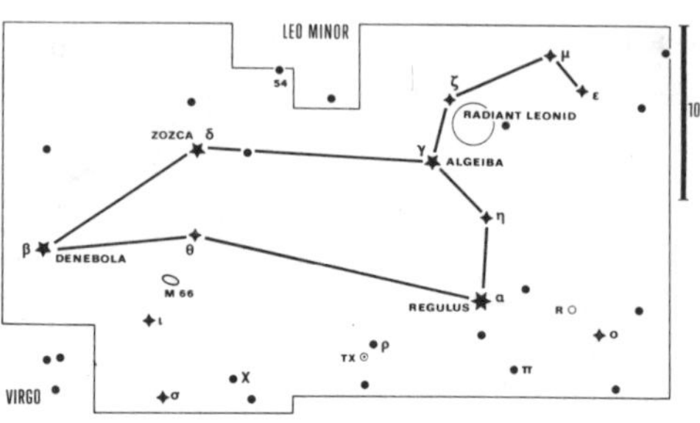

The renowned prophetess of Delphi, Pythia, once advised the hero Hercules to enter the service of the King of Mycenae. Mycenae is an ancient city on the Peloponnese peninsula – the ruins of the royal castle survive to this day. Hercules took her advice and went to Mycenae. There he was given 12 labours. The first was to slay a lion which lived in the nearby mountains. It was exceptionally large, and terrorized the surrounding countryside. Hercules found it in a cave (shown to tourists to this day) and carried it off to the King of Mycenae. The king was taken aback not only by the huge lion, but also by Hercules' strength. He therefore set him the next task, to kill the Hydra, hoping that the beast would make an end of the hero. But Hercules was successful once again. Because of his courage he soon became famous and immortal. He was taken among the gods on Olympus and attained the sky as a constellation. His victims the Lion, the Hydra and the Crab are up there with him to commemorate his deeds of valour.

Leo is a constellation of the Zodiac, reminiscent in shape of the king of beasts lying down. It is to the south of the Great Bear. In spring it can be seen in the evening. To see it in November, you would have to get up early in the morning. Between November 11 and 20 the meteors of the Leonid shower fly from the constellation's star ζ. They are the remains of Tempel's comet. Every 33 years the Earth passes through the densest stream of their particles. In 1833, 46,000 Leonid meteors were sighted in a single hour – what you might really call meteoric rain!

The bright star Regulus is one of the four 'guardians of the skies', or 'royal stars'. They were the stars that at the dawn of history marked the points of the summer solstice (Regulus), the vernal equinox (Aldebaran), the winter solstice (Fomalhaut) and the autumnal equinox (Antares). The royal stars divided up the path of the Sun (the ecliptic), and thus marked off the four seasons. So when the Sun was between Regulus and Antares, for instance, it was summer, and so on. It would be difficult to find a peasant today who still told the seasons by the stars, but in ancient times every herdsman, farmer and seafarer used the stars to determine the time of year and of the day.

Virgo

(The Virgin)

Virgo, Virginis, Vir

During the Golden Age on Earth men had no need of laws, and everyone honoured justice and good faith of his own accord. The goddess of justice and order, Astraea, daughter of Zeus, came down among men and taught them order, justice and law. It was said to have been a wonderful age in which to live. But men later became selfish, thinking only of their own interests, and invented weapons with which they slew their brothers and robbed the innocent. The Earth ran red with blood, and the gods deserted it. The last to remain was Astraea, in the hope that she might avert the wrath of Zeus and the destruction of mankind. But eventually she, too, deserted the Earth and returned to live among the other gods and goddesses. Now she dwells among the stars as the constellation of the Virgin.

For the inhabitants of the valley of the Tigris and the Euphrates this constellation represented the goddess Ishtar, daughter of heaven and queen of the stars. In Egypt it was called Isis, who was the mother of Horus, the solar deity, and wife of Osiris, god of the underworld.

At the moment when the Sun, at the autumnal equinox, passes through the constellation of Virgo, summer ends and autumn begins. Virgo is the sign of the Zodiac where the ecliptic passes through the equator (at the autumnal equinox), approximately in the middle, between the stars η Vir and β Vir. (On the map the point is marked by a cross.) The nearby γ Vir (Arich, Porrima) is a beautiful binary star.

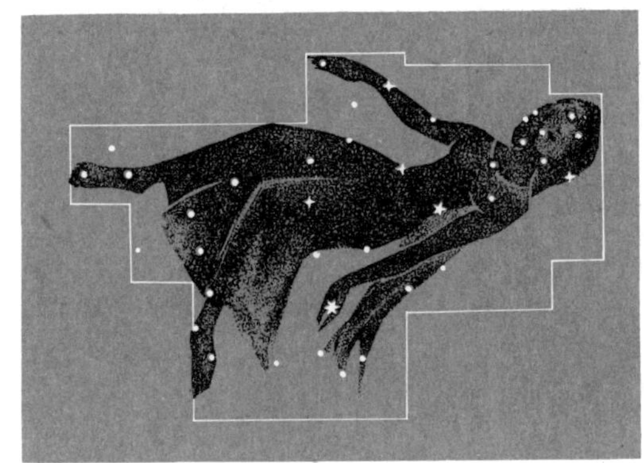

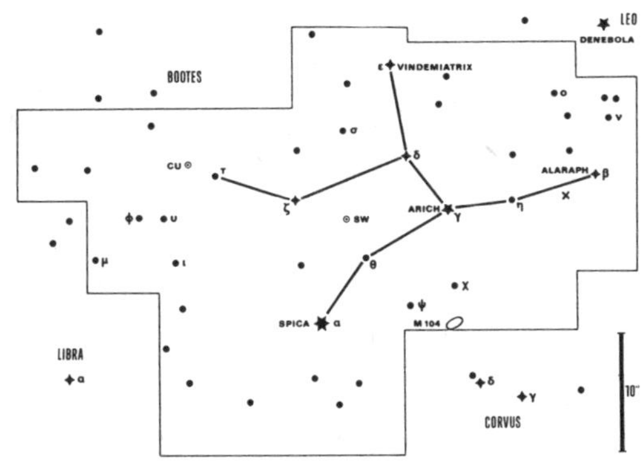

Among the most interesting objects in Virgo is a large cluster of galaxies which is moving away from us at a speed of 1200 km (750 miles) per second. The rate of recession is greater, the farther away any galaxy we observe. The cluster of galaxies in Virgo is significant in that it marks the approximate centre of our supergalaxy (Figure 142). It is around 60 million light years away. The local group of galaxies to which the Milky Way, the Magellanic Clouds, galaxy M 31 in Andromeda and 20 other galaxies belong, is at the southern edge of the supergalaxy. We therefore orbit the cluster of galaxies in Virgo. The diameter of the supergalaxy is about 150 million light years.

Virgo also contains the object known as 3C 273 (Figure 142 number 7). Through a telescope it looks like a faint star, at first sight no different from an immense number of others in the Milky Way. But spectral analysis shows that it is something quite different from the stars. Radio telescopes, for instance, have detected intensive radio radiation emitted from it. Though this object is much smaller than the Galaxy, its radiation is about one hundred times stronger than all the latter's 150,000 million stars put together. We now know many hundreds of such objects. Due to the expansion of the Universe, they are all receding from us at an enormous rate. They are known as *quasistellar objects*, or *quasars* for short. 3C 273 was the first quasar astronomers discovered, through its radio- and X-radiation. Quasars are the most distant objects so far ascertained in the Universe. It is likely that galaxies developed from quasars. If so, then quasars – like the 3C 273 in Virgo – were embryos of galaxies.

THE CONSTELLATIONS OF SUMMER

Summer is a very good time for observing stars. The nights in the northern hemisphere are short, but warm, and the Milky Way shines in all its glory. Its silvery band runs from the south, high above the eastern horizon, disappearing behind the northern horizon. It is easy to make out in it a clear triangle whose apexes are made up of the trio of bright stars: Deneb, in Cygnus; Altair, in Aquila; and Vega, in Lyra. The constellation Cygnus is shaped like a cross, and it is sometimes called the Northern Cross. If you take twice the distance between Deneb and Vega in the direction of Vega, you will find the typical quadrangular shape of Hercules. At the southern edge of the Milky Way there are two constellations of the Zodiac. The more westerly of these is Scorpio, with the red supergiant Antares, and to the east, where the Milky Way is brightest, there is Sagittarius. A radio telescope reveals the centre of our Galaxy far beyond the stars of Sagittarius. To the east of Sagittarius is the inconspicuous constellation of Capricorn. Between Scorpio and Virgo in the Zodiac lies Libra. Between Hercules and Scorpio are the elongated constellations of Ophiuchus and Serpens. High above the western horizon is Boötes – the shaft of the Plough points towards it. Between Boötes and Hercules is the small, regular constellation of Corona Borealis. Virgo sets below the western horizon. Between Aquila and Cygnus there are three small constellations: Delphinus, Vulpecula and Sagitta. Between Aquila and Sagittarius is Scutum.

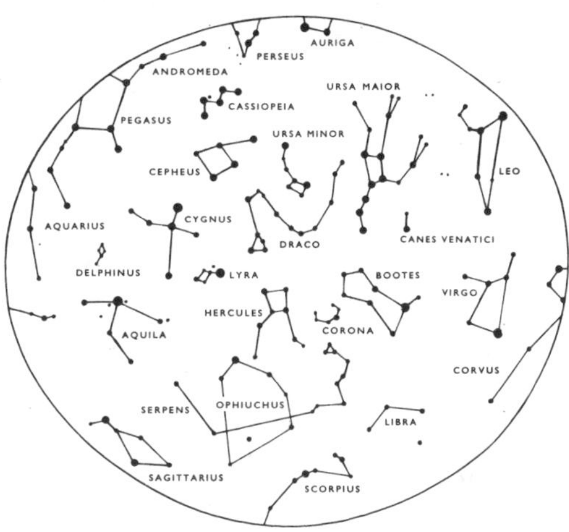

Libra
(The Scales)
Libra, Librae, Lib

Libra is one of the small constellations which commemorates the technical achievements of the ancient nations instead of the usual gods and heroes. It is a rather indistinct group of stars which lies on the ecliptic, and is thus one of the signs of the Zodiac. Not all the constellations of the Zodiac are named after animals, even though the Greek name zōidiakós (from *zôion* – animal) suggests this. The Zodiacal constellations are as follows: Aries (the Ram), Taurus (the Bull), Gemini (the Heavenly Twins), Cancer (the Crab), Leo (the Lion), Virgo (the Virgin), Libra (the Scales), Scorpio (the Scorpion), Sagittarius (the Archer), Capricorn (the Goat), Aquarius (the Water-Bearer) and Pisces (the Fishes) – Figure 105.

The Zodiac is the strip of sky along the ecliptic in which not only the Sun, but also the planets and the Moon are to be found (Figure 105). It stretches for 9° on either side of the ecliptic (the ecliptic is the apparent annual path of the Sun's centre in the sky) and is divided into 12 equal parts 30° long. The resulting rectangles (18° wide and 30° long) are called the signs of the Zodiac. More than 2000 years ago they were named after the constellations which were in them at the time. But because the vernal equinox is moving along the ecliptic in the opposite direction to the motion of the Sun, the signs of the Zodiac are also changing their position. In 36 years the vernal equinox shifts half a degree, which is a lunar diameter. Thus in 2000 years the signs of the Zodiac (in the sense of the rectangles they represent) have moved to places other than those occupied by their respective constellations. For any of us born under Libra according to the calendar (i.e. between September 23 and October 23), the Sun was actually in Virgo. But an early Greek or Roman born under Libra was actually born when the Sun was in that constellation.

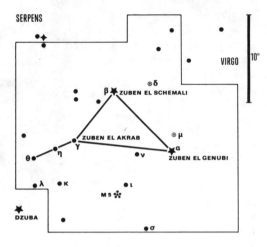

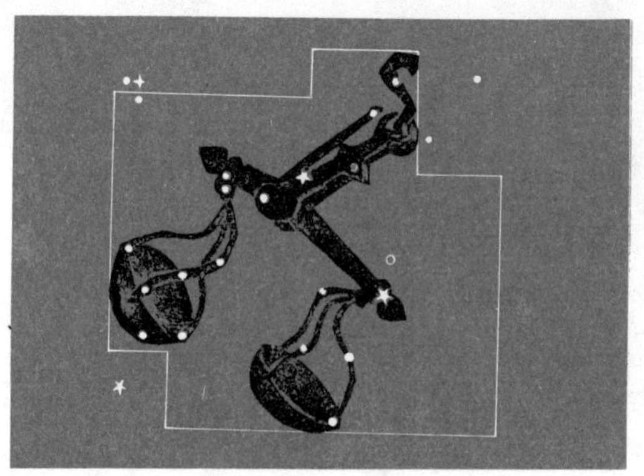

Few people really believe in astrology these days, but at least you will know why horoscopes always refer to the birth 'sign' rather than 'constellation'.

Scorpio
(The Scorpion)
Scorpio, Scorpii, Sco

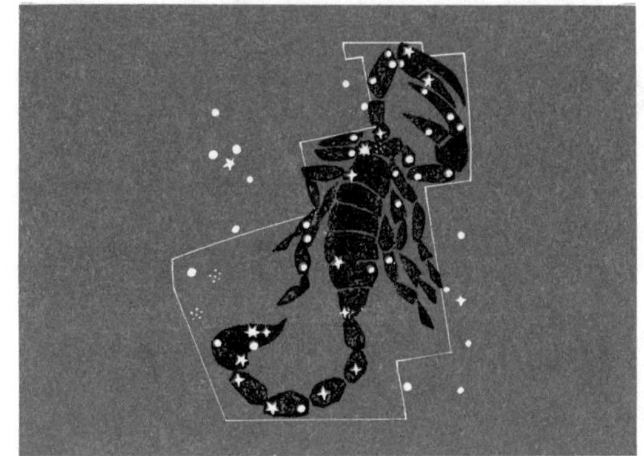

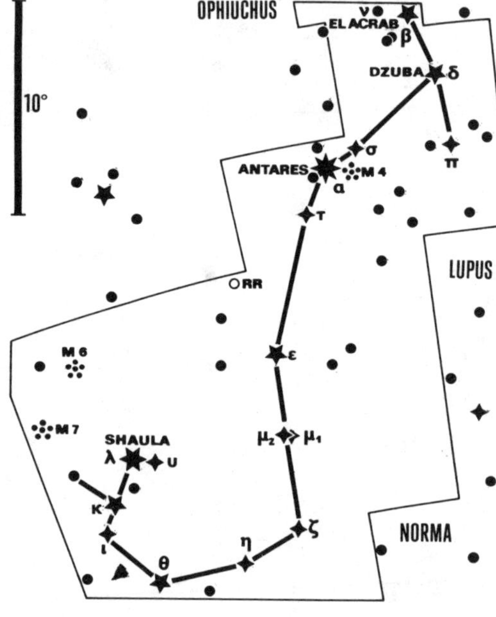

The group of stars between Sagittarius and Libra looks rather like a scorpion preparing to strike – a fact which did not go unnoticed even in prehistoric times, when it first received its name. It rises when Orion sets, for the scorpion is the sworn enemy of the hunter Orion. It was said that the goddess Hera herself released a huge scorpion from the underworld to sting Orion to death. She wished to avenge the insulting things Orion had said about her. Orion died after being stung, but on the intercession of the goddess Artemis he and his dog Sirius were lifted up to the heavens and placed opposite the scorpion. The grateful Hera had raised the scorpion to stardom in return for its services.

This Zodiacal constellation remains very close to the horizon, so that in Europe it is only partly visible. There are many interesting objects in it. One of the most important is the *association* of glowing stars. Associations are very young groups of stars which come into being at the same time and immediately after their birth start to disperse among the other stars of the Galaxy. They are in fact a sort of very extensive and sparsely distributed open stellar cluster. The life of associations cannot be longer than 10 to 20 million years. The association in Scorpio is about four million years old, and over 100 glowing stars have been counted in it. It is 4000 light years away from us and is about 100 light years in diameter.

A very striking star found in Scorpio is Antares, which is similar to the planet Mars. That is how it got its name: anti-Ares (*Ares* being the Greek name for the god of war the Romans called Mars). Its true diameter is 300 times greater than that of the Sun. If the Solar System was placed in such a way that the Sun was in the very centre of Antares, the Earth would orbit 50 million km (31 million miles) inside this supergiant. A white star, seeming greenish in contrast to the red Antares, orbits the latter, and is visible with even a small telescope. It is a very powerful radio emitter. The red Antares and its white companion are 400 light years away from us.

In the year 134 BC a supernova visible even by day exploded in the constellation of Scorpio. It was this that convinced the Greek astronomer Hipparchus that the sky and the stars were not unchanging, as was generally held at the time. He therefore made a list of the brightest stars (a total of 1080 of them) in order to study their alterations. In his catalogue the stars were divided into 48 groups, or constellations, which have survived to this day.

Sagittarius
(The Archer)
Sagittarius, Sagittarii, Sgr

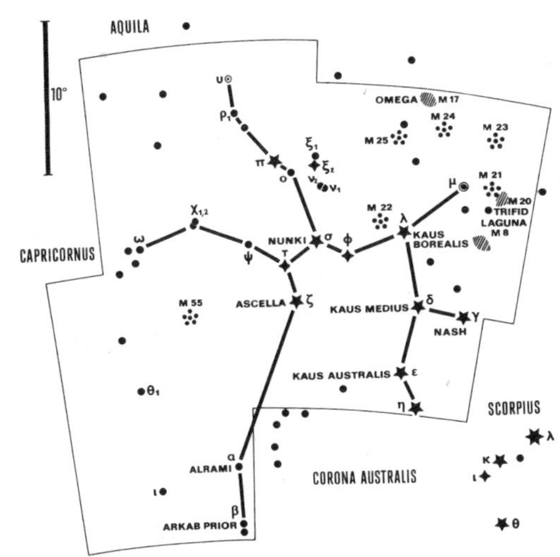

Apart from beautiful goddesses and bold heroes, Greek mythology also had its share of monsters. Among them were the centaurs, half man, half horse. Only a few of these were well-inclined towards men, and one of them was the protégé of the Muses and inventor of archery, who was raised to star rank for his services and became the constellation of Sagittarius.

Sagittarius is a fine Zodiacal constellation of the summer skies, very rich in binaries, variable stars, star clusters and nebulae. Out of the huge number of interesting objects we shall select three beautiful nebulae: the Trifid, Laguna and Omega nebulae, the last also being known as the Horseshoe nebula. They lie at approximately one-tenth of the distance in, to the centre of the Galaxy.

The constellation of Sagittarius is in the part of the Milky Way with the brightest clouds of stars. The luminous clouds of stars in Sagittarius (Figure 130) are only a small part of one arm of our Galaxy. This arm, called the Sagittarius arm, is 10,000 light years away from us (Figure 128-Z). The central part (nucleus) of the Galaxy is 20,000 light years beyond the luminous star clouds of Sagittarius. Unfortunately, light from the galactic centre can never reach us, since it is absorbed by clouds of interstellar matter. We are thus deprived of a fine sight, since the nucleus of the Galaxy would otherwise shine in the sky to such an extent that it would throw shadows comparable to those created by the Moon.

All we know about the nucleus of the Galaxy is learnt through infra-red or radio waves, since these pass through interstellar matter much more readily than light. Thanks to these types of radiation, we now know that in the centre of the Galaxy there exists a small nucleus about 7 light years across. The density of stars in the nucleus is around one hundred million times greater than that in our own part of the Galaxy. Apart from stars in the nucleus of the Galaxy there is matter in a form unknown on Earth, which is in a state of stormy evolution. From time to time there are huge explosions there, whose causes are unknown. But they are probably due to gravitation, which can release from matter up to half its rest energy. A very massive black hole with enormous gravitation may be the cause of stormy processes in the galactic nucleus.

Capricorn
(The Goat)
Capricornus, Capricorni, Cap

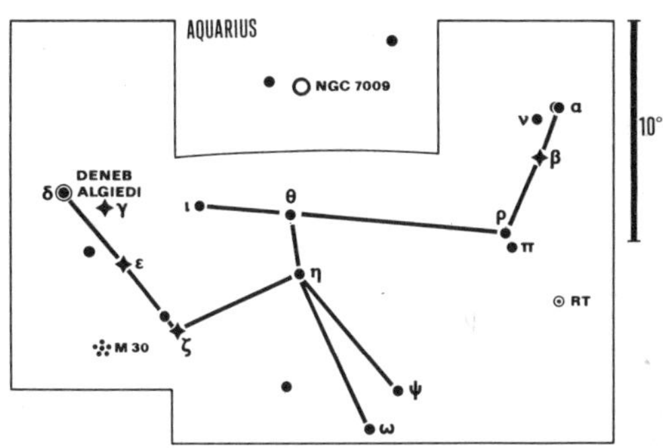

The Greek god of forest, field, herds and herdsmen was Pan (the Roman god Faunus). Since his appearance inspired fear in humans, he preferred to wander alone in the woods and mountains, passing the time by playing his pipes. He enjoyed music and dancing. Pan was often depicted as a goat with a fish's tail, a form he is said to have taken when he fled over land and water from his sworn enemy, the giant Typhon.

Today, as a result of precession, the southernmost constellation of the Zodiac is Sagittarius, which is therefore where the Sun is at the time of the winter solstice. Previously, more than 2000 years ago, it was Capricorn which was furthest south, and to this day the southern tropic is called the Tropic of Capricorn. The names of things in the Universe and on Earth often remain in use for thousands of years.

Cygnus
(The Swan)
Cygnus, Cygni, Cyg

According to the ancient legend Zeus would turn into a swan when he wished to descend from Mount Olympus and move among men. It was in this form that he visited the Spartan queen, Leda, whose beauty had made a great impression upon him. In order to get close to her, he had a huge eagle pursue him, and the queen rescued him. Leda became the mother of the beautiful Helen, cause of the Trojan War.

The constellation appears in the form of a swan flying southwards along the Milky Way. The white star Deneb makes up a striking triangle in the summer sky along with Vega in Lyra and Altair in Aquila. Cygnus can be seen in the evening from June to January. It is sometimes also known as the Northern Cross.

The Solar System and the surrounding stars are moving in the direction of Cygnus at a rate of 230 km (143 miles) per second as a result of the rotation of the Galaxy (Figure 128 number 14). That is the speed at which the Sun orbits the galactic nucleus, completing one orbital revolution every 200 million years (a *galactic year*).

The supergiants Deneb (from the Arabic word *dhanab,* meaning a tail) and Sadir have a luminosity of 10,000 times that of the Sun. The star 61 Cygni (making up a parallelogram with α, γ and ε) is famous as having been the first star whose distance was determined (in 1837). It has a small companion, whose mass is about one-hundredth that of the Sun. Such bodies have more in common with Jupiter than with the Sun.

Close to star ε is the Veil Nebula (Figure 156), catalogued as NGC 6992–5 and NGC 6960. The fine filaments of this nebula are luminescent, and so far are expanding at a rate of around 100 km (60 miles) per second. It is the remains of a supernova which exploded in Cygnus about 50,000 years ago, and is 2500 light years away. Because of the speed at which it is moving, its collisions with interstellar gas cause it to emit radiation.

The newly discovered binary Cygnus X-1 is an immense source of X-rays. One component is a glowing supergiant, while the other, invisible one, is probably a black hole with a mass about 15 times that of the Sun. From the huge supergiant a tremendous avalanche of hot gases rushes towards the black hole. Before being irretrievably swallowed, they collide at high speed, thus emitting X-rays.

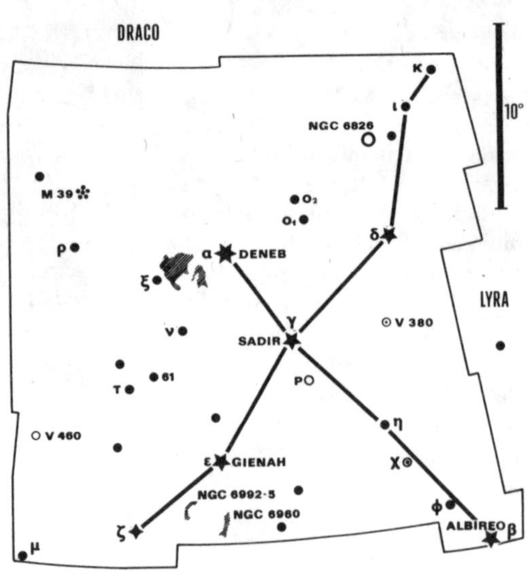

THE CONSTELLATIONS OF AUTUMN

A prominent grouping which helps you to get your bearings in the autumn sky is the large square of Pegasus, comprising the stars α, β and γ Persei and α Andromedae. If we add to these β and γ Andromedae and Algol in Perseus (also called the Winking Demon), we get a figure similar to the Great Bear. Beneath Andromeda are Triangulum and Aries. The open V at the south-eastern corner of Pegasus is the inconspicuous constellation of Pisces. The southern sky is occupied by the large constellations of Cetus and Aquarius.

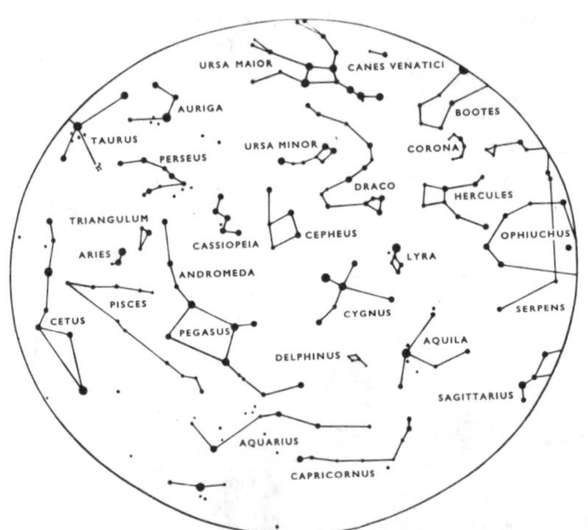

Aquarius

(The Water-Bearer)
Aquarius, Aquarii, Aqr

On old Babylonian monuments Aquarius is depicted as a kneeling man pouring water from a vessel on his shoulder. In Egypt he was a symbol of the Nile floods and the rainy season. The Egyptians believed that Aquarius caused the flooding of the Nile by using a huge vessel to transfer water from springs to the river.

In Greek mythology Aquarius represented the god Zeus pouring water on to the Earth to punish the sins of mankind. Previously, it was said, men had been very righteous and happy, living in what was called the Golden Age, when it had always been Spring on Earth. Then came the Silver Age, when Zeus divided the year into four seasons and men had to live in conditions of alternating hot and cold. Next came the Bronze Age, followed by the Iron Age, the worst of all. Men were very wicked, slaying each other in wars, truth and honour almost disappeared altogether, and violence and lies were supreme. Then Zeus grew very angry and poured a great deluge of water on to the Earth, in which most of mankind drowned. Only Prometheus' son Deucalion and his wife Pyrrha were saved, on Mount Parnassus, for they were just and good people. They then founded a new human race.

The constellation of Aquarius takes up a large section of the sky below Pegasus, but its stars are not very bright. The brightest of them, Sadalmelek (which in Arabic means 'happy star of the king') is a supergiant with a luminosity 10,000 times that of the Sun. It is about 1100 light years away from us.

In 1846 the Berlin-based astronomer Johann Galle found the planet Neptune close to the star ι. Its position had already been calculated by the French astronomer Urbain Leverrier from the irregular movement of the planet Uranus. The then unknown planet at the periphery of the planetary system attracts Uranus, thus making it diverge slightly from its regular elliptical orbit around the Sun. From the magnitude and direction of this deviation Leverrier was able to calculate the position of the then unknown planet. Since Galle found it in the constellation of Aquarius, it was given the name of the Roman god of the waters – Neptune.

Close to the star η is a radiant of the meteoric shower known as Eta Aquarides. The meteors of this shower are caused by a stream of meteoroids left behind by Halley's comet. These remnants of Halley's comet glow in the Earth's atmosphere and we see them as meteors. They most often fall around May 5. The Earth's orbit again passes through the shower around October 20. At that time the meteors fall in large numbers from the constellation of Orion, and are accordingly called Orionids. Halley's famous comet was last visible in late 1985 and early 1986.

The planetary nebula called Helix is one of the largest in our skies.

With a small telescope it can be seen as a flattened misty cloud. The glowing star in its centre which illuminates it is too weak to be seen without a powerful telescope. The planetary nebula was thrown off many thousands of years ago when a star similar to our Sun became extinct.

Pisces
(The Fishes)
Pisces, Piscium, Psc

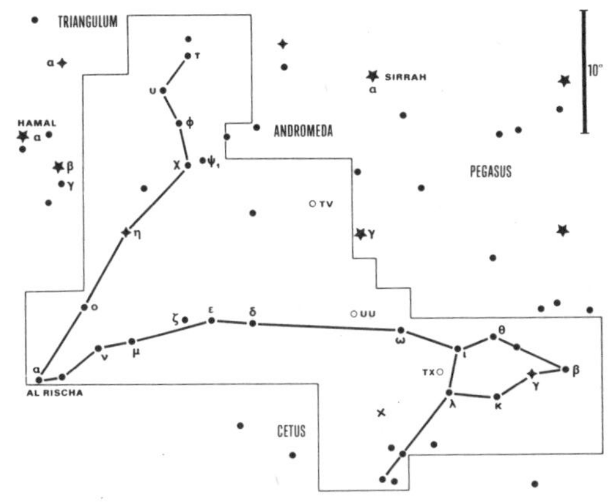

According to the ancient myth the goddess Aphrodite (Roman Venus) and her son Eros were walking along the banks of the Euphrates when the fearful giant Typhon suddenly appeared before them. In their alarm the two of them leapt into the river and changed into a pair of fishes. As a reminder of Aphrodite's clever escape the two fishes were raised up to the autumn heavens and placed close to Cetus, Delphinus and Piscis Austrinis. The constellation was previously called Aphrodite and Eros, but since the Babylonians, Assyrians and Persians also saw the figures of two fishes there, the name Pisces has remained.

The vernal equinox is in the constellation of Pisces. It is the intersection of the ecliptic and the celestial equator. The point in the sky is marked by a cross at the bottom of our map. In the northern hemisphere, spring starts when the Sun in its apparent annual motion crosses the vernal equinox. A circle passing through both celestial poles (the points on the northern and southern hemispheres around which the stars appear to turn) is the equivalent of a terrestrial meridian, and is called a *circle of declination.* The circle of declination which passes through the vernal equinox is just as important for the taking of celestial bearings as the Greenwich Meridian here below. From it, astronomers calculate *right ascension* of the Sun (similar to longitude). Similarly, the celestial equator is used to measure *declination* (the equivalent of latitude), positive declination being northwards and negative southwards.

But our planet behaves like a giant spinning top (Figure 414). Its axis describes a circular motion we call precession. It causes the celestial poles to shift, and the equator also moves. Therefore its intersection with the ecliptic, the vernal equinox, moves westwards along the ecliptic. The vernal equinox is therefore moving in the opposite direction to the Sun. The point is moving away from Aries, the position it occupied before the time of Christ. But the vernal equinox is still denoted by the symbol of Aries, ♈, even though it has long since moved to Pisces, and is on its way to Aquarius. The whole belt of 12 Zodiacal sings (Figure 105) is moving with it, of course.

The constellation of Pisces is poor in stars – in that direction we are looking perpendicularly through the disc of our Galaxy. For the same reason every body in the planetary system is clearly visible there. In particular the area of ε, ζ and δ Piscium is suitable for the discovery of any new minor planets.

Aries
(The Ram)
Aries, Arietis, Ari

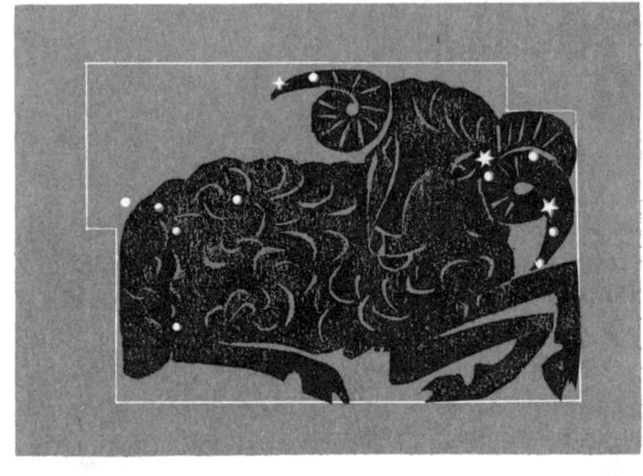

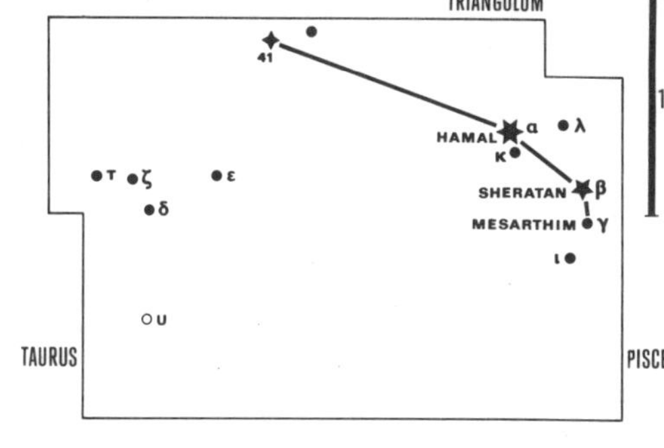

Legend has it that Athamas, King of Thebes, had two children. Their stepmother disliked and ill-used them. When the messenger of the gods, Hermes, discovered this, he sent the children to distant Colchis on the shores of the Black Sea, giving them a ram with a golden fleece, which was able to fly very quickly. The ram in the autumn skies is, according to legend, that very flying ram whose golden fleece was sought in Colchis by the Argonauts. Not only the ram, but also some of the heroes of the tale (Castor and Pollux), and even the ship *Argo* itself, made it to the sky at the end of their adventures.

The boy Phrixus and the girl Helle sat on the ram's back and held on to its golden fleece. But Helle did not cling tightly enough, and fell into the sea. From that time the part of the sea where she fell was named the Hellespont in her honour (now the Dardanelles).

Phrixus arrived safely in Colchis and was most warmly received by the king. The ram was sacrificed to Zeus, and the fleece was given to the King of Colchis in gratitude. He was very proud of his rich gift, and afraid lest someone should steal it, so he hid it in a sacred cave guarded by a dragon which never slept. Nevertheless, Jason and his fellow Argonauts later came to Colchis, where they took possession of the fleece and sailed back to Greece with it.

In the days of ancient Greece Aries was the first constellation of the Zodiac. It contained the vernal equinox, which has since moved into the neighbouring constellation of Pisces (to the west). But to this day the vernal equinox is denoted by the symbol of Aries, ♈ .

THE CONSTELLATIONS OF WINTER

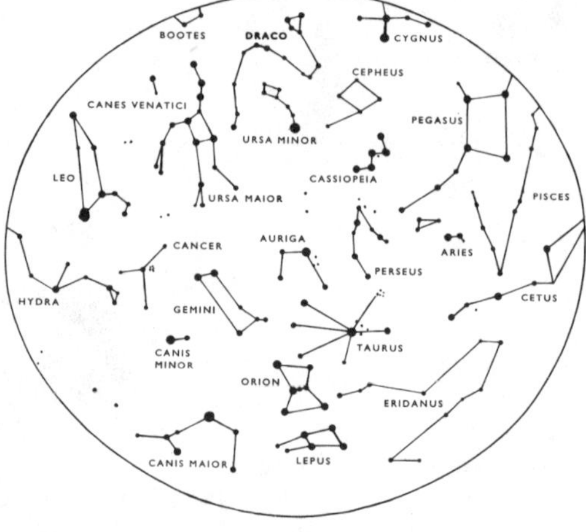

The most beautiful view of the sky is during the long winter nights. The winter sky is darker than the summer one, and there are many fine constellations in it. Perhaps the finest, because of its symmetry, is Orion. It is easy to find using the map. Between the red Betelgeuse α at the top and the blue Rigel β at the bottom is Orion's belt of three bluish stars.

If you extend the line of Orion's belt you come to Aldebaran, in Taurus, on one side and Sirius, in Canis Major, on the other. The line joining Rigel and Betelgeuse runs towards Gemini (Castor and Pollux). An extension of the line Bellatrix (γ Ori) – Betelgeuse points towards Procyon in Canis Minor.

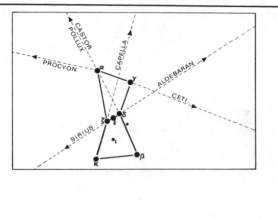

The brightest starts in the winter sky form a hexagon, in the centre of which is Betelgeuse, and whose apexes are, in order: Rigel, Aldebaran (in Taurus), Capella (in Auriga), Castor and Pollux (Gemini, the Heavenly Twins), Procyon (in Canis Minor), and Sirius (in Canis Major).

Orion

Orion, Orionis, Ori

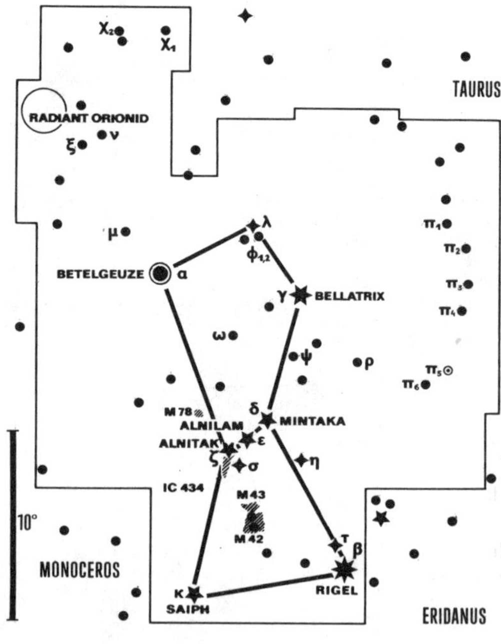

The constellation of Orion is without doubt one of the finest in the whole sky. It was known 3000 years before Greek civilization flourished. In those days the Mesopotamians called it *Uru-anna,* meaning light of the sky. It was this Mesopotamian name from which the name Orion was derived, and it has remained with us to this day. Is it not fascinating that men still use a name already in existence in the cradle of civilization?

Ancient myths described Orion as a strong, handsome hunter whose father was Poseidon, king of all waters. His mother was a huntress and a companion of the goddess of the hunt, Artemis. From his father he received the power to move in the depths of the ocean, but it got him into all manner of mischief. For instance, he pursued the beautiful nymphs, the Pleiades, so long that they begged Zeus to turn them into some sort of animal. Zeus did as they asked, and they became doves, and later became the group of stars still known as the Pleiades. On another occasion the boastful behaviour of Orion offended the goddess Hera, who as a punishment had him killed by a giant scorpion. To this day Orion and Scorpio cannot abide each other – when the latter appears over the eastern horizon, the former disappears in the west.

Most of the stars in Orion are white or slightly blue, i.e. hot and young – stars born relatively recently. They make up a stellar association. The remains of the original material from which this formed can be seen in the Orion Nebula (M 42) in Orion's sword, which is visible with the naked eye (Figure 148). Stars are still being formed there.

In the second half of October the Earth's orbit crosses an extensive stream of particles (meteoroids) which came from the nucleus of Halley's comet. All these particles have parallel orbits, and they fall to Earth from the direction of the star ξ Ori. When they enter the atmosphere they get hot and evaporate at a height of 100 km (60 miles). We see them as meteors flying from the point near the star ξ Ori. The point is called the radiant of Orionids and is marked by a circle on our map. The greatest number of the Orionids falls around October 20, when the Earth is passing through the thickest part of the stream of particles.

Near the star ζ Ori in Orion's belt is the luminous gaseous nebula IC 434, against which a cloud of dark interstellar matter can be seen. Due to its shape it is called the Horse's Head.

Taurus
(The Bull)
Taurus, Tauri, Tau

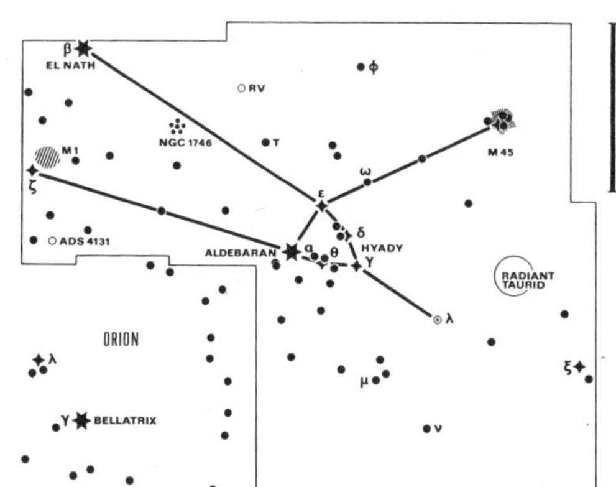

According to Greek mythology, Zeus wished to abduct Europa, the beautiful daughter of the Phoenician king. In order to deceive her, he changed into a snow-white bull and mixed with the royal herds. Europa took a liking to the tame young bull. One day she mounted it to take a ride. But the bull left the herd, plunged into the sea, and swam with Europa on its back to the island of Crete. That is the reason only the head and shoulders of the creature can be seen in the sky: the rest is beneath the waves.

There are two distinct groups of stars (stellar clusters) in Taurus with mythical names. The Hyades and the Pleiades were sea nymphs, daughters of the giant Atlas, who held up the sky to prevent it from falling to Earth.

The Pleiades are 400 light years away from us, and they are 50 million years old. They are veiled in a cloud of interstellar matter which is the remains of the matter from which they were created (Figure 125). Near the star ζ Tau is the Crab Nebula, visible through a telescope as a slightly luminous cloud. It is the debris from the explosion of a huge supernova, of which we have already spoken. The neutron star in the Crab Nebula turns at the rate of about 30 times a second, and at each turn it gives out a short flash. The pulsar in the Crab Nebula is one of the fastest of those so far discovered, pulsing at the rate of 30 times a second (Figure 155).

Gemini
(The Heavenly Twins)
Gemini, Geminorum, Gem

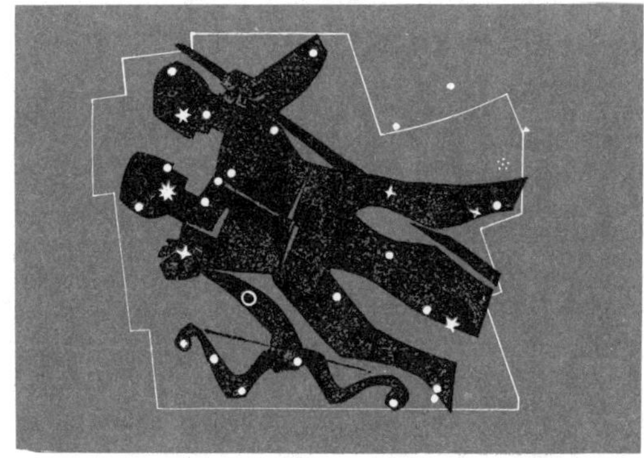

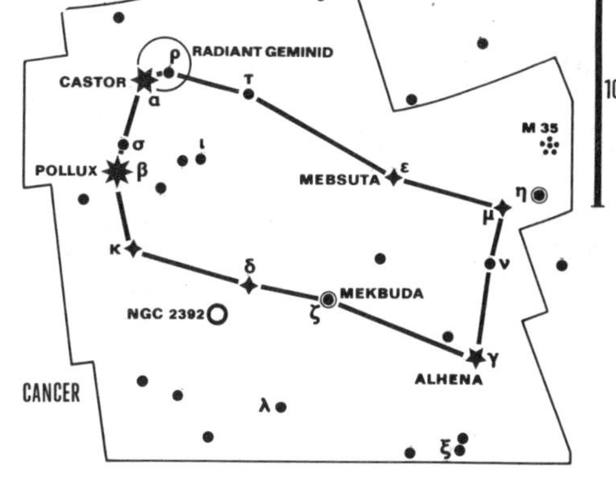

Gemini (Castor and Pollux) were the sons of Zeus and the Queen of Sparta, Leda. Helen, of Trojan War fame, was their sister. The brothers were very brave; they took part in the *Argo* expedition to faraway Colchis to fetch the Golden Fleece. So attached were they to each other that when one died the other no longer wanted to live. After their death their father Zeus brought them up to heaven, where they shine to this day.

The constellation of Gemini lies partly in the Milky Way, towards its outer edge (in Figure 128, to the left of the point X). The brightest of its stars, Castor and Pollux, are most dissimilar. The brighter and closer star, Pollux is an isolated orange giant, 30 light years away from us. Castor is one of the most interesting stars in the sky. A telescope reveals it to be a triple star. Two blue stars (Castor A and Castor B) orbit each other once every 340 years. The more distant reddish dwarf (Castor C) takes several thousand years to orbit the other two. But a spectrograph attached to a telescope shows that all three of these stars are binaries. Thus Castor is in fact a sextuple star consisting of three close pairs. If any of these has a planet with intelligent beings on it, they must have the wonderful sight of six suns in the sky – two light red, and four bright blue (Figure 122).

THE CONSTELLATIONS AROUND THE SOUTH POLE

This book is written for readers in the northern hemisphere, but we should also like to mention those constellations which are only visible from the southern hemisphere.

The sky appears to turn about the northern and southern poles of the Earth, like a huge, hollow globe spinning on a pair of pivots. You can find the North Pole with the help of Polaris, which is very close to it. But there is no bright star near to the South Pole which we might call the Southern Pole Star, and by means of which we could determine where that pole is. It can be found using the Magellanic Clouds. These make up an equilateral triangle with the South Pole. You can also use the Small Magellanic Cloud and the Southern Cross. If you imagine the line joining the star γ of the Southern Cross to the Small Magellanic Cloud, the pole is about one-third of the way from the cloud. Both Clouds – our nearest galaxies – are conspicuous luminous patches and cannot be mistaken. On our map you can find the southern pole with the help of the stars γ and α of the constellation Southern Cross. The southern pole lies on the line connecting the stars γ and α, at a distance four times greater than the distance between these two stars.

If our planet and the Sun were suddenly to disappear, we should find ourselves in a completely dark space with only the starry sky around us. There would be no top or bottom, no up or down, since there would no longer be gravity. The sky would no longer have either a north or a south pole, for without the Earth there would be no axis by which to determine them. We should, however, see all the constellations in the sky just by turning our heads, for the Earth would no longer cover half the sky, or the Sun outshine the stars in the daytime. We should find ourselves in the middle of a huge, hollow and velvet-black globe, on whose wall we would be able to observe stars in every direction. The Milky Way would form a continuous band of light all around us. Astronauts who go beyond the bounds of the Solar System will be able to see such a sight.

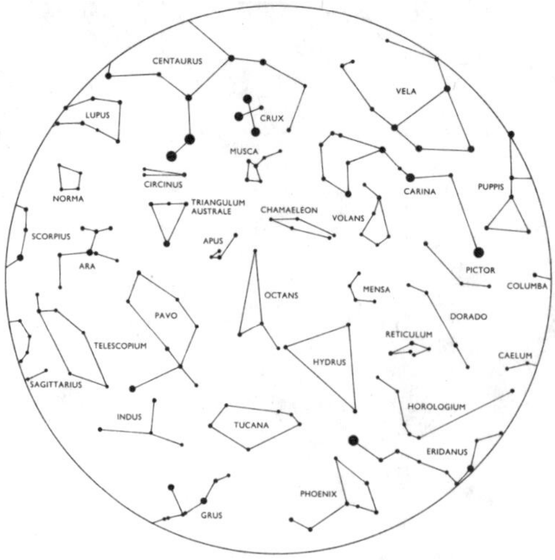

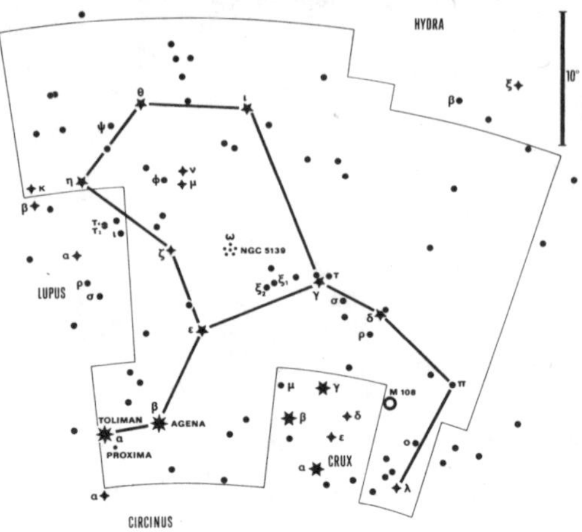

In the southern part of the Milky Way lies the famous Southern Cross. If you should ever travel to the southern hemisphere, do not forget that the Southern Cross is to be found under the belly of Centaurus, between its front and hind legs. The southern sky also contains a 'false' cross (Crux Falsa), which is prominent, and very similar in shape to the Southern Cross. The false cross led to the wrecking of many a ship in the days when seafarers were reliant on the stars. The constellation of the Southern Cross can be found on the flags of several southern states (Brazil, Australia and New Zealand).

To the north of the Southern Cross is the striking constellation of Centaurus. To the south of the cross is Musca. Near the brightest star of Centaurus (called Toliman or Rigil Kent, which means Centaur's foot) are the indistinct constellations of Circinus and Triangulum Australe. The brightest star in Centaurus (α Centauri = Toliman = Rigil Kent) is the third brightest star in the heavens (after Sirius, in Canis Major, and Canopus in Carina). It is a triple star. Its two brightest components (yellow and orange) are similar to our Sun, and are $4\frac{1}{4}$ light years away from us. The third, very faint component, called Proxima Centauri, is about 1 light month nearer than the two bright ones. It is at the moment the nearest star after the Sun (*proxima* means closest). But in a few thousand years it will move to the other side of the bright components of Toliman, which it orbits once in many thousands of years, and it will no longer be the nearest.

The distance to our nearest neighbours in space – the triple star α Centauri – is about 100 million times the distance of the Moon from the Earth. As yet it is not known whether or not α Centauri has planets. But we can easily imagine what the Universe would look like from there. The patterns of stars would be slightly shifted – the closer ones more, the more distant ones less so. The position of the globular stellar clusters and galaxies would be the same as that from here on Earth. The constellations would not look very different from a planet close to α Centauri. Only in Perseus would we see an additional yellow star in the vicinity of the double stellar cluster. If we had the patience and measured its proper motion, we should find that it does not move in a straight line, but in a wavy one. One oscillation lasts 11 years. We could also determine the mass of the invisible companion of that yellowish star. The star is the Sun, its companion Jupiter. From a planet of α Centauri we should not even detect the presence of the other planets – their mass is too small.

The band of the southern Milky Way is cut into by the large constellation of the ship *Argo*. It is so huge that it had to be divided into three: Carina (The Keel), Puppis (The Poop) and Vela (The Sails). Below Scorpio in the Milky Way you will find Lupus (The Wolf), Norma (The Ruler) and Ara (The Altar).

The line joining three bright stars facilitates orientation in the southern sky: Canopus (α Carinae), Achernar (α Eridani) and Fomalhaut (α Piscis Austrini). On this line you will find Pictor (The Painter), Dorado (The Swordfish), Reticulum (The Net), Horologium (The Clock), Phoenix and Sculptor. Around the South Pole are the constellations Octans (The Octant), Apus, Hydrus with the Small Magellanic Cloud, Mensa and Chamaeleon. The constellation Pavo is also easy to find by its solitary bright star. It is one of the group named after birds: Apus (bird of paradise), Grus (crane), Phoenix and Tucana (toucan).

The original 48 constellations inherited from bygone days have been recently supplemented by many more, especially in the southern sky. The borders between many of them were not well established, and constellations have often been introduced for small groups of poorly visible stars. The whole thing was set in order (in 1930) by the International Astronomical Union, which defined precise borders between constellations. The sky was divided into a total of 88 constellations, and each was given a Latin name to be used throughout the world. The stars in a given constellation are designated according to brightness by Greek letters and the genitive case of the constellation's name. Some stars have their own name, handed down from ancient times, in addition to their astronomical designations.

There are very many stars to be seen in the sky, however, especially if looked at through a telescope. Astronomers have therefore drawn up extensive catalogues giving the brightness, exact position and other information relating to stars. Among the best-known catalogues are the Bonner Durchmusterung (BD), containing 458,000 stars, the Cordoba Durchmusterung (CD) with 614,000 southern stars, and the Henry Draper Catalogue (HD) with 225,300 stars. The type of spectrum of each star is also given.

CONCLUSION

We have looked in some detail at the structure and evolution of the Universe. We have seen that the Earth is quite an inconsequential part of the Universe, and that not even the Sun, our star, is particularly special. If the Universe or the Earth were dismantled into smaller and smaller parts, they would end up as protons, electrons and neutrons. All things can be described as systems of these elementary particles, and all processes which take place in the whole Universe can be attributed to the action of nuclear, electrical and gravitational forces. But we should be aware that our knowledge of the structure and evolution of the Universe is far from being perfect. Some of our knowledge may well turn out to be incorrect. In dozens of observatories and laboratories on board satellites and other spacecraft, and deep below the surface of the Earth, scientists are carrying on intensive research and investigation. Not just for the sake of it, but because all knowledge has its significance, and even if mankind in the 20th century is sometimes unaware of the importance of his discoveries, generations to follow may well find it useful. Let us hope that they will make use of this knowledge to create a better world in which to live.